Advances in Intelligent Systems and Computing

Volume 1435

The series “Advances in Intelligent Systems and Computing” contains publications on theory, applications, and design methods of Intelligent Systems and Intelligent Computing. Virtually all disciplines such as engineering, natural sciences, computer and information science, ICT, economics, business, e-commerce, environment, healthcare, life science are covered. The list of topics spans all the areas of modern intelligent systems and computing such as: computational intelligence, soft computing including neural networks, fuzzy systems, evolutionary computing and the fusion of these paradigms, social intelligence, ambient intelligence, computational neuroscience, artificial life, virtual worlds and society, cognitive science and systems, Perception and Vision, DNA and immune based systems, self-organizing and adaptive systems, e-Learning and teaching, human-centered and human-centric computing, recommender systems, intelligent control, robotics and mechatronics including human-machine teaming, knowledge-based paradigms, learning paradigms, machine ethics, intelligent data analysis, knowledge management, intelligent agents, intelligent decision making and support, intelligent network security, trust management, interactive entertainment, Web intelligence and multimedia.

The publications within “Advances in Intelligent Systems and Computing” are primarily proceedings of important conferences, symposia and congresses. They cover significant recent developments in the field, both of a foundational and applicable character. An important characteristic feature of the series is the short publication time and world-wide distribution. This permits a rapid and broad dissemination of research results.

Indexed by DBLP, INSPEC, WTI Frankfurt eG, zbMATH, Japanese Science and Technology Agency (JST).

All books published in the series are submitted for consideration in Web of Science.

For proposals from Asia please contact Aninda Bose (aninda.bose@springer.com).

More information about this series at https://link.springer.com/bookseries/11156

Simona Dzitac · Domnica Dzitac ·
Florin Gheorghe Filip · Janusz Kacprzyk ·
Misu-Jan Manolescu · Horea Oros
Editors

Intelligent Methods Systems and Applications in Computing, Communications and Control

9th International Conference on Computers Communications and Control (ICCCC) 2022

Springer

Editors
Simona Dzitac
University of Oradea
Oradea, Romania

Domnica Dzitac
New York University Abu Dhabi
Abu Dhabi, United Arab Emirates

Florin Gheorghe Filip
Romanian Academy
Member of the Romanian Academy
Bucharest, Romania

Janusz Kacprzyk
Polish Academy of Sciences
Systems Research Institute
Warszawa, Poland

Misu-Jan Manolescu
Agora University of Oradea
Oradea, Romania

Horea Oros
University of Oradea
Oradea, Romania

ISSN 2194-5357 ISSN 2194-5365 (electronic)
Advances in Intelligent Systems and Computing
ISBN 978-3-031-16683-9 ISBN 978-3-031-16684-6 (eBook)
https://doi.org/10.1007/978-3-031-16684-6

This Springer imprint is published by the registered company Springer Nature Switzerland AG
The registered company address is: Gewerbestrasse 11, 6330 Cham, Switzerland

Preface

The volume contains a selection of 37 papers presented at the ninth edition of the International Conference of Computers, Communications and Control-ICCCC. The initial idea of the conference which was founded in 2006 by Ioan Dziţac, Florin Gheorghe Filip and Mişu Jan Manolescu was to set the stage for presenting papers containing scientific and practical results concerning the design and deployment of systems that integrate the three technologies: *computers*, *communications* and *control* at large. It is worth noticing that the conference was intended by its founders to be an opportunity offered to academia people and industry practitioners to exchange ideas and discuss possible future common projects in a friendly environment.

Since 2006, the conference has been organized every even year by *Agora University of Oradea*, under the aegis of the *Information Science and Technology* Section of Romanian Academy. The published conference proceedings and website witness a successful series of events.

From the very beginning, the conference was envisioned as a public and open event associated with the *International Journal of Computers, Communications and Control* (IJCCC), a publication which was meant to be a perennial means to archive and broadly disseminate selected articles describing the results obtained in the well-defined domain of solutions integrating the three technologies.

Over the years, outstanding scholars and successful professionals from industry presented high-quality papers, which generated the need of diversifying the set of core topics to be addressed. *Soft computing* (with particular emphasis on *fuzzy sets and systems* and associated solvers of complex and complicated problems), *decision-making models* and *computer-based support systems* have found their place in the sections of the conference. More recently, *green* and *trustworthy computing* has been accepted as a core topic of the conference in line with the current developments in the international scientific and technological landscape. Consequently, the ninth edition of the conference was organized into five sections

1. Integrated Solutions in Computer-Based Control;
2. Advanced Control Systems Integrating Computers and Communications;
3. Soft Computing Including Fuzzy System Approach;
4. Decision-Making and Support Systems;
5. Trustworthy and Green Design.

The ninth edition of ICCCC was dedicated in memory of Prof. Ioan Dziţac, the conference founder and the general chair of the previous eight editions. On this occasion, the Dziţac Family (wife Simona and daughter Domnica) awarded the "Ioan Dziţac Prize" of EUR 500 for the most appreciated scientific work presented at ICCCC 2022 by a promising young researcher, more precisely to the PhD student of the University of Granada, Spain—José Ramón Trillo. The conference general chair was prof. Simona Dziţac.

The programme included four key note speeches, five invited talks and 41 regular papers grouped in the above five sections. The papers were authored by researchers from 18 countries (Canada, Chile, China, Colombia, Japan, Israel, Italy, Lithuania, Mexico, Poland, Romania, Russia, Serbia, Spain, Switzerland, UAE, UK and USA).

The keynote speeches were held by high reputation scholars such as

1. Acad. Gintautas Dzemyda, Vilnius University, Lithuania (*Geometric Multidimensional Scaling: A New Approach for Data Dimensionality Reduction*);
2. Prof. Enrique Hererra-Viedma, University of Granada, Spain (*On Fuzzy and Linguistic Decision-Making: Scenarios and Challenges*);
3. Acad. Janusz Kacprzyk, Polish Academy of Sciences, Warsaw, Poland (*AI-enabled Decision Support Systems: An Effective And Efficient Solution For Complex Problems*);
4. Prof. Gaston Lefranc H., Pontificia Universidad Católica de Valparaíso, Chile (*Review of Trends in Manufacturing Systems Based On Industry 4.0: The Opportunities*).

The invited speakers were

1. Prof. Razvan Andonie, Central Washington University, Ellensburg, USA (*Computational Semiotics in Deep Learning*);
2. Dr. Valeriu Beiu, "Aurel Vlaicu" University of Arad, Romania (*Reflective Optics from Stars to Dust*);
3. Prof. Yezid Donoso, Universidad de los Andes, Bogota, Colombia (*Methodologies for Solving Complex Multi-Objective Combinatorial Problems in Engineering: An Evolutionary Approach Applied to Computer Networks*);
4. Prof. Alfred Bruckstein, Technion, Haifa, Israel (*Guiding Swarms with Pheromone Gradients or Broadcast Signals*);
5. Prof. Valentina E. Balas, "Aurel Vlaicu" University of Arad, Romania (*Advances in Soft Computing and Applications*).

Fifty-three regular papers were submitted for evaluation. Forty-one papers were accepted by the *International Programme Committee*. Five of the papers presented at ICCCC 2022 have obtained the "Best paper award". A number of 37 final versions of the accepted papers are included in the five sections of the volume as follows:

- Section 1. *Integrated Solutions in Computer-Based Control*: eight papers;
- Section 2. *Advanced Control Systems Integrating Computers and Communications*: seven papers;
- Section 3. *Soft Computing Including Fuzzy System Approach*: eight papers;
- Section 4. *Decision-Making and Support Systems*: nine papers;
- Section 5. *Trustworthy and Green Design*: five papers

The study of the papers contained in the volume can be useful for researchers, consultants and postgraduate students in *computer science and design*, *applied informatics*, *automation* and *industrial engineering*.

The editors are grateful to Springer for accepting to publish the proceedings of ICCCC in the *Advances in Intelligent Systems and Computing* Series and ensuring a high-quality technical version of the volume.

June 2022 The editors

Contents

Integrated Solutions in Computer-Based Control

Automatic System Based on Riesz MV-algebras, for Predictive Maintenance of Bearings of Industrial Equipment Using Temperature Sensors

Dan Noje[1(✉)], Radu Catalin Tarca[2], Nicolae Pop[3], Alin Octavian Moldovan[1], and Ovidiu Gheorghe Moldovan[4]

[1] Doctoral School of Industrial Engineering, University of Oradea, Universității, 1, 410087 Oradea, Romania
dan.noje@gmail.com

[2] Department of Mechatronics, University of Oradea, Universității, 1, 410087 Oradea, Romania
rtarca@uoradea.ro

[3] Institute of Solid Mechanics of the Romanian Academy, Sector 1, 15 Constantin Mille, 10141 Bucharest, Romania

[4] Department of Mechatronics, University of Oradea, Universității, 1, 410087 Oradea, Romania
omoldovan@uoradea.ro

https://doctorat.uoradea.ro/en

Abstract. Predictive maintenance systems take a more and more important place in increasing the reliability of industrial equipment, of productivity and reducing production costs. There's a constant focus to develop new system architectures and in using the latest numerical methods and algorithms in developing more and more reliable predictive maintenance systems. Based on recent results in using Shepard local approximation operators defined in Riesz MV-Algebras for IoT devices signals processing, in this article we propose a new architecture for a predictive maintenance system by adding an extra layer (denoted Data Validation Layer). This new layer is processing the acquired signals by sensors such that to provide to the predictive algorithms a complete and validated data set. The proposed architecture was validated by implementing and testing a predictive maintenance system that is monitoring the bearings of industrial equipment using temperature sensors.

Keywords: Predictive maintenance · IoT · Industrial equipment · Sensors · Riesz MV-Algebras

1 Introduction

Predictive Maintenance (PrM) isused throughout the industry for several reasons mainly in order to reduce cost, minimize break-down times, improve the quality of the process and also it can have a major impact in environment protection [29].

S. Dzitac et al. (Eds.): ICCCC 2022, AISC 1435, pp. 3–19, 2023.
https://doi.org/10.1007/978-3-031-16684-6_1

Many industries adopt PrM and rely on forecasting abilities of different system in order to ensure adequate safety and reliability of all systems [25]. In today's industry, smart products and components (equipped with embedded smart devices) are wirelessly networked and remotely monitored in real-time [8]. Maintenance strategies are responsible for the performance and the proper operation of manufacturing systems, turning into increasingly more relevant for the sustainability of manufacturing processes [14]. Recent developments in information technology, data transmission and information processing, such as IoT (Internet of Things) technology, allow the development of PrM systems to be efficient and costeffective. This makes new PrM techniques being adopted in industries where this was either not technically possible or not cost-effective [25] . PrM tools are used to determine when specific maintenance is required. The tool relies on continuous observation of the machine, characteristics of a machine or process, and this permits maintenance to be performed only when is needed [25]. The PrM [18,32] is an effective tool in reducing operating costs and the frequency interval for maintenance activities on industrial equipment [26,28]. PrMcan be a key factor in the Industry 4.0 concept when the integration of sensors and sensorial systems target the cost reduction of maintenance, the implementation zero-waste production, and the reduction of down time caused by failures [17,25]. In developing PrM strategies and tools, the IoT is taking a more and more important role [1,24,30]. A significant part of the literature focuses on methods of acquiring sensor data and interpreting this data in order to identify potential problems in the production process. In general, these techniques are generally called conditionbased maintenance techniques for industrial equipment and processes [15]. These techniques are divided in three categories, each category uses signals from existing process sensors, uses signals from test sensors or generates a test signal that is sent towards the equipment and the response is evaluated [15].

Another important aspect of PrM and data acquisition is the why the data is used and processed. Specifically if we look at the IoT based PrM systems, machine learning (ML) models are employed to evaluate the acquired data and for the decision making process [3]. The most advanced PrM techniques focuses on using Artificial Intelligence (AI) techniques in combination with proven measurement and data processing strategies [11] in order to obtain best results in term of efficiency and cost efficiency. Accurate and reliable data measurement and acquisition for PrM is the general goal [4]. In order to determine de accuracy and the results of data analysis using the AI models, generally Artificial Neural Networks models are used [10,13]. This paper addresses the issue of data integrity in PrM systems. As PrM becomes increasingly complex, PrM uses more and more artificial intelligence algorithms [5] so, the need for the transmitted data to be correct or to be compensated if some of them are missing or damaged is becoming more and more important. There are many causes of data integrity problems, such as data acquisition problems or signal transmission errors caused by electromagnetic interference or other causes. Random errors can also be a factor in distorting the information and integrity of the data which is then considered in the processing for the decision-making system of the PrM component. If an IoT-based PrM system is considered, the importance of data integrity is even greater. It is known that not all phenomena in the real world can be described accurately, i.e., you cannot answer yes or no to all questions. Consequently, classical logic cannot cope with these types of problems. The concept of multivalent algebra, in short MV algebra, introduced by C.C. Chang [6,7] allows the description of the inaccuracy associated with real phenomena. MV-algebras are seen as algebraic correspondents of a multivalent logic, just as Boolean algebras are the algebraic correspondents of classical logic. Based on this algebraic structure, vectorial MV-algebras [21] were introduced to provide instruments to develop approximation methods. Later, Riesz MV-algebras [9] who are a particular type of vectorial MV-algebras were introduced and studied [2]. Based on the Riesz MV-algebra structure of IoT signals [22,23] a new architecture of an automatic system for PrM of bearings of industrial equipment is proposed and presented. The data used in this research is related to the bearing housing temperature of an electric motor. Several research identified bearing housings as critical points for data collection in PrM systems [10–12,16,31]. In most application the drive system is a key point for PrMand the evolution of bearing temperature (along with vibration) is generally monitored and used as a reliable information regarding system health and performance.

2 Materials and Methods

As mentioned, the new proposed architecture of an automatic system for PrM is based Riesz MV-Algebras of IoT signals, in fact is using Shepard local approximation operators defined on these Riesz MV-algebras [22,23].To validate the new architecture was decided to implement and test an automatic system for PrM of bearings of industrial equipment. Considering this, we have to map the acquired signals by the temperature sensors with algebraic structures such that to be able to use the proposed mathematical methods.

2.1 Autonomous Systems

2.2 Mathematic Fundamentals Used in IoT Signal Processing

In [23] was considered that the data sent by an IoT device on regular time intervals is a time-based function

$$f[0,n] :\rightarrow [0, 2^t - 1] \tag{1}$$

0 is considered the moment when the first signal was sent, and that 1 is the length of the time interval when a new signal is transmitted by an IoT device and t is the number of bits used to store the values in computer memory. In other words, in our case, the function f gives the bearing temperature at a specific moment.

In [19] as proved that the structure $([0, 2^t - 1], \oplus, \neg, 0)$ is an MV-algebra, if the following definitions are used:

$$x \bigoplus y =_{def} min(2^t - 1, x + y) \quad (2)$$

$$\neg x =_{def} 2^t - 1 - x \quad (3)$$

$\forall x, y \in [0, 2^t - 1]$

If is considered the external operation $\bullet : R_+ \times [0, 2^t - 1] \rightarrow [0, 2^t - 1]$, defined as follows:

$$a \bullet x =_{def} min(2^t - 1, a \cdot x). \quad (4)$$

$\forall a \in R_+ \; and \; f \forall \, x \in [0, 2^t - 1]$

In [20] was proved that the structure $([0, 2^t - 1], \oplus, \neg, 0, \bullet)$ is a vectorial MV-algebra and it is easy to see that it is also a Riesz MV-algebra. In [23] the Shepard local approximation operator was introduced as a function

$$S : [0, n] \rightarrow A \quad (5)$$

defined as follows:

$$S(f, x) = \oplus_{x_i \in B(x,r)} \frac{K\left(\frac{|x - x_i|}{r}\right)}{\sum_{x_i \in B(x,r)} K\left(\frac{|x - x_i|}{r}\right)} \bullet f(x_i) \quad (6)$$

where $\oplus$ and $\bullet$ are the Riesz MV-algebra operations,
$B(x, r) = \{y \in [0, n] \, || x - y| \leq r\}$ and a kernel, which is a strictly decreasing function $K : [0, 1] \rightarrow R_+$.

In the above definition the Shepard local approximation operator, xi are the time intervals when new data was received, and x are the time intervals when no data was received and that has to be approximated. For more accurate monitoring process, we can consider that the data sent by an IoT device on regular time intervals is a two variable function:

$$[0, n] \times [0, m] \rightarrow [0, 2^t - 1] \quad (7)$$

where the first variable represents the time dependency as in the one variable case, and the second variable is the bearing revolution speed. 0 for the second variable is when the bearing is not moving at all, and m is considered as maximum revolution speed. If we consider this approach, the definition of the generalized Shepard local approximation operator should be used [22]. In this case, the function f gives the bearing temperature at a specific moment and dependent on a specific revolution speed.

2.3 Automatic System Based on Riesz MV-algebras, for PrM of Bearings of Industrial Equipment Using Temperature Sensors

As it is described in [30] usually an IoT-based PrM system consists in five different components:

- The Data Acquisition Layer consists in sensors placed in specific points of the industrial equipment;
- The Data Communication Layer, is the layer that transfer the collected data to a data storage;
- The Data Storage, which is the layer that store the data in structured way such to be suitable for analysis processes and decision making;
- The Predictive Analysis Layer implements predictive algorithms, mostly based on machine learning. They have to generate maintenance tasks in case of a specific incident may occur;
- The Predictive Maintenance Software (PrMSw) is the layer that send notifications to the maintenance team regarding the necessary maintenance tasks that have to be performed and also provide all required information needed for the task's accomplishment. The PrMSw should also provide reports.

It is known that in the process of data acquisition may occur errors. Those errors may be generated by noises that are altering the signal or by discontinuity in data acquisition. Another known fact is that the predictive analysis layer needs a complete and validated set of data in order to deliver accurate predictions. Considering these facts, a new architecture of an IoT-based predictive maintenance system is proposed. A new layer to be added is proposed. This new layer should be placed between the data acquisition layer and the data communication layer, and we call it Data Validation Layer. The Data Validation Layer has the role collect all the signals acquired by the sensors to filter the altered values by noises and to fill in the missing signals. This way, complete set of validated signals is provided to the Data Communication Layer. The signal processing algorithm used in the Data Validation Layer is the one presented in Fig. 1.

If we consider the bearing temperature as a time-based function, it is obvious that is a continuous function. Also, if we consider that the temperatures are stored in memory, the possible values are in the range $[0, 2^t - 1]$, where t represents the number of bits used to store the value. Considering the results presented in [19], and using the operations as defined in (1), (2) and (3), the structure $([0, 2^t - 1], \oplus, \neg, 0, \bullet)$ is a Riesz MV-algebra, means that the temperature values acquired by sensors from bearings has a Riesz MV-algebra structure. Considering this, we can conclude that:

- If we consider that the bearing temperature is a time-based function, the approximation algorithm introduced in [23] can be used. This algorithm is based on the Shepard local approximation operator.
- In case that we consider that the bearing temperature is a time and bearing revolution speed function, in the approximation algorithm, the Generalized Shepard local approximation operator [22] can be used.

If we plan to optimize the transmission of data, the values collected in the buffer will only be transmitted from time to time, depending on the available connectivity. Also, the buffer should not be emptied because we still need some

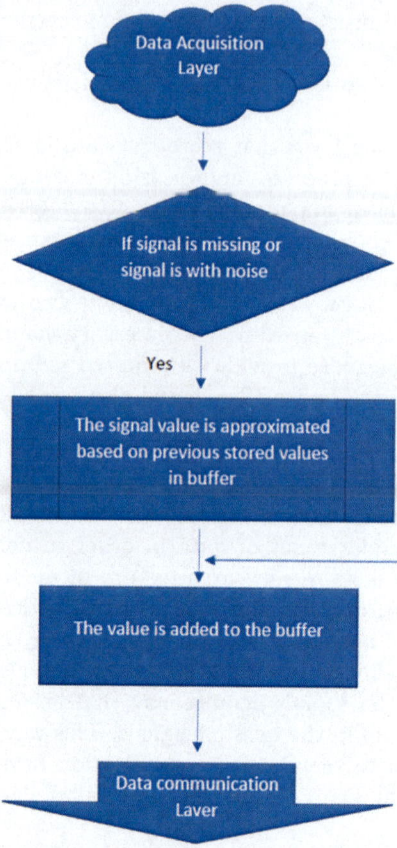

Fig. 1. The signal processing algorithm used in the Data Validation Layer

validated values for further computations. The number of values that should be kept in the buffer depend on the approximation parametrization.
An interesting fact is that the approximation algorithm based on Shepard local approximation operator can be used not only to generate the approximated signal value, but also to determine if the signal is altered by noises.
The algorithm to determine if the signal is altered by noise, has the following sequence:

1. We consider a signal value Sg;
2. A new signal value NSg is approximated based on previous stored values in buffer;
3. If $|Sg - NSg| \geq$ Accepted error, the Signal value is considered altered by noise.Machine learning or experiments can be used to set the Accepted error.

3 Results

The experimental implementation of a predictive maintenance system as described above, was implemented on an electric motor. The electric motor for which the data acquisition on the temperature evolution was carried out is a three phase motor with a power of 1.5 kW. The motor shaft is mounted in two zones by means of two radial ball bearings. A bearing type 6206ZZ is mounted on the front side and a bearing type 6205ZZ on the rear side. In order to monitor the temperature development in the two bearing housings, two temperature sensors are mounted for each bearing. The sensors used are LM 35 type sensors produced by National Instruments. The sensors have a guaranteed accuracy (at +25 °C) of 0.5 °C, the operating temperature range is −55 ° to +150 °C and the operating voltage is between 4 and 30 V. The heat transfer was ensured by using a thermally conductive paste applied between the motor housing and the temperature sensor, as seen in Fig. 2.

Fig. 2. The temperature sensors setup

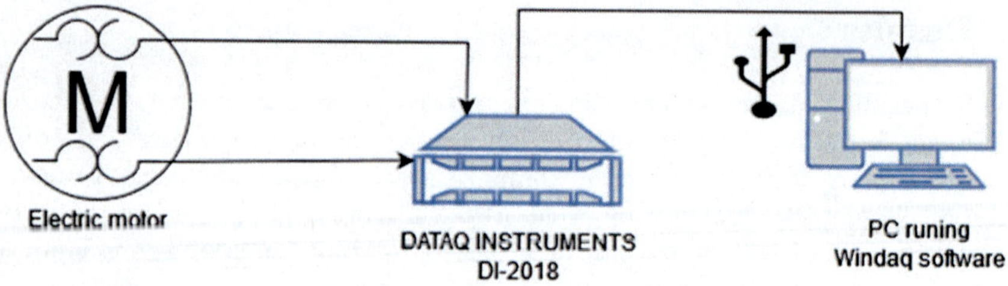

Fig. 3. The setup used for temperature acquisition.

Data acquisition from the four temperature sensors was done using a DATAQ INSTRUMENTS acquisition board model DI-2018 and Windaq application (Fig. 3). The acquisition system used has a resolution of up to 16 bits and a 220 kHz sample throughput rate. To acquire data on the temperature evolution in the bearings, 100 samples/second were acquired for each channel at a sampling interval of 10 ms. The data acquisition time was 30 min. The acquisition was made when the motor was under normal operating conditions, i.e., it was forced ventilated by means of a fan mounted on the motor shaft (Fig. 4).

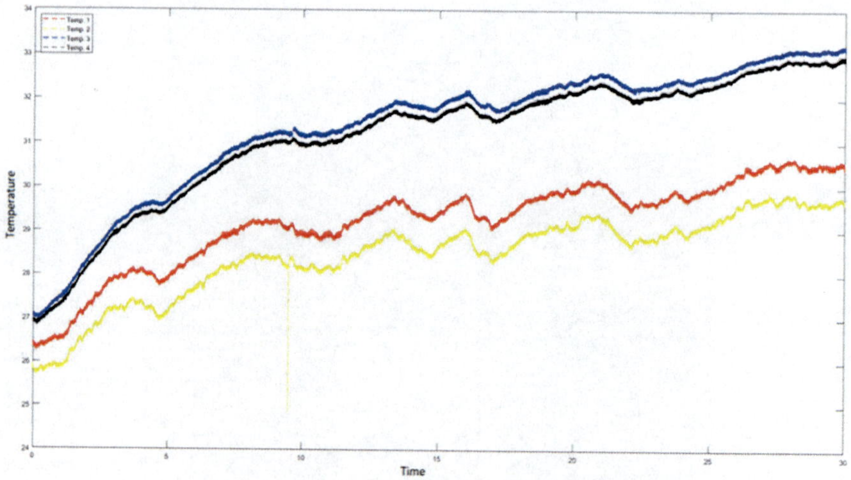

Fig. 4. Acquired temperatures over 30 min period.

In the implementation was considered that the bearing temperature is a time-based function and thus the approximation algorithm introduced [26] was be used. The implemented algorithm (Algorithm 1) has approximated each missing value based on the previously acquired temperatures that were in the proximity of radius r of the missing value.In this case we have to compute each missing value during the acquisition time. In this situation we can say that each time

we encounter a missing value, the previous one exists, no matter if it is acquired or approximated one. Considering these, we decided to use the parametrization presented in [23], that provided the best results in case each second value is missing, situation that correspond to our assumption that each value has at least one acquired value before it.
The used parametrization is:

- the selected kernel was Exponential kernel: $K(u) = e^{-\lambda u^2}$
- $r = 3$ *and* $\lambda = 10$.

In the experiments were used the four sets of approximately 360000 values of acquired temperatures by the four sensors.The acquired temperatures were stored in a buffer and transferred to the Data Validation Layer once the buffer is full. After the buffer content is transferred, the buffer is emptied, but the last r values are kept, such that to be able to approximate the temperature in case the first value is missing after the buffer content is transferred.

The PrMSw used was PriDesk [27] a collaborative tool that ensure traceability and process automation and a powerful reporting module. The integration was realized using secure connection to its API. In order to reduce the complexity of the implementation and since the focus of this research is to determine a suitable implementation of the Data Validation Layer, we decided to create a maintenance ticket in case the bearing temperature has exceeded a certain previously set threshold temperature. To prevent the flood of the human operators with unnecessary tickets, a new ticket is created not for each temperature that exceeds the threshold temperature, but only in the situation that we switch from a normal operating temperature to a temperature above the threshold one. Since all the tests were very similar, for exemplification we present the results of the tests performed on a set of 50 consecutive acquired temperatures with the help of first sensor.

We mention again that each every 10 ms a new temperature was acquired. In order to simulate the case of a malfunction of the sensor, we randomly eliminate some of the acquired temperatures and generate a new value instead. We considered two experiments. In the first experiment, of which obtained results are presented in Fig. 5, we chose to consider that from time to time a single value is missing. The approximated values are marked in Fig. 5 using a dark purple point over the orange line.

In the second experiment, of which obtained results are presented in Fig. 6, we chose to consider that from time-to-time multiple values are missing. The approximated values are marked in Fig. 6 using a dark purple point over the orange line.

4 Discussion

Analysing the results of the performed experiments, we noticed that the approximated values tend to keep the trend of previous acquired temperatures, thus the approximation isn't good enough. This led us to the idea to use another

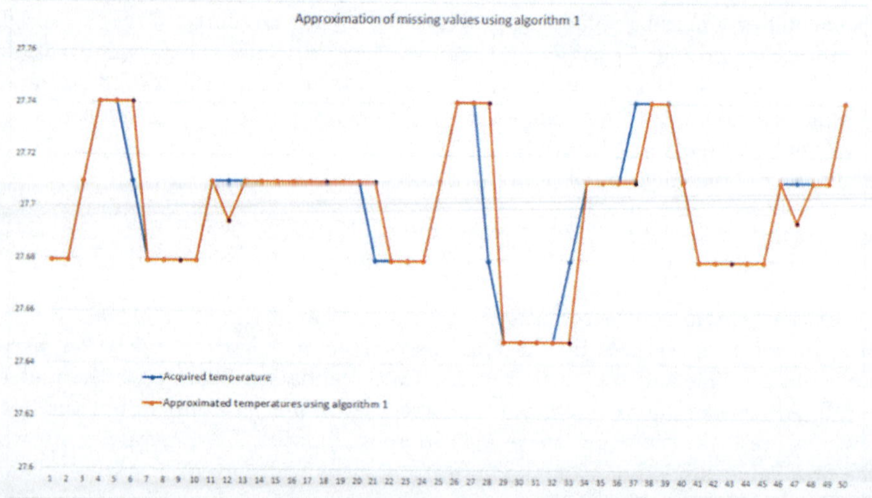

Fig. 5. Approximation of missing values using Algorithm 1 in experiment 1 for singular missing values.

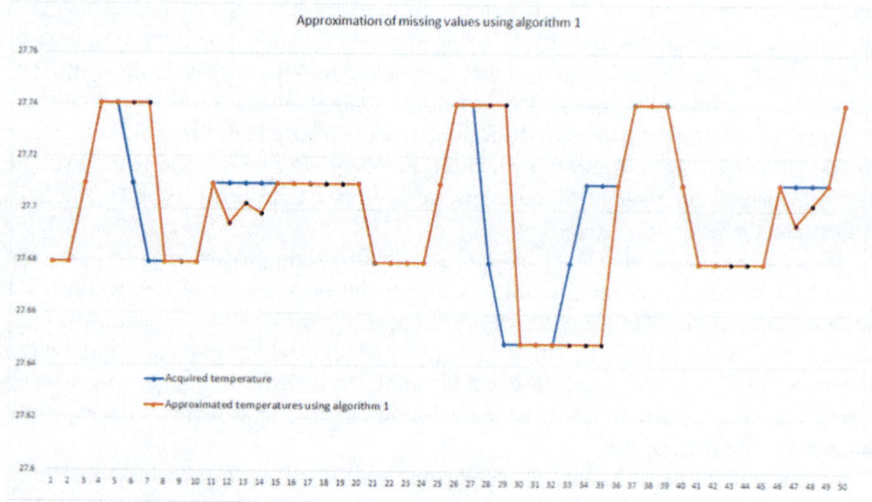

Fig. 6. Approximation of missing values using Algorithm 1 in experiment 2, for multiple missing values.

algorithm (Algorithm 2) that compute each missing value based on both previous and next acquired temperatures that are in the proximity of radius r of the missing value. To implement this algorithm, we have to compute the missing values just after the temperature buffer is full. In order to avoid the possibility to be in the situation that the buffer starts or ends with missing values, in which situation the approximation algorithm fails, in the Data Validation Layer, we implemented a variable length buffer considering the following constrains:

1. If the new buffer that is created starts with missing values, the last r values from the previous buffer will be kept as start in the new one. The r value is the radius used to determine the values in the proximity of the missing one that will be approximated;
2. If the new buffer that is created ends with missing values, it will be considered as ending after the last subset of existing values. In this case, we try to keep no more than r values from this subset end together with the missing values as start of the next buffer. In order to avoid having this subset duplicated it will be used only to approximate the missing values and will not be parsed to the Data Storage Layer.

Considering the above premises, we have run the same two experiments that have been used to test Algorithm 1:

- in the first experiment, of which obtained results are presented in Fig. 7, we chose to consider that from time to time a single value is missing.
- in the second experiment, of which obtained results are presented in Fig. 8, we chose to consider that from time-to-time multiple values are missing.

Since we intended to compare the two algorithms in the same conditions, we selected the same values as missing as for the Algorithm 1. In both Fig. 7 and Fig. 8, the approximated values computed using Algorithm 2 are marked using a red point over the gray line.

As seen in the Fig. 5 and Fig. 7, in most of the cases the Algorithm 2 gives us a better result than Algorithm 1 in the case of missing singular values. This statement is enforced by the maximum difference between the acquired temperature and the approximated one:

- using Algorithm 1, the maximum difference was 0.06103;
- using Algorithm 2, the maximum difference was 0.01526.

Also, in the case of second experiment, when multiple values were considered missing, Algorithm 2 gave better results than Algorithm 1, as can be seen in Fig. 6 and Fig. 8. For this experiment the maximum difference between the acquired temperature and the approximated one was:

- using Algorithm 1, the maximum difference was 0.09155;
- using Algorithm 2, the maximum difference was 0.04069.

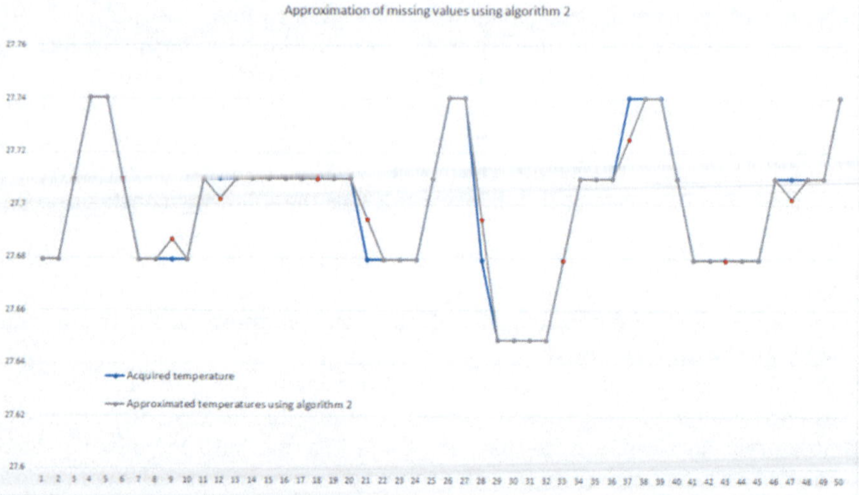

Fig. 7. Approximation of missing values using Algorithm 2 for singular missing values.

After the two experiments, we can conclude that Algorithm 2 is providing better results no mater in which situation we are, thus is more suitable for our purpose.

In [23], for multiple missing values, and using the Exponential Kernel, were obtained better results using the following parametrization:$r = 5$ and $\lambda = 10$. This led us to the idea to experiment this parametrization for our collected temperatures too in the case of multiple missing values. For consistency, we performed this experiment using both algorithms. If in the case of Algorithm 1, the result was not surprising, and with this parametrization it gave worst results, as seen in Fig. 9. In the case of Algorithm 2 not always we got better results using this new parametrization, as seen in Fig. 10.

In this case we noticed that there are two different situations:

- when the temperature values are having a smoother variation. In this situation the new para-metrization confirms the results of [23];
- when the missing temperature is near a local minimum or local maximum, because in our situa-tion the temperature variation is quite big next to these points, the results of [23] are not confirmed.
- in conclusion, we can state that the obtained results using Algorithm 2 are better than Algorithm 1, thus we can conclude that Algorithm 2 is more suitable to be used to develop the Data Validation Layer. During the experiment, new ideas raised to further develop the algorithms used by the Data Validation Layer and also to develop new IoT devices to detect in a more accurate way if the bearing gives any signs that may start to perform abnormally. As new research directions to further develop the newly introduced automatic

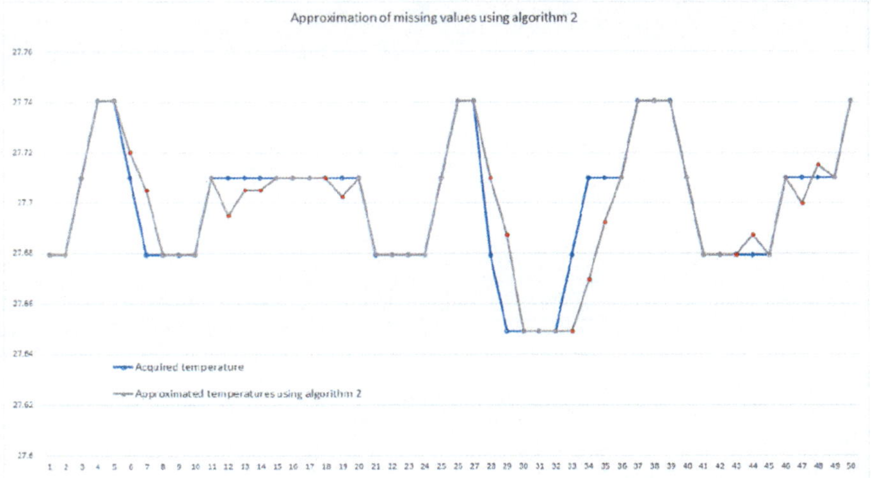

Fig. 8. Approximation of missing values using Algorithm 2 for multiple missing values.

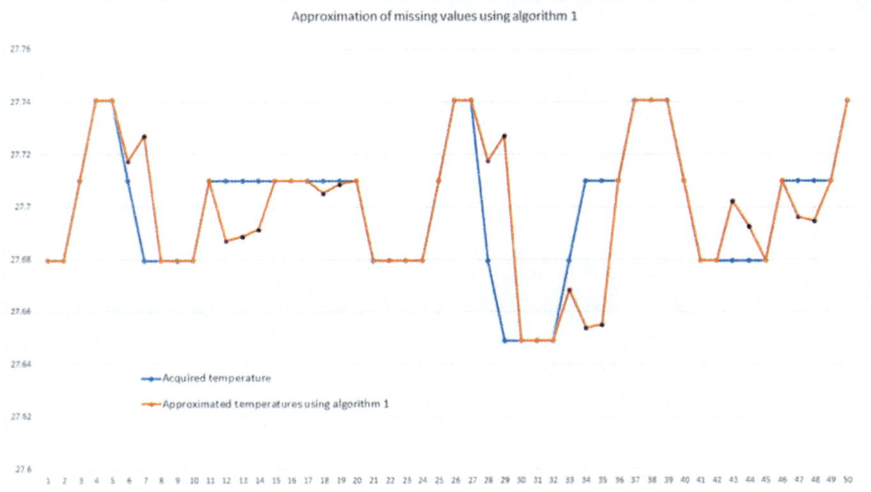

Fig. 9. Approximation of missing values using Algorithm 1 using parametrization $r = 5$ and $\lambda = 10$ for multiple missing values.

system based on Riesz MV-algebras for PrM of bearings of indus-trial equipment using temperature sensors, we mention:

- considering the results of the two different parametrizations used by the Algorithm 2, we came to the conclusion that we could probably get better results if we would develop a new algorithm that uses a mix of parametrizations for

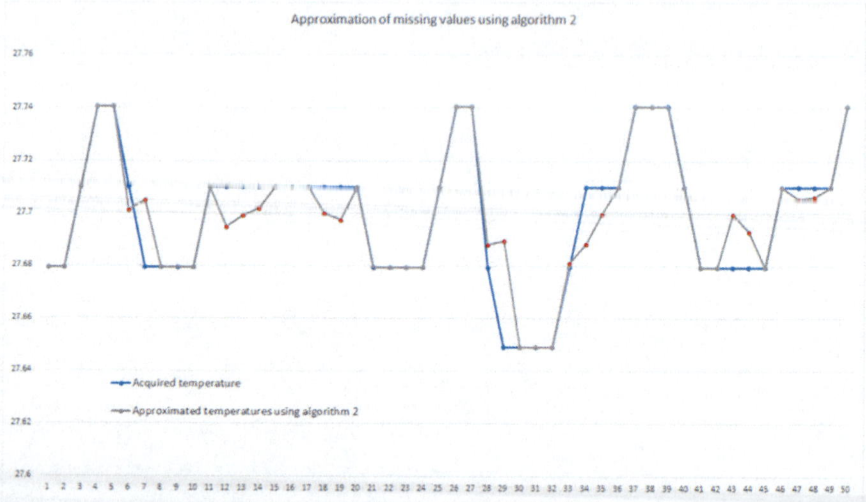

Fig. 10. Approximation of missing values using Algorithm 2 using parametrization $r = 5$ and $\lambda = 10$ for multiple missing values

the radius r depending if we are or we are not in the region of a local minimum or maximum;
- we have considered that a signal is altered by noise if the difference between the acquired value and the approximated value is at least x °C, but more accurate detection algorithms can be used. Thus, a research idea would be either to develop new algorithms to detect signals altered by noises or to identify existing algorithms that are suitable for this purpose;
- to develop applications considering the bearing temperature as a time and bearing revolution speed function. In this case we should use the generalized approximation algorithm described in [22];
- to develop new IoT devices starting from other types of sensors, like vibration or sound detection sensors.

5 Conclusions

This research started from the structure of an IoT-based predictive maintenance system as it is described in [30] and from the necessity to provide a complete set of validated data to the Predictive Analysis Layer such that all decisions regarding the actions that have to be made to be as accurate as possible. In order to achieve this goal, we proposed to add a supplementary layer, called Data Validation Layer, in the structure of the analyzed IoT-based predictive maintenance system. This supplementary layer roles to verify if the acquired data by the IoT devices is altered by noises or is missing. If the data is altered by noises, it is considered as missing. In both situations, a new value is generated using approximation algorithms and inserted in the acquired date stream. To demonstrate the viability of this new architecture, we performed experiments by eliminating some acquired temperatures of the bearings of an electric motor, values that were considered either altered by noises or not acquired, with new values generated using Shepard local approximation operator [23]. Were considered two implementations, the first one is using for approximation only the validated data acquired before the one that should be generated and a second implementation that is using for approximation the validated data acquired both, before and after the one that should be generated. The second implementation has provided better results, but from our point of view both implementations provided good enough results to validate the scope of the experiments. Also, out of the performed experiments, we concluded that based on the possible variation paten of the expected acquired values, several parametrizations of the Algorithm 2 can be used. As further development of the Data Validation Layer, can be considered:

- testing and/or developing new methods to detect the acquired data by sensors that are affected by noise;
- testing and/or developing new methods to approximate the acquired data by sensors that were not validated (affected by noise or missing);
- to optimize the data transfer, we can consider compressing the data contained in the processing data buffer and after it is received by the Data Storage Layer to be decompressed.

Funding Information. The research has been funded by the University of Oradea, within the Grants Competition "Scientific Research of Excellence Related to Priority Areas with Capitalization through Technology Transfer: INO - TRANSFER - UO", Project No. 326/21.12.2021.
This project is funded by the Romanian Ministry of Research, Innovation and Digitisation through Programme 1 - Development of the National Research and Development System, Subprogramme 1.2 - Institutional Performance - Projects for funding the excellence in RDI, Contract No. 29 PFE / 30.12.2021 with University of Oradea.

References

1. Adila, A.S., Husam, A., Husi, G.: Towards the self-powered internet of things (IoT) by energy harvesting: trends and technologies for green IoT. In: Proceedings of the 2018 2nd International Symposium on Small-scale Intelligent Manufacturing Systems (SIMS), pp. 1-5. IEEE, Cavan (2018)
2. Bede, B., Di Nola, A.: Elementary calculus in Riesz MV-algebras. Int. J. Approx. Reason. **36**, 129–149 (2004)
3. Cakir, M., Guvenc, M.A., Mistikoglu, S.: The experimental application of popular machine learning algorithms on predictive maintenance and the design of IIoT based condition monitoring system. Comput. Ind. Eng. **151**, 106948 (2021). https://doi.org/10.1016/j.cie.2020.106948

4. Cardona, M., Cifuentes, M., Hernandez, B., Prado, W.: A case study on remote instrumentation of vibration and temperature in bearing housings. JLPEA **11**, 44 (2021). https://doi.org/10.3390/jlpea11040044
5. Carvalho, T.P., Soares, F.A.A.M.N., Vita, R., Francisco, R.D.P., Basto, J.P., Alcalá, S.G.S.: A systematic literature review of machine learning methods applied to predictive maintenance. Comput. Ind. Eng. **137**, 106024 (2019). https://doi.org/10.1016/j.cie.2019.106024
6. Chang, C.C.: Algebraic analysis of many valued logics. Trans. Amer. Math. Soc. **88**, 467–490 (1958)
7. Chang, C.C.: A new proof of the completeness of the Lukasiewicz axioms. Trans. Amer. Math. Soc. **93**, 74–80 (1959)
8. Csokmai, L.S., Ţarcă, R.C., Bungău, C., Husi, G.A.: Comprehensive approach to off-line advanced error troubleshooting in intelligent manufacturing systems. Int. J. Comput. Commun. **2014**(10), 30 (2014). https://doi.org/10.15837/ijccc.2015.1.1561
9. Di Nola, A., Flondor, P., Leustean, I.: 2004. MV-Modules J. Algebra **261**, 21–40 (2003)
10. Esim, E., Yıldırım, Ş: Drilling performance analysis of drill column machine using proposed neural networks. Neural Comput. Appl. **28**(1), 79–90 (2016). https://doi.org/10.1007/s00521-016-2322-8
11. Falekas, G., Karlis, A.: Digital twin in electrical machine control and predictive maintenance: state of the art and future prospects. Energies **14**, 5933 (2021). https://doi.org/10.3390/en14185933
12. Faysal, A., Ngui, W.K., Lim, M.H., Leong, M.S.: Noise eliminated ensemble empirical mode decomposition scalogram analysis for rotating machinery fault diagnosis. Sensors **21**, 8114 (2021). https://doi.org/10.3390/s21238114
13. Fontes Godoy, W., Morinigo-Sotelo, D., Duque-Perez, O., Nunes da Silva, I., Goedtel, A., Palácios, R.H.C.: Estimation of bearing fault severity in line connected and inverter fed three phase induction motors Energies, **13**, 3481 (2020). https://doi.org/10.3390/en13133481
14. Franciosi, C., Voisin, A., Miranda, S., Iung, B.: Integration of I4.0 technologies with maintenance processes: what are the effects on sustainable manufacturing? IFAC-PapersOnLine, **53**, 1-6 (2020). https://doi.org/10.1016/j.ifacol.2020.11.001
15. Hashemian, H.M.: State-of-the-art predictive maintenance techniques. IEEE Trans. Instrum. Meas. **60**, 226–236 (2011). https://doi.org/10.1109/TIM.2010.2047662
16. Hou, Z., Lv, X., Zhuang, S.: Optimized extreme learning machine-based main bearing temperature monitoring considering ambient conditions' effects. Energies **14**, 7529 (2021). https://doi.org/10.3390/en14227529
17. Li, D., Landström, A., Fast-Berglund, Å., Almström, P.: Human-centred dissemination of data information and knowledge in industry 4.0. Procedia CIRP **84**, 380–386 (2019). https://doi.org/10.1016/j.procir.2019.04.261
18. Mobley, R.K.: An Introduction to Predictive Maintenance, 2nd edn. Butterworth-Heinemann, Amsterdam, New York (2002), ISBN 978-0-7506-7531-4
19. Noje, D., Bede, B.: The MV-algebra structure of RGB model. Stud. Univ. Babes-Bolyai, Informatica **56**, 77–86 (2001)
20. Noje, D.: Using Bernstein polynomials for image zooming. In: Proceedings of the Symposium Zilele Academice Clujene, Computer Science Section, pp. 99–102 (2002)
21. Noje, D., Bede, B.: Vectorial MV-algebras. Soft Comput. **7**(4), 258–262 (2003)

22. Noje, D., Tarca, R., Dzitac, I., Pop, N.: IoT devices signals processing based on multi-dimensional shepard local approximation operators in Riesz MV-algebras. Int. J. Comput. Commun. **14**, 56–62. https://doi.org/10.15837/ijccc.2019.1.3490
23. Noje, D., Dzitac, I., Pop, N., Tarca, R.: IoT devices signals processing based on Shepard local approximation operators defined in Riesz MV-algebras. Informatica **31**(1), 131–142 (2020). https://doi.org/10.15388/20-INFOR395
24. Parpala, R.C., Iacob, R.: Application of IoT concept on predictive maintenance of industrial equipment. In: MATEC Web Conference, vol. 121, p. 02008 (2017). https://doi.org/10.1051/matecconf/201712102008
25. Pech, M., Vrchota, J., Bednář, J.: Predictive maintenance and intelligent sensors in smart factory: review. Sensors **021**(21), 1470 (2021). https://doi.org/10.3390/s21041470
26. https://www.avnet.com/wps/portal/abacus/solutions/markets/industrial/predictive-maintenance-iot/
27. https://pri-desk.ro/
28. https://www.scnsoft.com/blog/iot-predictive-maintenance-guide
29. Selcuk, S.: Predictive maintenance, its implementation and latest trends. Proc. Inst. Mech. Eng. Part B: J. Eng. Manuf. **231**, 1670–1679 (2017). https://doi.org/10.1177/0954405415601640
30. https://www.softwareadvice.com/resources/iot-predictive-maintenance/
31. van den Hoogen, J., Bloemheuvel, S., Atzmueller, M.: Classifying multivariate signals in rolling bearing fault detection using adaptive wide-kernel CNNs. Appl. Sci. **11**, 11429 (2021). https://doi.org/10.3390/app112311429
32. Zerhouni, N.: From Prognostics and Health Systems Management to Predictive Maintenance 1: Monitoring and Prognostics. ISTE Ltd/ Wiley Hoboken, NJ (2016)

Smart Contracts: A Valuable Technology in ERP e-Commerce Systems and for Customer Experience

Maria Georgiana Stoica(✉)

Department of Economic Informatics and Cybernetics, Bucharest University of Economic Studies, Romana Square 6, 010374 Bucharest, Romania
stoicamaria15@stud.ase.ro

Abstract. In the past few years, more and more e-commerce companies are progressively looking for business opportunities and ways to evolve in terms of their informatic system in such a crowded market and gain a unique customer experience to earn the client's trust. This need is exacerbated, taking into account the competition generated by the major market leaders, by constantly introducing innovations in planning the business resources, and by the results-oriented for stimulating the competition. This study theoretically systematizes how blockchain technology and smart contracts are slowly becoming a necessity in the ERP systems from the big companies that offer online shopping services, whether B2B or B2C, providing extra security for its clients. The models defined in this study highlight the technical efficiency in a resource system planning of three companies, market leaders in Romania, in terms of supply-chain, secure payments, and forecast, for comparing the costs of the change when using blockchain technology. Moreover, for a better shopping experience in online stores, the efficiency and effectiveness scores of applying these technologies to the customer review section, so that frauds are analyzed and can be significantly reduced by making in this way a secured and more reliable system. The study analysis of these aspects shows that the efficiency of minimizing frauds and improving the technology of ERP development by creating a safe and stable environment varies substantially between models and leads to a new era in e-commerce that is so needed nowadays.

Keywords: Blockchain · e-commerce · ERP

1 Introduction

With the rise of globalization and the perpetual change of individual's lifestyle according to nowadays needs, in the e-commerce domain, there is a growing demand to cover all the customer needs, but also an evolution towards a technology that gives business profit and minimization of business costs. A decisive factor in the significant emphasis of the business performance analysis in the e-commerce business is the current pandemic, which has risen in the last 3 years the need to increase IT and ERP systems, due to the huge demand of orders and online transactions. Due to this, large volumes of customers

S. Dzitac et al. (Eds.): ICCCC 2022, AISC 1435, pp. 20–33, 2023.
https://doi.org/10.1007/978-3-031-16684-6_2

are directed to the major market leaders, because as a buyer, the decision is to choose the most stable, secure, and reliable IT product, through which he can meet his need. From the company's point of view, they are facing the satisfaction and maintenance of the shop characteristics that customer is attracted, but also the fact that a high evolution is needed to keep competitors at distance and to face with technology which continues to evolve and surprise through its new applications. Given these aspects, the blockchain is considered a remarkable technological step forward that leaves a significant touch on everyday business, especially in the field of e-commerce.

According to the World Economic Forum [11], this technology has the potential to become a powerful tool for transactions, so it helps users gain trust, boost business and drastically reduce corruption. More specifically, if we take as an example the big market leaders in e-commerce such as Amazon, Carrefour, or IBM, they have already approached blockchain technology since 2019 in their solutions for the supply chain, but also for verifying the authenticity of customer goods. More specifically, through the blockchain technology applied to SCM, companies can document and keep track of production updates in a single shared register, which has the role of providing transparency to the data, and more importantly, the data source will always be correct. In this way, from the stage of production to the stage of delivery of the product to the client, companies can query the status and location of the ordered product at any time. This helps to combat issues such as counterfeit goods, breaches of compliance, delays, and waste [9]. In contrast to these international models, this technology has not yet been adopted nationally in Romania, in the field of e-commerce.

Having all this information about the business-level technological models of the big e-commerce giants, the current study simulates and make a detailed analysis of the integration of blockchain technology and smart contracts on the SCM, payments, and statistical forecasts at the level of company costs, but also it will be explaining a method to avoid fraud at the level of client reviews. The study will detail two perspectives, namely: the company and the client. Also, for an accuracy of the data and the evolution of this study, we will analyze the aspects of 3 companies from Romania, eMAG, Fashion Days, and Altex, and we will analyze the growths and changes that will take place on users as well as businesses in case integration of blockchain technology and smart contracts at ERP level in e-commerce companies.

2 Background

I will start by reviewing the existing literature and research on technologies applied to e-commerce systems and how the strategy of the resource planning system can influence the value of system security and customer confidence. Then I will identify the reasons why a customer is influenced to shop online and the issues that need to be improved technologically. I will consider 3 online stores in Romania, namely eMAG, Fashion Days, and Altex and I will follow the aspects that I specified in correlation with the international models that have adopted smart contracts at the ERP level to minimize business costs and client confidence. I will refer to the Amazon e-store as a reference for its successful integration of blockchain technology.

Several studies are researching the impact of smart contracts and how blockchain technology influences ERP systems or more, the supply chains. In a study, it is making detailed references on how to achieve real-time inventory visibility, full visibility, and real-time updates for the price, product catalog, price breaks, and many others aspects. In comparison with this study, it does not offer a clear architecture or design of such a service, but theoretically presented aspects regarding the automation of many expensive processes and integrations that can be done in real-time, such as Demand and Supply Management, Logistics, and Manufacturing [16]. An important thing in this study is the advantages offered for stakeholders in a business that integrates smart contracts, but also how the origin of the products can be found, and more importantly the state, the origin of the ingredients or materials that make up the product. The authors also propose to record this data through a source map.

Another notable journal study provides general information on the implications of blockchain in industry 4 and details the supply chain, but for the field of construction. This time, it is presented an architectural model of creating a database that manages the supply chain and uses blockchain. By comparison to this study, it does not detail the entire level of ERP, but only details the process of data management and the benefits of their transparency [17]. Also, in another article in the International Journal of Information Management, the study assesses profitability geographically in India but does not cover all aspects of an SCM, such as supplier, buyer performance, integration times, and technology cost minimization [18]. However, the demographic profit, Reliability, and AVE values brought by the blockchain integration in the SCM are summarized.

A more practical study, which was applied to a range of 236 people, which, although not in the field of eCommerce, presents concrete and measurable information on the profitability of the blockchain brought at the level of SCM, improving in 6 dimensions this aspect. It is also concluded to improve the reliability of delivery and mass customization which significantly increases the profitability of companies [19]. The dimensions analyzed would be CRM, information quality, company improvement, product quality, SCM, and mass customization. Interestingly, it was concluded that among many other good practices in using blockchain, one would be integrated into it when the main players in the supply chain are mistrust [19].

2.1 Empirical Regularities on National and International e-Commerce Trends

Empirical research at the national level, more precisely in Romania, results that any improvement in e-commerce technologies inevitably leads to a sharp increase in employee productivity, in this way facilitating costly processes [13]. At the same time, there is a progression in the efficiency of the IT system, which leads to an increase in the competitiveness of companies and the acceleration of interdependencies between national economies and markets. For this reason, how e-commerce is constructed is a decisive factor in the activity of economic organizations, so it leads to better management of forecasting strategies and estimation of decisions taken at the level of the planning system [2].

The 3 online stores, eMAG, Fashion Days, and Altex, leaders on the Romanian national market, have similar technologies, but none of them use blockchain in the resource planning system. EMAG and Fashion Days employ ERPs closely related to a Business Intelligence solution. The first two are part of the same company, so they use similar technologies in their approach to resource planning systems. To maintain its market-leading position in sales, the first two e-stores have chosen Hostway as their hosting provider and now operate on more than 40 IBM and Dell servers. Also, in addition to Altex, the latter do both B2B and B2C trade, so they offer other companies services through which they can sell and supply their products through eMAG. In terms of customer satisfaction for the supply chain, Altex manages to deliver the ordered products with an average greater than eMAG. This is due to the technologies approached in the supply chain management system so that any improvement in the technology results in faster delivery of orders. There is also no specialized track and trace system, especially at the Altex e-shop. This has, as a consequence for the customer, the dissatisfaction with the need to track the goods, but also an inefficient forecast for the company because it is not possible to accurately track the duration and route of the package to the destination. To avoid fraud at a high level in the supply chain, blockchain technology is a great success.

Analyzing the mentioned e-commerce websites, both eMAG, Altex, and Fashion-Days, use as payment method PayU, which is one of the oldest electronic payment processors since 2006 in Romania. After completing an order, a request is made to the server, where an API is called from inside the ERP to PayU, where the payment will be processed, and it will return with an answer to the order completion page.

Most e-stores in Romania use robust payment gateway solutions and services. Similar is the payment gateway service provider, PayU, because it simplifies the connection between the customer and the business by making payments and ensuring that sensitive information is kept secure and nothing else more. The question is how this guarantee can be improved and how e-commerce can evolve into future technology, by transforming the business into the future, and the answer is blockchain.

Why blockchain? Even though the existing traditional payment systems are still effective, they have major limitations, such as delays in cross-border transactions, lack of process automation, and involvement of intermediaries. Moreover, there is a growing need to improve payment systems in areas such as the supply chain and retail, where, of course, it is essential to maintain a transparent balance between the flow of goods and money. Blockchain-based payments offered by Gateway services, such as Mastercard Blockchain, Samsung Blockchain Platform, Solana Pay, and many more, have become increasingly popular due to the promising future that it brings, in which digital currencies predominate and money digital moves through the internet as data, which means without censorship and intermediaries charging each transaction. These services are known for their speed and low transaction costs, and Solana is one of the best performing digital currencies in recent years, with has great potential when it comes to solving some of Ethereum's most pressing issues [14]. Blockchain-based payments get so much faster transaction processing times and offer consumers payment with a wide range of digital currencies from their e-wallet directly to the merchant's account, immediately settling the measured costs in cash fractions. The Solana Pay payment gateway guarantees the processing of over 50,000 transactions per second [14] which is amazing compared to about 100 payments per second as offered by a regular payment API, which means that Blockchain solutions are 500 times faster than traditional ones. This incorporates consumer loyalty but also the confidence of the integration of this technology by business people in their companies and thus becomes real weapons for the future of e-commerce platforms and businesses.

What companies want is to secure their data, to effectively collect customer data, such as name, age, buyer profile, and this can be strengthened and improved by integrating blockchain solutions into e-commerce systems to provide complete security and wallet protection. A blockchain-based currency requires consumers to disclose sensitive data.

Moreover, considering a blockchain encryption algorithm that does not fail, the obvious conclusion is that by implementing this technology of the future, fraud is minimized, and a secure, stable, and fast ERP is created, which offers security and customer confidence and certainly, the conversion rate, from the visitor status of the e-commerce platform to buyer status, will be higher than 3%, as estimated on average [15].

$$Conversion\ Rate = Number\ of\ Customers/Number\ of\ Visitors;$$

Therefore, taking into account the payment method, Blockchain technology offers more secure and credible solutions and services for e-commerce businesses, combining transparency in transactions with more reliable management of consumer data. After all, the benefits of blockchain for e-commerce extend to cheaper business processes, greater data security, and a substantially improved customer experience.

In contrast to Romania's national models, Amazon has reinvented e-commerce supply chain management by implementing smart contracts in SCM. The company focuses on high response capacity, combined with warehouse automation and robotics, which results in one-day order delivery and even offers two-hour delivery with Prime Now technology. Due to this, Amazon is attracting more and more customers from various geographical areas and gaining their trust by massively fulfilling orders. In September 2019, Amazon purchased 100,000 electric delivery vans, and in 2020, the company took advantage of low prices during the coronavirus pandemic to expand its means and vehicles for air transportation. As estimates for the future, Amazon aims to deliver drone packages to customers with Amazon Prime Air, to minimize the length of the goods delivery process [12].

In this way, Amazon is becoming an international e-commerce ideal, thanks to its well-developed ERP system, but also for the innovation of acceptance and integration of new technologies, such as the implementation of smart contracts in the SCM system and beyond. Romania is lagging slowly towards this business ideal, but there is still a lot of work to be done to reach this IT potential, as the current technology we have thus limited the resource planning systems as well as the satisfaction index of customers, keeping on a steady trend.

2.2 Correlation Between SCM and Customer Satisfaction on Regular ERP Systems

In the digital economy, the study of buyers focuses primarily on technologies that can change the demand curve and, secondly, on interaction with buyers to stimulate demand in their favor. Therefore, I will further present through a predictive analysis performed between the independent variable that often influences the customer towards the purchase of a product, the quality, and the dependent variable the customer's expectations for the ordered product.

To emphasize some measures for the client expectations I analyzed an existing dataset from xlstat.com [8], which is a web platform that collects data. The results are based on real-life data, where 250 customers of mobile phone operators have been asked several questions to be able to model their loyalty. Each latent variable is related to one or more manifest variables that are measured [8]. In this case, the manifest variables questions are on a 0–100 scale. For example, for the Satisfaction latent variable, the five manifest variables are: the product can be trusted in what it says and does, it is stable and functional, it is delivered fast, the quality is good, the reviews are not fake.

It can be seen in Table 1 that the independent variable, namely the quality of the products, has the most important effect on customer satisfaction, followed by the perceived value. The impact of expectations is insignificant. At the same time, it is observed that the independent variable given by the quality perceived at the client's level in an e-commerce system has a contribution of approximately 64.8% to R^2.

Table 1. Equation of the model: Quality = 0.556741866775216 * Expectation

ANOVA								
Latent variable	Value	Standard error	t	Pr > |t|	f^2	Critical ratio	The lower bound (95%)	Upper bound (95%)
Intercept	0.000	0.053	0.000	1.000		−0.580	0.000	0.000
Expectation	0.557	0.053	10.555	0.000	0.449	10.723	0.433	0.648

As such, approximately 35.2% of the variation in customer satisfaction is explained by other external factors, which may be the payment methods, the trust in the respective electronic store, the influence of the reviews. Therefore, according to Fig. 1, it will be analyzed to what extent the change in technology brought to an IT system related to the e-store, can influence the value of 35.2% of foreign factors and how we can increase buyers' confidence and ERP performance in that company.

3 An Empirical Strategy for Building Client's Trust by Implementing Smart Contracts

I will start by presenting and theorizing the ideal model of an ERP system based on building smart contracts on the blockchain and in what proportion it is more significant than the traditional model. Also, a comparative analysis was made between the proposed system and the existing ones, that will illustrate the benefits of the blockchain solution and is being considered the impact of its implementation on the supply chain, payments, and how significantly improves economic forecasting.

3.1 A Theoretical Blockchain Model for ERP System Benefits

From managing payments and reviewing audits to tracking inventory and assets, blockchain technology allows for a much larger supply chain platform than ever before.

At the same time, the electronic payment model, which would be an intermediate step in the SCM process, is an important step that needs to be changed by implementing smart contracts, to increase customer confidence, but also to reduce costs for the currently used model nationwide. The companies mentioned in the previous chapter, namely 3D Secure technology [1].

Regardless of the intended application, blockchain technology offers so many advantages to the current system that Romania also tends to adapt to this already existing technology at the international level in the field of e-commerce.

Some of the most important advantages over the regular supply chain system are increased transparency, greater scalability, better security, and increased overall innovation. The biggest advantage of the blockchain is its transparency. By documenting the process of migrating a product along the entire supply chain, manufacturers can thus reduce withdrawals by sharing logs with regulators and all relevant parties in the event of a problem [5]. In this sense, it will increase customer confidence, but also better management at the company level, and last but not least, a powerful tool in building future statistical and economic forecasts at the data level.

Table 2. Comparison between blockchain and traditional SCM in eCommerce [7]

Comparative item	Blockchain-based SCM	Traditional SCM
Data integrity	Tamper proof	No temper proof
Data security	Decentralized protection	Centralized protection
Data verification	Archived	Not archived
Government regulations	Using authentication protocol	No auth protocol

I have summarized above in Table 2 a comparative detailed analysis about the benefits of Blockchain for a traditional e-commerce system in the supply chain management sector.

Blockchain can provide greater security than current strategies. Virtually any number of participants can access information from any number of contact points, providing flexibility as well as security throughout the system. Security is an advantage of blockchain technology, as a permanently shared registry with hard-coded rules could reduce or eliminate the need for audits required by internal systems and business processes [7].

Finally, enhanced innovation is a huge advantage over our current technology, with new and specialized uses being able to be created from Blockchain technology to provide an extra layer of encrypted protection, which tends to be an easily scalable and transparent architecture.

3.2 Proposed Model for e-Commerce ERP System

In this section, I am proposing a model, based on blockchain and distributed systems, attached to any e-commerce company for supply chain management and the visibility of payments in the system.

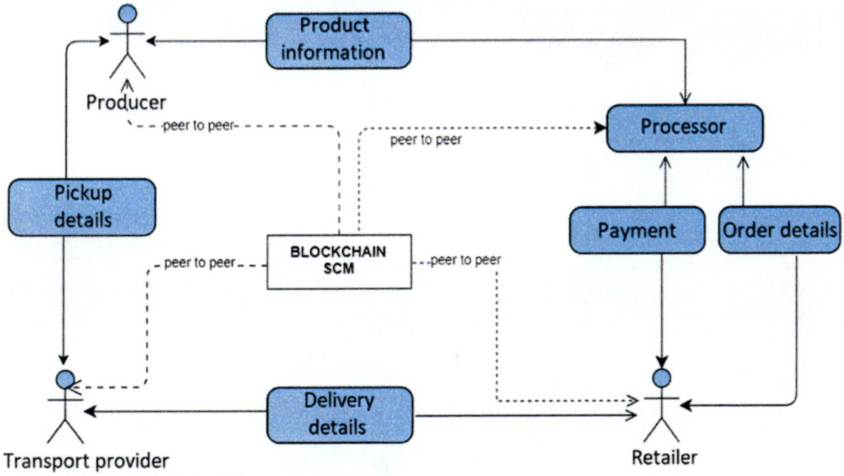

Fig. 1. A supply chain management blockchain architecture

Figure 1 depicts at the architectural level the links that are created in supply chain management by applying blockchain technology. Basically, at the SCM level, the visibility increases at the level of each step of the product migration from the manufacturer, to the seller and then to the customer. During this time a thorough check is made and each recorded process is stored from the beginning of each event until it transits in each node representative of the SCM process. The payment process is visible and immutable, being added to the blockchain, and this aspect favors the minimization of frauds that can be made at the level of payments between producer and seller [6].

In an e-commerce system that conducts either a B2B or B2C business, the relationship between the company and the supplier is as important as the business and customer relationship. Thus, the architecture presented above strengthens the link between the manufacturer and the seller, and also between the e-commerce business and the producers from whom the supply is made. This is possible through the visibility and transparency of the data provided to the mentioned actors, thus gaining performance on the trust between them.

Next, I analyze the activity at the blockchain level, a process visible in Fig. 1, above, and I will explain the benefits of each process built in the proposed algorithm for supply and order management.

In this way, Fig. 2 diagrams the exchange of messages at the level of micro-services between the service of Manufacturer, Transport, Transport points, Delivery, and Customer service. Specifically, each micro-service communicates with the other micro-services by exchanging messages.

At the ERP level, the micro-services are intended to be implemented through the Apache Kafka service-level architecture as is shown in Fig. 3 below. Single Kafka clusters can even be deployed over multiple regions and globally. Mission-critical deployments without downtime or data loss are crucial for Blockchain and many other use cases. Architecture patterns for the distributed, hybrid, edge and global, Apache Kafka

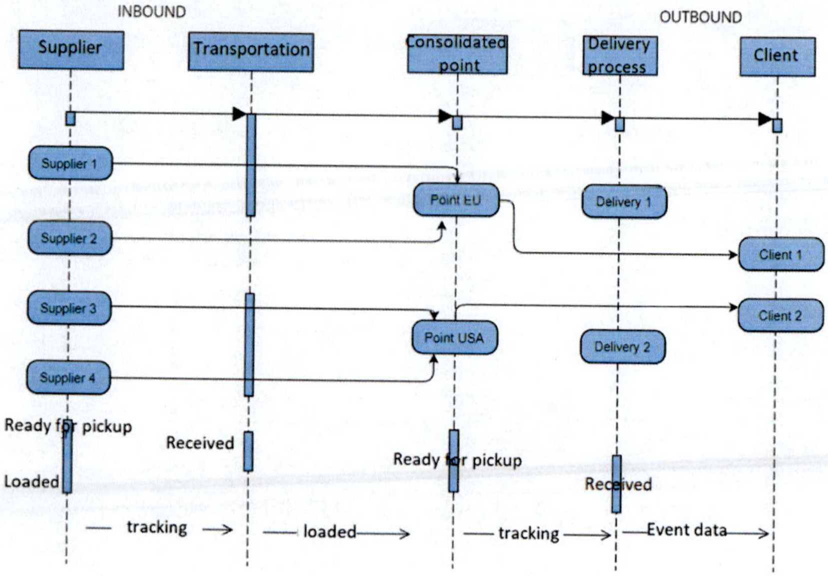

Fig. 2. Activity diagram for blockchain process in SCM

deployments explore these topics in detail. Kafka will be run as a cluster of one or more micro-services, which can span multiple data centers or cloud regions, such as the Client-side or the Consolidated point region. They continuously import and export data through event streams, making it possible to integrate this architecture with blockchain.

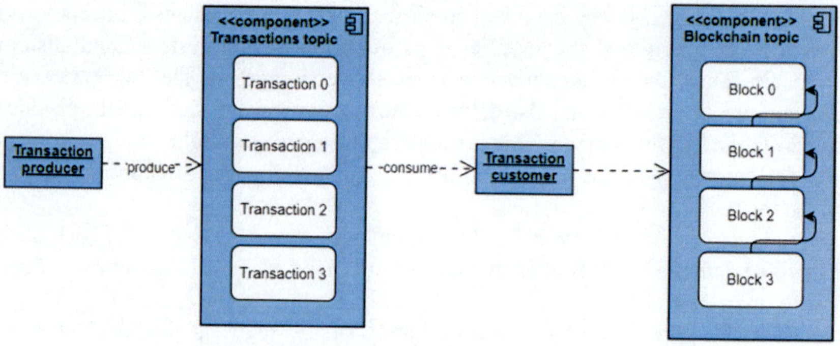

Fig. 3. Transaction's component diagram for integration Kafka with blockchain

Also, for the infrastructure, a big step is to use or build a Kafka-native blockchain infrastructure, as in Fig. 3, so the deployment will be sequentially by the topics and the blocks in the consumer and producer Kafka API protocol. It tends to get high scalability and high-volume data processing in real-time for mission-critical deployments is a crucial differentiator of using Kafka compared to traditional blockchain deployments. Kafka enables a blockchain for real-time processing and analysis of historical data with one single platform. Messages are alternatively encrypted and hashed consecutively. The library methods are called inside a Kafka application's message producer code to wrap messages and called inside the application's customer code to unwrap messages. Since blockchains must be entirely successively requested, Kafka blockchain points must either have a single partition, or consumers for each segment must participate to arrange and sequence the records.

At the same time, the integration of the two technologies significantly minimizes data loss, because if one of the Kafka servers loses the connection or no longer works in normal parameters, the other servers will take over the work to ensure continuous operations without any data loss.

Therefore, it will greatly increase the visibility and transparency of data in supply chain processes and payments made between the main actors of the SCM process, having as indisputable effects the increase of technical performance as well as the improvement of operational processes.

3.3 Blockchain as a Fake Reviews Solution

In addition to the benefits presented above, blockchain also has the power in e-commerce business to confirm whether the reviews brought by its supposed customers are true or fake. With the ability to create a digital map of the customer, which cannot be modified, the blockchain can thus check whether or not users who have left a review have purchased on the site before publishing them. More specifically, when buying something new or trying out a new service, many of us head online to check what other customers think. Over 80% of adults in the United States occasionally read online customer reviews before buying something for the first time. But only half of those who read reviews think they give an accurate description of the product, according to the Pew Research Center [10]. So, the blockchain has the power to track customer information such as if he ordered before and if the review is about what he ordered, and if it is relevant.

Customer reviews are especially important in an e-commerce business because they have a significant influence when a customer chooses whether or not to buy a product. Beyond these benefits, blockchain technology, used in conjunction with machine learning, can match a review with a real and appropriate user. In this case, the services can significantly reduce fraud and allow people to convey their opinion about their purchases, the quality of orders, the duration of transport, or the quality of the products purchased [4].

The system runs on a blockchain with powerful algorithms that take into account customer preferences for features such as location, language, product or brand preferences, cost, and other useful factors to provide real, useful, and appropriate product reviews. Which customers follow. False reviews can be prevented using a verified digital identification, supported by blockchain immutability and security. By combining a biometrically authenticated ID with a payment source, you can create an unquestionable way to authenticate reviews. An example is implemented by Amazon which consists of biometric verification of your digital identity when buying a product. One day after you have purchased with the funded decentralized identity, the courier delivers the package to you, by pinning your digital ID to allow contactless delivery. After this, the customer leaves a sincere review of your digital identity verified by payment [12].

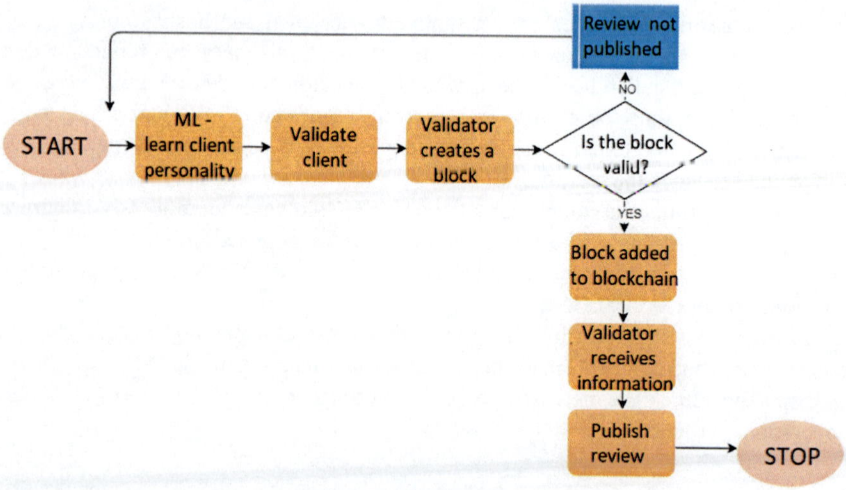

Fig. 4. The logic schema for blockchain review algorithm

Figure 4 illustrates this, and so the algorithm is proposed by combining blockchain technology with machine learning to increase data accuracy so that a real profile of each client will be created by ML and verified by blockchain whether or not it is a client.

4 The Effects of Smart Contracts and ERP Performance Evaluations

Blockchain gives companies the ability to have different types of information in their resource planning system, which also gives the parties easy access to information. For example, in the traditional supply chain process, the transit of products until they are delivered to customers is not easily monitored. Blockchain integration would provide real-time product traceability.

Once a smart contract is integrated into an organization's ERP system, it optimizes operations, internal data control, and business processes, such as business-to-business transactions. For example, as discussed in previous chapters, smart contracts can help to automatically verify a transaction between companies, specifically in the SCM between the manufacturer and the seller [6]. When integrating the blockchain into an organization's ERP system, this should ideally be done with minimal disruption to normal business processes and avoid hampering the processing and verification of data and event algorithms.

Blockchain within the supply chain significantly progresses straightforwardness, quality, security, and handle keenness through the secure record. Essentially, within the fabricating segment, Blockchain empowers unwavering quality and works as a trust-based layer without go-between intercession. Whereas companies inevitably target utilizing open blockchains for commerce operations, actualizing private blockchains is less demanding to control and oversee security and protection. The Blockchain and ERP integration makes and ports unused data into the dispersed record and draws existing information from endeavor frameworks [3]. Blockchain ERP combination empowers businesses to control information sharing firmly.

The decentralization property of Blockchain gives companies the potential to join together a huge supply chain organize. Blockchain in ERP combinedly makes a difference to make strides in the robotization supply chains. So, each company can utilize its inside ERP framework whereas counting one rule-enforced blockchain arrange. This brings up a few unmistakable focal points. For illustration, it diminishes the visibility issue inside the supply chain. Each member within the whole framework can follow the item travel from fabricating to last conveyance on the rack.

Not at all like ERP frameworks, Blockchain ERP arrangement is additionally cheaper as all the members like producers, retailers, and wholesalers can utilize the same arrangement. So, the trade rules and the information right now dwelling with diverse companies would be accessible in one organization [1]. And each exchange would naturally be recorded on the dispersed record innovation. Companies would have perceivability into the complete supply chain. In any case, they can still record and keep the private information important. They don't have to be share commerce insights.

Last but not least, having transparency of the data from the ERP system of the e-commerce companies, makes possible the economic forecast and the accurate estimation of the company's statistical forecasts.

5 Conclusions and Future Research

In this paper, I have discussed what blockchain technology has brought to an e-commerce company, more precisely at the ERP level, including excellent security performance, helped to reduce frauds, improved performance at the SCM level, and improved economic forecasts due to transparent data brought through the smart contract, but it also brings high energy costs. Considering the DCS theorem [21], energy costs in enterprise permissioned BC platforms are significantly lower compared to pure decentralized solutions like Ethereum or Bitcoin which employ Proof of Work as the consensus mechanism. The energy cost is blurred by the profit from the implementation and integration of this solution, as the performance and security costs increase significantly, and with this slope, the customer's confidence and gain increases.

I have also proposed an architecture model at the ERP level, more precisely for better supply chain management, by integrating the Apache Kafka architecture with Blockchain technology, a more secure payment method that avoids online fraud, and a model for an ideal solution for creating real customers, starting from the proposed intelligent contract integrated with machine learning for the review section in an electronic store. I analyzed the benefits and costs of implementing the model within the ERPs of the e-Commerce stores in Romania, eMAG, Fashion Days, and Altex, which are based on a traditional ERP, but not much different from the other. Thus, it was found that the change brings the maximization of the number of customers through the numerous performances brought to the IT system. It is a cost-effective decision to adopt this solution, compared to the model that Amazon has adopted, which is mentioned in Sect. 2 of this paper.

Figure 5 shows that the use of the blockchain in an ERP-type system may bring excess costs but compensates for the fact that the performance of the system increases significantly, this result having the effect of satisfying the customer with a secure, trustworthy product. For the widespread adoption of smart contracts by e-commerce markets, blockchain companies should ensure that they reduce the operational complexity and costly transit process of new technology with an existing traditional ERP system.

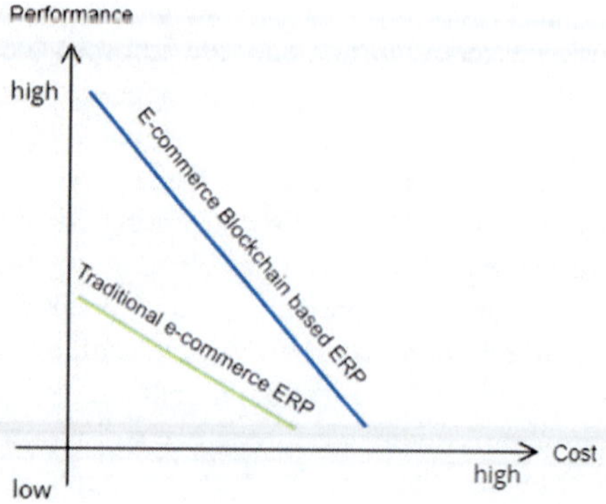

Fig. 5. Technology cost correlated to the performance of ERP system

As future developments, I would like to pursue the application of this technology at the level of CRM and how it could improve the entire business logic of the company because the blockchain provides huge access to data through its transparency at the system level.

Acknowledgement. This paper was co-financed by The Bucharest University of Economic Studies during the Ph.D. program.

References

1. Faulkner, S.: Benefits of Blockchain in E-commerce, case studies, and resources (2021)
2. 5 Ways Blockchain is transforming eCommerce. https://hackernoon.com/5-ways-blockchain-is-transforming-ecommerce-b90b4dd61f1c. Accessed 26 Jan 2022
3. Treiblmaier, H., Sillaber, C.: The impact of blockchain on e-commerce: a framework for salient research topics. Modul University Vienna, Am Kahlenberg (2021)
4. Stallone, V., Wetzels, M., Klaas, M.: Applications of Blockchain Technology in marketing—A systematic review of marketing technology companies. Zürich University of Applied Sciences, Winterthur, Switzerland (2021)
5. Banerjee, A.: Blockchain technology: supply chain insights from ERP, principal consultant, Chap. 3. Infosys Limited, India (2018)
6. Sundarakani, B., Ajaykumar, A., Gunasekaran, A.: Big data-driven supply chain design and applications for blockchain: an action research using case study approach. The University of Wollongong in Dubai, United Arab Emirates (2019)
7. Sunmola, F., Burgess, P., Tan, A.: Building blocks for blockchain adoption in digital transformation of sustainable supply chains. University of Hertfordshire, UK (2021)

8. Consumer satisfaction analysis with PLSPM. https://help.xlstat.com/s/article/consumer-satisfaction-analysis-in-excel-with-plspm. Accessed 02 Feb 2022
9. 3 Immediate Blockchain uses for Ecommerce Companies. https://www.practicalecommerce.com/3-immediate-blockchain-uses-for-ecommerce-companies. Accessed 22 Jan 2022
10. Pew Research Center. https://www.pewresearch.org/fact-tank/2020/03/05/most-americans-rely-on-their-own-research-to-make-big-decisions-and-that-often-means-online-searches. Accessed 04 Feb 2022
11. World Economic Forum. https://www.weforum.org/platforms/shaping-the-future-of-technology-governance-blockchain-and-distributed-ledger-technologies. Accessed 12 Jan 2022
12. Amazon SCM-Blockchain. https://aws.amazon.com/blockchain/blockchain-for-supply-chain-track-and-trace/. Accessed 20 Jan 2022
13. E-commerce in Romania, Statistics and Facts. https://www.statista.com/topics/7418/e-commerce-in-romania/#dossierKeyfigures. Accessed 03 Jan 2022
14. Solana Pay Gateway. https://www.analyticsinsight.net/solana-pay-a-payments-platform-that-will-overtake-paypal/. Accessed 21 Mar 2022
15. Conversion Rate. https://adoric.com/blog/what-is-a-good-conversion-rate-2020/. Accessed 21 Mar 2022
16. Banerjee, A.: Blockchain Technology: Supply Chain Insights from ERP. Infosys Limited, India (2018)
17. Kumar, S.A., Prasath, K.: Smart contracts and supply chain management using blockchain. J. Eng. Res. (2022). ACMM Special Issue, India
18. Karamchandani, A., Samir, K., Rajiv, K.: Perception-based model for analyzing the impact of enterprise blockchain adoption on SCM in the Indian service industry. Int. J. Inf. Manag. India **52**, 102019 (2019)
19. Karamchandani, A., Samir, K., Kumar, S., Srivastava, A.: Analysing perceived role of blockchain technology in SCM context for the manufacturing industry. Int. J. Prod. Res. **59**, 3398–3429 (2021)
20. Al-Farsi, S., Muhammad, M.R., Bakiras, S.: Security of blockchain-based supply chain management systems: challenges and opportunities. MDPI, Basel, Switzerland (2021)
21. Slepak, G., Petrova, A.: The DCS theorem (2018). https://doi.org/10.48550/arXiv.1801.043352018

A Nursing Mobile Robot for COVID19 Patients Using a RPI and Mobile Device

Víctor Lomas-Barrie[1,2,3,4], Mario Peña-Cabrera[1], Mauricio Reyes-Castillo[2], T. Alcantara-Concepcion[3], and Gastón Lefranc[4](✉)

[1] Departamento de Ingeniería en Sistemas Computacionales, IIMAS, Universidad Nacional Autónoma de México, Mexico City, Mexico
{victor.lomas,mario.pena}@iimas.unam.mx

[2] Laboratorio de Robótica, CIDI, Universidad Nacional Autónoma de México, Mexico City, Mexico
mauricio.reyes@cidi.unam.mx

[3] Dirección General de Cómputo y Tecnologías de la Información y Comunicación, Universidad Nacional Autónoma de México, Mexico City, Mexico
talcantarac@unam.mx

[4] Pontificia Universidad Católica de Valparaíso, Valparaíso, Chile
gaston.lefranc@pucv.cl

Abstract. The article shows the design and implementation of a caterpillar structure mobile platform, which includes integrated electronic communication and control devices, as well as 3D printed parts for custom mechanical and additive manufacturing. The objective is to help in the care of patients with COVID19 within the restricted areas of the hospital. PWM technique is used for motion control due to its proven use of robust and reliable performance for service robot applications. The article shows the design and implementation of a caterpillar structure mobile platform, which includes integrated electronic communication and control devices, as well as 3D printed parts for custom mechanical and additive manufacturing. The objective is to help in the care of patients with COVID19 within the restricted areas of the hospital. PWM technique is used for motion control due to its proven use of robust and reliable performance for service robot applications. The implementation is done with a commercial mechanical caterpillar platform with the architecture and electronic components manufactured by 3D printing are described in order to show the scope and technique used to control the operation. The article aims to test the efficiency and reliable performance of a mobile platform using PWM control techniques and Wi-Fi communication implemented in a RaspBerry Pi hardware electronic device as an auxiliary nursing robotic system in hospital areas restricted by COVID19. Assisted remote control is web-based, so any mobile device can perform actions. MQTT WebSocket is used to send and receive commands and telemetry.

Keywords: Mobile robot · Nursery · COVID19 · 3D manufacturing · MQTT · Wi-Fi

S. Dzitac et al. (Eds.): ICCCC 2022, AISC 1435, pp. 34–46, 2023.
https://doi.org/10.1007/978-3-031-16684-6_3

1 Introduction

The global and digitized society today finds in mobile robotics a field of development in continuous growth with important expectations in the near future. Trends such as the delivery of packages with drones by Amazon and DHL, or the granting of licenses for autonomous cars are proof that every day robotics is and will be more present in contemporary everyday life. However, there are applications that by their nature need to be very robust and require the complementary intervention of a human due to the risk involved, whether to interact with human lives, living beings or in high-risk areas, this makes the mobile robotics human assistance systems a very powerful tool.

Mobile robots are used for many purposes, but one of particular relevance, due to the COVID19 Pandemic, is to assist health personnel in hospitals with patients infected with the SARS-CoV-2 virus. Basic tasks performed by a mobile platform such as carrying food, medicine and establishing a remote communication with a patient is the subject of this article. This paper deals with the design and implementation of a commercial caterpillar platform to provide it with a remote-control system and visual feedback for its movement. This configuration allows to control such a robot system with greater precision and smoothness movements, as well as to guarantee that the distance travelled and the direction of the mobile is correct by means of a remote-control system through Wi-Fi networks. Which guarantees to operate correctly [1].

The crisis caused by the pandemic due to Covid-19, researchers from developing countries such as Mexico and Chile, are dedicated to finding local solutions, which are shared, as a way to relieve collapsed medical systems. There are projects similar to the one described in this document, such as the development of a kiosk-platform called Charlie in 2014 focused on helping to monitor the vital signs of older adults in hospitals.

The mobile robot was developed at the Korean company Yujin Robotics, with the initial goal of serving in restaurants and supporting teachers in schools. Charlie is 1.2 m tall, has a 24V LiPo battery and is made up of a camera, a touch screen, speakers, microphones, contact and laser sensors, and two computers. The robot adapts with more components or different software depending on the environment it is in; therefore, it is possible to add a sphygmomanometer, oximeter, glucometer, or more cameras and microphones. It is described that it also has blood pressure measurement and an entertainment module with music, videos and games [2].

The Smooth project has been developed in Denmark since 2017, with a prototype that has been designed to support the transport of materials for older adults; is a developing robot with three goals: to be a garbage or laundry collector, to serve drinks, and to be a guide, walker, or wheelchair that supports the movements of the adult in action [3, 4]. In [5], a mobile robot is incorporated, assistant to the medical staff, who cares for older adults. In this case, PR2 has the ability to manipulate objects; with a retail price of USD 400,000 [6]. The study contains the opinions of 14 nurses, who expressed fear of being replaced by robotic assistants, and who prefer the robot to perform support tasks to save time, such as having a third hand. In this same sense, other studies were found, such as the one by Turja et al. in 2018 [7]. The autonomous service robot for hospitals (Relay) of (Swisslog) [8], used for the transport of medicines, laboratory samples and other medical materials; and uses laser navigation, computer vision, and has a touch screen and wireless communications.

On the other hand, applications have been reported in telemedicine [9] and with the aim of promoting sustained research in robotics to address infectious disease risks [10]. Finally, a study carried out in 2017 [11] addresses the problem by determining the characteristics to consider when designing a nursing assistant robot, similarly to the situation presented in 2020 with COVID-19, the study was carried out after detecting the needs of patients and nurses in the face of MERS infection in 2015 in South Korea. From this situation, the need to develop a support system for nursing staff became evident.

Several aspects should be considered based on these works: the mobile robot should have a pleasant and simple interaction, with dimensions that must be in accordance with the place where it moves, and be able to navigate avoiding obstacles, and be in harmony with the environment, for example, it should be considered that the lights, the noise it produces, the temperature and the alarms do not disturb the environment in which it is located.

The objective of this paper is to help in the care of patients with COVID19 within the restricted areas of the hospital.

2 The Mobile Robot's Description

Mobile robots have characteristics that make them suitable for a certain task, with a physical structure with traction and steering systems that are in accordance with the requirements of speed, maneuvers and terrain characteristics. Robot sensors and a control system, with obstacle avoidance, are added to achieve the precision and speed necessary to perform a reliable navigation to go from a point to its destination. This implies that the mechanical, electronic and computer systems must be more complex. There are three basic systems from which different configurations can be obtained: drive and steering on independent axle, drive one direction on the same axle and steering and drive on all axles [12]. There are three basic systems from which different configurations can be obtained: drive and steering on independent axles [13], drive and steering on the same axle [14] and drive and steering on all axes [15].

In the mobile robot described in this document, a pair of tracks with independent motors are used, which resembles the traction and steering system in the same axis, also called differential traction. It must have the necessary information to allow the mobile robot to carry out three fundamental tasks: estimate its position and orientation, keep the map of the environment updated and detect possible obstacles.

For the estimation of the position and the updating of the environment map, characteristics such as precision, spatial resolution and range are required, in the detection of obstacles, the time between observations must be less, to have the information already processed as quickly as possible. In the case of this robot-nurse project, the above characteristics are not essential since the navigation operation is controlled remotely by a human operator who has visual feedback through a camera located on the platform itself, this is due to the type of application and the rapid design and construction required to have the impact in the time required by the COVID19 Pandemic, in this way the control of the robot is easier and requires less technical knowledge to be operated within a hospital environment in an area confined for patients with COVID19.

The mobile robot Ayúdame 1.0 is a platform for the support and location of Food, Medication and Equipment (Fig. 1), developed jointly by the Research Institute in

Applied Mathematics and Systems (IIMAS), PUCV and the Industrial Design Center, both from the National University of Mexico (UNAM). The objective is to serve as an assistant in nursing processes in confined areas in hospitals with COVID19 patients in a tele-operated assisted way up to 50 [m]. The design must be practical, robust, stable and capable of carrying a hospital-type tray with food, medicine and other necessary materials. In addition, the platform will be able to bring the means to establish a video call with the patient. Navigation is visual by means of a camera that transmits to a mobile device, in addition to other telemetry data, including an electronic compass.

The Ayúdame 1.0 system has the following parts: An electromechanical system, an electronic system, a control system, remote command, support and serving tray.

2.1 The Electromechanical System

The mobile robot Ayúdame 1.0 uses the caterpillar-type commercial platform (T'Rex Tank) with differential traction based on six support wheels and one traction on each side and with a total load capacity of up to 8 [kg].

Fig. 1. The mobile robot Ayúdame 1.0

The crawler drive allows to control the movements with greater precision and smoothness, as well as guarantee that the distance traveled and the direction of the mobile is correct by means of a remote-control system.

The chassis is made of anodized aluminum, and the sprocket wheels, the bearing wheels, the track and the suspension are made of anodized zinc. The width is 265 [mm] and the length is 355 [mm]. The mobile structure has independent suspension to improve traction and absorb shocks, it has a height of 70 [mm] necessary to make the chassis very operational and manageable on rough terrain and not rub against the floor. The chassis has a cavity to transport batteries and components, while its lid serves as a base for transport or to integrate other mechanical components, which in the case of Ayúdame 1.0 is a structure for transport hospital-type trays to carry food, medicine or some material necessary for the patient.

The tracks are driven by 2 independent 12 [V] direct current motors with a nominal current consumption without load of 1.3 [A], with load of 4 [A] and stall of 11 [A]. Both motors are controlled by a STMicroelectronics VNH5019 dual H-bridge and the electrical power is supplied by a lithium-ion polymer (LiPo) 3S35C battery of 5,200 mAH.

2.2 Electronic System

Figure 2 shows a connection diagram of the different components that make up the control system of the mobile robot AYUDAME 1.0. The main component of the electronic system is the small format computer, the Raspberry Pi 4 (RPI) with 4 GB of RAM. The communication processes, control of the caterpillar traction motors, reading of the current consumed by the motors, reading and processing of values coming from an inertial measurement unit, are carried out by the RPI, as well as the detection of objects and surfaces to through two ultrasounds. Sensors and panning movements of the camera. The operating system used is Raspbian 10 (Buster) with Linux Kernel 4.19 in its Lite version.

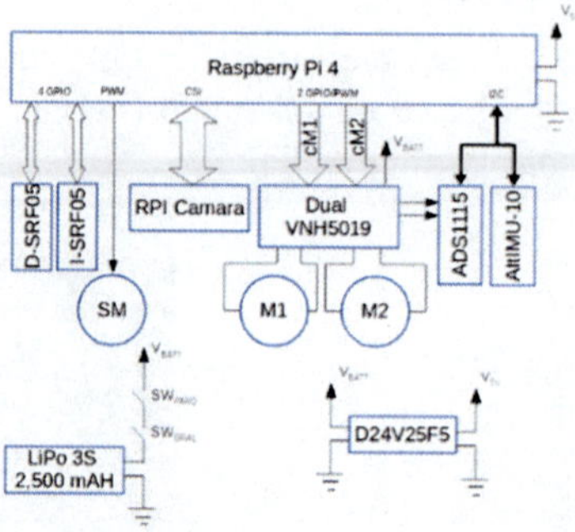

Fig. 2. Diagram of electronic components

Port and pin monitoring is described in Table 1. General Purpose Port (GPIO) monitoring is done through the Python library (Gpiozero). Three PWM channels are used, one of them implemented by software is to control the panning of the servo motor (SM) that moves the camera, the other two DMA channels are used to ensure the maximum frequency of 20 [kHz] to regulate the speed of the camera. VNH5019 circuit, which controls the traction motors (M1 and M2). 3 pins are needed to control the direction and enable of each motor (cM1 and cM2).

The current consumed by the motors (cM1 and cM2), the VNH5019 dual IC provides two analog channels, one for each H-bridge, which are read by the first and second channels of the MCP3008 analog-to-digital converter. Then, they are read sent by the RPI through the I2C protocol. The reading of the IMU AltIMU-10 electronic compass is captured on this same channel. Two GPIO pins are used to drive two SRF05 sensors (left, I-SRF05 and right, D-SRF05) and two to listen for echo return to detect an object in front of the platform. The platform stops when an object is detected within a distance of 15 [cm] and a signal is constantly sent to the remote control to avoid collisions. The battery is connected to a regulated power supply (D24V25F5) that powers the Raspbery Pi, the AltIMU-10 and the MCP3008 with 5 [V]. An emergency stops button (SWPARO) and a general switch (SWGRAL) cut the power supply when necessary.

Table 1. Port pin assignment for RPI

Port(pin)	Type	Device
GPIO[2, 3]	I2C	MCP3008
GPIO[2, 3]	I2C	AltIMU-10
GPIO[23, 24]	I/O	D-SRF05
GPIO[22, 27]	I/O	I-SRF05
GPIO[17]	PWM	SM
GPIO[12, 13]	PWMDMA	VNH5019
GPIO[5, 6, 14, 15, 16, 18]	I/O	VNH5019
CSI	BusCSI	Camera

2.3 Control Systems

The architecture of the control system of the mobile platform Ayúdame 1.0 is implemented in the RPI software and has a remote control (Fig. 3). The Control Server is developed in Python 3 with the purpose of managing address commands and telemetry data propagation to and from the control motors M1 and M2, and SM, as well as current draw feedback, reading of SRF05 sensors and the AltIMU. -10. It sends and receives data through a broker-provided MQTT service installed and running on the RPI using WebSockets over port 9001, as the remote uses JavaScript. (Figure 3) shows the Wi-Fi network and the set of files loaded by the remote-control browser.

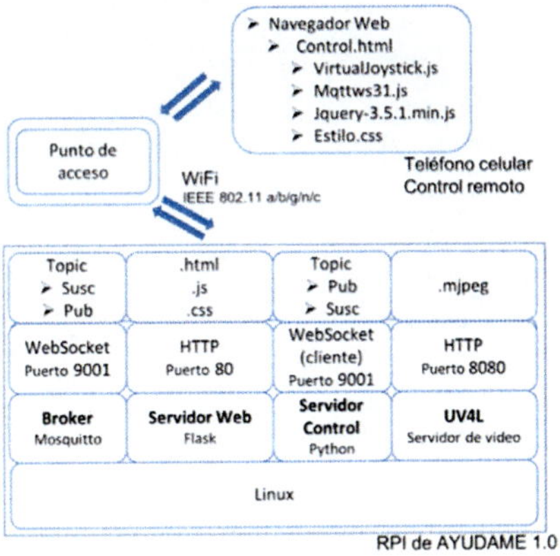

Fig. 3. Control software architecture

Risk mitigation actions prevent the Wi-Fi signal from being weak, the platform automatically stops when a defined threshold is reached, and a visual alert is displayed on the remote.

The Broker handles three themes (Fig. 4) "telemetry" (a), "motion" (b), and "commands" (c), using [forward x] and [forward y] which are used to "go to left or right", and "forward or backward", respectively.

Digital video streaming is used with a RPI V 2 Sony IMX219 camera with a resolution of 8 [Mpixels] and night vision function, for which UV4L (User space Video4Linux) is used. The transmission format is jpeg over the HTTP protocol on port 8080 and provides RESTful services for configuring service parameters. The maximum video resolution is 1080p, however, it can be scaled to lower resolutions so as not to saturate the transmission channel.

This service is configured dynamically, following a scheme of risk mitigation actions that constantly monitors the speed and quality of the Wi-Fi connection with the access point. If both decrease, it activates and lowers the resolution. The same happens in the opposite case.

The web server runs an application using Python's Flask. The application allows the client to view the web-based interface from a browser on the remote. This service publishes .html, .js and.css files on HTTP port 80, necessary to draw and articulate the behavior of the software for remote control of the platform.

1) Assisted acceleration and deceleration: Ayúdame 1.0 have a weight of 6.4 [kg] composed of 3.7 [kg] of the chassis, 1 [kg] of battery and electronic components and 1.7 [kg] of the body support and the tray. An example of a typical payload consisting of food (soup, dinner plates and cutlery, fruit and drink) (Fig. 5) is estimated to have a maximum weight of 1.6 [kg]. This weight is considered the maximum since other medical supplies, such as medicines, syrups, syringes, cotton, alcohol and medical instruments that may be transported on this platform will be of lesser weight.

```
{
  "sensores": {
    "IM1":        [corriente_M1],
    "IM2":        [corriente_M2],
    "FM1":        [falla_M1],
    "FM2":        [falla_M2],
    "COMPASS":[grados],
    "onMov":      [true_en_mov],
    "WiFiNivel":  [nivel_senal_wifi],
    "WiFiLink":   [link_senal_wifi}],
    "WiFiDisp":   [listado_wifi_disp]
  },
  "servicios":{
    "SC":         [servidor_control],
    "SW":         [servidor_web],
    "ssh":        [servicio_ssh],
    "WIFIID":     [WIFI_ID],
    "WIFIPASS": [WIFI_contrasena]
  }
}
```

a)

```
{
  "x": [avance_en_x],
  "y": [avance_en_y],
  "angCamara: [angulo_camara]
}
```

b)

```
{
  "servicios":{
    "rSC":        [reiniciar_s_control],
    "rSW":        [reiniciar_s_web],
    "rssh":       [reiniciar_s_ssh],
    "sWIFIID":    [establece_WIFI_ID],
    "sWIFIPASS": [establece_WIFI_con]
  }
}
```

c)

Fig. 4. JSON Syntaxis

Due to the vertically offset center of gravity, objects may fall or spill liquid if the operator abruptly actuates the virtual joystick on the remote control. To prevent this, the acceleration and deceleration system performs countermeasures. Then the position change is transmitted directly from the remote to the RPI, the Control Server takes the data, stores it and waits for the next data to arrive, it is updated every 500 [ms]. If the difference in positions is greater than 30%, it initiates a smooth acceleration or deceleration action. This is by commanding 10% by 10% pulse width increments that activate the motors until reaching cruise speed.

2) Configuration of Wi-Fi parameters: The hospital infrastructure guarantees a stable Wi-Fi network signal in all the spaces where the mobile robot moves, having the same SSID and the same authentication, however, at some point it will be necessary register a new network, for which an application was developed for Android devices for RPI Bluetooth and the mobile device and can configure the parameters.

2.4 Remote Control and Teleoperation

Ayúdame 1.0 teleoperation is carried out through a Wi-Fi IEEE 802.11 a/b/g/n/c wireless network supported by RPI 4. Control can be carried out from any smartphone, tablet or personal computer connected to the same network. An interface to teleoperate the platform, might be loaded with a web application and browser supported by JavaScript ES6. (Figure 3) shows the Wi-Fi network.

Figure 5 is the mobile robot and Fig. 6 is a snapshot of the graphical interface of Ayudame1.0 remote control is showed, and a virtual joystick can be seen to control forward, breaking and direction actions. The remote control receives data from telemetry and are displayed in the main window, can be accessed by others menus in screen. The video coming from the camera is projected on background.

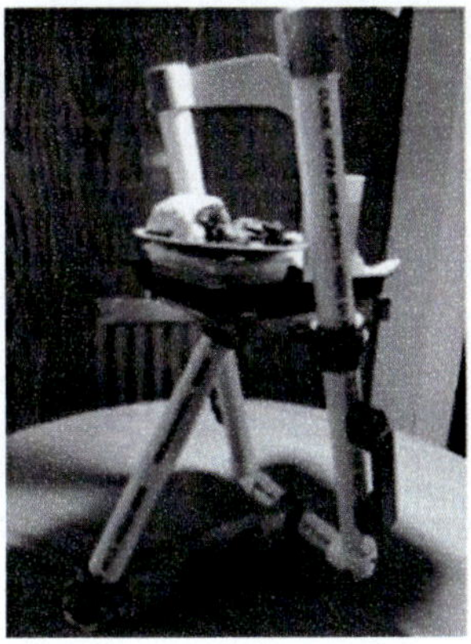

Fig. 5. Ayúdame 1.0 with payload

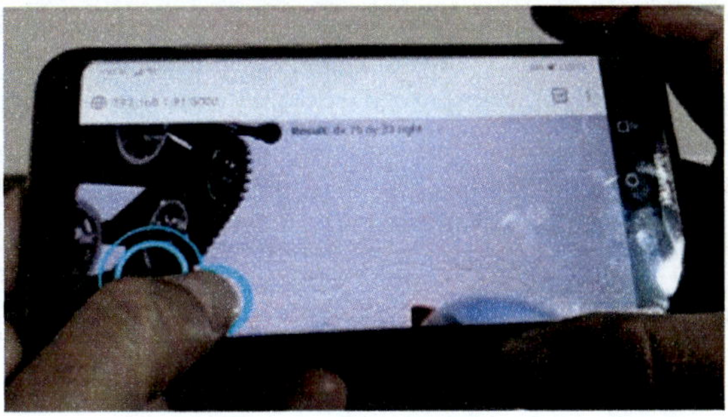

Fig. 6. Remote control

2.5 Support and Service Tray

The structure is based on commercial low-cost and 3D printed elements Fig. 7. Advantages of this technology allows the customization of products, modular manufacturing, industrialization of low-production parts, quick weight and resistance verification of the elements, as well as, the products finish without making major changes in production [16].

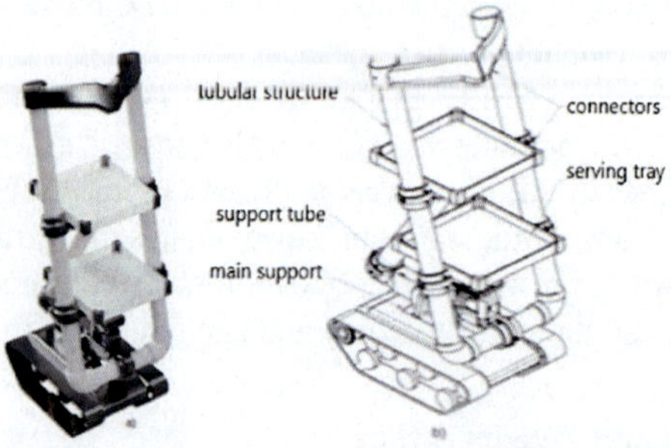

Fig. 7. a) 3D model of mobile robot and b) Isometric of the base

The proposed system is modular and easily replicable with the fused deposition molding method (FDM) and additive manufacturing process [12, 17]. The manufacturing process can be configured considering the following factors: a) capacity of the 3D printing equipment, b) available material, c) precision, d) volume of the parts to be printed and e) manufacturing time.

The additive manufacturing process facilitates the modeling of three-dimensional figures with complex geometry difficult to produce in other industrialized processes.

The use of polymers in additive manufacturing has applications in medical and biomedical areas [13], and the manufactured products can be subjected to disinfection and sterilization processes to be used in health centers [18]. One of the materials used for printing three-dimensional parts with applications in medical areas is the Polylactic Acid (PLA) filament [19]. Therefore, PLA is applied in structural parts of Ayúdame 1.0.

Commercial pipes such as polyvinyl chloride (PVC) or aluminum pipe were used, both materials offer advantages and disadvantages in relation to weight, costs, flammability, durability and toxicity under extreme conditions [20].

The structure has the following elements: a) main support, b) support tube, c) tubular structure d) connectors and e) service tray. (Figure 7) (b), in this way, it allows to place or remove trays and increase the height up to the robotic platform used.

The three-dimensional design was made with Fusion 360 3D parametric solid modeling package of AutoDesk1. The modeling is available for viewing, downloading and modification at the following link: https://a360.co/2KRdHPB.

3 Evaluations and Results

Two types of tests were done: 1) aspects of powertrain control and capabilities, and 2) Wi-Fi networks teleoperation capacity and efficiency.

3.1 Powertrain

Twenty tests were performed to test the capabilities and limitations of the electronic system and determine the maximum drive current the system may support. 2 scenarios are considered; 1) the platform is at full stop and reaches cruising speed; and, 2) the platform is at cruise power. Table 2 shows the results of these scenarios with no payload and when carrying (1.6 [kg]). Test for cruise speed which represents 75% of the PWM work cycle, reached approximately 0.87 [m/s] on a very common linoleum floor used in public hospital.

To test battery life and how many trips could the platform may support, 20 trips of 50 m (round trip) were done with a very similar used route on a COVID19 patient floor. The outbound trips were made without a payload and the return trips with a payload. Measurements are shown in (Fig. 8) and can be observed the maximum limit of trips for each charge in twenty-first trip when the voltage was close to the battery breakdown.

Finally, the assisted acceleration and deceleration algorithm was tested to prevent liquid spillage on the tray. A glass of water was freely placed on the tray. Different test conditions were done: 20 starting tests (A) from full stop to cruising speed, and 20 braking tests (F), from cruising speed to full stop. Three types of effusion were qualitatively classified as: mild (DL), medium (DM) and severe (SD). Table 3 shows the results.

Table 2. Powertrain current conditions

Power condition	No payload		Payload	
	I_{M1}[A]	I_{M2}[A]	I_{M1}[A]	I_{M2}[A]
Cruise stop	7.6	7.4	9.2	9.1
Cruise	5.4	5.3	6.8	6.8

Table 3. Effusion liquid test

	DL		DM		DS	
	A	F	A	F	A	F
Yes	3	4	1	2	0	1
No	17	16	19	18	20	19
[%]	85	80	95	90	100	95

3.2 Teleoperation

A communication between a local network connected by Wi-Fi with one or more access points can have several factors that affect data transport. The most critical parameters are intermittent or total disconnections and the reduction of the data transmission speed. The first case is a good infrastructure issue rather than the design of AYUDAME 1.0; however, the reconnection is done automatically and a complete shutdown occurs if it occurs. On the other hand, for the second case, there are different causes that originate it, for example, the number of clients connected simultaneously, the distance from the access point to the mobile platform, the distance from the access point to the remote control, or obstacles such as walls.

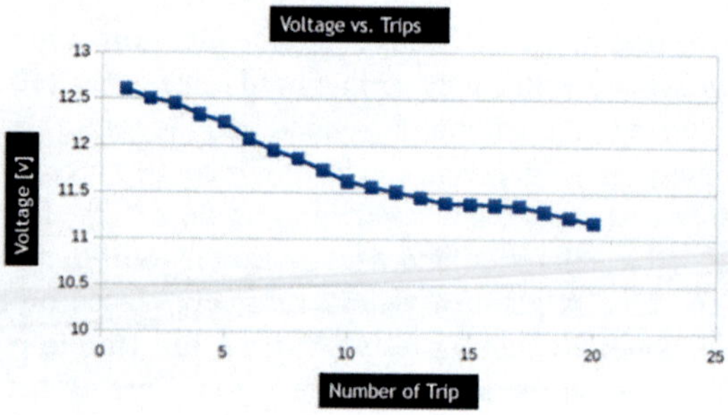

Fig. 8. Graph of voltage vs. number of trips

Therefore, reduction of speed transmission over the Wi-Fi network limits AYUDAME 1.0 platform operation. Video is the most critical service which requires Wi-Fi acceptable speed signal stability. Telemetry can also be affected but MQTT message packets are much smaller in size.

Table 4 shows the result of 20 tests performed by decreasing the connection speed and if Ayudame1.0 video risk mitigation (MRV) and telemetry risk mitigation (MRT) actions were activated correctly, as well as the visual alert on remote control.

Table 4. Video and telemetry quality test at different speed transmission

Speed [Mb/s]	MRV%	MRT%	Alert%
867	0	0	0
216	0	0	0
17	0	0	0
1.7	85	0	100
0.2	100	50	100

4 Conclusion

A mobile robot based on a low-cost pl commercial mobile platform and printed in 3D is presented, to be used in restricted areas of hospital care due to COVID19.

The technique used is robust and reliable for service robot applications. PWM technology is used for navigation and motion control. Test results with and without payload, the mobile robot is able to make an average of 20 times. The sudden movements caused by overcoming inertia and friction mean that the objects placed on the tray can fall, come out or spill the liquid from a glass of water. As a solution, the assisted acceleration and deceleration technique is used, which is simple and robust.

The wireless communication between the AYUDAME 1.0 platform and the remote control is carried out through a Wi-Fi network, a necessary and mandatory infrastructure that complies with international and national standards to operate TCP/IP in most hospitals in the world, allowing the use of a large number of secure and robust applications. A sensitive point was the use of roaming to switch between base stations or access points and not lose communication, however, it is guaranteed in the hospital infrastructure. In case of reconnection to another network, the Android application AYUDAME1.0_WiFi allows you to configure the parameters through the Bluetooth of the Raspberry and the mobile device.

On the other hand, the remote control based on a web browser allows the mobile platform to be operated remotely from any mid-range device onwards. This clearly eliminates the need to create specific applications for specific versions of the operating system, which also limits the number of devices where it can be installed due to the specific use of libraries. The graphical interface of the remote control has great capabilities to increase its complexity but at the same time be intuitive, sensitive to touch and with accessible menus, this is because it is designed with web technology, which allows a very flexible way of creating windows and buttons functional. Likewise, the use of the MQTT protocol through WebSocket proved to be an excellent combination for the transport of data and telemetry in this type of application, due to the persistence of the communication channel, and the simplicity and secure way of sending in contrast to the RESTful services, which are in applications with RPI.

The video service turned out to be a very effective support tool as it reaches a resolution of 1080p and allows for colorful images with good contrast. The base and tray are very solid, inexpensive and easily reproducible structural elements and do not add marginal weight to the system. The materials used are not toxic or medically contraindicated and it has been proven that they travel in risk areas due to COVID19.

The height of the first tray is 53 [cm] and the second tray is 81 [cm], which is enough for a standing person to pick up items without bending down.

References

1. Lefranc, G.: Some trends in robotics and automation. In: 2021 IEEE CHILEAN Conference on Electrical, Electronics Engineering, Information and Communication Technologies (CHILECON), pp. 1–8 (2021), https://doi.org/10.1109/CHILECON54041.2021.9703003
2. Ahn, H.S., et al.: Design of a Kiosk type healthcare robot system for older people in private and public places. In: Brugali, D., Broenink, J.F., Kroeger, T., MacDonald, B.A. (eds.) SIMPAR 2014. LNCS (LNAI), vol. 8810, pp. 578–589. Springer, Cham (2014). https://doi.org/10.1007/978-3-319-11900-7_49
3. Juel, W.K., Haarslev, F., Ramírez, E.R., et al.: SMOOTH robot: design for a novel modular welfare robot. J. Intell. Robot. Syst. **98**, 19–37 (2020). https://doi.org/10.1007/s10846-019-01104-zW

4. Krüger, N., Heidtman, C. (2020) https://smooth-robot.dk/en/project/
5. Mitzner, T.L., Tiberio, L., Kemp, C.C., Rogers, W.A.: Understanding healthcare providers' perceptions of a personal assistant robot. Gerontechnology **17**(1), 48–55 (2018)
6. Garage, W.: Willow Garage Inc., pp. 2–3 (2010). http://www.willowgarage.com/blog/2010/09/07/pr2-pricing-
7. Turja, T., Van Aerschot, L., Sarkikoski, T., Oksanen, A.: Finnish health care professionals' attitudes towards robots: reflections on a population sample. Nurs. Open **5**(3), 300–309 (2018)
8. Swisslog Healthcare: Relay Autonomous Service Robot (2020). https://www.swisslog-healthcare.com/en-us/products-and-services/transport-automation/relay-autonomous-service-robot
9. Koceska, N., Koceski, S., Zobel, P.B., Trajkovik, V., Garcia, N.: A telemedicine robot system for assisted and independent living. Sensors (Switz.) **19**(4), 834 (2019)
10. Yang, G.Z., et al.: Combating COVID-19 – the role of robotics in managing public health and infectious diseases. Sci. Robot. **5**(40), eabb5589 (2020)
11. Han, J., Kang, H.J., Kwon, G.H.: Understanding the service scape of nurse assistive robot: the perspective of healthcare service experience. In: 2017 14th International Conference on Ubiquitous Robots and Ambient Intelligence (URAI) (2017)
12. Frandsen, C.S., Nielsen, M.M., Chaudhuri, A., Jayaram, J., Govindan, K.: In search for classification and selection of spare parts suitable for additive manufacturing: a literature review. Int. J. Prod. Res. **58**(4), 970–996 (2020)
13. Adamowski, J.C., Simões, M.G., Gozman, F.G.: Desenvolvimento de un robo móvel. Escola Politecnica de la Universidade de Sao Paulo (1990)
14. Cox, I.J.: Blanche—An experiment in guidance and navigation of an autonomous robot vehicle. IEEE Trans. Robot. Autom. **7**(2), 193–204 (1991)
15. Connell, J., Viola, P.: Cooperative control of a semi-autonomous mobile robot, pp. 1118–1121 (1990)
16. Berman, B.: 3-D printing: the new industrial revolution. Bus. Horiz. **55**(2), 155–162 (2012)
17. González-Henríquez, C.M., Sarabia-Vallejos, M.A., Rodriguez-Hernandez, J.: Polymer for additive manufacturing and 4D-printing: materials, methodologies, and biomedical applications. Prog. Polym. Sci. **94**, 57–116 (2019)
18. Rutalaand, W.A., Weber, D.J.: Disinfection and sterilization in health care facilities. Bennett & Brachman's Hospital Infections: 6th edn, no. May (2013)
19. Sin, L.T., Rahmat, R.R., Rahman, W.A.: Polylactic Acid: PLA Biopolymer Technology and Applications, pp. 1–341 (2012)
20. Dixit, V., Kalra, P., Ruan, A., Chesebro, B.B.: Plastics in HealthCare: Crisis or Opportunity? (2020)

Electric Vehicles Between Recycling and Sustainable Development - @.ro

Razvan Marcus, Cosmin Rus(✉), Monica Leba, and Marius Risteiu

University of Petrosani, Universitatii 20, 332006 Petrosani, Romania
cosminrus@upet.ro

Abstract. Even if until now electric vehicles are not 100% clean, they are becoming an increasingly viable alternative to classic transport solutions, bringing in addition new innovations designed to help within the general concept of sustainable development, especially within large urban areas. In addition to the classic transport function, through new technologies (V2V, V2G), an electric vehicle can be successfully transformed into an active tool in achieving the goals of a Smart City. This paper presents several alternatives both as an effective part of control hardware for an electric vehicle and algorithms implemented on this hardware. The special characteristics that can be implemented in an electric vehicle are highlighted by presenting the achievements of such vehicles. The original note is the conversion of a vintage off-road vehicle from Romania into an electric vehicle. It is noteworthy that many of the hardware components used in the construction of various electric vehicles are recycled so it can be said that the first step in terms of the concept of sustainable development has already been achieved. All electric vehicles designed or converted at the University of Petrosani benefit from LoRa type communications (a communication network still little used in Romania but which has proven its reliability) and are intended for use inside an urban area and can successfully fulfil the role of the platform for monitoring the air quality parameters in a certain area precisely in view of the fact that it does not pollute directly.

Keywords: Electric conversion · BMS · BLDC · V2X · Smart City · Smart Grid

1 Introduction

We are in an era where more and more emphasis is being placed on using new technologies in order to protect the environment but without affecting the comfort we have gained in recent decades. So, if we are talking about the equipment in our own home or about the car in the garage, there are more and more concerns regarding the reduction of energy consumption and if possible, we can try a relative balance between what is produced and what is consumed. In recent years, the idea of recycling certain parts of equipment, devices or products in order to limit consumption is also becoming more and more pronounced. In addition to the basic idea surrounding the concept of recycling, there is also the desire of more or less nostalgic engineers in terms of bringing back

S. Dzitac et al. (Eds.): ICCCC 2022, AISC 1435, pp. 47–62, 2023.
https://doi.org/10.1007/978-3-031-16684-6_4

to life some equipment or devices with a certain tradition behind them. Here, too, new concepts appear, including restoration or conversion. If about the restoration things are very well known instead about the conversion and especially the conversion of some vehicles with internal combustion engines into electric vehicles things are in a rather incipient phase at the level of our country.

This paper presents in a relatively detailed form the stages of such a conversion of a historic 4 × 4 ARO vehicle as well as computer-based command and control methods in terms of the fact that all new systems brought into the vehicle structure are managed through well-established algorithms. This whole conversion process started with the idea of bringing back to life one of the most prestigious brands of Romanian 4 × 4 vehicles (ARO) as well as finding ways to make a vehicle that fits perfectly in terms of cars market dynamics from today (non-polluting directly vehicles and interconnected both with each other and with the transport infrastructure). All these ideas and concepts fit perfectly into the overall framework of a Smart City concept and so each idea that structures the whole set can easily be based on these bases. So, if we are talking about interconnected vehicles and intelligent infrastructure, first and foremost we must raise the issue of creating a Smart Grid in a city or a settlement just to be able to build the necessary dependencies. Smart Grid must be the first thing done in a Smart City, starting from the premise that a good foundation and a rigorous management of each action will lead in the end to a reliable system that will have minimal implementation costs. Smart Grid is the foundation of any Smart City and includes all the starting points for the realization of each part of the entire system. An electric vehicle in a Smart City built on a Smart Grid will integrate perfectly from all points of view, taking into account that all the ideas behind each achievement are designed to blend perfectly. The success of the development of electric vehicles and the growing demand for their purchase on the background of greater autonomy of power systems and the development of charging infrastructure also highlights the possibility of using these vehicles as an integral part of a Smart Grid system in the context of urban development of the Smart City concept in which other innovative concepts can be integrated, as shown in the paper by Rus et al. [1]. In addition to presenting conversions of electric vehicles, this paper aims to present new research developed around the idea of Smart City in Petroșani, starting from electric vehicles that can serve both as a means of transport and as a system for monitoring the parameters of electric vehicles. Medium, but which can also be integrated into the electricity grid, so that in times of high consumption it can supply electricity when needed and in times of over-production to charge its batteries (the new concept of prosumer). Although at present in Petroșani there are not many electric vehicles that can become part of the electricity network, this idea is applicable especially in a public or private installation with the possibility of scaling in the future.

2 Literature Review

As it appears in some papers and as Findl [2] and Thornton [3] states, the first electric vehicle was built in 1888 by the English inventor J.K. Starley though in the United States, Fred M. Kimball of Boston is supposed to have been the first builder in the same year. Other papers describe the fact that the first electric vehicle was built in the United States in 1892. Although Findl [2] also provides some significant data, it is not possible to say with certainty who invented the first electric vehicle, the electric car industry has known a rapid growth as early as 1892. The first large companies to produce such vehicles were Electric Carriage and Wagon Co., Studebaker Brothers Manufacturing Co. and Oldsmobile. By 1912, more than 100 companies worldwide produced electric vehicles, and by 1912, more than 10,000 such vehicles were produced in the United States alone. Internal combustion vehicles gained more and more popularity in the early 20th century and were more preferred by male drivers. At that time, the lack of noise, the limited driving speed and the high dependence on the electrical network of electric vehicles were more associated with female drivers. Manufacturers noticing this trend began to introduce in the design of electric vehicles different shades and colored prints in terms of paint but this philosophy proved to be fatal and for this reason in addition to the many advantages of using internal combustion vehicles (high autonomy, low fuel price, many refueling facilities).

In the early 1930's, the only company that still produced electric vehicles was the Detroit Electric Car Company. The further development of internal combustion engines and the low price of fossil fuels (gasoline or diesel) have led to a significant increase in the market for vehicles with internal combustion engines. For almost half a century, from the 1930s to the 1970s, gasoline or diesel-powered vehicles dominated the world car market and no concern was taken into account, both in terms of pollution and regarding the amount of fuel available. But the years have passed and there are more and more signs of the impact of carbon monoxide resulting from the burning of fuel by vehicles with internal combustion engines. Deterioration of the ozone layer in the atmosphere was observed as well as the appearance in the large urban agglomerations of smog (high concentration of pollutants above an inhabited area). Ever since the publication of Findl's article [4] (1977), there has been the problem of reorienting the automotive industry towards electric vehicles in the context of the fossil fuel crisis. We are the victims of our own success, stated in the introduction of Dave and Dong's work [4], which presents a completed project for the conversion of an internal combustion engine vehicle into a fully functional electric vehicle that complies with all safety regulations. This conversion project presents the conversion of a vehicle with a thermal engine into an electric vehicle in which two people could travel, has a range of about 60 km between battery charges, complies with safety standards. Its electric motor offered a 90% efficiency and the controller offered the possibility to select several types of driving (normal, low consumption or sport) as well as offered a brake control. The charger could charge the battery up to 144VDC from the 115VAC network. The battery that offered a range of about 80 km consisted of a dozen 12VDC batteries. The coupling between the electric motor and the vehicle transmission was made of a low-carbon steel gear (which offered a certain flexibility in operation). Following the tests (on public roads, after all the approvals were obtained and the vehicle was put into circulation) extraordinary results were obtained for that moment. Many papers show how CO_2 emissions can be reduced if two work scenarios are followed. Thus, if the number of electric vehicles on the streets were increased and the electricity needed to charge their batteries from renewable sources was obtained, a decrease of about 58% in carbon emissions could be obtained. Francis et al. [5] shows how vehicles with internal combustion engines produce over 900 million tons of CO2 annually (which means about 15% of global CO2 emissions). Although there have been major concerns about the development of the electric car industry since the 1970s amid concerns about excessive pollution, about 16 million vehicles with internal combustion engines are produced each year. Although electric vehicles do not directly pollute, the question remains whether electric vehicles do not actually cause pollution. In an attempt to answer this question, Rievaj and Synák [6] present significantly relevant data on the whole process of generating electricity in many European countries and correlate these data with the number of electric vehicles on the market. It shows how in countries such as Sweden and Iceland, based on geothermal energy sources and other renewable resources, there are premises for truly obtaining 100% clean electric vehicles. Instead, the issue of recycling the batteries of these vehicles remains in an even more urgent discussion, but also in this process, the Nordic countries are also starting to develop different strategies.

There are currently many scientific papers that present in detail the entire process of converting an internal combustion engine vehicle into an electric vehicle. Dankov, Madjarov and Prodanov [7] show the conversion of a small city vehicle (Opel Corsa) into an electric vehicle. Particular emphasis is placed on the software management part of all components and especially on the management of vehicle battery charging. Francis et al. [5] presents a special implementation model of a conversion of a classic vehicle into an electric one. Thus, the use of a BLDC motor capable of developing a maximum torque of about 16Nm and running at a torque of 6 Nm at 3500 rpm is presented. It also shows the use of a power regeneration system using the braking system. The prototype developed and presented by Francis et al. [5] has a range of about 108 km at an efficiency of about 75% of the entire system. A special feature of this prototype is the implementation of a semi-autonomous driving system that can be used to park the vehicle through an Android application. The paper shows how to reduce the costs that may occur in the whole process by obtaining an electric vehicle that requires only 20% of the total original costs required to operate. The use of a system of electricity production from renewable resources necessary to charge the batteries of the electric vehicle can reduce the operating costs of the vehicle to almost zero and a amortization of the production costs in a short time.

3 Evolutionary Background of Some Electric Vehicle Conversion Projects

The author's concerns regarding the conversion of internal combustion engine vehicles into electric vehicles began several years ago when an electric kart was built. The motor chosen in the first stage consisted of two BLDC motors recovered from a hoverboard (two-wheeled mobile electric device). The control part and the control part were taken from the hoverboard and a suitable drive system was designed accordingly. Tests were made with this variant of motorization controlled by means of control units but although a high efficiency (about 85%) and a low power consumption were found, due to the control mode which supposed that the driver synchronized the motor almost perfectly, left with the right motor in the control levers (left, right, front, rear) could not be obtained a perfect control according to safety regulations (Fig. 1).

Fig. 1. The first prototype of an electric vehicle (electric kart) using hoverboard motors

Due to the fact that the steering axle was removed (steering column removal) it was not possible to obtain an optimal control over the steering only from the acceleration or deceleration of one of the drive wheels and so a second constructive variant was tried which although had a lower yield offered a greater stability and safety in operation. The second motor variant was the one with a DC motor recovered from a forklift vehicle and adapted accordingly. The power of the motor was 1 kW, 1420 rotations/min and cosφ of 0.82, protection class IP 44. The power supply of this motor was made with the help of three 12 V, 66 Ah batteries. The control part of the motor is made analogically using the control and command interface by means of a potentiometer with which the motor speed is adjusted and with the help of two contacts with which the operating direction can be selected (forward - reverse) (Fig. 2).

Another conversion project was that of an ATV (All-Terrain Vehicle) with a classic thermal engine (250 cm^3 2-cylinder petrol engine) in an electric vehicle that also had an autonomous vehicle functionality as shown in the paper made by Rus, Mija, and Leba [8]. The electric motor chosen is 2.2 kW three-phase asynchronous motor, 1420 rpm and cosφ of 0.82, protection class IP 44. The power supply of this motor is made with the help of a 12 V, 240 Ah, 1200 A battery. The battery is connected to the circuit via a 3000 W 12 V – 220 V voltage inverter. The purpose of this inverter is to enable the power supply of a Yaskawa VS-606V7 frequency converter with a power of 4 kW. The three-phase motor is controlled from the three-phase output of the frequency converter (Fig. 3).

A revolutionary new driving concept for this electric vehicle has been proposed and realized. This driving algorithm was developed in the first phase to reduce the power

Fig. 2. Electric kart with DC motor from a lifting machine

Fig. 3. Electric ATV with asynchronous motor

consumption used in propulsion but later this driving algorithm was used to lay the foundations of an autonomous (driverless) driving model.

A deep-learning algorithm has also been developed which, based on images taken from a webcam mounted on the functional platform of the ATV, can allow the realization of an autonomous driving scenario. Rus et al. [9] presents a special use of such an electric vehicle that can bring a major benefit and greater accuracy in retrieving data using environmental quality sensors, data taken from the environment are not affected by the vehicle's polluting emissions (an electric vehicle does not directly emit polluting emissions).

This vehicle (ATV smart electric) was made mainly of recycled components and parts, and the problem was to make a power supply system from renewable sources, so that the vehicle is truly environmentally friendly. Another significant part that was worked on was the integration of this vehicle in a network of sensors using the LoRa communication protocol (which allows long distance data transfer with low power consumption). Thus, in addition to the fact that their own LoRa communication modules were developed, the integration of the entire system in a Vehicle to Grid (V2G) concept was also achieved by being able to determine the position of the vehicle (ATV) in an enclosed space without being need to use GPS as well as various environmental data (pollution factors) could be retrieved. Thus, it was possible to create a smart electric vehicle that was integrated in the Smart City concept, as shown in the paper by Rus et al. [10].

Another special project of the research team is the conversion of a Buggy type vehicle into an electric one. To make this conversion we will use our own custom BLDC motor shown in Fig. 4. The operation of the BLDC motor is based on the supply of windings on the stator and on the fact that the motor switching is done electronically. One of the methods used to control the BLDC motor is done using at least 3 Hall sensors with which the rotor position is read.

Fig. 4. BLDC motor custom made

Hall sensors generate a digital signal based on the polarity of the magnetic pole in front of the sensor. For a pole N the signal is on logic level 1, and for a pole S the signal is on logic level 0. These signals are usually 3.3 V or 5 V TTL signals. The sensors detect magnetic poles on the rotor, and depending on the binary combination obtained based on the signals, the correct winding supply sequence is established. The power supply sequence is maintained until a new binary combination appears from Hall Sensors.

To achieve and maintain this process it is necessary to use a microcontroller, which has the following main objectives:

- reading signals from Hall sensors;
- processing the signals from the Hall sensors and establishing the supply sequence, which ensures the correct energization of the windings;
- generation of control signals for the power circuit (Fig. 5).

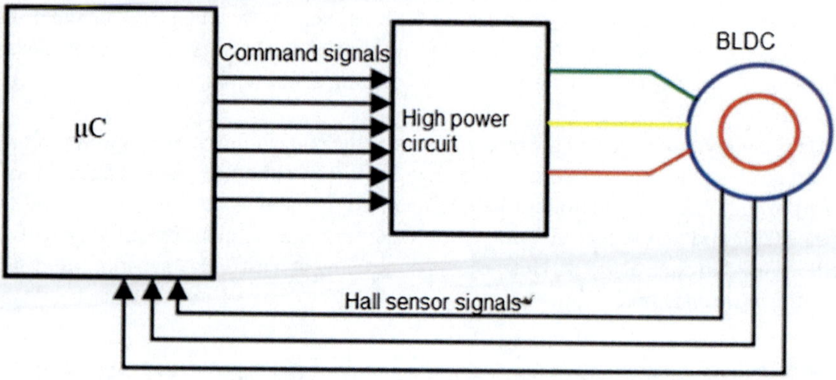

Fig. 5. Block diagram of the electronic control module

Because there are three Hall sensors, there are a total of 8 possible binary combinations based on digital signals. Given that the Hall sensors are placed at a 120° angle and the magnetic poles on the rotor are distributed in pairs, two of the 8 binary combinations are eliminated: binary combination 000 and binary combination 111. The remaining six binary combinations are repeats periodically in a certain order, order depending on the direction of rotation of the motor. Completely traversing the six binary combinations and properly feeding the windings is called an electric rotation. The number of electrical rotations required to perform a mechanical rotation is directly proportional to the number of pairs of magnetic poles on the rotor. In this case the BLDC motor with 5 pairs of poles, results in a number of 5 complete electric rotations on one mechanical rotation. There are six control sequences for the six binary combinations. At each control sequence, the windings are fed differently.

Figure 6 shows the three motor windings marked A, B and C in the star connection and the winding supply for the six control sequences. The direction of the arrow indicates the negative polarization of the windings relative to zero, and the unmarked part with the arrow indicates the positive polarization of the windings relative to zero.

Table 1 shows the clockwise control sequence for each binary combination of Hall sensor outputs.

The table shows that a Hall sensor changes its state only after three control sequences, which results in a magnetic pole passing through a Hall sensor after three control sequences. When a magnetic pole has been detected three times by the sensor in the same state, the winding of the sensor is not supplied. When an S-pole is detected by the sensor, the winding of the sensor that sensed the pole is negatively fed for two sequences, resulting in a repulsive force due to the identical polarity of the two magnetic fields. Otherwise, when an N-pole is detected by the sensor, the winding of the sensor that sensed

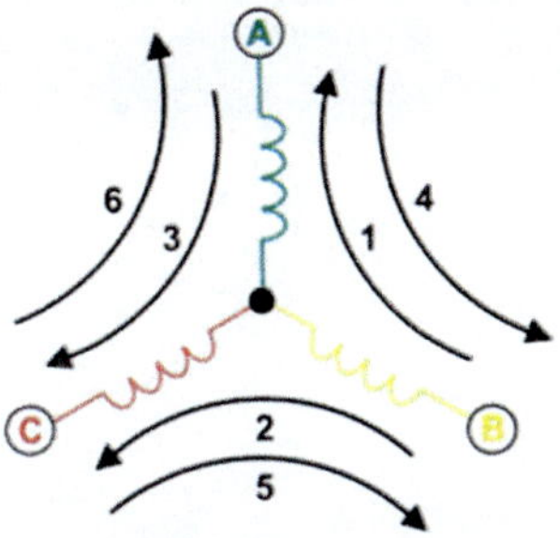

Fig. 6. Winding feed sequences

Table 1. Clockwise sequence of combinations of Hall sensor outputs

Sequence	Hall 1	Hall 2	Hall 3	Phase A	Phase B	Phase C
1	0	1	1	-V	+V	NC
2	0	1	0	NC	+V	-V
3	1	1	0	+V	NC	-V
4	1	0	0	+V	-V	NC
5	1	0	1	NC	-V	+V
6	0	0	1	-V	NC	+V

the pole is positively fed for two sequences, thus resulting in a repulsive force due to the identical polarity of the two magnetic fields. Table 2 shows the corresponding counter clockwise control sequence for each binary combination in the Hall sensors.

Table 2. Counter clockwise control sequence for combinations of Hall sensor outputs

Sequence	Hall 1	Hall 2	Hall 3	Phase A	Phase B	Phase C
1	0	0	1	+V	NC	-V
2	1	0	1	NC	+V	-V
3	1	0	0	-V	+V	NC
4	1	1	0	-V	NC	+V
5	0	1	0	NC	-V	+V
6	0	1	1	+V	-V	NC

The differences from the previous case are that the order in which the combinations of the Hall sensors occur is reversed, and therefore the control sequences are reversed. When a magnetic pole is first detected by the sensor, the winding of the sensor is not supplied and the force is attractive in the case of the next two control sequences and the polarities of the magnetic fields are different. The entire BLDC motor control algorithm was developed using a dedicated microcontroller also using an Arduino development board that acts as a link between the data at the system input (accelerator pedal position, movement direction) and effective control system (Fig. 7).

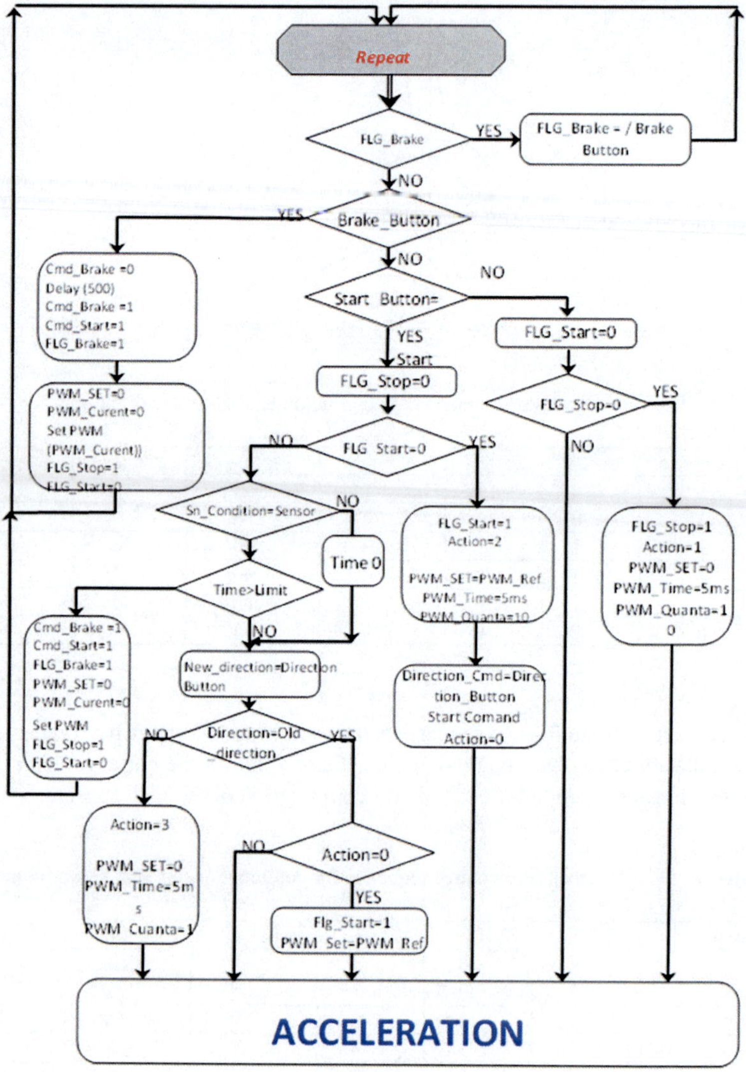

Fig. 7. Control logic diagram

As can be seen in the logic diagram above the control algorithm consists in a loop that monitors the state of the BLDC motor and the control signals received from the car user in order to give the PWM pulses to move forward or backwards, to accelerate or decelerate, or to stop the motor.

4 Conversion into an Electric Vehicle of the Historical Romanian Vehicle ARO

Successfully obtaining satisfactory results in terms of converting small vehicles with internal combustion engines into electric vehicles, we moved on to the next stage when the conversion of a historic Romanian 4 × 4 vehicle began. ARO (Auto Romania), one of the symbols of Romania before 1989, was a manufacturer of off-road vehicles in Câmpulung-Muscel, Romania. Production of off-road vehicles began in 1957 and more than 330,000 vehicles were produced, with most of the production being exported to over 110 countries worldwide. These vehicles have not been manufactured in Romania since 2003, but they still remain a special landmark in terms of domestic vehicle production and are still highly sought after by lovers of Romanian classic vehicles.

This system is designed to be a research stand for various adapted hardware elements, but also for testing different driving algorithms. In October 2021, an ARO 243 was purchased, checking beforehand that its mechanical platform (body, chassis and accessories) is in relatively good condition, taking into account the age of the vehicle. The parts related to the classic motorization were disassembled (internal combustion engine, exhaust and other elements that were present on the old structure).

The process of changing all mechanical parts (steering, brake, suspension) has begun and work has begun on repairing the body affected by time and rust (Fig. 8).

Fig. 8. Reconditioning of mechanical and body systems

After almost completely (at this moment) the reconditioning of the entire mechanical system as well as the repair of the body, because this vehicle is to be put back into public circulation after receiving the necessary approvals, the installation process of the new electric motor has started. The basic system developed contains an 80 kw BLDC motor, powered by a 25 kw battery bank, at a voltage of 400VDC. We used as components the construction of a standard electric vehicle, a Nissan Leaf electric motor (Fig. 9) that proved to be very reliable and safe in operation (greatly reducing the risk of "leakage" of electricity that could occur in the case of building one's own motor).

Fig. 9. The electric motor mounted on the mechanical platform of the ARO 243

The connection between the electric motor and the gearbox was made using an elastic clutch with claws, thus benefiting from higher tolerances than in the case of a rigid coupling and thus obtaining a perfect alignment between the motor and the gearbox, as shown in Fig. 10.

Fig. 10. The elastic coupling between the electric motor and the ARO 243 gearbox

For the charging part we use a BMS (Battery Management System), which has the function of charging by stabilizing the cells in the batteries without being destabilized. This project uses Orion BMS 2 [11] which protects and monitors a battery by monitoring multiple sensors and using multiple outputs to control battery charging and discharging. The BMS measures the inputs from the socket cell voltage, the total pack voltage socket, has a current sensor with hall effect and thermistors. The BMS controls the current flow (charging or discharging) by sending data on current limits (via CANBUS or analog reference voltages) or by simply turning digital signals on/off as needed. The BMS needs the intervention of a user to provide external controls that respect the current limits (charging/discharging) set to protect the batteries. Balancing the battery cells of the electric vehicle is done by the BMS with the help of shunt resistors. The Orion 2 BMS can also be interfaced with most existing charging stations. Orion BMS 2 programming is relatively simple and intuitive using a computer that uses the Orion BMS 2 Utility software (Fig. 11), the entire programming process is done through the CANBUS communication protocol.

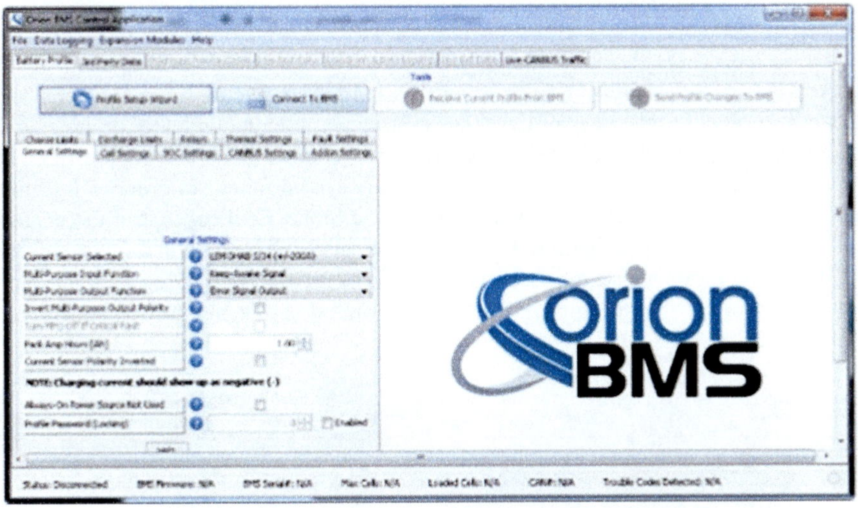

Fig. 11. Orion BMS 2 Utility software interface [11]

The control of the entire loading process can also be achieved through a small development board (Raspberry PI) which, in addition to the advantage of having extraordinary portability, has sufficient computing power to meet this goal, thus being able to achieve a integrated computer-based control solution. The current entering and leaving the battery (charging/discharging) is measured every 8ms using the external sensor with hall effect. The Hall effect current sensor is attached to a wire that carries all the current from the battery and converts the measured amperage into two analog voltages between 0–5 V. One channel is used to measure lower amperages to ensure a high resolution for low currents, and the other channel is used to measure higher currents. These values of the analog voltages at the output of the current sensor are used by BMS for battery protection. Figure 12 shows the correlation between the analog voltage supplied at the output by the current sensor and the actual measured current.

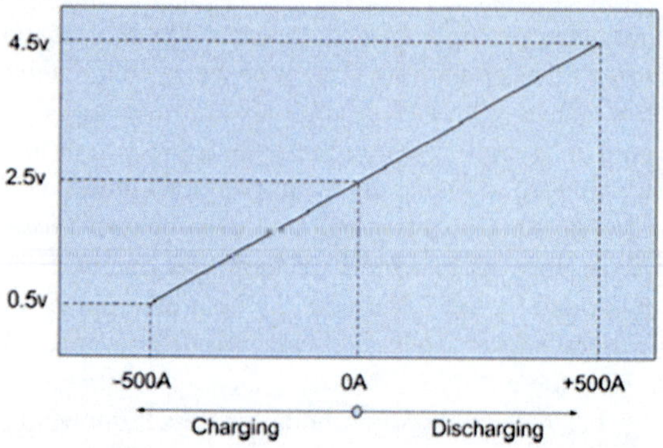

Fig. 12. Battery charge/discharge correlation diagram based on analog values of a current sensor [11]

Orion 2 BMS [11] is one of the most universal battery management systems on the market. Due to its versatility, it can be easily used for electric cars, electric ships and network storage. In this way, all CAN bus messages can be individually designed and the addresses assigned. With various construction versions that allow control of 24 to 180 cells, the optimal size for any project can always be found. In order to integrate our electric vehicle, ARO, in the whole chain of a Smart Grid concept in the context of a Smart City, our own charging station was designed and built, which is currently in the functional prototype stage in tests (Fig. 13). This prototype charging station is built around a SmartEVSE controller and allows the generation of a 44 Ah charging current.

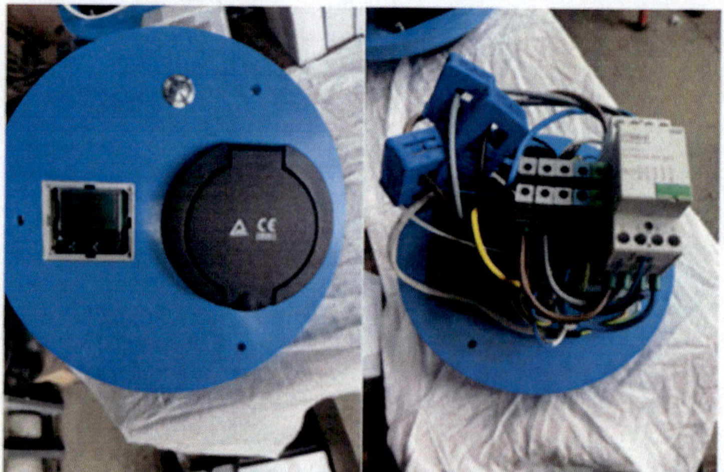

Fig. 13. Own charging station built using a SmartEVSE controller

Even if the question remains whether or not electric vehicles pollute, there are more and more achievements that demonstrate that satisfactory benefits can be obtained from this technology.

By including in the entire hardware and software structure a LoRa communication system [10] designed and built in the laboratories of the University of Petrosani, it will be possible to obtain an advanced control system that integrates the calculation part related to the vehicle control system and the communications made with the help of LoRa technology. LoRa is a wireless communication technology that uses an original radio modulation technique. This specific modulation technique allows the creation of connection points between two systems that will be able to transmit small amounts of data over relatively long distances, thus being able to use a reduced bandwidth and at the same time minimizing energy consumption. Some of the special applications that can use the LoRa communication protocol would be the realization of communication systems between electric vehicles (V2V) as well as between them and the transport infrastructure (V2G).

5 Conclusions

In the current context of concerns about the fate of this planet in terms of climate change, truly clean solutions must be found as soon as possible, especially in the car industry, because we see that each year the number of vehicles on the streets increases and as their number increases, so does pollution. This paper briefly presents various achievements of electric vehicles designed to be a viable alternative to classic vehicles. All the achievements presented are in fact the result of an extensive process of recycling some vehicles or parts of a vehicle that at one time only occupied a space without being functional. The electric kart, the electric ATV and the electric buggy vehicle were the starting points for a major conversion process into an electric vehicle of a historic Romanian vehicle, ARO 243, rebuilt almost from scratch with many recycled parts. This electric vehicle is intended to be approved as a street vehicle in the near future and can be put into public circulation.

All these achievements so far presented in this paper are based on the principle of reusing different equipment or parts and on obtaining functional platforms with which to take data on environmental factors, especially in large urban areas. If until now using small electric vehicles developed at the University of Petrosani could make tests or monitor various environmental parameters only on the university campus, once the conversion of the vehicle ARO 243 and its approval by the Romanian Road Transport Authority (ARR), it will be possible to move to a new level, being possible to build a mobile laboratory in terms of various possible applications in the road industry. The integration of these vehicles in the Smart Grid and Smart City concepts is very simple to do because from the beginning it was wanted to make electric vehicles that have different smart functionalities that allow them to communicate with each other (V2V) or with infrastructure (V2G).

Thus, through the LoRa communication protocol and other specific tools used in the case of making the electric ATV, it has reached the point where the electric vehicle can determine a certain driving trajectory on its own. The conversion of the historic

ARO 243 vehicle into an electric vehicle also brought multidisciplinary challenges, so much work was done on the mechanical part as well as on the design part, wishing to obtain in the end a completely restored vehicle, but which would not lose its originality that made him known and appreciated. All these achievements lay the foundations for special projects that have major implications both in terms of environmental protection strategies and in terms of developing new concepts of Smart Grid and Smart City.

References

1. Rus, C., Leba, M., Ionica, A.: Road traffic control in the context of an environmental friendly smart city. In: 17th International Technical-Scientific Conference on Modern Technologies for the 3rd Millennium, pp. 351–356 (2018)
2. Findl, E.: Physics and the electric car. Phys. Teach. **15**(2), 72–77 (1977)
3. Thornton, R.: Design considerations for an electric car. SAE Technical Paper (1970)
4. Dave, J., Dong, J.: Conversion of an existing car to a rechargeable electric vehicle. In: 2009 Annual Conference & Exposition, pp. 14–367 (2021)
5. Francis, J., Narayamparambil, A.B., Johnson, A., Mathew, J., Sankar, V., George, J.M.: Conversion of internal combustion engine car to semi-autonomous electric car. J. Phys.: Conf. Ser. **2070**(1), 012203 (2021). IOP Publishing
6. Rievaj, V., Synák, F.: Does electric car produce emissions?. Zeszyty Naukowe. Transport/Politechnika Śląska (2017)
7. Dankov, D., Madjarov, N., Prodanov, P.: Description of the conversion process and results on the Opel Corsa with an internal combustion engine into an electric car. In: 2021 XXX International Scientific Conference Electronics (ET), pp. 1–5 (2021)
8. Rus, C., Mija, N., Leba, M.: Autonomous electric ATV using IPM based inverter control and deep learning. In: Rocha, Á., Adeli, H., Reis, L.P., Costanzo, S., Orovic, I., Moreira, F. (eds.) WorldCIST 2020. AISC, vol. 1159, pp. 746–755. Springer, Cham (2020). https://doi.org/10.1007/978-3-030-45688-7_73
9. Rus, C., Marcus, R., Pellegrini, L., Leba, M., Rebrisoreanu, M., Constandoiu, A.: Electric cars as environmental monitoring IoT Network. In: IOP Conference Series: Materials Science and Engineering, vol. 572, no. 1, p. 012091. IOP Publishing (2019)
10. Rus, C., Leba, M., Marcuş, R., Pellegrini, L., Costandoiu, A.: LoRa communication and geolocation system for sensors network. In: MATEC Web of Conferences, vol. 305, p. 00043. EDP Sciences (2020)
11. Orion 2 BMS Operation Manual. https://www.orionbms.com/manuals/pdf/orionbms2_operational_manual.pdf. Accessed 31 Mar 2022

Conversion of a Hybrid Flexible Manufacturing Cell to Industry 4.0

Valente Váquez-Velazquez[1], Mario Peña-Cabrera[1], Víctor Lomas-Barrié[1], and Gastón Lefranc[2](✉)

[1] Instituto de Investigaciones en Matemáticas Aplicadas y en Sistemas (IIMAS), Universidad Nacional Autónoma de México, Mexico City, Mexico
valente@ciencias.unam.mx, mario.penia@unam.mx, victor.lomas@iimas.unam.mx

[2] Pontificia Universidad Católica de Valparaiso, Valparaíso, Chile
gaston.lefranc@pucv.cl

Abstract. A flexible manufacturing cell implemented in a hybrid way is presented, that is, integrating a real physical part with a simulated one. The cell is made up of a Kuka robotic manipulator, a vision system, sensors, actuators and containers for parts and products. The simulated part is completely connected to the real part and vice versa. The cell is converted with industry 4.0 criteria, which is implemented with a wireless communications system and is controlled through a hierarchical system. The cell is tested by performing a part pick and place task on a conveyor belt.

Keywords: FMS · Flexible manufacturing cell · Industry 4.0 · Computer vision · Robotics · Simulation

1 Introduction

For several years, flexible manufacturing systems (FMS) have been used to make a wide variety of products to meet market demand. This requires a certain level of automation and artificial intelligence allowing to improve production. When FMS is combined with Industry 4.0 technology, the production is accelerated.

FMS focuses on the automation of cells that function as a system, consisting of processing, automated material handling, and material and product storage workstations. FMS has flexible manufacturing cells, flexible assembly cells, quality control cells, storage cells for parts, tools and finished products. The proper used software allows the different tools, materials and processed parts to be identified and distinguished; and allows quick changes to operating instructions and physical cell reconfiguration. A computer manages the cell, and is connected to the FMS system computer, through a hierarchical communication network [1].

In the cells, different machines are used to handle parts and tools, such as drills, lathes, etc., to manufacture products. It also has the flexibility to route the movement of

S. Dzitac et al. (Eds.): ICCCC 2022, AISC 1435, pp. 63–71, 2023.
https://doi.org/10.1007/978-3-031-16684-6_5

materials and parts; as well to route work from one machine to another, according to a given process.

In the evolution of the automation of manufacturing processes, the concept of Industry 4.0 is reached, which includes the exchange of information wirelessly within industries. This criterion uses new technologies in production plants that include: Internet of Things IOT, Industrial Internet of Things (IIoT), Smart Manufacturing, Cloud Computing, Cognitive Computing, Artificial Intelligence and Cyber-Physical Systems (CPS). This means that all devices and technology used are wirelessly connected using actuators, sensors and monitors to visualize all processes, allowing a solid overview of operations in a company. In recent years, FMS has improved automation by using cobots (cooperative robots) for routing and some machine process steps [2].

Manufacturing companies should enhance FMS with digital transformation technologies such as AI, generative design, and 3D printing. 3D printing represents an example of such a transformation which in many cases includes digital design of the final product and production of parts, showing 3D printing true flexibility, even if it costs around 10 times more.

To understand the impact of the Industry 4.0 concept on manufacturing processes, it is important to understand the evolution from Industry 1.0 to 4.0.

- In Industry 1.0, mechanical production uses steam engines (18th century).
- Industry 2.0 begins in the 20th century with the development of machines working with electricity, simplifying mass production.
- Industry 3.0 begins with the incorporation of electronic devices and systems in the 20th century. They use machines with electronic processors and software.
- Industry 4.0 began in 1990, when the Internet and telecommunications transformed the way the information is connected and exchanged. This technology merges the physical world, the virtual world and cyber-physical systems (CPS).

A flexible manufacturing system (FMS) is a production method that is designed to easily adapt to changes in the type and quantity of the product being manufactured. Machines and computerized systems can be configured to make changes of variety of parts and handle levels of production [3].

This work presents the conversion of a flexible manufacturing cell using industry 4.0 criteria, implemented with a wireless communications system and controlled through a hierarchical system. The cell used is hybrid, that is, it integrates a real physical part with a simulated part.

2 The Flexible Manufacturing Cell

The general architecture of the flexible manufacturing cell (FMC) is showed in Fig. 1.

FMC is composed of a central computer, a vision system (CVS), a robotic manipulator (RM), a conveyor belt (CB) and a simulator system (SS).

The FMC information is handled by the computer who controls the cell based on a hierarchical structure approach: for example, the pick and place movement of the

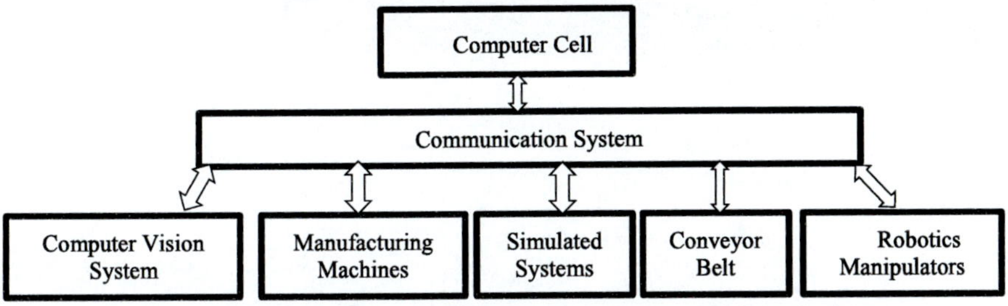

Fig. 1. General architecture of the flexible manufacturing cell FMC

robotic manipulator is a function of the CVS vision system as well as the movement of the conveyor belt (Fig. 2) [4, 5].

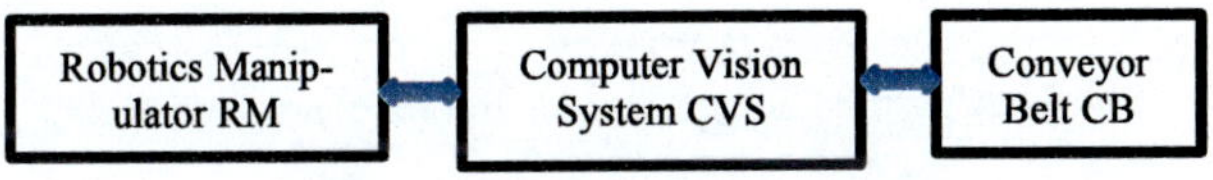

Fig. 2. FMC hierarchical structure

The CVS is executed when CB movement is activated. If a part is within the camera's field of view, CVS generates the descriptor vector called Boundary Object Function vector (BOF), optical flow and Kanade methods are used to determine the position, orientation and pick points of the object and tracking of trajectory starts. This have already been tested in previous works to get the coordinates of the object [15]. The coordinates are sent to the RM, who picks up the part and then the process is confirmed (Fig. 3).

Figure 4 shows the real FMC that uses a Kuka robotic manipulator, a conveyor belt and parts made by 3D printer. In Fig. 5 FMC is simulated using CopeliaSim [6].

Both can work alone or work in coordination, that is, if the real FMC picks and places tools or parts, the simulated FMC performs the same task. On the other hand, if the simulated FMC is to pick and place tools or parts, the real FMC does the same job.

In addition, we can carry out hybrid work, that is, a real machine can be replaced by its simulated counterpart. In the example, we can replace the real CVS using the CVS from the simulation. This idea has been used before in the control of mobile robots using software on a computer [5]. The operating algorithm of the robotic manipulator is written in Python.

The coordinate system used in the Kuka manipulator (Xk, Yk, Zk, αk, βk, γk), coincides with the simulator coordinate system (Xc, Yc, Zc, αc, βc, γc), (1):

$$(Xk, Yk, Zk, \alpha k, \beta k, \gamma k) = (Xc, Yc, Zc, \alpha c, \beta c, \gamma c) \quad (1)$$

To do this, the system simply review and compare the angles of the real system (Fig. 6a) with the virtual one (Fig. 6b).

The images of (Fig. 6) show that the X axis of the FMC is a mirror with respect to that of the simulator. It is possible to send and/or receive the pick or place coordinates if

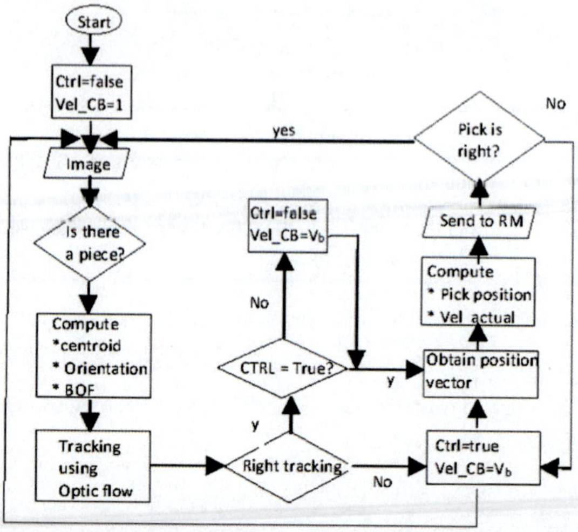

Fig. 3. Algorithm implemented in the computer of the cell

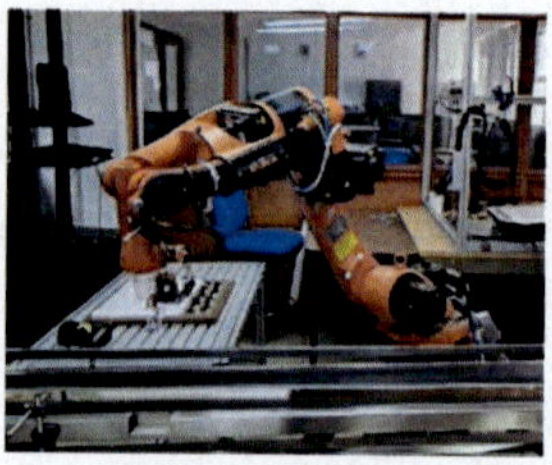

Fig. 4. Real FMC

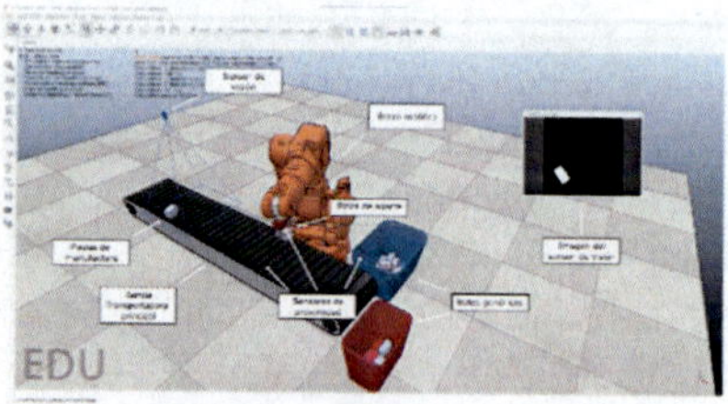

Fig. 5. FMC simulation CoppeliaSim

the proper calibration is done, since the robot's movement depends on its own reference system. As for the rotation angles α, β and γ, the only difference is that the axis of rotation about the Z axis in the physical system is the α coordinate, while in the virtual system it is the γ coordinate. Therefore, Eq. (2) becomes:

$$(Xk, Yk, Zk, \alpha k, \beta k, \gamma k) = (Xc, Yc, Zc, \gamma c, \beta c, \alpha c) \tag{2}$$

(a) Cartesian axes of real FMC (b) Cartesian axes of FMC simulated

Fig. 6. Cartesian axes of the physical and virtual FMC.

In the FMC example, the actual manufacturing parts are simple geometry objects printed in ABS plastic in 3D [7]. In this work grasping process is achieved using rectangular prisms, since it is a simple way to hold them (Fig. 7). An opening movement through a SG902 servo motor takes place with a designed circuit to this purpose, where a gripper close or open based on a variable received through the communication protocol MQTT, the system also has an infrared sensor model QRD1114 3 whose function is to send a status flag on whether the piece is correctly grasped.

An effector of the robotic gripper manipulator is chosen, because the angle of grip is important. The effector is 3D print, and similar to the one presented in [8] in Fig. 8.

(a) 3D image of a part. (b) Diagram of the part (mm).

Fig. 7. Manufacturing part.

Fig. 8. Clamp model as 3D end effector (Atthariq, 2021).

3 The Hybrid Flexible Manufacturing Cell

As an example, a solution which combines the virtual system with the physical one is described, that is, it starts the artificial vision CVS inside the simulator [11] and sends the grip positions to both the Kuka and the virtual manipulator; both controllers have the same scheduled routes. These paths are created by the robot by tracking the part along with its movement from the starting point [12, 13] (Fig. 9).

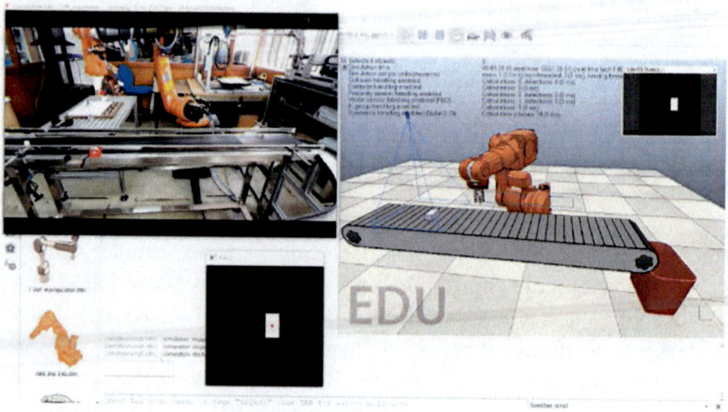

Fig. 9. Tracking of a manufacturing part using computer vision

This configuration works robustly in case of failures, since there have been no problems when testing. This part tracking is achieved, even if there are parts that are not fully supported on the web and/or in any part orientation [14].

The process to obtain the BOF vector is not performed when more than one piece is presented in the camera's field of view [15].

4 Converting Hybrid Flexible Manufacturing Cell

To improve the automation of FCM, some Industry 4.0 concepts are used to implement an information exchange wirelessly within the manufacturing cell, MQTT (Internet of Things IOT protocol) and Cyber-Physical Systems (CPS) are used to create 3D parts. This criterion complements the uses of Cloud Computing and Artificial Intelligence.

This means that all devices and technology are wirelessly connected and use actuators, sensors and monitors to visualize all processes, allowing a robust overview of cell operations. The new architecture of the FCM is (Fig. 10):

In the Hybrid FCM, information is sent wirelessly via MQTT, from the virtual system to the physical system. In this example, the machine vision CVS inside the simulator sends the grip position information to both the Kuka and the virtual manipulator. The information sent is the position and orientation of the hilt, the size of the item, and how the effector should choose the part. This information depends on the type of effector used, since it is not the same if a vacuum suction cup or a similar one is used.

The wireless communication system is based on the Standard for IoT Messaging, the MQTT protocol [9], which needs an intermediary to manage all the information and devices that are connected to the network, so the intermediary used is Mosquito [10]. This agent is perfect as it is lightweight, secure and suitable for use on a variety of

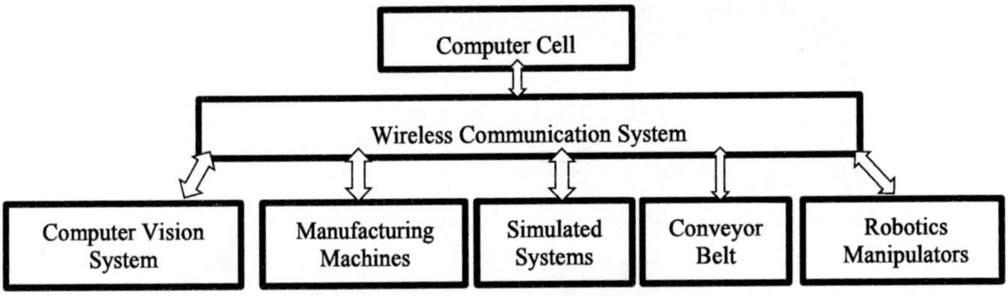

Fig. 10. New architecture of the FCM

devices, and is open source. This type is fine for this purpose, due to its potential use and low cost, specifically with a Raspberry Pi 4 Model B, it allows to easily configure a Mosquito runner inside a Raspberry.

5 Evaluations

The FMC cell is converted with industry 4.0 criteria, using a wireless communications system, controlling a hierarchically system processors. The wireless communication system is based on the MQTT protocol.

To test the FMC conversion it is configured as hybrid cell, where the artificial vision CVS inside the simulator sends the grip position information to both the Kuka and the virtual manipulator. So, the information sent is: the position of the grip, the orientation of the grip, the size of the object and how the effector should pick the piece.

The vision system obtains the BOF vector, uses the optical flow and the Lucas-Kanade algorithm to determine the position, orientation and pick points, which have already been tested in previous works. The robot manipulator works well because it is using the results of the previous work.

The cell is tested by picking and placing parts on a conveyor belt. During the tests, the system behaves robustly in the event of failures, since there have been no problems, nor when the following circumstances occur: due to not achieving correct monitoring of the manufacturing part; for having pieces that are not fully supported by the band; or the part is in any orientation; or when more than one piece is presented in the camera's field of view, provided that the process of obtaining the BOF vector is not being carried out. The test of pick an object is shown in Fig. 11.

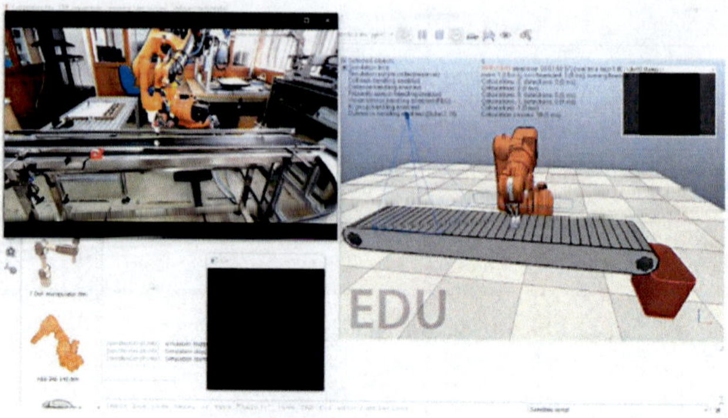

Fig. 11. Manipulator pick an object

6 Conclusion

The conversion of a flexible manufacturing hybrid cell using industry 4.0 criteria has been presented. The Hybrid Cell integrates a real physical part with a simulated part, which is the CVS computer vision system, which sends the grip positions to both the Kuka and the virtual manipulator; both have the same scheduled routes. The cell is made up of a Kuka robotic manipulator, a vision system, sensors and containers for parts and products.

To convert the cell, a wireless communications system is implemented and controlled through a hierarchical system. The cell is tested with the task of picking and placing parts on a conveyor belt.

The vision system obtains the BOF vector, uses optical flow and kanade methods to determine the position, orientation and pick points, which have already been tested in previous works.

The robot manipulator works well because the results of the previous work are being used.

During the tests carried out on the cell, they allow to say that the cell can track and take or place a piece of manufacture; you can do it even when the pieces are not fully supported by the band; or be with any orientation.

When more than one part is present in the camera's field of view, it cannot determine its position, if the BOF vector calculation has not yet started.

In general, the developed hybrid flexible manufacturing cell is based on highly optimized, functional and robust artificial vision algorithms: BOF and Lucas-Kanade optical flow, this makes the system inherit the same characteristics, thus being an excellent solution, and contribute original, for pick and place systems within Industry 4.0. In addition, various artificial intelligence methods are used in Kuka robots, conveyor belts, and other parts of the cell.

The main future work is definitely to transfer the main algorithm to a specific electronic architecture, such as an FPGA. This would give advantages accelerating computation, making the operation speed grow a significant percentage. The characteristic vector

BOF is indispensable for the training and execution stages of whatever computational method is used.

References

1. Lefranc, G.: Some trends in robotics and automation. In: 2021 IEEE CHILEAN Conference on Electrical, Electronics Engineering, Information and Communication Technologies (CHILECON), pp. 1–8 (2021). https://doi.org/10.1109/CHILECON54041.2021.9703003
2. Sullivan, M.: Flexible Manufacturing Systems Combined with Industry 4.0 Can Further Progress Toward Mass Customization (luxresearchinc.com) (2020)
3. Hughes, I., Millán, G., Cubillos, C., Lefranc, G.: Colony of robots for exploration based on multi-agent system. Int. J. Comput. Commun. Control **9**(6), 103–130 (2014)
4. Leighton, F., Osorio, R., Lefranc, G.: Modelling, implementation and application of a flexible manufacturing cell. Int. J. Commun. Comput. Control **1**, 278–285 (2011)
5. Rojas, D., Millan, G., Passold, F., Osorio, R., Cubillos, C., Lefranc, G.: Algorithms for maps construction and localization in a mobile robot. Stud. Inform. Control **23**(2), 189–196 (2014). ISSN 1220-1766
6. CoppeliaSim Release 4.2.0 is out (April 6th 2021). https://www.coppeliarobotics.com/
7. Markforged 3d printer. https://marforged.com
8. Atthariq, L.: Gripper servo sg90 (2021). https://grabcad.com/library/gripper-servo-sg90-1
9. R MQTT: The Standard for IoT Messaging. https://mqtt.org/
10. Mosquito Version 2.0.13 released, Eclipse Mosquitto. Obtenido de: https://mosquitto.org/
11. Osorio-Comparán, R., et al.: Recognition and simulation of a vision system for a KUKA KR-5 industrial manipulator. In: 2021 IEEE CHILEAN Conference on Electrical, Electronics Engineering, Information and Communication Technologies (CHILECON), pp. 1–6 (2021). https://doi.org/10.1109/CHILECON54041.2021.9702984
12. Osorio-Comparan, R., et al.: Vision system for a Kuka KR-5 industrial manipulator. In: 2019 IEEE CHILEAN Conference on Electrical, Electronics Engineering, Information and Communication Technologies (CHILECON), pp. 1–7 (2019). https://doi.org/10.1109/CHILECON47746.2019.8988027
13. Peña-Cabrera, M., Amezquita, L., Lomas-Barrié, V., Osorio, R., López, J.I., Lefranc, G.: Artificial vision in embedded system for rigid parts recognition. In: 2019 IEEE CHILEAN Conference on Electrical, Electronics Engineering, Information and Communication Technologies (CHILECON), pp. 1–6 (2019). https://doi.org/10.1109/CHILECON47746.2019.8988010
14. López-Juárez, I., Castelán, M., Castro-Martínez, F.J., Peña-Cabrera, M., Osorio-Comparan, R.: Using object's contour, form and depth to embed recognition capability into industrial robots. J. Appl. Res. Technol. **11**, 05–17 (2013)
15. López-Juárez, I., Ríos-Cabrera, R., Peña-Cabrera, M., Méndez, G.M., Osorio, R.: Fast object recognition for grasping tasks using industrial robots. Comput. y Sistemas **16**(4), 421–432 (2012). ISSN 1405-5546

A New Way of Collecting Traffic Data Inside a Smart City Based on IoT Devices and Communication Protocols Designed for Preserving Perfect Anonymity

Nicolaie Popescu-Bodorin[1,2] and Mariana Nagy[1,3](✉)

[1] Applied Computer Science Testing Laboratory, Bucharest, Romania
bodorin@live.com
[2] Spiru Haret University, Bucharest, Romania
[3] Aurel Vlaicu University of Arad, Arad, Romania
mariana.nagy@uav.ro
http://lmrec.org/bodorin/

Abstract. The basic schemes of new, generic, anonymity-preserving, radio frequency based IoT devices and communication protocols are presented here. Their usage is illustrated in the particular context of avoiding traffic jams in Smart Cities, while also exemplifying traffic data as IoT big data that can be transmitted, collected, processed and interpreted with important social benefit, without necessarily bounding it to the identity of a person or device, hence without creating any room for user's and device's privacy to be compromised and/or abused. We defined these new schemes by descending along the OSI (Open System Interconnection) model down to the point where the goal of preserving anonymity is achievable by changing the transmission protocol rules, and also the structure and functionality of the IoT devices in an appropriate manner for the purpose.

Keywords: Smart City · IoT · FR-IoT · Privacy · Anonymity · RFID · Anonymous radio frequency device · Anonymity preserving communication protocols

1 Motivation, Terminology and Literature Review

In the IoT communication based on the OSI (Open System Interconnection) model, by the nature of the communication protocols, one can find out, eventually, who is using what device, under what IP address, in what location, at what time - and this is what we - ironically - still call 'privacy' today, when in fact, it is exactly the lack of that.

Since we insist that in various situation - some of them closely related to the Smart City concept - the data should be collected while preserving the anonymity of its source user and device, one could legitimately ask how much

S. Dzitac et al. (Eds.): ICCCC 2022, AISC 1435, pp. 72–84, 2023.
https://doi.org/10.1007/978-3-031-16684-6_6

(social) benefit can be provided by using anonymized data only. To illustrate an answer, the case of aggregating and processing data for modelling and controlling the traffic flow is considered here. Other particular use cases well-answering for the same question can be easily derived/identified by the reader, for sure.

The basic schemes of new, generic, anonymity-preserving, radio frequency (RF) based IoT systems (devices and communication protocols) are presented here as an incremental development on top of the initial RFID patent of Mario Cardullo [3]. Their usage is illustrated here for the particular context of avoiding traffic jams in Smart Cities, while also exemplifying traffic data as IoT big data that can be transmitted, collected, processed and interpreted with important social benefit, without necessarily bounding it to the identity of a person or device, hence not only without creating any room for user's and device's privacy to be compromised and/or abused, but also without allowing for data to be collected otherwise than as anonymized data in the first place, not as private/personal data. We defined these new schemes by descending along the OSI model down to the point where the goal of preserving anonymity is achievable by changing the transmission protocol rules, and also the structure and functionality of the IoT devices in an appropriate manner for the purpose.

The remaining of the current section establishes the ground for positioning our approach in comparison to the systems discussed in the literature review.

1.1 Anonymity vs. Privacy

Just to avoid any possible confusion, the terms privacy and anonymity are considered here as they are defined in Oxford [5] and Cambridge [6] online dictionaries. Various well-known privacy-related incidents [7] warn us all about the possible consequences of collecting private data - on one hand, and on giving up using such (apparently too stiff, but very) clear and precise dictionary definitions - on the other. Nevertheless, these definitions allow us for easy translations as formal knowledge within a formal grammar later used for configuring the flow of data inside and between the IoT devices such that the system as a whole to act as a membrane [8] whose permeability to sensitive identity data is always null. We aim to describe the target system as a formal grammar with specified target properties by using a dictionary of data flow methods defined over a vocabulary of data and data structures. The flow methods describe how the devices work internally on data, and how do they communicate data, whereas the vocabulary of data and data structures provides both an abstract and a hardware/physical description of the data and data types that can be internally stored by the IoT devices or transmitted within the communication stream. The big advantage of designing the newly proposed anonymity-preserving system in this manner is that whenever our search ends up with finding a target formal grammar satisfying the required properties, it is formally checked already, hence it is - by design - unable to accidentally break the rules concerning the continuous preservation of anonymity. As a result, no private data is collected in the first place, and the risk of someone or something using such data in relation with an identity of an user or device is entirely avoided. For a large range of applications, ensuring the anonymity is better than subsequently managing and guarding the privacy.

1.2 Our Contribution in the Context of the Literature Review

The contribution of this paper is incrementally innovating on top of the original and oldest RFID patent [3] by redesigning the flow of RFID data and the memory of the transponder to effectively anonymize users, RFID data, and the transponder itself. In what follows, the literature review is presented in contrast to our approach.

The most unusual fact in our approach is the anonymization of the transponder, because until now, in the majority of previously proposed RFID applications (many refhere) the transponder was always marked with some identity data - the usage of RFID technology being very much related with logistics [1] or other similar areas where the unique identity of the transponder is primordially important. Still, for many other applications this is unnecessary, and the transponder can store and use only temporary, changeable, anonymous identities - established and given by the transceiver/antenna unit or, on the contrary, randomly and locally generated by the transponder itself.

Since 1973, when the RFID technology has been introduced for the first time [3,4], the range of RFID applications expanded continuously [13,14].

Alarming in our view, the trend is to only stronger and stronger bound the RFID data to the identity of a specific user and/or device, and that explains why RFID technology and devices are not so largely preferred when it comes for the users to agree on their usage [18,22] - fact witnessed by the discrepancy between the number of RFID-related patent documents indexed by Google (136'658 today, [2]) and the number of RFID applications that we can see already implemented and in use around us - the latter being incredibly low by comparison. In the quest of correcting this situation, other authors proposed decoupling IoT devices from the user identity [9] by calling for a third party or by introducing/considering an additional storage module (similar to a third party authorization/authentication provider) in the RFID data flow to maintain the association between the user identity and the device (s)he is using, a module ranging in size and technology from a small chip on the transponder and/or antenna unit [], to an actual SGBD [19,20] used to operate an authentication protocol. Still, even with this solution the user identity and actions can be tracked, subject to querying for - and eventually obtaining - the association data from that third party. It could be taken as a progress, for sure, but we think much more can be done, at least for some specific applications, simply by altering the design of the RFID system to allow only anonymous users and anonymized RFID data in the first place. Other ways of increasing the privacy of RFID data were proposed mainly based either on Blocker-Type methods for protecting customers' privacy in RFID systems [15], or on increasing power consumption on the transponder to allow low-power cryptography on it [21].

One sensitive issue in the context of using RFID technology with supply chain and logistics is optimizing the locations of the RFID antennas [10] - possibly a valid concern also for many other applications relying on anonymous RF devices, but not for the task of sensing and managing the traffic jams, where the streets intersections are by default the best option for installing the antennas.

Antenna and transponder coverage is also an important matter for any RFID infrastructure in general [11]. Still, in the context of traffic sensing, the transponders are mobile devices installed on cars, and therefore, the inevitably decreasing distance from transponders to antennas is mitigating well the coverage risks.

From a different perspective, we aim our proposed system to be compatible to the still-to-come technology and also to the latest and most robust developments in the RFID field, allowing for low-cost consumable tags, miniaturization, and avoiding directionality and mutual antenna-transponder orientation issues [12], therefore we will formulate our approach in terms of simple hardware and data-flow schemes enabling further integration with other designs and technologies. Regarding the transponder-to-antenna operational distance, we target only the long-range passive [23] or active [24] mode against any other, whereas when it comes to choose how passive [16] or active [17] the transponder should be, we aim to offer compatibility for the entire range starting with UHF passive re-writable transponders [24], up to active transponders, with some enhancements for the latter ones, and with the precaution that in case of the passive transponders, UHF mode is mandatory to ensure that enough energy (necessary for many read/write operations) can be transferred on the transponders.

2 Proposed System

The neighbouring application field matching our approach is Vehicle - to - Infrastructure (V2I) communication segment of IoT underlying previous attempts in traffic management [26,27], real-time traffic-adaptive intersection signal scheduling [28], traffic flow optimization [29–31], traffic tool collection [32], vehicle tracking [33], and automated car-parking [25]. The major difference between all of these works from above and our approach is that we focus exclusively on the Communication Layer within the RFID paradigm, where, by analogy with the structure of OSI model, we define a Data Link Layer divided into two sub-layers, namely MAC (Media Access Control) data link layer, and LLC (Logical Link Control) data link layer, both of them with enhanced functionalities for preserving device's and user's anonymity, and isolating sensitive RFID data within a closed-loop communication specified inside these two sub-layers.

2.1 Terminal, Network and Central Control/Storage Devices, and the Events on Them

In the system architecture that follows to be discussed here, the terminal devices are custom-built anonymous RF transponder-alike units, further denoted ATU (Anonymous Transponder Unit), whereas the network device is the RF public antenna/receiver unit, further denoted PAU (Public Antenna Unit), fixed as location for a given exploitation period, but in general movable, (re)configurable, and compatible, in principle, with both wireless and wired metropolitan network infrastructure in the context, by nature a public, non-anonymous, non-private network device providing:

- wireless UHF RF communication interface for the nearby ATUs (installed on cars - in the particular case of controlling the traffic),
- master MAC management service (a DHCP-alike service) provided on the MAC data link layer for the terminal devices entering its operational range - which are the MAC slaves,
- so much needed isolation (firewall-alike filtering) for the RFID identity (MAC/sensitive/private) data exchanged with the terminal devices (ATUs).

Any ATU works in strictly closed loop with a PAU when it comes for the MAC layer, whereas any PAU is a subsidiary of the Smart City network infrastructure and is responsible for routing the flow of anonymized payload data (coming from ATUs on LLC layer) into the metropolitan network to Central Control and Storage Unit (CSSU) for further processing. At least in the case of controlling traffic jams, it's not the privacy of the PAU concerning us, but only the anonymity of the ATUs that are, in a way, also public devices because their communication can be intercepted by nearby devices other than the designated PAUs.

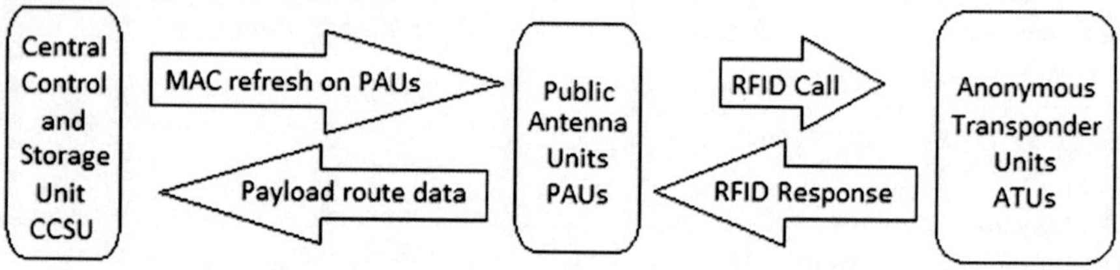

Fig. 1. System structure and main data flow

For both antenna (PAU) and transponder (ATU) devices, we consider, by design, the following list of events: the power-related events (Power-On, Power-Off), Fig. 1 MAC-related events (get/set MAC address and frame ID), command-related events (get/set a command code from or to the frame), and the LLC-related events (get/set/send payload data, get/set data frame integrity-related fields used for error detection and correction, check frame integrity, rewrite/resend the received frame if needed), memory-full event on the ATU, and the externally-oriented event on the PAU to send ATU-submitted route to the storage. From all these events, for simplicity, the current paper will ignore those concerning error detection and correction, since this is a well-known separate topic, and therefore, by this simplification, without reducing the generality, it is considered here that any received frame is correctly mirroring the frame that has been sent.

2.2 Making the Effort of Intrusive PAU- and ATU-Listeners Useless

The CCSU is responsible to change the MAC addresses of all the PAUs accordingly to established refresh rate, propagation strategy and anti-collision policy, and to register these changes and the time they occur to be latter used as triplets (MAC address, time stamp, location) for decoded the routes submitted by the ATUs to PAUs and forwarded by PAUs to the CCSU. Refreshing the MAC addresses on the PAUs ensures that an intrusive listener system can only acquire and make sense of the correspondence MAC-to-location only if listening all PAUs within the city and only if submitting these data to its own CSSU. In other words, in order for an intrusive listener system to make sense of the MAC addresses of the PAUs, it has to be built at the same scale as the system it is listening to, hence it will require a big financial investment, time, and resources to be installed, while its deployment couldn't go unnoticed. Even such a system would be successfully installed, the data it will capture will be a copy of the total data the official system captures, i.e. the collection of all MAC addresses for all the PAUs along the time, and the collection of anonymized routes of the given length k reflecting the number of consecutive intersections each ATU can store before reaching memory-full event and sending its data to a PAU. Since the total data doesn't allow for the identification of the drivers and cars, it can just be published for research purposes, making any effort to deploy an intrusive listener system just a non-sense.

2.3 ATU-PAU Inner Communication Protocol Layers

Without establishing now a certain length for them, the messages exchanged by ATUs and PAUs are symbolically represented here as digital binary frames in which the fields can be recognized conventionally by their individual length and possibly - as an additional precaution - by their position within the frame.

SOF	FID	SID	RID	CID	(AID_1,TS_1)	...	(AID_k,TS_k)	FCS	EOF

Fig. 2. Frame structure

For simplicity, this additional precaution will be left aside, and therefore, we can consider the classical situation of the frames formed with length-specified binary words (displayed in the scheme above), first five of which in our case here are: the Start-Of-Frame word (SOF), the Frame ID (FID), the Sender ID (SID), the Destination ID (DID), and the Command ID to reach the destination (CID). The subsequent content of the frame is the payload field - a collection of data pairs (data-word, time-stamp), followed by the Frame Check Sequence (FCS) and the End-Of-Frame word (EOF).

For the purpose of capturing, analyzing and managing the traffic, the payload is set to contain a history of the car movement relative to a fixed number (k) of previously encountered PAUs as an array of pairs (AID, TS) - i.e. (Antenna ID, Time Stamp) pairs.

The sender and the destination within the frame can temporary identify a PAU or an ATU, depending on the situation: in a RFID call (Fig. 1), the SID is the current MAC of the current calling PAU, whereas in an RFID response

(Fig. 1), the SID is the temporary MAC of an ATU replying to a RFID call made by a PAU. In a RFID call (Fig. 1), the RID is the current MAC of the current replying ATU, whereas in an RFID response (Fig. 1), the RID is the temporary MAC of the PAU that made the RFID call, as it was last time refreshed by the CCSU. In order to avoid providing useful feedback for intruder devices mimicking ATU response, the PAUs will acknowledge the RFID response to the sender regardless the source of the response is recognized as an ATU in the system or not.

2.4 MAC and LLC Data-Link Layers

In fact, we have here only one communication channel: the data-link layer used to communicate frames (Fig. 2). Inside a frame, however, different fields encode data for few different purposes: frame container (SOF, EOF), frame authentication (FID), MAC data (SID, RID), command (CID), payload (the array of pairs AID-TS), and frame integrity (FCS). All of these are actually interdependent.

Logical Link Control layer can be seen for the moment as the collection of rules expressing this interdependence, whereas MAC data and the rules for manipulating it - as MAC data-link layer.

Still, the MAC data within a frame comes there as a result of some services running far up above the level of the frame: CSSU refreshes MAC addresses of the PAUs, which together with some timestamps are further carried to ATUs (as payload pairs) by a PAU RFID call, and then stored in the local memory of the ATUs until the memory-full event triggers the memory to be copied to a frame and sent from the ATU to the PAU as RFID response (Fig. 1). Once received by PAU, the frame is truncated (by eliminating MAC address of the sender) and only the payload data - the route encoded as a collection of pairs (AID, TS) - is submitted to the CCSU (Fig. 1).

Hence, as an important difference from how MAC layer is defined in OSI [19], MAC layer here is a convention to describe a parallel distributed asynchronous lossy communication protocol and processing centered on MAC addresses of the ATUs and PAUs. It happens that, in our system here, MAC manipulation occurs in parallel, distributed in different sub-systems, without them waiting to each other, whereas some small amounts of payload data can go lost - accidentally not recorded in the CCSU, while the MAC of the sender ATUs is on purpose never sent nor recorded in the CCSU.

2.5 MAC Address Refresh on the ATUs

Of course, a frame is communicated in radio frequency and can be intercepted - at least by curious students wishing to understand, "spy", and test the system locally, or by any other people using locally the right gadgets - and therefore, the anonymization of data within the frame is capitally important. Capturing the anonymized data within the frame has to be, by design of the system, far less useful and has to provide far less meaningful data to someone than what data a visual observer can capture by simply eye-witnessing the passing car(s).

In fact, we aim that the content of the frame to be useless and meaningless for the intrusive listeners outside the system. This precaution is conditioning the MAC data-link layer on transponders in three ways: at power-on event, set-payload data event and send-payload data event, the slave MAC management service on the transponder (ATU) randomly initiates its MAC address that will be latter used by the system as SID or RID - as needed. This way, it's impossible that consecutive ATUs or intrusive listeners to trace cars movement. Until downloading the payload data from ATU to PAU, the MAC address of the ATU is only needed for maintaining ATU-to-PAU communication, so it can be changed after the car passes each intersection.

Due to above restrictions, the ATUs in our proposed system are not really IoT devices, because the only ones addressing them are the PAUs. Only the PAUs are IoT devices, and on their side, when they send the traffic data to the storage, the MAC address of the transponder is ignored and not transmitted, being replaced in the storage by an auto-number record-ID for the submitted route. This additional precaution avoids the possibility that an intrusive listener to pair the data it intercepted on the street with some data in the storage (CCSU), even in the case when a penetration bridge would exist from the listener to the storage.

2.6 Central Control and Storage Unit - CCSU

The CCSU is required as a component of the system given the power, network, memory and computational limits imposed on PAUs and ATUs. On the CCSU side, there is no transponder identification data (ATU MAC address) to arrive and to be stored there. All that it's stored on CCSU is the fact that an unknown car traveled on the recorded k-step timed route in a given time interval. Not even the total number of cars on the roads in a specific day can not be retrieved from the storage without authorization to perform expensive intelligent processing estimating a result by inevitably using some hypotheses to reverse a projection operator (given by the elimination of the transponders MAC addresses when the frame data is sent by PAU to the CSSU). It is only the intensity of road occupancy and its evolution along the time that can be easily derived from the data collected by the system and stored on CCSU.

2.7 Avoiding (Fake) Data Injection from Intruders Mimicking ATUs

Relative to the task of collecting some data - which is our case here, one can attack the system either by copying the data it intercepts (this case was discussed above) or by streaming fake data into the system, hence another type of attack against the proposed system will be to inject fake data into MAC or LLC layers of the communication.

We introduced the Frame ID (FID) field into the frame structure exactly for avoiding this kind of attack. Any frame will be received by the PAUs without any reject response, but only those frames for which the FID field is right will be submitted to the CCSU.

Our FID authentication mechanism can't and won't be disclosed here, but one can see that as a mechanism similar to the so called subliminal authorization channel for remotely authenticating stand-alone unconnected independent token devices when accessing online banking interfaces.

3 The Proposed System as an Abstract Machine

The first approach to describe the system and prove its formal correctness is to consider the algorithm governing the functionality of the system and show that the standby state (2, 13, 23) is the equilibrium point to which the system is inevitably and always returning regardless the interaction perturbing this state.

Anonymity Preserving RFID Communication Algorithm (Bodorin, Nagy):

while True **do** Run Parallel Distributed Asynchronus Threads:
- 1. on CCSU:
  - 2. On Running:
    - 3. Refresh MAC addresses table and of online PAUs (Sect. 2.2)
    - 4. Listen for AID Request and Payload Route Data (Sect. 2.2)
  - 5. On AID Request:
    - 6. Provide caller PAU with requested MAC/AID
  - 7. On Route Data Arrival:
    - 8. Parse Submitted Payload Route Data
  - 9. On Submitted Route Data Valid:
    - 10. Store Submitted Payload Route Data (Sect. 2.6).
  - 11. On Submitted Route Data Invalid:
    - 12. Log the details
- 13. on PAU_k:
  - 14. On Restart:
    - 15. Retrive AID_k from CCSU
  - 16. On RFID Response (Fig.1):
    - 17. Parse Frame RFID Response (Fig.2)
    - 18. Acknowledge Frame Response Received (Sect. 2.7)
  - 19. On Frame Authentication Successful (Sect. 2.7):
    - 20. Send Anonymized Payload Route Data to CCSU (Sect. 2.6)
  - 21. On Frame Authentication Failed (Sect. 2.7):
    - 22. Send Frame to CCSU Intrusion Log (Sect. 2.6)
- 23. on ATU_n:
  - 24. On Restart:
    - 25. Self-renew the MAC (Sect. 2.5)
    - 26. Initiate In-Memory Frame Storage.
  - 27. On RFID Call:
    - 28. **If** Memory Full:

29. Send Frame RFID Response to the calling PAU (Fig.1)
30. **Else**
31. Collect the current payload route pair (AID, TS) - Fig.2
32. **EndIf.**
33. Self-renew the MAC (for the reasons explained in Sect. 2.5).

end while

The route state of the system is (2), and by equivalence (3, 4), then becomes (3, (5, 7)), (3, (6, 7)), (3, (6, (9,11))), (3, (6, (10, 12))), (3, (6, 4)). The 6th state transitions to (3) if the PAU provided with the MAC has no interaction with any ATU until the next MAC refresh, and therefore, in this case, the state (3, (6, 4)), changes to (3, (3, 4)), i.e. to (3, 4).

The alternative is that an ATU-PAU interaction takes place, and consequently, the 6th state transitions to (27), then to (31), (29), (16), (17), (18), (19, 21), (20, 22), (9, 11), (10, 12), and finally (4) again. Therefore, also in this case, the state (3, (6, 4)), changes to (3, 4), i.e. to (2).

On the other hand, the state (2) is equivalent with (2, 13, 23), the equivalence expressing the system standby state continuously waiting for CCSU-PAU or/and ATU-PAU interaction to occur. In conclusion, the system is stable in its standby state (2, 13, 23), regardless the possible read/write intrusions.

Next target is to prove that the system is collecting anonymized data as intended, unaffected by read/write intrusions eventually occurring. Therefore, in a second approach to describe the system and prove its formal correctness, the data content of a frame will be considered as the abstract state of the device sending or preparing to send that frame, whereas the collection of events on the devices mirrors the transitions of the automata describing the system.

On the temporal axis, the state of an ATU is initiated empty on restart:

SOF	FID	MAC/SID	NaN	NaN	(NaN, NaN)	...	(NaN,NaN)	FCS	EOF

On each RFID call event, until the route/memory is full, the frame collects the payload route data step by step, as the ATU interacts with k successive PAUs on the road:

SOF	FID	SID	RID	CID	(AID_1, TS_1)	...	(AID_k, TS_k)	FCS	EOF

Since the route/memory is now full, the same frame is transmitted to the current PAU, which on its turn, after checking frame validity, send the payload data to CCSU:

Nan	Nan	Nan	Nan	Nan	(AID_1, TS_1)	...	(AID_k, TS_k)	Nan	Nan

whereas the ATU re-initiate the frame:

SOF	FID	MAC/SID	NaN	NaN	(NaN, NaN)	...	(NaN, NaN)	FCS	EOF

Only the route is stored in CCSU while the frame content on ATUs is re-initiated, and therefore, users and devices anonymity is guaranteed.

4 Conclusion and Future Work

The basic schemes of new, generic, anonymity-preserving, radio frequency based IoT devices and communication protocols were presented here. Their usage was illustrated in the particular context of capturing traffic data in Smart Cities. Traffic data was exemplified as IoT big data that can be transmitted, collected, processed and interpreted with important social benefit, without necessarily bounding it to the identity of a person or device, hence without creating any room for user's and device's privacy to be compromised and/or abused.

Future work will explore by analogy the possibility of defining a similar system but based on anonymized WiFi communication and devices.

References

1. Casella, G., Bigliardi, B., Bottani, E.: The evolution of RFID technology in the logistics field: a review. Procedia Comput. Sci. **200**, 1582–1592 (2022). https://doi.org/10.1016/j.procs.2022.01.359. ISSN 1877-0509
2. Google. RFID Patents. https://patents.google.com/?q=RFID&oq=RFID. Accessed 30 Jan 2022
3. Cardullo, M., Parks, W.: Transponder Apparatus and System, US Patent 3‘713‘148, 1970(S)/1973(G)/1990(E)
4. Walton, C.: Electronic recognition and identification system, US Patent 3‘816‘708
5. Oxford Dictionary Online. privacy. anonymity. Accessed 30 Jan 2022
6. Cambridge Dictionary Online. privacy. anonymity. Accessed 30 Jan 2022
7. Google. privacy scandals. Accessed 30 Jan 2022
8. Grigoras, M.V., Morogan, L.: Password cracking using P-systems with active membranes. In: 7th International Conference on Security for Information Technology and Communications, SECITC 2014, Bucharest, Romania, 12–13 June (2014)
9. Wilson, S.: Decoupling identity from devices in the internet of things, U.S. Patent US10164966B2 (2018)
10. Fuertes, G., Alfaro, M., Soto, I., Carrasco, R., Iturralde, D., Lagos, C.: Optimization model for location of RFID antennas in a supply chain. In: Proceedings of the International Conference on Computers Communications and Control, ICCCC, Oradea, 8–12 May 2018 (2018). https://doi.org/10.1109/ICCCC.2018.8390459

11. Carbunar, B.: Coverage problems in wireless sensor and RFID systems. Ph.D. thesis, Center for Education and Research in Information Assurance and Security, Purdue University (2005). https://citeseerx.ist.psu.edu/viewdoc/download?doi=10.1.1.159.408&rep=rep1&type=pdf
12. Dobrykh, D., Yusupov, I., Ginzburg, P., Slobozhanyuk, A., Filonov, D.: Self-aligning roly-poly RFID tag. (Nature's) Sci. Rep. **12**. ISSN 2045-2322 (online), article 2140/2022. https://doi.org/10.1038/s41598-022-06061-6
13. Duroc, Y., Tedjini, S.: RFID: a key technology for humanity. Comptes Rendus Physique **19**(1–2), 64–71 (2018). https://doi.org/10.1016/j.crhy.2018.01.003
14. Sobur, M.A., Masud, M.A., Chowdhury, N.R.: Design and prototyping of a security locker system for public places using RFID technology. Int. J. Inf. Tecnol. **14**, 579–585 (2022). https://doi.org/10.1007/s41870-021-00835-3
15. Blaskiewicz, P., Klonowski, M., Majcher, K., Syga, P.: Blocker-type methods for protecting customers' privacy in RFID systems. In: Proceedings of the International Conference on Cyber-Enabled Distributed Computing and Knowledge Discovery, pp. 89–96 (2013). https://doi.org/10.1109/CyberC.2013.23
16. Zhang, J., Periaswamy, S.C.G., Mao, S., Patton, J.: Standards for passive UHF RFID. GetMobile: Mob. Comput. Commun. **23**(3), 10–15 (2019). https://doi.org/10.1145/3379092.3379098
17. Li, M., Mori, T., Noguchi, H., Shimosaka, M., Sato, T.: Use of active RFID and environment-embedded sensors for indoor object location estimation. In: Proceedings of the International Universal Communication Symposium, pp. 93–99 (2009). https://doi.org/10.1145/1667780.1667799
18. Spiekermann, S.: RFID and privacy: what consumers really want and fear. Pers. Ubiquit. Comput. **13**(6), 423–434 (2009). https://doi.org/10.1007/s00779-008-0215-2
19. Wang, X., et al.: A new RFID ultra-lightweight authentication protocol for medical privacy protection in smart living. Comput. Commun. **186**, 121–132 (2022). https://doi.org/10.1016/j.comcom.2022.01.014. ISSN 0140-3664
20. Kumar, V., Kumar, R., Khan, A.A., Kumar, V., Chen, Y.C., Chang, C.C.: RAFI: robust authentication framework for IoT-based RFID infrastructure. Sensors Spec. Issue Secur. Priv. Mob. Comput.: Challenges Solutions **22**(9), 3110 (2022). https://doi.org/10.3390/s22093110
21. Xiao, M., Shen, X., Yang, Y., Wang, J.: Low power implementation of hummingbird cryptographic algorithm for RFID tag. In: Proceedings of the IEEE International Conference on Solid-State and Integrated Circuit Technology, pp. 581–583 (2010). https://doi.org/10.1109/ICSICT.2010.5667322
22. Tewari, A., Gupta, B.B.: Security, privacy and trust of different layers in Internet-of-Things (IoTs) framework. Future Gener. Comput. Syst. **108**, 909–920 (2020). https://doi.org/10.1016/j.future.2018.04.027
23. Byondi, F.K., Chung, Y.: Longest-range UHF RFID sensor tag antenna for IoT applied for metal and non-metal objects. Sensors **19**(24), 5460 (2019). https://doi.org/10.3390/s19245460
24. Costa, F., Genovesi, S., Borgese, M., Michel, A., Dicandia, F.A., Manara, G.: A review of RFID sensors, the new frontier of internet of things. Sensors **21**, 3138 (2021). https://doi.org/10.3390/s21093138
25. Nguyen, A.Q., Tran, H.A., Tran, T.H., Dao, N.P.: Implementation of a WiFi-based V2V-V2I communication unit for low speed vehicles. In: Proceedings of the International Conference on Advanced Technologies for Communications, pp. 79–82 (2021). https://doi.org/10.1109/ATC52653.2021.9598224

26. Milanes, V., Villagra, J., Godoy, J., Simo, J., Perez, J., Onieva, E.: An intelligent V2I-based traffic management system. IEEE Trans. Intell. Transp. Syst. **13**(1), 49–58 (2012). https://doi.org/10.1109/TITS.2011.2178839
27. Lourenço, M., Gomides, T.S., de Souza, F.S.H., Meneguette, R.I., Guidoni, D.L.: A traffic management service based on V2I communication for vehicular ad-hoc networks. In: Proceedings of the Latin America Networking Conference, pp. 25–31 (2018). https://doi.org/10.1145/3277103.3277132
28. Alexander, W., Hong, X., Hainen, A.: V2I communication-enabled real-time intersection traffic signal scheduling. In: Proceedings of ACM South-East Conference, pp. 26–33 (2017). https://doi.org/10.1145/3077286.3077573
29. Barrachina, J., et al.: A V2I-based real-time traffic density estimation system in urban scenarios. Wirel. Pers. Commun. **83**(1), 259–280 (2015). https://doi.org/10.1007/s11277-015-2392-4
30. Zhang, Y.: RFID-based tracking in supporting real-time urban traffic information. In: Proceedings of International Joint Conference on INC, IMS and IDC, pp. 657–659 (2009). https://doi.org/10.1109/NCM.2009.372
31. Benzaman, B., Sharma, D.: Discrete event simulation of a road intersection integrating V2V and V2I features to improve traffic flow. In: Proceedings of Winter Simulation Conference, Article no. 249, pp. 1–12 (2017). https://doi.org/10.1109/WSC.2017.8248026
32. Fong, S.L., Abu-Bakar, A.A.A., Ahmed, F.Y.H., Jamal, A.: Smart transportation system using RFID. In: Proceedings of International Conference on Software and Computer Applications, pp. 579–584 (2019). https://doi.org/10.1145/3316615.3316719
33. Pandit, A.A., Mundra, A.K., Talreja, J.: RFID tracking system for vehicles (RTSV). In: Proceedings of International Conference on Computational Intelligence, Communication Systems and Networks, pp. 160–165 (2009). https://doi.org/10.1109/CICSYN.2009.50
34. Day, J.D., Zimmermann, H.: The OSI reference model. Proc. IEEE **71**(12), 1334–1340 (1983). 10.1109/PROC.1983.12775

What UAE Software Students Think About Software Testing: A Replicated Study

Luiz Fernando Capretz[1](✉), Saad Harous[2], and Ali Bou Nassif[2]

[1] Western University, London N6G5H5, Canada
lcapretz@uwo.ca
[2] University of Sharjah, Sharjah, United Arab Emirates
{harous,anassif}@sharjah.ac.ae

Abstract. Software testing is vital to improve software quality. However, software tester role is stigmatized, partly due to misperception and partly due to the treatment of the testing process within the software industry. The present study analyses this situation aiming to explore what might inhibit an individual from taking up a software testing career. In order to investigate this issue, we surveyed 132 senior students pursuing degrees in information systems, information and communication technology, computer science, computer engineering, software engineering, and other closely-related disciplines at three universities in the United Arab Emirates: two publicly funded and one top-notch private university. The students were asked to describe the PROs and CONs of taking up a career in software testing and to ponder the likelihood that they would take up the career themselves. The study identified 7 main PROs and 9 main CONSs for pursuing a testing career, and indicated that the role of software tester is perceived as a social role, which may require as many soft skills as technical prowess. The results also show that UAE software-related students have a stronger negative attitude towards software testing compared to their counterparts in other countries where similar investigations have been carried out in the past three years.

Keywords: Software career · Human dimension of software testing · Software testing · Human factors in software engineering · Software engineering education · Computer science curriculum

1 Introduction

As software systems are becoming more pervasive, they are also becoming susceptible to failures, resulting in potentially lethal combinations. There have been catastrophic failures such as Therac-25 [1], Ariane 5 [2], and most recently the Boeing 737-Max crisis [3], which led to the loss of capital and life. Many more similar incidents are happening now all over the world as we write this article.

This research was conducted at New York University in Abu Dhabi in 2020, when L.F. Capretz was a visiting professor of Computer Science.

S. Dzitac et al. (Eds.): ICCCC 2022, AISC 1435, pp. 85–93, 2023.
https://doi.org/10.1007/978-3-031-16684-6_7

Software testing is critical to prevent software failures. Therefore, research has indeed been carried out in testing, but these works are largely limited to the process [4] and technology [5, 6] dimensions; the human dimension has not been sufficiently addressed. Even though there are reports about inadequacies of testing professionals and their skills [7], only a few studies have tackled the education and curriculum dimensions directly [8, 9].

The role of tester does not appear as one of the preferred roles among the population of software practitioners, according to previous study results [9–11]. Some studies point out the need for reversing people's perceptions regarding this role by using career progression [12] and other incentives to reinforce the crucial insights that a tester brings to the software project [13].

It has been pointed out that human and social aspects play a significant role in software testing practices [14, 15]. Attention to human factors in software testing in an academic setting has been encouraged by Hazzan and Tomayko [16], Shah and Harold [17], and Capretz [18]. Therefore, we decided to explore the human dimension of this endeavor. We started with the basic problem that plagues the testing profession, the shortage of talent, by asking why students are reluctant to consider testing careers.

If very few bright individuals voluntarily choose testing careers, this robs the industry of good testing and delivery of quality products. To change this situation, it is necessary to analyze the reasons for such apathy towards testing careers.

In this research, we studied the chances of senior software-related degree students taking up software testing careers and their reasons. To that end, we conducted surveys with senior software-related programs at three universities in the United Arab Emirates endeavoring to expose actual reactions of students to the role of tester in the software industry.

2 Methodology

The objectives of the study were explained to students, and their responses to the survey were sought. They were assured that their responses would not influence course grades in any way and were offered an option of not disclosing gender and GPA. A fresh and simple survey questionnaire, which is presented in Appendix A. The first author conducted replicated research using the same questionnaire in Canada, India, China, Cuba, Pakistan, Malaysia and Brazil, thus ascertained that the students neither had any queries nor reported any lack of clarity in the questions.

The first two questions were open ended questions: 1) What are three PROs (in order of importance) of pursuing a career in software testing; and 2) What are three CONs (also prioritized) of pursuing a career in software testing. The third question asked participants to indicate their intentions of pursuing a career in software testing. They were given the option to answer with either "certainly not," "no," "maybe," "yes," and "certainly yes." Participants were also invited to share the reasons behind their responses. Lastly, participants were asked to provide demographic information about themselves.

The research is descriptive, diagnostic, cross-sectional, and mixed. Descriptive research describes the characteristics of a population being studied and does not explore the reasons for those characteristics. Diagnostic research studies determine the frequency with which something occurs or its association with something else. We studied the event not over time but at a cross-section, making the study cross-sectional. We used a qualitative method of seeking open-ended responses about the PROs and CONs of testing careers and quantitative method, we asked categorical questions about the probability of choosing testing careers, making the study a mixed one too.

Our study analyzed the reasons that UAE students did not choose testing careers. These students are seniors pursuing a degree in information systems, information and communication technology, computer science, computer engineering, software engineering, and other closely-related disciplines at three universities in the UAE. One hundred and thirty two (132) students voluntarily participated in the survey during the Fall/2019 and Spring/2020 semesters.

3 Findings

Similar statements were combined during the refining of data. We found 7 main PROs and 9 main CONs in total; these statements are listed below.

The most important reasons considered as PROs for taking up a testing career among the surveyed individuals are presented in Table 2, along with their frequencies. The most frequent, with a 38% of respondents pointing that software testers earn more monetary benefits.

The second-most cited PROs, with 35%, is that testing tasks particularities allows software testers to learn about the system in hand, new technologies, and testing tools because the testing tasks provide a full background of the project scope, modularization, and integration strategy in a short period of time.

The third-most cited reason (31%) is the perception that software testing increases the software product quality thus worth doing despite the costs and efforts involved.

The fourth-most popular motivator is the perception that software tester deliver an important job, crucial to improve to guarantee that the software works properly and satisfies customer needs. Also, 12% of respondents mentioned job opportunities for software testers. Only a scarce 14% of individuals noted that software testing is a thinking job that requires critical thinking and creativity.

In contrast, when asked about the CONs for taking up a testing career, results are shown in Table 3. Respondents gave most importance to the following reasons: (a) it is a tedious job for 50% of respondents, (b) it is a complex task that requires expertise for 34%; and (c) 16% of the respondents pointed out the perception that other team members may become upset facing tester's findings, i.e. software failures. Lastly, 7% of the subjects noted that in the labor market the role of tester is a role for which wages are lower than the average wages for other roles; that view is reinforced by the perception that software testers are treated as second-class citizens within a software project (6%).

Regarding the third question, Table 1 shows the responses to the actual chances of respondents taking up a testing career according to their personal preferences.

Table 1. Chances of students taking up software testing careers.

Responses (132)	Numbers	Percentage
Certainly not	9	7%
No	22	17%
May be	70	53%
Yes	24	18%
Certainly yes	7	5%

3.1 The PROs of a Software Testing Career

The PROs responses are analyzed and presented in Table 2. Since we exclude PROs that were less than 5% (too small to consider) of the total, and the students could list up to three reasons, hence the total in the column is more than 100%.

Table 2. Percentage of salient PROs.

PROs	Percentage
Learning opportunities	35%
Important job	26%
Easy job	14%
Thinking job	11%
More job opportunities	12%
Monetary benefits	38%
Increase product quality	31%

We analyzed the responses to our prompt and found some of the following principles to be common themes cited for PROs:

1. Learning opportunities – Testers can learn about different products, technologies, techniques, and languages as well as domains such as retail, finance, embedded systems, etc. Testing activities provide the full background of a project's scope and architecture in a short period of time and may span all project stages. Testers can also hone their softer skills due to more interactions with developers and customers, although many of them being difficult ones.
2. Important job – Testers are accountable and responsible for the product quality. In that sense, testing is perceived as an important part of the software life cycle.
3. Easy job – This refers to students' belief that testing does have well-defined and easy processes, etc. There is also a perception that testing does not require strong technical skills.

4. Thinking job - This encompasses views about testing such as being challenging, creative, innovative, and requiring logical and analytical thinking.
5. More job – This states that more testing jobs are available; due to the higher demands, the jobs are secure and stable.
6. Monetary benefits – Testing jobs pay good salary packages, as much or more than for software developers, thus it is a lucrative career.
7. Increase product quality – Some students felt that proper software testing will improve the quality of a software product, thus considering a worthwhile endeavor in terms of associated time and cost.

3.2 The CONs of a Software Testing Career

The CONs responses are analyzed and presented in Table 3. Since the students could list up to three CONs and we excluded CONs categories that were less than 5%, therefore the total percentages is more than 100%.

Table 3. Percentage of salient CONs.

CONs	Percentage
Second-class citizens	6%
Career limitations	16%
Complexity/Expertise needed	34%
Tedious/Time consuming tasks	50%
Prefer development	12%
Less monetary benefits	7%
Anti social/Find others' errors	16%
Stressful job	5%
Detailed-oriented skills	5%

We also analyzed the responses to our prompt and found some of the following principles to be common themes for CONs:

1. Second-class citizen – This points out that testers are not involved in decision-making. Testers often take the blame for poor quality software while developers reap the reward for good quality. They perceive an evident lack of support from management in terms of unrealistic schedules, poor allocation of resources, and inadequate recognition.
2. Career development – Students believe that there is limited growth in the testing field. Some also believe that testers' jobs are dead-end careers and that they are the first ones to lose their jobs during business downturns.
3. Complexity – Testers often face complex situations such as different versions of software products, platform incompatibilities, defects not getting reproduced, and

testing not given sufficient time. At the same time, testers are often uniquely held responsible for product quality. This also includes the fact that testers need to look at business and technology artifacts and understand many abstractions. The lack of clarity around requirements adds to the difficulties. This factor also includes dealing with different versions and vendors of third party software, problems with testing tools, development environments or a weak infrastructure, and the ask exhaustively for patience.

4. Tedious – This refers to the repetitive nature of testing; respondents also used descriptors such as, monotonous, boring, not-challenging, and time-consuming.
5. Prefer development – Some students believe that they would miss opportunities for professional development by taking up testing careers. While some testers do develop test automation systems, the students seem to consider that as different from actual development activity. Some also think that they lose learning opportunities that are available to developers because they do not code anymore.
6. Less monetary benefits – Some students believe that testers' jobs do not have monetary benefits on par with developers.
7. Finding the mistakes of others – Fundamentally, it is not pleasant to find mistakes in others' work and present them. As testers have to report anything that adversely impacts the value of the product, which may not go well with some stakeholders; testers thus risk generating an anti-social role.
8. Stressful job – Students mentioned that software testers are gatekeepers of the software quality, therefore accountable for faulty software releases. Some think that tester must be on stand-by all the time, just in case a software fails and actions must be taken, thus negatively affecting the work-life balance.
9. Detailed-oriented skills – Some students just mentioned that they have no interest in the field of software testing because it requires concentration for several hours, every day, looking for details or edge conditions to test software.

4 Threats to Validity

This study has some inherent limitations. The subjects come from two large public universities and one top-ranked private university in the United Arab Emirates only. Although respondents represent a sample of currently active senior students throughout the UAE, including expatriates, their origins were not recorded.

5 Discussions and Implications

The study analyzed the opinions of 132 senior students enrolled in information science, information and communication technology, computer science and software-related disciplines in the UAE. Opinions were elicited on whether they would choose testing careers and what they felt were the advantages and drawbacks (PROs and CONs) of a testing career.

We found considerable variations in the student-perceived advantages of a career in testing. Among the key PROs of learning opportunities, monetary benefits, and job opportunities, percentages narrowly vary from 31% to 38%. With 18% of students indicating that they would like to be a software tester, and only 6% saying they would certainly take up a career as software tester, it is evident that a software testing profession among students is as unpopular as it is in other countries [19]; where only 7% of the students would not like to be software tester. These results do concur with prior studies [20, 21], which point to the tester role as one of the least popular roles compared to others, such as project manager, analyst, designer, programmer, and maintenance.

Not surprisingly, there are few outliers found in this investigation that are worth mentioning: (a) five respondents saw in software testing an opportunity to beat or challenge the developer; (b) stressful job was mentioned by only 5% of the subjects in this study, consistent with previous studies where this issue appeared less than 5%; and (c) five students alarmingly asserted that software testing is too expensive, and a waste of time and effort because the tester may not find any error in the software. Such students have an alarming misconception about the role of testing in the software life cycle, and surely need to take a course on software verification and validation.

This study has many implications for colleges, especially for information technology, information science, computer science, software engineering programs, and information and communication technology programs. Since testing courses can improve the perception of testing careers, universities can introduce them into their curricula. They can regularly review the curricula by consulting their alumni and researchers. Since testing offers additional jobs, a testing course can help colleges improve placement prospects of their students.

The program curricula, software testing and verification and validation courses need to reflect the understanding that testers need to provide correct information to various stakeholders, and appreciate that testing is 'applied epistemology' grounded in 'social psychology'. Course instructors must dispel beliefs such as, 'testing is just mechanically running tests and comparing outputs with expected results'. Instead, testing instructors should explain the importance of testing and the philosophy behind it, and they should impress upon the students that any testing assignment and design of test cases can be very creative.

Instructors should educate tester who can understand different domains and the needs of users in these different domains. These testers must be able to understand the developer mindset, and to anticipate mistakes that developers may make. These testers must do their testing jobs creatively and efficiently under challenging constraints, and they must report finding wisely and tactfully to all stakeholders.

6 Conclusions

As previously emphasized, software testing is a human activity [22, 23] and testers, who willingly take up testing careers, can influence the quality of the final product.

Nevertheless, software testing appears to be a neglected area in the software industry [24]. There are not enough testing specialists, and test schedules are squeezed as development overruns occur and delivery milestones are considered non-negotiable. Many times, testing is perceived as a nuisance that is sandwiched between development and deployment, when in fact it is a critical activity that needs to be performed in parallel to design and development activities, as advocated by software verification and validation. Therefore, the software industry has been facing a shortage of qualified software testers which has led to quality problems.

This study offers useful insights and helps educators to come up with an action plan to change the outlook towards testers in their programs; educators must put the software testing profession under a new light. Enthusiasm for software testing could increase the number of software engineers deciding on testing as a career of their choice, could increase the quality of software testing, and improve the overall productivity and turnaround time of software development activity.

Appendix A – Survey Questions

1. What are the three PROs (in the order of importance) for taking up testing career?
   a) b) c)
2. What are three CONs (in the order of importance) for taking up testing career?
   a) b) c)
3. What are chances of my taking up testing career?
   Certainly Not No Maybe Yes Certainly Yes
   Reasons:
4. Gender (optional):
5. GPA (optional):

References

1. Leveson, N.G., Turner, C.S.: An investigation of the Therac-25 accidents. Computer **26**(7), 18–41 (1993)
2. Nuseibeh, B.: Ariane 5: who dunnit? IEEE Softw. **14**(3), 15 (1997)
3. Johnston, P., Harris, R.: The Boeing 737 Max saga: lessons for software organizations. Softw. Qual. Prof. **21**(3), 4–12 (2019)
4. Kaner, C., Bach, J., Pettichord, B.: Lessons Learned in Software Testing. Wiley, Hoboken (2008)
5. Kasurinen, J., Taipale, O., Smolander, K.: Software test automation in practice: empirical observations. In: Advances in Software Engineering, p. 18 (2010)
6. Graham, D., Fewster, M.: Experiences of Test Automation: Case Studies of Software Test Automation. Addison-Wesley, Boston (2012)
7. Florea, R., Stray, V.: Software tester, we want to hire you! An analysis of the demand for soft skills. In: Garbajosa, J., Wang, X., Aguiar, A. (eds.) XP 2018. LNBIP, vol. 314, pp. 54–67. Springer, Cham (2018). https://doi.org/10.1007/978-3-319-91602-6_4
8. Astigarraga, T., Dow, E.M., Lara, C., Prewitt, R., Ward, M.R.: The emerging role of software testing in curricula. In: IEEE Transforming Engineering Education: Creating Interdisciplinary Skills for Complex Global Environments, pp. 1–26 (2010)
9. Waychal, P., Capretz, L.F..: Why a testing career is not the first choice of engineers. In: 123rd Annual Conference of the American Society for Engineering Education, New Orleans, Paper ID #14994, p. 6 (2016)
10. Waychal, P., Capretz, L.F., Pardeshi, S.N.: Analyzing popularity of software testing careers in Canada. In: 125th ASEE Annual Conference, Salt Lake City, ID #21627, p. 6 (2018)

11. Deak, A., Stålhane, T., Cruzes, D.: Factors influencing the choice of a career in software testing among Norwegian students. Softw. Eng. **796**, 1–8 (2013)
12. Deak, A.T., Stålhane, T., Sindre, G.: Challenges and strategies for motivating software testing personnel. Inf. Soft. Technol. **73**, 1–15 (2016)
13. Santos, R.E.S., Magalhaes, C., Correia-Neto, J., Silva, F.Q.B., Capretz, L.F.: Would you like to motivate software testers? Ask them how. In: Empirical Software Engineering and Measurement, Toronto, Canada, pp. 95–104 (2017). https://doi.org/10.1109/ESEM.2017.16
14. Capretz, L.F., Varona, D., Raza, A.: Influence of personality types in software tasks choices. Comput. Hum. Behav. **52**, 207–214 (2015). https://doi.org/10.1016/j.chb.2015.05.050
15. Rooksby, J., Rouncefield, M., Sommerville, I.: Testing in the wild: the social and organizational dimensions of real world practice. Comput. Support. Coop. Work **18**(15–16), 559–580 (2009)
16. Hazzan, O., Tomayko, J.: Human aspects of software engineering: the case of extreme programming. In: Eckstein, J., Baumeister, H. (eds.) XP 2004. LNCS, vol. 3092, pp. 303–311. Springer, Heidelberg (2004). https://doi.org/10.1007/978-3-540-24853-8_45
17. Shah, H., Harrold, M.J.: Studying human and social aspects of testing in a service-based software company: case study. In: Proceedings of ICSE Workshop on Cooperative and Human Aspects of Software Engineering, pp. 102–108 (2010)
18. Capretz, L.F.: Bringing the human factor to software engineering. IEEE Softw. **31**(2), 102–104 (2014). https://doi.org/10.1109/MS.2014.30
19. Waychal, P., Capretz, L.F., Jia, J., Varona, D., Lizama, Y.: Practitioners' testimonials about software testing. In: IEEE International Conference on Software Analysis, Evolution and Reengineering, pp. 582–589 (2021). https://doi.org/10.1109/SANER50967.2021.00070
20. Capretz, L.F., Waychal, P., Jia, J., Basri, S.B.: Comparing the popularity of testing career among Canadian, Malaysian, Chinese, Indian, and Malaysian students. In: 3rd International Conference on Advanced Information Science and Systems (AISS 2021), Sanya, China, p. 7 (2021). https://doi.org/10.1145/3503047.3503091
21. Capretz, L.F., Waychal, P., Jia, J.: Comparing the popularity of testing careers among Canadian, Chinese, and Indian students. In: IEEE/ACM 41st International Conference on Software Engineering (ICSE Companion), Montreal, Canada, pp. 258–259 (2019). https://doi.org/10.1109/ICSE-Companion.2019.00103
22. Cerioli, M., Leotta, M., Ricca, F.: What 5 million job advertisements tell us about testing: a preliminary empirical investigation. In: The 35th Symposium on Applied Computing (2020)
23. Capretz, L.F., Ahmed, F.: Making sense of software development and personality types. IEEE IT Prof. **12**(1), 6–13 (2010). https://doi.org/10.1109/MITP.2010.33
24. Capretz, L.F., Waychal, P., Jia, J., Varona, D., Lizama, Y.: International comparative studies on the software testing profession. IEEE IT Prof. **23**(5), 56–61 (2021). https://doi.org/10.1109/MITP.2020.3031862

The Impact of COVID-19 on the Air Transportation Network

Sergey Shvydun[1,2](✉)

[1] National Research University Higher School of Economics, Myasnitskaya Street 20, 101000 Moscow, Russia
shvydun@hse.ru

[2] V.A. Trapeznikov Institute of Control Sciences of Russian Academy of Science, Profsoyuznaya Street 65, 117342 Moscow, Russia

Abstract. The COVID-19 pandemic have changed our lives and caused severe social, economic and environmental disruption worldwide. This paper examines the impact of coronavirus outbreak and subsequent wave of lockdowns and travel restrictions on air transportation from the network perspective. Using the air traffic data from 2019 to 2021, we examine how the structure of the network has changed both globally and locally. We quantify the similarity of air transportation network for adjacent years and compare it to the pre-pandemic period. The connectivity configuration of countries is also considered. Finally, we identify the most important participants of the air transportation industry and reveal countries with the largest change of centrality during the pandemic.

Keywords: Air transportation · Network analysis · COVID-19

1 Introduction

The coronavirus outbreak caused a large impact on society, economy and the environment. The social disruption is devastating: according to the World Health Organization (WHO), there has been more than 300 million of confirmed cases while 5.5 million of people has died from COVID-19 by the end of 2021 [1]. It is also estimated that around 768 million of people were undernourished in 2020 (≈18% increase compared to 2019), which is equivalent to 10% of the world's population [2]. The coronavirus pandemic also affected all sectors of the national economies and businesses. Estimates indicate the pandemic reduced the annual global economic growth in 2020 by 3.4%, with a recovery of 5.5% for 2021 [3].

Air transportation, which is one of the world's most important industries, suffered major losses during the pandemic. Massive wave of lockdowns and travel restrictions had a severe impact on movement of citizens, international workers, tourists, students, etc. For instance, the UN World Tourism Organization (UNWTO) states that the crash in international tourism could cause a loss of more than $4 trillion to the global gross domestic product (GDP) during 2020–2021 [4]. This paper aims to analyze how the

S. Dzitac et al. (Eds.): ICCCC 2022, AISC 1435, pp. 94–104, 2023.
https://doi.org/10.1007/978-3-031-16684-6_8

movement of people has changed during 2019–2021 by tracking the changes in the air transportation network.

The impact of COVID-19 on the air transportation has been examined in many studies. A detailed survey of recent works is provided in [5]. However, we did not find any study that provides a detailed analysis of transportation network from the network perspective. In this paper, examine how the structure of the network has changed both globally and locally. To quantify the evolution of the network, we compute the similarity of air transportation network for adjacent years and compare it to the pre-pandemic period. Additionally, we analyze the connectivity configuration of countries to identify phase transitions of nodes. Finally, we detect the most important participants of the air transportation industry and reveal countries with the largest change of centrality during the pandemic.

The structure of the paper is organized as follows. First, we describe the air traffic dataset, present some data preparation steps and perform its preliminary analysis. Second, we analyze the evolution of the air transportation in terms of the network properties. Next, we quantify the connectivity configuration and centrality of nodes. The final Section concludes.

2 Data Description

To analyze the development of air transportation during the COVID-19 pandemic we used the crowdsourced air traffic data from the OpenSky Network [6]. It is updated monthly and contains information about 81,954,814 flights seen by the OpenSky Network from January 2019 to December 2021. The dataset includes information about the origin and destination airports for each flight as well as the date of the last airborne message received by the OpenSky Network.

2.1 Data Preparation

The construction of the air transportation network from initial data relies on some data preparation process. First, one should note that the initial data contains information about flights from all the airports including large, medium and small airports, heliports, airbases and seaplane bases. Since we focus on the movement of people, we considered only large and medium airports according to OurAirports classification [7]. Large airport is defined as "land airport with scheduled major airline service with millions of passengers/year, or major military base". Medium airport is defined as "land airport with scheduled regional airline service, or regular general aviation or military traffic". Finally, we excluded airbases and all the airports that do not have International Civil Aviation Organization (ICAO) or International Air Transport Association (IATA) airport codes. As a result, we have obtained a list of 2043 airports from 136 countries.

Second, it was observed that some flights have the same origin and destination. Such flights may occur by various reasons: technical problems of the aircraft, adverse weather conditions, etc. We excluded all these flows since they do not lead to the movement of people. We also excluded all the routes between airports within 100 km distance. As a

Table 1. Countries with the highest number of inbound and outbound flights during 2019–2021.

ID	Country	Outbound flights	Inbound flights
USA	United States of America	13,157,229	12,374,490
DEU	Germany	1,355,388	1,365,566
GBR	United Kingdom	1,198,832	1,214,938
AUS	Australia	766,694	762,967
ESP	Spain	764,565	754,087
FRA	France	744,481	749,015
ITA	Italy	710,265	705,535
CAN	Canada	610,565	623,090
IND	India	528,377	532,843
RUS	Russian Federation	389,849	386,352
JPN	Japan	377,417	360,998

result, our data include more than 26.5 million flights since 2019. The list of countries with the highest number of inbound and outbound flight is presented in Table 1.

The countries with the largest domestic flights are the United States (≈12.37 million), Australia (≈0.66 million), Germany (≈0.4 million), India (≈0.39 million) and Canada (≈0.33 million). The countries with the number of largest international flights include the United Kingdom (≈1.93 million), Germany (≈1.91 million), the United States (≈1.59 million), Spain (≈1.11 million), France (≈1.04 million).

The final part of data preparation is the data aggregation. The initial dataset contains information about individual flights and there can be multiple flights between the same airports during the day. Thus, there is a need to aggregate this information to define the intensity of connections among the airports. One may consider different levels of aggregation: daily level, weekly level, monthly level. This paper considers the monthly level since fluctuations for daily or weekly levels tend to occur due to different schedule at the airport while we focus on capturing the effect of the pandemic. Thus, we constructed 36 networks that correspond to air transportation network at a particular month.

2.2 Descriptive Statistics

Let us examine the aggregated statistics to understand the impact of coronavirus pandemic on air transportation industry. The total number of flights dynamics is illustrated in Fig. 1.

According to Fig. 1, the volume of flights increased over time until February 2020. The peak was in October 2019 (≈1.03 million flights) while the total number of flights averaged approximately 870,000 thousand flights per month. In March 2020 the WHO declares COVID-19 a pandemic and by that time many countries imposed lockdowns. In the following month, the volume of flights dropped to about 229,000 thousand flights (≈77% decrease compared to February 2020). However, since May 2020 we observe a

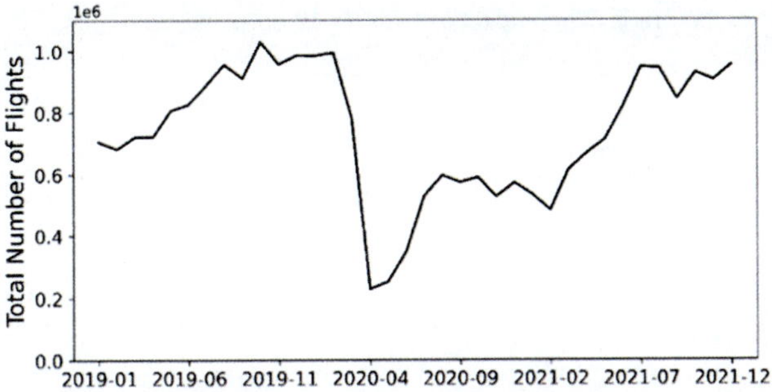

Fig. 1. Dynamics of total flights, 2019–2021.

gradual increase in the number of flights, which is now comparable to the volume in the middle of 2019. To assess how much the air transportation industry has recovered, we provide information about the structure of flights in Fig. 2.

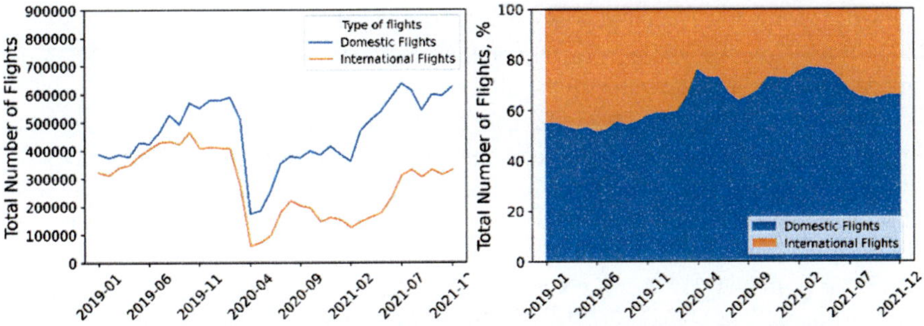

Fig. 2. Structure of flights (in absolute and in percentage), 2019–2021.

Although the total number of flights at the end of 2021 is comparable to the values before the pandemic, Fig. 2 shows that the airline industry has not recovered yet. It is seen that the total number of international flights reached 329,000 flights per months by the end of 2021, which is still 20% less than the volume of flights at the end of 2019. On the contrary, we observe the increase of domestic flights, which now account for 624,000 flights (≈8.1% increase compared to the maximal value in 2019). Thus, we can see that airlines have increased domestic flights as many international destinations are not allowing incoming passenger traffic. At the end of 2021, domestic flights account for 66% of all the flights (≈55% in 2019). Finally, there is also a relative jump of domestic flights in April 2020 as international routes dropped significantly.

Next, let us consider how the pandemic affected domestic and international flights for countries from Table 1. In Table 2 we compare the change in the volume of domestic flights for countries with > with respect to three periods: the largest flow before the pandemic (pre-pandemic) and the smallest flow since March 2020 (largest drop) as well as the largest flow before the pandemic (pre-pandemic) and the largest flow at the end of

Table 2. The impact of COVID-19 on domestic flights

ID	Country	Pre-pandemic vs. largest drop	Pre-pandemic vs. now
USA	United States of America	−65%	10%
DEU	Germany	−87%	−42%
GBR	United Kingdom	−84%	16%
AUS	Australia	−84%	−24%
ESP	Spain	−92%	66%
FRA	France	−93%	34%
ITA	Italy	−89%	−18%
CAN	Canada	−71%	15%
IND	India	−97%	−45%
RUS	Russian Federation	−84%	19%
JPN	Japan	−76%	159%

2021 (now). Note that Table 2 does not include the Netherlands as there are no domestic flights in the country according to the database.

It is shown in Table 2 that India and European countries had the largest drop of domestic flights compared to pre-pandemic values, which can be explained by strict lockdowns in these countries. If we compare pre-pandemic values with the latest data, one can observe that some countries increased the volume of domestic flights (Japan, Spain, France, Russian Federation, the United Kingdom, Canada and the United States). Interestingly, the largest increase was for Japan that increased the number of domestic flights from 6827 in January 2020 to 17720 in December 2021. Finally, we can see that India have still not recovered the volume of domestic flights.

Similarly, Table 3 describes how the pandemic changed the international flows for countries from Table 1.

Table 3 shows that the pandemic affected international flights more compared to domestic flights. As the COVID-19 pandemic forces lockdowns, most countries were not available to visit, thus, the volume of international flights has decreased by >90% for most of them. It is also observed that countries with the largest flights have not still recovered from the pandemic.

Table 3. The impact of COVID-19 on international flights

ID	Country	Pre-pandemic vs. largest drop	Pre-pandemic vs. now
USA	United States of America	−78%	−7%
DEU	Germany	−86%	−32%
GBR	United Kingdom	−91%	−35%
AUS	Australia	−82%	−66%
ESP	Spain	−95%	−30%
FRA	France	−91%	−10%
ITA	Italy	−95%	−34%
CAN	Canada	−88%	−29%
IND	India	−91%	−63%
RUS	Russian Federation	−91%	−44%
JPN	Japan	−84%	−40%

3 Network Analysis

Since the network evolution is demonstrated as a set of snapshots for different times $G_t = (V_t, E_t)$ where V_t is a set of nodes at time t and E_t is a set of edges at time t, our goal is to examine the properties of the linkages for each network, identify its structure as well as its most important elements and analyze its evolution.

3.1 Network Characteristics

Let us construct the air transportation network for each month and provide its descriptive statistic. First, we can observe that the total number of countries varies from 93 (May 2020) to 112 (May 2021). One can observe that the total number of nodes decreased by 10% in April-May of 2020. Similarly, the total number of edges averaged 2800 before the pandemic, dropped by more than 25% in April 2020, then increased to 3000 by the end of 2021.

Next, we computed density and centralization indices to evaluate the network structure in general. Density is a ratio of the number of actual edges to the number of possible edges in a network. It is observed that the air transportation network is relatively dense (0.23–0.31) as countries have direct flights to 27 other countries in average. Centralization determines an extent to which the network depends on certain nodes [8]. The failure of these nodes has the potential to disrupt the entire system. Centralization measure $C_D(t)$ at time t is based on degree centrality and computed as

$$C_D(t) = \frac{|V_t| \cdot \max_i d_i(t) - \sum_i d_i(t)}{(|V_t| - 1) \cdot (|V_t| - 2)}, \tag{1}$$

where $d_i(t)$ is a degree of node i at time t. If nodes have the same degree, then $C_D(t) = 0$. On the contrary, $C_D(t) \to 1$ if one node completely dominates the network with respect to degree centrality (e.g. star network). One can observe that the air transportation network is decentralized (≈0.49), thus, there is no single node that connects the entire system. As for other network characteristics, the clustering coefficient averages 0.71, the network diameter is 4 while the assortativity of the network is low and negative (accounts for – 0.17), which indicates a weak relationship between nodes of different degree.

3.2 Network Similarity

Next, one can examine how the structure of edges is changed with respect to time. If we compare adjacent months, it is observed that the network preserves 85% of connections in average, so it does not tend to vary. However, we also noticed the shock to the industry caused by the pandemic as only 65% of connections were preserved in April 2020. Additionally, to account for the edge weights, we computed the weighted version of the Jaccard index which is defined as the ratio of intersection and union of edge sets corresponding to two networks. We measured the similarity between graph G_t and graph G_{t-1} (adjacent months) as well as graph G_t and the average graph for the pre-pandemic period $G_{pre\text{-}pandemic}$ (September 2019–February 2020). The results are presented in Fig. 3.

According to Fig. 3, we can see how much the pandemic affected the air industry. From the beginning of 2019 to February 2020 the air flows almost do not change. There is a high similarity between air transportation networks corresponding to adjacent months (≈0.88). However, the spread of the pandemic and subsequent lockdowns disrupted the air system as more than 70% of flows disappeared. In the following months, we can observe high values of Jaccard index which reflects to the fact that air transportation industry is recovering slowly from the slump experienced in March 2020. Moreover, it is also seen that air transportation network is becoming more similar to the network of pre-pandemic period although the level of similarity is still quite low (≈0.7).

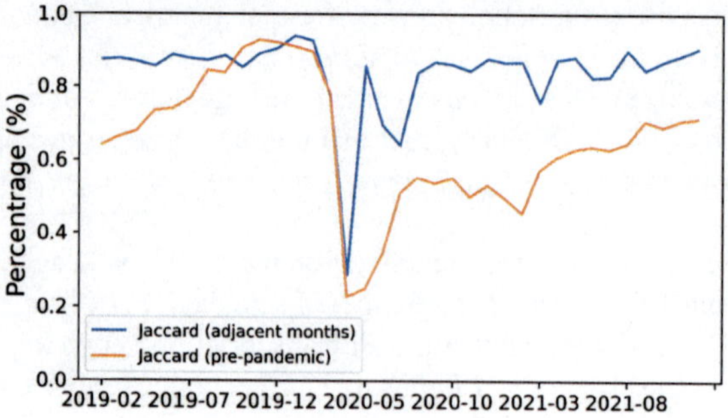

Fig. 3. Weighted Jaccard index (pre-pandemic period and adjacent months)

3.3 Nodes Connectivity

Let us consider how the pandemic affected individual nodes (countries). Since the coronavirus outbreak has led to massive wave of lockdowns and travel restrictions in many countries, the connectivity of certain nodes has significantly decreased. On the other hand, some countries did not impose strict lockdowns, thus, their connectivity has not decreased much. We want to capture this feature. Thus, in this paper we quantify the diversity of the connectivity configurations that a node has. In order to do it, we firstly calculate the probability of visiting each node in the network starting from a particular node i using the Personalized Random Walks with Restarts [9]. Next, we compare the obtained probability distribution for each node to the uniform distribution using the Jensen-Shannon (JS) divergence [10]. The choice of the uniform distribution is explained by the fact that we want to compare the real connectivity of the node with ideal case when all nodes can be reached with equal probability (≈no travel restrictions). By definition, JS $= 0$ indicates that two probability distributions are identical while JS $= 1$ indicates that node i is not connected to other nodes in the network (≈strict lockdown, travel restrictions, etc.).

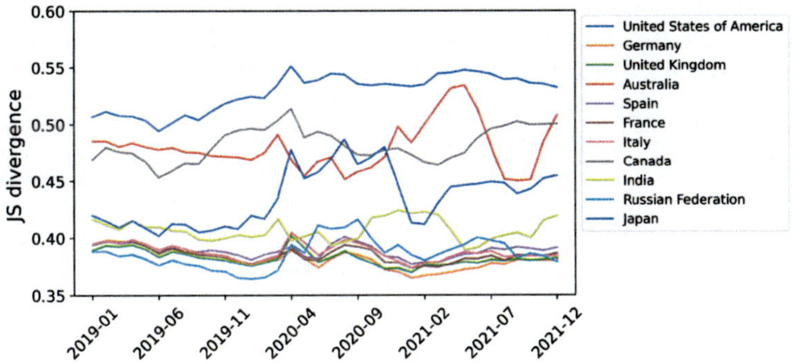

Fig. 4. Jensen-Shannon (JS) divergence for countries with the largest amount of flights.

Figure 4 shows the dynamics of JS divergence for the countries we largest number of flights. First, we can observe that JS divergence is almost identical for European countries (Germany, United Kingdom, Spain, France and Italy). The connectivity configuration for these countries remains stable during 2019–2021 except for a slight increase in April 2020. Second, the USA and Canada shows similar dynamics of JS divergence although their connectivity configuration is less similar to the uniform distribution. Third, we observe the increase of JS measure for some countries during the pandemic. For instance, there is an increase of the JS measure for Australia in March 2020 as the government closed the borders. In the following months, the JS divergence increased due to its low connectivity to other countries and the rise of domestic flights. Similarly, there is an increase of JS measure in April 2020 for Japan which can be explained by the entry ban imposed by the government.

Next, let us examine the largest change of nodes connectivity during 2019–2021. We investigated countries with more than 100,000 flights in total. As a results, we observe that the largest difference of JS divergence was for Thailand (increase by 0.327 in October 2020), Republic of Korea (increase by 0.28 in May–August 2021), New Zealand (increase by 0.14 after April 2020). There might be many reasons for such rapid increase of JS measure, which should be investigated further.

Finally, it was found that countries with the lowest JS divergence include United Arab Emirates, Qatar, Thailand and Turkey. The connectivity structure of these countries are very diverse.

3.4 Nodes Centrality

Next, one can examine how the pandemic affected the centrality of nodes. In this paper, we are focused on identification of countries that play a major role in the air industry as well as countries that are the leaders of international transportation. The importance of nodes in the network is usually determined by applying various centrality metrics. In Sect. 2 we provided the results for weighted in-degree and out-degree centrality indices (see Table 1) which do not take into account the global structure of the network. Since our paper deals with directed weighted networks, we calculate the PageRank [11] centrality. The PageRank algorithm is based on random walks in a graph and it assigns high values to nodes that have a higher probability of being visited. An intuitive interpretation of PageRank centrality for the air transportation network is that the most visited nodes correspond to countries which are highly engaged in air transportation.

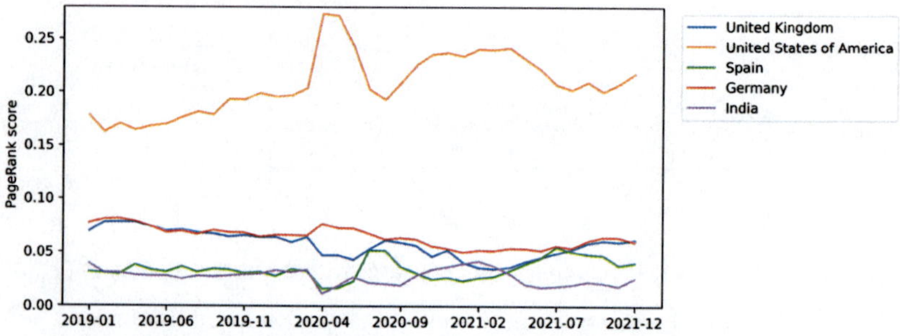

Fig. 5. PageRank score (TOP-3 countries) with domestic flights, 2019–2021.

First, we consider air transportation network with domestic flights. Figure 5 shows the dynamics of PageRank centrality for the countries included in the TOP-3 for one of the snapshots. As it is shown, the USA, Germany and the United Kingdom are the most important countries in air industry according to the PageRank centrality. We have already shown that these countries have the largest volume of inbound and outbound flights, consequently, they are the key participants in the network. There is also an increase of influence of the USA in March 2020, which can be explained by the drop of international flights and the major role of the USA in domestic flights. Next, we also observe the increase of influence for Spain in summer 2020–2021 as it is one of the most popular holiday destinations. Interestingly, we also see India in TOP-3 in February 2021.

Additionally, we compared the average PageRank centrality of countries before the pandemic to the same measure at the end of 2021. One should note that the highest increase of centrality was observed for Turkey, Spain and the United States while the largest decrease was for India, Australia and Malaysia.

Next, we consider air transportation network without domestic flights to identify the most important international players. Figure 6 shows the dynamics of PageRank centrality for the countries included in the TOP-3 for one of the snapshots.

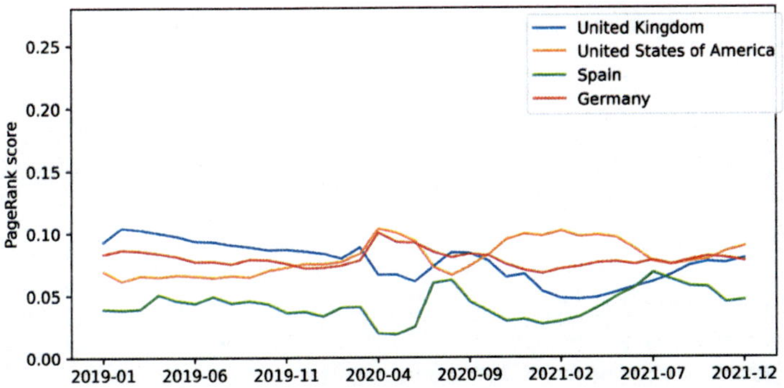

Fig. 6. PageRank score (TOP-3 countries) without domestic flights, 2019–2021.

According to Fig. 6, the USA, Germany and the United Kingdom are still the most important countries in the international transportation. However, on the contrary to Fig. 5, we can notice the decrease of importance for the USA as we excluded domestic flights from consideration. Interestingly, we can see the increase of influence for the USA and Germany and the decrease of influence for the United Kingdom and Spain in March 2020. As for other countries, we also observe the increase of the PageRank centrality for Hong Kong, Korea in March 2020 as well as for United Arab Emirates by the end of 2020. Finally, we examined countries with the largest increase/decrease of the PageRank centrality after the pandemic started. It was found that the main beneficiaries are Turkey and Spain while countries with the largest decrease of centrality include United Kingdom, Canada, Malaysia and Thailand.

4 Conclusion

The coronavirus outbreak caused a severe impact on society, economy and the environment while the air transportation is one of the most affected industries. In this paper we examined the effect of COVID-19 from the network perspective. Although it seems that air transportation industry is close to recover from the drop in March 2020 in terms of the total flights, we observe that many countries have not regained the volume of international flights. However, we also notice the rapid increase of domestic flights in some countries as one of the side effects of the pandemic.

Next, we examined similarity of air transportation network for adjacent years and compare it to the pre-pandemic period. Although air transportation is becoming more similar to the network of pre-pandemic period, the current level of similarity is still quite low. We also investigated the connectivity structure of countries. As a result, we identified countries with diverse or poor connectivity during the pandemic and found some correspondence with the introduction of lockdowns and travel restrictions. Finally, we identified the most important participants of the air transportation industry and revealed countries with the largest change of centrality during the pandemic.

Acknowledgement. The article was prepared within the framework of the HSE University Basic Research Program and funded by the Russian Academic Excellence Project '5–100'. The centrality analysis of the network (Sect. 3.4) was supported by grant of President of Russian Federation No. MK-3867.2022.1.6.

References

1. Covid19.who.int. 2022. WHO Coronavirus (COVID-19) Dashboard. https://covid19.who.int/. Accessed 31 Jan 2022
2. World Economic Forum. 2021. This is how COVID-19 has affected world hunger. https://www.weforum.org/agenda/2021/07/world-hunger-malnutrition-un-report-covid-coronavirus-pandemic-sdgs. Accessed 31 Jan 2022
3. World Bank. Global Economic Prospects, World Bank, Washington, DC, January 2022
4. UNCAD. Covid-19 and Tourism: Assessing the Economic Consequences. https://unctad.org/webflyer/covid-19-and-tourism-update. Accessed 31 Jan 2022
5. Sun, X., Wandelt, S., Zheng, C., Zhang, A.: COVID-19 pandemic and air transportation: successfully navigating the paper hurricane. J. Air Transp. Manag. **94**, 102062 (2021)
6. Strohmeier, M., Olive, X., Lübbe, J., Schäfer, M., Lenders, V.: Crowdsourced air traffic data from the OpenSky Network 2019–2020. Earth Syst. Sci. Data **13**, 357–366 (2021)
7. OurAirports database. https://ourairports.com/. Accessed 31 Jan 2022
8. Freeman, L.C.: Centrality in social networks i: conceptual clarification. Soc. Netw. **1**, 215–239 (1979)
9. Haveliwala, T.H.: Topic-sensitive pagerank: a contextsensitive ranking algorithm for web search. IEEE TKDE **15**(4), 784–796 (2013)
10. Cichocki, A., Amari, S.: Families of alpha- beta- and gamma- divergences: flexible and robust measures of similarities. Entropy **12**, 1532–1568 (2010)
11. Brin, S., Page, L.: The anatomy of a large-scale hypertextual Web search engine. Comput. Netw. **30**, 107–117 (1998)

Advanced Control Systems Integrating Computers and Communications

Linear and Nonlinear Observers Developed for Direct Current Electric Drive Systems

Alexandra-Iulia Szedlak-Stinean(✉), Radu-Emil Precup, and Raul-Cristian Roman

Department of Automation and Applied Informatics, Politehnica University of Timisoara, Bd. V. Parvan 2, 300223 Timisoara, Romania
{alexandra-iulia.stinean,radu.precup,raul.roman}@aut.upt.ro

Abstract. This paper suggests four estimation approaches, namely two linear and two nonlinear ones: Extended Luenberger State Observer (ELSO), Kalman Filter Observer (KFO), Extended Kalman Filter Observer (EKFO) and Sliding Mode Observer (SMO), for electric drive systems (EDSs) with direct current (DC) motor. The application considered in this paper is a simple mechanism, a mechatronic system that does not require complex hardware and also exhibits attractive features such as high efficiency, low manufacturing cost, simple construction and operation and linear torque-speed characteristics. The performance and the efficiency of the four estimation approaches are validated through results obtained by digital simulation. To better highlight the performance, a comparative study of all four estimation approaches is performed considering the values for the mean square error (MSE) regarding two simulation scenarios – step reference input and staircase reference input – obtained for two state variables. The results show that all these estimation approaches work efficiently for such types of processes.

Keywords: DC electric drive system · Extended Luenberger state observer · Kalman filter observer · Extended Kalman filter observer · Sliding mode observer

1 Introduction

The increasing complexity of industrial applications requires a permanent modernization of industrial automation and has led to new requirements in relation to electric drives and related automatic control solutions. These requirements actually refer to the assignment of increasingly high performance in relation to the operation in different modes – starting, speed control, braking, reversing – and the development of the control system in relation to these requirements [1–6]. Electric drive systems (EDSs) are a way of converting electricity into mechanical energy and in terms of performance are often superior to hydraulic and pneumatic. EDSs are encountered in a large number of applications mainly due to the facilities they can provide, through the development of control systems. The choice of the type of electric drive is determined by the operating conditions and operating regimes [5–8]. The performance specifications imposed to the operation of the EDSs influences the mathematical modeling of the processes, and in order to meet the

S. Dzitac et al. (Eds.): ICCCC 2022, AISC 1435, pp. 107–119, 2023.
https://doi.org/10.1007/978-3-031-16684-6_9

performance specifications, a mathematical model as complete and correct as possible is required. Taking this into consideration, Direct Current (DC) motors have a relatively simple construction, are robust, have high reliability, can adjust speed in relatively wide areas and ensure fast speed reversal.

In the context of the quality requirements expressed as performance specifications imposed to the operation of EDSs, there are currently several problems, two of which are highlighted: one is to make the control systems robust to variations in parameters and under the influence of disturbances and the other one consists in making the control system adaptable to such changes. The main objectives of these systems are to provide the most accurate following of the reference, smallest possible settling time, reduced overshoot, zero adjustment error (as low as possible) in relation to certain types of references and disturbances and thereby a static character of the system. Given these objectives, the analysis of the state-of-the-art of control theory and applications related to DC EDSs shows that the solution which involves implementation of robust control techniques based on observers leads to high system performance. Two types of observers can be used: linear and nonlinear. From the category of linear observers, the most common are full-order observers known as the Luenberger-type observers and reduced-order observers. Among the nonlinear ones we mention: extended Kalman filters, sliding mode observers, fuzzy and neural network observers [9–15].

Several approaches for the estimation of state variables from obtainable measurements have been created and publicized due to the increased attention given by researchers to the idea of developing appropriate estimation approaches, techniques and methods. In this context, in [9], Luenberger demonstrated the fact that the development of an observer for a system with m outputs can be replaced with the development of m individual observers for single-output subsystems as an effect of a special canonical form presented in the paper for multiple-output systems. Later on, Thau introduced in [10] a main topic in literature from the control systems point of view, the dilemma of developing observers for nonlinear control systems. Aghannan and Rouchon proposed in [16] a novel design approach of asymptotic observers for a certain type of nonlinear mechanical systems: Lagrangian systems with position measurements. Marx et al. suggested in [17] an approach for state-estimation of Takagi-Sugeno descriptor systems influenced by unknown inputs. The developed observers are presented in usual form and not in descriptor form so that it facilitates their implementation. Adequate existence conditions for the unknown input observers are offered and strict linear matrix inequalities are solved in order to establish the observer gains. Lendek et al. proposed in [18] the division of a linear system into a series of basic subsystems and the employment of a Kalman filter to independently obtain a states estimation of these subsystems. In [19], Spurgeon presented an analysis of both linear and nonlinear Sliding Mode Observer (SMO) paradigms, and in [20] Mercorelli proposed a two-stage SMO in order to lessen the effect of uncertainties and disturbances for sensorless control in automotive applications. These observers use the equivalent command to estimate the equivalent disturbance. In general, their structures are closely related to the mathematical models used in the controlled process. Later on, in [21] Habibullah and Lu designed an Extended Kalman Filter Observer (EKFO) with the interest of improving the prediction model of sensorless finite-state predictive torque control for induction motor drives, and in [22]

Magnis and Petit proposed a method to estimate the angular speed of a rigid body using single vector measurements. Several acceleration and speed observers for brushless DC drives have been recently proposed by Szedlak-Stinean et al. in [23].

The principle contributions of this paper are: (i) the development and the implementation of the four estimation approaches – Extended Luenberger State Observer (ELSO), Kalman Filter Observer (KFO), EKFO and SMO – designed for the EDS with DC motor; (ii) the creation of Matlab files and Simulink models to test the considered approaches; (iii) the digital simulation and validation of the linear and nonlinear observers and (iv) a comparative study of all four estimation approaches in order to assess their performance and efficiency.

This paper further includes: Sect. 2 presents the linear mathematical model (MM), the block diagram and the parameters of the EDS with DC motor. Section 3 explains the design and the implementation of the four estimation approaches. Section 4 offers the digital validation and comparison of the proposed approaches. In Sect. 5 drawn conclusions and directions for future research are specified.

2 Electric Drive System with DC Motor

2.1 Mathematical Model of the EDS with DC Motor

The principal hypotheses after which the mathematical model (MM) of the process is established are the following:

(i) the excitation flow provided by the independent excitation circuit Φ_e is constant;
(ii) in the design of the proposed approaches, the nonlinearities owing to the diverse constructive elements are neglected;
(iii) to simplify the writing, it is accepted that the parameters of the inductor and the smoothing coil are merged, being written in the form: $R_a = R_a + R_{sc}$, $L_a = L_a + L_{sc}$;
(iv) the moment of inertia of the driven mechanism, J, is constant and
(v) the measuring transducers have a linear static characteristic.

Taking into account these aspects, the behavior of the EDS with DC motor is expressed by the following equations [1, 7, 8, 15]:

$$\begin{cases} \frac{di_a(t)}{dt} = -\frac{1}{T_a} i_a(t) - \frac{k_e}{L_a}\omega(t) + \frac{1}{L_a} u_a(t), \\ \frac{d\omega(t)}{dt} = \frac{k_m}{J} i_a(t) - \frac{k_f}{J}\omega(t) - \frac{1}{J} m_s(t), \\ \frac{du_a(t)}{dt} = -\frac{1}{T_\mu} u_a(t) + \frac{k_E}{T_\mu} u_c(t), \end{cases} \tag{1}$$

where: i_a, u_a – the armature current and voltage [A, V], u_c – the control voltage [V], $T_a = L_a/R_a$, L_a, R_a – the electrical parameters [s, H, Ω,], ω – the angular speed [rad/s], k_e, k_m, k_f – the electromagnetic coefficients [V/rad/s, N·m/A, N·m/s], k_E – the actuator gain [V/V], J – the moment of inertia [kg·m^2], and m_s – the load torque [N·m].

The characteristic parameters related to EDS with DC motor used in the informational characterization of the controlled process are: (i) inputs: u_c – the control voltage and m_s

– the load torque, (ii) outputs: ω – the angular speed and i_a – the armature current, and (iii) state variables: $\{\omega, i_a, u_a\}$.

The state-space MM of the EDS with DC motor is obtained by inserting the state variables $x_1(t) = i_a(t)$, $x_2(t) = \omega(t)$, $x_3(t) = u_a(t)$:

$$\begin{cases} \begin{bmatrix} \dot{x}_1(t) \\ \dot{x}_2(t) \\ \dot{x}_3(t) \end{bmatrix} = \begin{bmatrix} -\frac{1}{T_a} & -\frac{k_e}{L_a} & \frac{1}{L_a} \\ \frac{k_m}{J} & -\frac{k_f}{J} & 0 \\ 0 & 0 & -\frac{1}{T_\mu} \end{bmatrix} \begin{bmatrix} x_1(t) \\ x_2(t) \\ x_3(t) \end{bmatrix} + \begin{bmatrix} 0 & 0 \\ 0 & -\frac{1}{J} \\ \frac{K_E}{T_\mu} & 0 \end{bmatrix} \begin{bmatrix} u_c(t) \\ m_s(t) \end{bmatrix}, \\ y(t) = \begin{bmatrix} x_1(t) \\ x_2(t) \end{bmatrix} = \begin{bmatrix} 1 & 0 & 0 \\ 0 & 1 & 0 \end{bmatrix} \begin{bmatrix} x_1(t) \\ x_2(t) \\ x_3(t) \end{bmatrix}. \end{cases} \quad (2)$$

In Fig. 1, the block diagram corresponding to the DC electric drive system is shown.

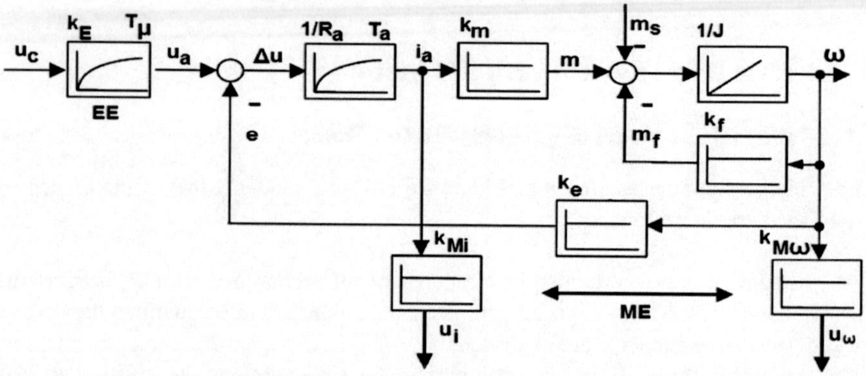

Fig. 1. Block diagram of the EDS with DC motor.

2.2 Numerical Values of the EDS with DC Motor

The numerical values of the EDS with DC motor used in the linear and nonlinear observers design are given in Table 1.

Table 1. Numerical values of system parameters.

Parameters	Notation	Value	Units
Rated power	P_{an}	1000	[W]
Rated voltage	u_{an}	200	[V]
Rated control voltage	u_{cn}	4	[V]
Rated capacity	η	0.82	
Rated speed	n	2000	[rpm]
Moment of inertia	J	0.0632	[kg·m^2]
Mechanical time constant	T_m	0.6	[s]
Electrical time constant	T_a	0.1	[s]
Execution element time constant	T_μ	0.001	[s]
Current transducer constant	kMi	0.7	[V/A]
Angular speed transducer constant	$kM\omega$	0.0239	[V/rad/s]

3 Linear and Nonlinear Observers

The evaluation of an observer design success is based on its ability to estimate the difficult-to-measure states with acceptable convergence rates and with approximately zero estimation errors. In order to achieve a comparative study on armature current and angular speed control performance, the four design proposals presented below will be taken into consideration. The design will use the dynamic model shown in (2).

3.1 Extended Luenberger State Observer (ELSO)

A solution to compute the state estimates $\hat{x}_1(t)$, $\hat{x}_2(t)$ and $\hat{x}_3(t)$ is to use two ELSOs having as predictive correction terms the armature current error $\Delta x_1(t) = x_1(t) - \hat{x}_1(t)$ and the angular speed error $\Delta x_2(t) = x_2(t) - \hat{x}_2(t)$.

For estimating the armature current, a gain matrix $L = \left[l_{1ia}\ l_{2ia}\ l_{3ia} \right]^T \in \Re$ is used (here, T represent the matrix transposition) and the dynamic model of ELSO is [9, 15, 23]:

$$\begin{cases} \underbrace{\begin{bmatrix} \dot{\hat{x}}_1(t) \\ \dot{\hat{x}}_2(t) \\ \dot{\hat{x}}_3(t) \end{bmatrix}}_{\dot{\hat{x}}(t)} = \underbrace{\begin{bmatrix} -\frac{1}{T_a} & -\frac{k_e}{L_a} & \frac{1}{L_a} \\ \frac{k_m}{J} & -\frac{k_f}{J} & 0 \\ 0 & 0 & -\frac{1}{T_\mu} \end{bmatrix}}_{\mathbf{A}} \underbrace{\begin{bmatrix} \hat{x}_1(t) \\ \hat{x}_2(t) \\ \hat{x}_3(t) \end{bmatrix}}_{\hat{\mathbf{x}}(t)} + \underbrace{\begin{bmatrix} 0 & 0 \\ 0 & -\frac{1}{J} \\ \frac{k_E}{T_\mu} & 0 \end{bmatrix}}_{\mathbf{B}} \begin{bmatrix} u_c(t) \\ m_s(t) \end{bmatrix} + \underbrace{\begin{bmatrix} l_{1ia} \\ l_{2ia} \\ l_{3ia} \end{bmatrix}}_{\mathbf{L}} (x_1(t) - \hat{x}_1(t)), \\ \hat{y}(t) = \underbrace{[1\ 0\ 0]}_{\mathbf{C}} \underbrace{\begin{bmatrix} \hat{x}_1(t) \\ \hat{x}_2(t) \\ \hat{x}_3(t) \end{bmatrix}}_{\hat{\mathbf{x}}(t)}. \end{cases} \tag{3}$$

For estimating the angular speed, a gain matrix $L = \left[l_{1\omega}\ l_{2\omega}\ l_{3\omega} \right]^T \in \Re$ is used and the dynamic model of ELSO is [9, 15, 23]:

$$\begin{cases} \underbrace{\begin{bmatrix} \dot{\hat{x}}_1(t) \\ \dot{\hat{x}}_2(t) \\ \dot{\hat{x}}_3(t) \end{bmatrix}}_{\dot{\hat{x}}(t)} = \underbrace{\begin{bmatrix} -\frac{1}{T_a} & -\frac{k_e}{L_a} & \frac{1}{L_a} \\ \frac{k_m}{J} & -\frac{k_f}{J} & 0 \\ 0 & 0 & -\frac{1}{T_\mu} \end{bmatrix}}_{\mathbf{A}} \underbrace{\begin{bmatrix} \hat{x}_1(t) \\ \hat{x}_2(t) \\ \hat{x}_3(t) \end{bmatrix}}_{\hat{\mathbf{x}}(t)} + \underbrace{\begin{bmatrix} 0 & 0 \\ 0 & -\frac{1}{J} \\ \frac{k_E}{T_\mu} & 0 \end{bmatrix}}_{\mathbf{B}} \begin{bmatrix} u_c(t) \\ m_s(t) \end{bmatrix} + \underbrace{\begin{bmatrix} l_{1\omega} \\ l_{2\omega} \\ l_{3\omega} \end{bmatrix}}_{\mathbf{L}} (x_2(t) - \hat{x}_2(t)), \\ \hat{y}(t) = \underbrace{[0\ 1\ 0]}_{\mathbf{C}} \underbrace{\begin{bmatrix} \hat{x}_1(t) \\ \hat{x}_2(t) \\ \hat{x}_3(t) \end{bmatrix}}_{\hat{\mathbf{x}}(t)}. \end{cases} \tag{4}$$

In both cases $A \in \Re^{3x3}$, $B \in \Re^{3x2}$, $C \in \Re^{1x3}$ are the continuous-time ELSO matrices and for m_s a step signal is utilized. The observer gain matrix L is calculated in terms of the pole placement method, considering the pair (A, C) observable. For the imposed poles $P_\sigma = \{p_1, p_2, p_3\} = \{-50, -60, -70\}$ the observer gains are given in Table 2, column 1.

3.2 Kalman Filter Observer (KFO)

KFO is the most popular and often utilized observer for linear systems and it offers a good approach for estimating the states while minimizing the mean square error [18, 21, 23]–[24]. The dynamical state-space model is defined by [23, 25]:

$$\begin{aligned} \dot{x}(t) &= Ax(t) + Bu(t) + Gw(t), \\ y(t) &= Cx(t) + Du(t) + Hw(t) + v(t), \end{aligned} \tag{5}$$

where: x – the state vector, u – the input, v, w – the white measurement noise and white process noise, Q, R, N – the process covariance, the measurement covariance and noise cross covariance and the elements of matrices A, B, C and D (here $D = 0$) are gained using (2). The values for G, H, Q, R and N are given Table 2, column 2. The KFO constructs the error covariance as stated by:

$$P = \lim_{t \to \infty} E\left(\left\{x(t) - \hat{x}(t)\right\}\left\{x(t) - \hat{x}(t)\right\}^T\right). \tag{6}$$

Taking this into consideration, the equations of the KFO are:

$$\begin{aligned} \dot{\hat{x}}(t) &= A\hat{x}(t) + Bu(t) + L(y(t) - C\hat{x}(t) - Du(t)), \\ \begin{bmatrix} \hat{y}(t) \\ \hat{x}(t) \end{bmatrix} &= \begin{bmatrix} C \\ I \end{bmatrix} \hat{x}(t) + \begin{bmatrix} D \\ 0 \end{bmatrix} u(t). \end{aligned} \tag{7}$$

As far as the KFO is concerned, in order to obtain the filter gain, a Riccati equation must be solved. The filter gain L, R_r and N_n are according to:

$$\begin{aligned} &L = (PC^T + N_n)R_r^{-1}, \\ &R_r = R + HN + N^T H^T + HQH^T, \ N_n = G(QH^T + N). \end{aligned} \tag{8}$$

3.3 Extended Kalman Filter Observer (EKFO)

EKFO is the nonlinear version of KFO, and it linearizes around an estimate of the current mean and covariance. It is deemed to be a relevant standard for well-defined transition models, for nonlinear estimation, global positioning systems and navigation systems [25, 26]. Because most engineering systems are nonlinear, it was attempted to use the EKFO for functions approximated by using a linearization process.

The dynamical model is represented by the equations [25, 27, 28]:

$$\begin{aligned} \dot{x}(t) &= f(x(t), u(t)) + w(t),\ \ w(t) \sim \mathcal{N}(0, Q(t)), \\ y(t) &= h(x(t)) + v(t),\ \ v(t) \sim \mathcal{N}(0, R(t)), \end{aligned} \tag{9}$$

where: x, u, w, v, Q and R are exactly the same vectors, variables and parameters as in the case of KFO, f – a function utilized for calculating the predicted state from the previous estimate and h – a function utilized for calculating the predicted measurement from the predicted state.

The EKFO predict-update equations are [25, 27, 28]:

$$\begin{aligned} \dot{\widehat{x}}(t) &= f(\widehat{x}(t), u(t)) + K(t)(y(t) - h(\widehat{x}(t))), \\ \dot{P}(t) &= F(t)P(t) + P(t)F(t)^T - K(t)H(t)P(t) + Q(t), \\ K(t) &= P(t)H(t)^T R(t)^{-1}, \\ F(t) &= \frac{\partial f}{\partial x}\Big|\hat{x}(t), u(t), \\ H(t) &= \frac{\partial h}{\partial x}\Big|\hat{x}(t), \end{aligned} \tag{10}$$

where: P – the error covariance matrix, K – the Kalman gain matrix, F, H – the state transition and the state observation matrices. The states of EKFO are revised employing the error $\hat{e}(t) = y(t) - h(\hat{x}(t))$ and also the K matrix. The parameters used in the EKFO design are given in Table 2, column 3.

3.4 Sliding Mode Observer (SMO)

These observers have the ability to minimize the error between the measured process output and the output of the observer that guarantees that the observer offers a set of state estimates that are exactly proportional to the actual output of the process.

For SMO design, the dynamic model given in (2) will be utilized, knowing that the matrix dimensions are $A \in \Re^{nxn}$, $B \in \Re^{nxm}$, $C \in \Re^{pxn}$, $p \geq m$. In this context, the following change of coordinates $x \to T_c z$, $T_c = [N_c^T\ \ C]^T$, was considered, where $det(T_c) \neq 0$ and the submatrix $N_c \in \Re^{nx(n-p)}$ covers the null space of C [18, 23, 29].

The matrices $T_c A T_c^{-1}$, $T_c B$ and $C T_c^{-1}$ are presented in (11) with the parameters given in Table 2, column 4.

$$T_c A T_c^{-1} = \begin{bmatrix} A_{11}\ A_{12} \\ A_{21}\ A_{22} \end{bmatrix},\ T_c B = \begin{bmatrix} B_1 \\ B_2 \end{bmatrix},\ C T_c^{-1} = \begin{bmatrix} 0\ I_p \end{bmatrix}. \tag{11}$$

Knowing that $T_c z = [z\ \ y]^T$, the dynamical model of the SMO is [18, 23, 29]:

$$\begin{aligned} \dot{\hat{z}} &= A_{11}\hat{z} + A_{12}\hat{y} + B_1 u + L\nu_i, \\ \hat{y} &= A_{21}\hat{z} + A_{22}\hat{y} + B_2 u - \nu_i, \end{aligned} \tag{12}$$

where $L = \begin{bmatrix} L_1 & L_2 \end{bmatrix} \in \Re^{(n-p)xp}$ is a gain matrix and $v_i = \begin{bmatrix} v_1 & v_2 \end{bmatrix}^T = M\,\mathrm{sgn}(\hat{y}_i - y_i),\ i = 2$ is an injection term in which $M \in \Re_+$.

The error dynamics is expressed as:

$$\begin{aligned} \dot{e}_z &= A_{11}e_z + A_{12}e_{yi} + L\,v_i,\ e_z = \hat{z} - z, \\ \dot{e}_{yl} &= A_{21}e_z + A_{22}e_{yl} - v_i,\ e_{yl} = \hat{y}_i - y_i. \end{aligned} \tag{13}$$

In order to reduce the sliding motion, the error system defined in (13) provides the relation: $\dot{e}_z = (A_{11} + LA_{21})e_z$, which reliant on L, describes a stable system, so $e_z \to 0$ and therefore $\hat{z} \to z$.

Table 2. Numerical values of ELSO, KFO, EKFO and SMO parameters.

ELSO			KFO				EKFO		
1			2				3		
l_{1ia}	l_{2ia}	l_{3ia}	G	$H{=}N$	Q	R	Q	R_{ia}	R_ω
-830.03	-147.9	$5.1\cdot10^5$	$\begin{bmatrix} 0.001 \\ 1 \\ 0.1 \end{bmatrix}$	0	10^{-9}	$\begin{bmatrix} 1\cdot10^{-10} & 0 \\ 0 & 1\cdot10^{-6} \end{bmatrix}$	10^{-9}	10^{-10}	10^{-6}
$l_{1\omega}$	$l_{2\omega}$	$l_{3\omega}$							
$6.6\cdot10^4$	-830.03	$-4.02\cdot10^7$							

SMO						
4						
T_c	$T_cAT_c^{-1}$	T_cB	CT_c^{-1}	M	L_1	L_2
$\begin{bmatrix} 0 & 0 & 1 \\ 1 & 0 & 0 \\ 0 & 1 & 0 \end{bmatrix}$	$\begin{bmatrix} -1000 & 0 & 0 \\ 1.6167 & -10 & -1.3051 \\ 0 & 12.77 & -0.0305 \end{bmatrix}$	$\begin{bmatrix} 50000 & 0 \\ 0 & 0 \\ 0 & -15.8198 \end{bmatrix}$	$\begin{bmatrix} 0 & 1 & 0 \\ 0 & 0 & 1 \end{bmatrix}$	1	10	3

4 Simulation Results

A conventional control structure was designed and tested for an EDS with DC motor with regard to four proposed observers. This conventional control structure uses, in combination with all these estimation approaches, a classical Proportional-Integral (PI) controller and the parameters of this controller are tuned using the Modulus Optimum method (MO-m) [30].

The PI controller is described by the parameters: $k_r = 3.3814$ and $T_r = 0.6$. For determining the performance of the suggested design approaches – ELSO, KFO, EKFO and SMO – tests are run that compare the actual values with the estimates. The validation of the control approaches will be done by numerical simulation scenarios; in this paper, the system's responses in two simulation scenarios were taken into consideration: (i) a step reference input and (ii) a staircase reference input. The digital results are given in Fig. 2 and Fig. 3.

A comparative study of all four estimation approaches is performed considering the value of the mean square error (MSE) with the purpose of evaluating the performance

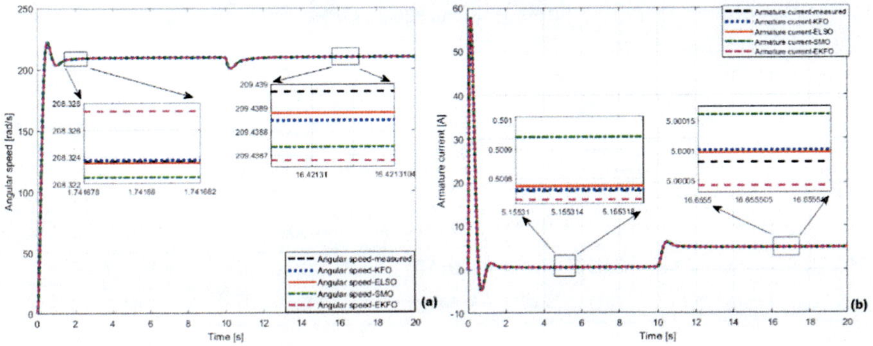

Fig. 2. Digital results (simulation) for DC electric drive system (step reference input): (a) angular speed versus time, (b) armature current versus time

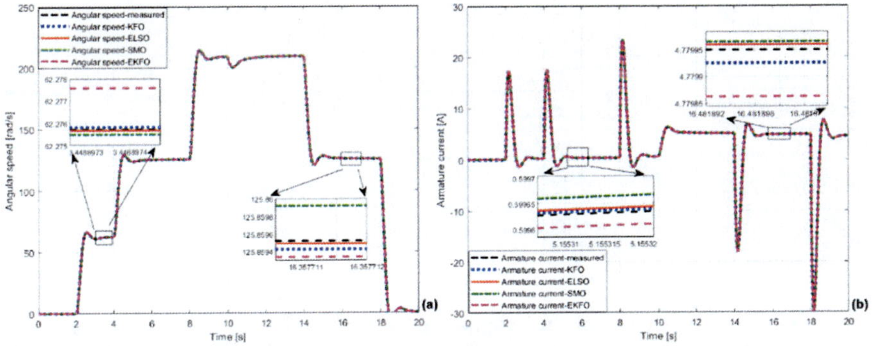

Fig. 3. Digital results (simulation) for DC electric drive system (staircase reference input): (a) angular speed versus time, (b) armature current versus time

and the efficiency of these estimation aproaches. The MSE values corresponding to these observers are shown in Table 3. The MSE values [23], that gather the control errors over the time horizon of N sampling intervals, examined as a global performance index, between the real system variable v_k and its estimation $\hat{v}_k$, are defined as (v – angular speed and armature current):

$$MSE = \frac{1}{N}\sum_{k=1}^{N}(v_k - \hat{v}_k). \tag{14}$$

The computational effort realized by the four estimation approaches, necessary for the future implementation in an adequate control system structure for different applications is shown in Table 4. The results presented in Table 4 were obtained after running the experiments on a personal computer with Intel(R) Core(TM) i5-1035G1 CPU 1.19 GHz processor, with 8 GB RAM, Windows 10 Pro and Matlab R2021b.

Table 3. Mean square errors.

Estimation approaches	Variables			
	$\omega(t)$		$i_a(t)$	
	Step	Staircase	Step	Staircase
MSE_{ELSO} (rad/s), (A)	$2.1381 \cdot 10^{-8}$	$1.1751 \cdot 10^{-8}$	$4.7856 \cdot 10^{-9}$	$4.0193 \cdot 10^{-9}$
MSE_{KFO} (rad/s), (A)	$1.2473 \cdot 10^{-8}$	$1.2472 \cdot 10^{-8}$	$4.50282 \cdot 10^{-10}$	$4.50281 \cdot 10^{-10}$
MSE_{EKFO} (rad/s), (A)	0.0201	0.0157	$1.3685 \cdot 10^{-5}$	$1.7216 \cdot 10^{-6}$
MSE_{SMO} (rad/s), (A)	$2.2440 \cdot 10^{-6}$	$9.6018 \cdot 10^{-7}$	$3.9860 \cdot 10^{-8}$	$2.4473 \cdot 10^{-8}$

Table 4. Computational effort.

Estimation approaches	Variables			
	$\omega(t)$		$i_a(t)$	
	Step	Staircase	Step	Staircase
ELSO (rad/s), (A)	0.7066 s	0.7321 s	0.7865 s	0.8130 s
KFO (rad/s), (A)	2.5054 s	2.8474 s	3.1410 s	3.5550 s
EKFO (rad/s), (A)	12.7210 s	13.3214 s	12.7980 s	13.6329 s
SMO (rad/s), (A)	0.7875 s	0.8094 s	0.7956 s	0.8345 s

Taking into consideration the graphics presented in Fig. 2 and Fig. 3, the MSE values given in Table 3 and also the computational effort in Table 4, the following set of conclusions is formulated:

(i) the best performances have been obtained by the KFO and by the ELSO due to the fact that they are linear observers and the EDS with DC motor is a linear system;
(ii) the ELSO is less efficient than KFO due to the fact that the process parameters influence the parameters of the observer gain;
(iii) the SMO provides better performance than the EKFO due to the fact that similarly to KFO has attractive noise resilience properties and also is insensitive to unknown inputs while in sliding;
(iv) in terms of execution times, the most efficient estimation approaches are ELSO, SMO, KFO and EKFO;
(v) the proposed observers are demostrated to be viable, efficient and guarantee a good reference tracking ability, and
(vi) the control system should benefit from the advantages of the continued analytical development and implementation of dynamic model-based observers that promise to improve the performance of different types of mechatronics systems.

5 Conclusions

This paper gave details with regard to the development and implementation of four observers for estimating the angular speed and the armature current for an EDS with DC motor. The proposed design approaches are utilized to estimate states in a linear process in the eventual presence of disturbances or noise. Due to the fact that observers can only be designed for observable systems, the first important consideration before designing the observer is to verify the observability condition of the system, which in turn has influence upon the observer's feasibility conditions. Although the observer gain and its estimation error dynamics are also significant, the evaluation of an observer designs success is based on its ability to estimate the difficult-to-measure states with acceptable convergence rates and with approximately zero estimation errors.

The validation of these estimation approaches was done by digital simulation results and the graphics are shown in Fig. 2 and Fig. 3. The graphics and also the MSE values demonstrated that these observers are viable and guarantee good reference tracking ability. The further development and improvement of the control approaches can be pursued independently.

These observers are important because they can be combined with different control approaches dedicated to various industrial and non-industrial processes including fuzzy logic-based controlled ones [31–35], robotics [36, 37], hybrid electric vehicles [38], power electronics [39], and networks [40, 41]. Future work will focus on adapting these observers to other significant cases, by extending the approaches presented in this paper to these illustrative applications.

Acknowledgment. This work was supported by grants from the Ministry of Research and Innovation, CNCS-UEFISCDI, project numbers PN-III-P1–1.1-PD-2016–0331, PN-III-P4-ID-PCE-2020–0269 and PN-III-P1–1.1-PD-2019–0637, within PNCDI III.

References

1. Isermann, R.: Mechatronic Systems: Fundamentals. Springer-Verlag, London (2005). https://doi.org/10.1007/1-84628-259-4
2. Koch, C., Radler, O., Spröwitz, A., Ströhla, T., Zöppig, V.: Project course. Design mechatronic systems. In: IEEE International Conference on Mechatronics, 69–72. Budapest, Hungary (2006). https://doi.org/10.1109/ICMECH.2006.252499
3. Hehenberger, P., Naderer, R., Schuler, C., Zeman, K.: Conceptual design of mechatronic systems as a recurring element of innovation processes. IFAC Proc Volumes **39**(16), 342–347 (2006). https://doi.org/10.3182/20060912-3-DE-2911.00061
4. Pabst, I.: An approach for reliability estimation and control of mechatronic systems. IFAC Proc. Volumes **39**(16), 831–836 (2006). https://doi.org/10.3182/20060912-3-DE-2911.00143
5. Tunsoiu, G., Seracin, E., Saal, C.: Acţionări electrice: Editura Didactică şi Pedagogică, Bucharest (1982). (in Romanian)
6. Kovacs, L.: Soluţii de conducere a sistemelor de acţionare electrică cu moment de inerţie variabil. Diploma project, "Politehnica" University of Timisoara, Timisoara, Romania (2000). (in Romanian)

7. Golnaraghi, F., Kuo, B. C.: Automatic Control Systems. 10th edn. McGraw-Hill Education, New York (2017)
8. Boldea, I., Nasar, S.A.: Electric Drives, 3rd edn. CRC Press, Boca Raton, Florida (2017)
9. Luenberger, D.G.: Observers for multivariable systems. IEEE Trans. Autom. Control **11**(2), 190–197 (1966). https://doi.org/10.1109/TAC.1966.1098323
10. Thau, F.E.: Observing the state of nonlinear dynamic systems. Int. J. Control **17**(3), 471–479 (1973). https://doi.org/10.1080/00207177308932395
11. Ackermann, J.: Der Entwurf linearer Regelngssysteme im Zustandsraum. Regelungs-technik **20**, 297–300 (1972). https://doi.org/10.1524/auto.1972.20.112.297
12. Friedland, B.: Advanced Control System Design. Prentice Hall, Englewood Cliffs (1996)
13. Gajic, Z., Lelic, M.: Modern Control Systems Engineering. Prentice Hall International, London (1996)
14. Stefani, R.T., Shahian, B., Savant, C.J., Hostetter, G.: Design of Feedback Systems, 4th edn. Oxford University Press, New York (2002)
15. Preitl, S., Precup, R.-E., Preitl, Z.: Structuri și algoritmi pentru conducerea automată a proceselor. 1, 2, Editura Orizonturi Universitare, Timisoara (2009)
16. Aghannan, N., Rouchon, P.: An intrinsic observer for a class of Lagrangian systems. IEEE Trans. Autom. Control **48**(6), 936–945 (2003). https://doi.org/10.1109/TAC.2003.812778
17. Marx, B., Koenig, D., Ragot, J.: Design of observers for Takagi-Sugeno descriptor systems with unknown inputs and application to fault diagnosis. IET Control Theor. Appl. **1**(5), 1487–1495 (2007). https://doi.org/10.1049/iet-cta:20060412
18. Lendek, Z., Babuska, R., De Schutter, B.: Distributed Kalman filtering for cascaded systems. Eng. Appl. Artif. Intell. **21**(3), 457–469 (2008). https://doi.org/10.1016/j.engappai.2007.05.002
19. Spurgeon, S.K.: Sliding mode observers: a survey. Int. J. Syst. Sci. **39**(8), 751–764 (2008). https://doi.org/10.1080/00207720701847638
20. Mercorelli, P.: A two-stage sliding-mode high-gain observer to reduce uncertainties and disturbances effects for sensorless control in automotive applications. IEEE Trans. Industr. Electron. **62**(9), 5929–5940 (2015). https://doi.org/10.1109/TIE.2015.2450725
21. Habibullah, M., Lu, D.D.-C.: A speed-sensorless FS-PTC of induction motors using extended Kalman filters. IEEE Trans. Industr. Electron. **62**(11), 6765–6778 (2015). https://doi.org/10.1109/TIE.2015.2442525
22. Magnis, L., Petit, N.: Angular velocity nonlinear observer from single vector measurements. IEEE Trans. Autom. Control **61**(9), 2473–2483 (2016). https://doi.org/10.1109/TAC.2015.2501358
23. Szedlak-Stinean, A.-I., Precup, R.-E., David, R.-C.: Speed and acceleration control of BLDC drives using different types of observers. In: 13th IEEE International Symposium on Applied Computational Intelligence and Informatics, pp. 229–236. Timisoara, Romania (2019). https://doi.org/10.1109/SACI46893.2019.9111608
24. Brown, R.G., Hwang, P.Y.C.: Introduction to Random Signals and Applied Kalman Filtering, 4th edn. Wiley, Hoboken (2012)
25. Mohinder, S.G., Andrews, A.: Kalman Filtering: Theory and Practice using MATLAB. Wiley, Hoboken, NJ (2015)
26. Julier, S.J., Uhlmann, J.K.: Unscented filtering and nonlinear estimation. Proc. IEEE **92**(3), 401–422 (2004). https://doi.org/10.1109/JPROC.2003.823141
27. Beyhan, S., Lendek, Z., Alci, M., Babuska, R.: Takagi-Sugeno fuzzy observer and extended Kalman filter for adaptive payload estimation. In: 2013 Asian Control Conference, pp. 1–6. Istanbul, Turkey (2013). https://doi.org/10.1109/ASCC.2013.6606241
28. Simon, D.: Optimal State Estimation: Kalman, H Infinity, and Nonlinear Approaches. Wiley, Hoboken (2006)

29. Shtessel, Y., Edwards, C., Fridman, L., Levant, A.: Sliding Mode Control and Observation. Springer, New York (2014). https://doi.org/10.1007/978-0-8176-4893-0
30. Åström, K.J., Hägglund, T.: Advanced PID Control. Systems, and Automation Society, Research Triangle Park, NC, ISA - Instrumentation (2006)
31. Precup, R.-E., Preitl, S.: Popov-type stability analysis method for fuzzy control systems. In: Fifth European Congress on Intelligent Technologies and Soft Computing (EUFIT 1997), vol. 2, pp. 1306–1310. Aachen, Germany (1997)
32. Vascak, J., Kovacik, P., Hirota, K., Sincak, P.: Performance-based adaptive fuzzy control of aircrafts. In: 10th IEEE International Conference on Fuzzy Systems, vol. 3, pp. 761–764. Melbourne, VIC, Australia (2001). https://doi.org/10.1109/FUZZ.2001.1009066
33. Precup, R.-E., Tomescu, M.L., Preitl, S.: Lorenz system stabilization using fuzzy controllers. Int. J. Comput. Commun. Control **2**(3), 279–287 (2007). https://doi.org/10.15837/ijccc.2007.3.2360
34. Dzitac, I.: Zadeh's centenary. Int. J. Comput. Commun. Control **16**(1), 1–13 (2021)
35. Precup, R.-E., David, R.-C., Roman, R.-C., Szedlak-Stinean, A.-I., Petriu, E. M.: Optimal tuning of interval type-2 fuzzy controllers for nonlinear servo systems using slime Mould algorithm. Int. J. Syst. Sci. (2021)https://doi.org/10.1080/00207721.2021.1927236
36. Precup, R.-E., Roman, R.-C., Teban, T.A., Albu, A., Petriu, E.M., Pozna, C.: Model-free control of finger dynamics in prosthetic hand myoelectric-based control systems. Stud. Inf. Control **29**(3), 399–410 (2020). https://doi.org/10.24846/v29i4y202002
37. Guechi, E.H., Lauber, J., Dambrine, M., Klančar, G., Blažič, S.: PDC control design for non-holonomic wheeled mobile robots with delayed outputs. J. Intell. Rob. Syst. **60**(3–4), 395–414 (2010). https://doi.org/10.1007/s10846-010-9420-0
38. Johanyák, Z.C.: A modified particle swarm optimization algorithm for the optimization of a fuzzy classification subsystem in a series hybrid electric vehicle. Technicki Vjesnik - Technical Gazette **24**(2), 295–301 (2017). https://doi.org/10.17559/TV-20151021202802
39. Iordache, C.A., Bodea, M.: Analysis and design of a high efficiency current mode buck converter with I2C controlled output voltage. Rom. J. Inf. Sci. Technol. **23**(2), 188–203 (2020)
40. Osaba, E., Del Ser, J., Camacho, D., Bilbao, M.N., Yang, X.S.: Community detection in networks using bio-inspired optimization: latest developments, new results and perspectives with a selection of recent meta-heuristics. Appl. Soft Comput. **87**, 106010 (2020). https://doi.org/10.1016/j.asoc.2019.106010
41. Priya, B., Solai Manohar, S.: Adaptive power control and duty cycle based medium access control protocol for cluster based wireless sensor network. Rom. J. Inf. Sci. Technol. **23**(1), 48–54 (2020)

Energy-Efficient Wireless Sensor Networks for Greenhouse Management

Attila Simo[1(✉)] and Simona Dzitac[2]

[1] Politehnica University Timisoara, 300223 Timisoara, Romania
attila.simo@upt.ro

[2] University of Oradea, 410087 Oradea, Romania
sdzitac@uoradea.ro

Abstract. Agriculture plays a crucial role in the economy of the country. The agricultural industry is getting more data-centric and requires precise, more advanced data and technologies than before, despite being familiar with agricultural processes. The agriculture industry is being advanced by various information and advanced communication technologies, such as the Internet of Things (IoT). Development and deployment of sensing technologies is one of the main steps in achieving sustainability in crop production through precision agriculture. The energy crisis will further accelerate this digitization process. The energy crisis will accelerate this digitization process even more, but at the same time it will help farmers consume energy more responsibly. To optimize electrical energy consumption, farmers need monitoring devices. In this paper the authors propose a low-cost, low-power, long range monitoring device development. The results obtained show the usefulness of such a monitoring device.

Keywords: Energy efficiency · Wireless sensor network · LPWA technologies · IoT

1 Introduction

Because the range of legacy non-cellular wireless technologies is restricted to a few hundred meters at best, they are not appropriate for connecting low-power devices spread over large geographical areas. As a result, the devices can't be placed or moved about at will, which is a requirement for many applications in agriculture, smart cities, transportation and logistics, and wildlife monitoring. For machine-type communication, legacy WLANs have smaller coverage areas and higher power consumption. A dense deployment of devices and gateways connected via multihop mesh networking [1] expands the range of these technologies.

Low Power Wide Area (LPWA) technologies offer unique sets of features including wide-area connectivity for low power and low data rate devices, not provided by legacy wireless technologies [2]. LPWA networks are unique because they make different tradeoffs than the traditional technologies prevalent in IoT landscape such as short-range wireless networks (Zig-Bee, Bluetooth, Z-Wave), legacy wireless local area

S. Dzitac et al. (Eds.): ICCCC 2022, AISC 1435, pp. 120–134, 2023.
https://doi.org/10.1007/978-3-031-16684-6_10

networks (WLANs), and cellular networks (GSM, LTE), etc. With a range of a few to tens of kilometers and battery life of ten years and beyond, LPWA technologies are promising for the Internet low-power, low-cost, and low throughput things. These features enable devices to spread and move over large geographical areas, thus IoT and M2M devices connected by LPWA technologies can be turned on anywhere and anytime to sense and interact with their environment [3].

IoT has become even more interesting as a result of the existence of LPWA networks. LPWA is a new communication paradigm that will work in tandem with traditional cellular and short-range wireless technologies to meet the different needs of IoT applications [2, 3]. The Internet of Things (IoT) has the potential to transform how we live and work. IoT has the potential to assist us in overcoming major global concerns such as energy crisis, resource depletion, and so on. To realize this vision, "things" must be able to perceive their surroundings and exchange that information with one another as well as with us in order to make intelligent decisions.

It's important to note that LPWA technologies accomplish long range and low power at the expense of low data throughput (usually in the tens of kilobits per second) and significant latency (typically in orders of seconds or minutes). As a result, it is evident that LPWA solutions are not intended to address all IoT use cases. LPWA technologies, in particular, are being evaluated for use cases that are delay tolerant, do not require high data speeds, and often require low power consumption and low cost.

Figure 1 depicts the wide range of applications that can use Low Power Wide Area technologies to connect their end devices across a variety of industries. Smart metering, smart grid management, smart buildings, smart agriculture, environmental monitoring, transportation and logistics, manufacturing process monitoring and control, and so on are among these sectors.

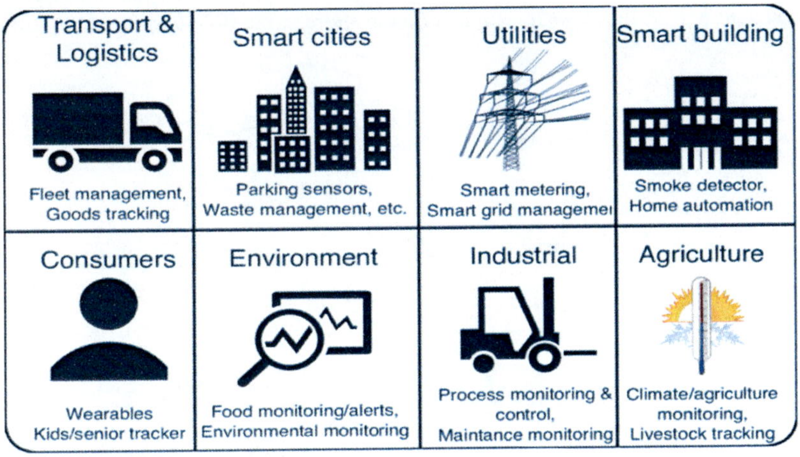

Fig. 1. Applications of LPWA technologies across different sectors

For LPWA connectivity, LoRa is a proprietary physical layer. A physical layer technology developed and marketed by Semtech Corporation that modulates signals in the

SUB-GHz ISM band using a patented spread spectrum approach. A unique chip spread spectrum technique, which spreads a narrow band input signal over a wider channel bandwidth, allows for bidirectional communication. The resulting signal resembles noise, making it more difficult to identify or jam [4].

The LoRa™ Alliance, a special interest group made up of commercial and industrial participants, proposed LoRaWAN, an open standard that defines architecture and layers above the LoRa physical layer. At the MAC layer, a basic ALOHA scheme is utilized, which, when combined with the LoRa physical layer, allows several devices to communicate simultaneously while using various channels and/or orthogonal codes (spreading factors). End devices can connect to any base station without incurring additional signaling costs. The network server, the primary unit of the LaRaWAN system that suppresses duplicate receptions, modifies radio access links, and transmits data to appropriate application servers, connects end devices through a back-haul. The data is then processed by application servers, which then conduct user-defined actions. To authenticate end devices with the network and protect the privacy of application data, the LoRaWAN standard employs symmetric-key cryptography [5, 6].

According to LoRaWAN, devices will have varying capabilities depending on application requirements. As a result, LoRaWAN divides end-devices into three categories (Fig. 2), each of which supports bidirectional communication but has differing downlink latency and power requirements [4]:

- *Class A* device achieves the longest lifetime but with the highest latency. It listens for a downlink communication only shortly after its uplink transmission.
- *Class B* device, in addition, can schedule downlink receptions from base station at certain time intervals. Thus, only at these agreed-on epochs, applications can send control messages to the end devices (for possibly performing an actuation function).
- *Class C* device is typically mains-powered, having capability to continuously listen and receive downlink transmissions with the shortest possible latency at any time.

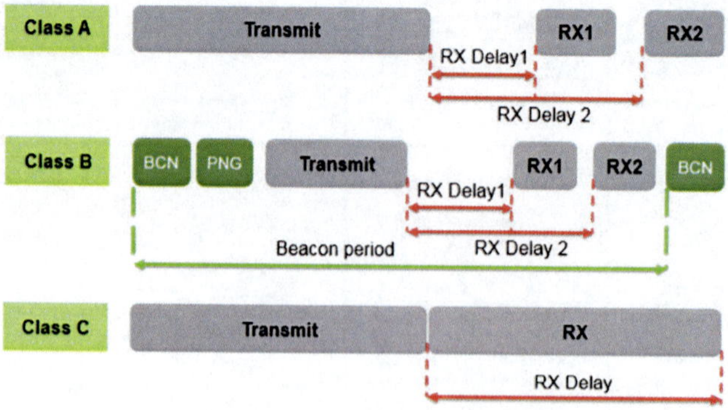

Fig. 2. End-device classes defined in LoRaWAN protocol

In recent years, Wireless Sensor Network (WSN) technologies have advanced rapidly. WSNs are used in a variety of applications, including medicine, agriculture, sports, and industry. Agriculture is one of the most advantageous facilities for WSNs to improve food crop yields and reduce farmer burden [7].

The benefits of smart agriculture include increased productivity, crop quality, efficient management and control of agriculture systems, and thus increased human income. IoTs comprised of wireless sensor network nodes are widely used for temperature, light, gas, smart agriculture, and smart cities applications [8]. WSNs are used in a variety of agricultural applications, including climate monitoring and forecasting crop health and product quality using soil nutrient data. WSNs can predict irrigation planning by monitoring weather conditions (such as temperature and humidity) and soil moisture. Other sensor nodes can be added to the existing WSN to improve the agricultural monitoring system's parameters and make the network more scalable. However, some obstacles have hampered the agricultural application of WSNs, including determining optimal deployment schemes, measurement periods, routing protocols, energy efficiency, cost, communication range, scalability, and fault tolerance [7, 8].

Many agricultural solutions are discussed in the literature, but the majority of these solutions are aimed at IoT device development for smart crop field monitoring system and automated irrigation system using wireless sensor networks [9], nutrient monitoring methods [10], low-cost remote monitoring and control network for household cultivation [11], a custom-built cost-effective device based on the MQTT IoT proto-col that sends data over network systems and gathers soil condition reports [12] or a low-cost, versatile, and robust data gathering device that can be deployed in a variety of operational scenarios [13].

There are a few other items that provide an examination of the demands in this sector, such as the necessity for monitoring equipment. All applicable approaches, from planting through harvesting, packing, transportation, and modern technologies available to farmers throughout the farming system are covered in [14]. Furthermore, this review article emphasizes the use of additional tools, such as unmanned aerial vehicles (UAVs), for crop monitoring and other beneficial actions, such as crop yield optimization. Furthermore, advanced IoT-based programs are discussed. Finally, based on our thorough assessment, we found advanced opportunities for IoT, which are critical instruments for sustainable agriculture. [15–17] provide an overview of the cutting-edge technology used in precision agriculture for soil evaluation and pollutant monitoring with respect to heavy metals in agricultural soil utilizing various sensors. Second, this publication shows how data collected by sensors is processed.

The energy crisis has gotten worse in the previous year, and its impacts are being seen in more and more places. Agriculture is no different. Heat is provided by electrical energy in many greenhouses. It is necessary to collect data as well as devise a strategy to reduce electrical energy usage without compromising product quality. This necessitates the use of devices, networks, and software applications.

The main contributions of this paper are summarized as follows:

- The potential of using long-range wireless communications protocols or technologies in agriculture was investigated, and the technology with the best power consumption and communication distance was identified.
- A low-cost device has been developed for monitoring the real-time electrical energy consumption in greenhouses, in order to obtain data for an efficiency plan.

- A LoRaWAN network has been developed for these monitoring devices.

Following the introduction presented within the 1st section, the 2nd one refers to the electrical quantities monitoring device prototype developing. The results and discussions are tackled within the 3th section. Finally, conclusions are synthesized within the 4th section.

2 Electrical Quantities Monitoring Device

The energy crisis in general, but the electrical energy crisis in particular, is also affecting farmers. In this context, it is important for them to have access to a monitoring device through which they can track some important parameters, in our case the electrical quantities. Based on these data they can impose measures to improve electrical energy consumption, thus, taking care of the environment but also of the value of the bills.

Our team proposed the construction of an electrical quantities monitoring device, with the following characteristics:

- *Low-cost* – the market offers such monitoring devices, but which for various reasons have quite high prices, and when they have to buy a few dozen pieces the budget is seriously affected. Which is why farmers give up buying them. They try different methods to solve the problem and end up seriously affecting the quality of their products. Farmers do not need approved/certified instruments, for them a device that shows the consumption trend is convenient.
- *Long-rage communication* – greenhouses, for example, are often far away from people's homes. In this case, if there is no Internet connection in the greenhouses, the data transmission from the devices to the application server and implicitly to the application, where people can see and analyze the data, is difficult. In these situations, LPWA technologies can help us. Specifically, we used LoRa technology and the LoRaWAN protocol to transmit data from devices.
- Low power - using LoRa it is our duty to make these devices with low power consumption. The prototype of our monitoring device is class C.

The parameters that can be tracked by the developed device are the following:

- voltage; electric current; active, reactive, apparent power; active, reactive energy; power factor.

For the data acquisition we used an 8-bit AVR RISC-based microcontroller running at 16 MHz, ATMEGA 16U2 Chipset - ATmega328P microcontroller (Fig. 3). This microcontroller has been connected to an inhouse developed board, but later we gave up this own developed board idea, realizing that the price of the device will increase considerably (also, we would have needed another board for data transmission). We turned our attention to Arduino, being a relatively cheap solution. We quickly found a board (OSEPP™ UNO R4 Plus – Fig. 4) with this type of microcontroller that also satisfies the rest of the project requirements.

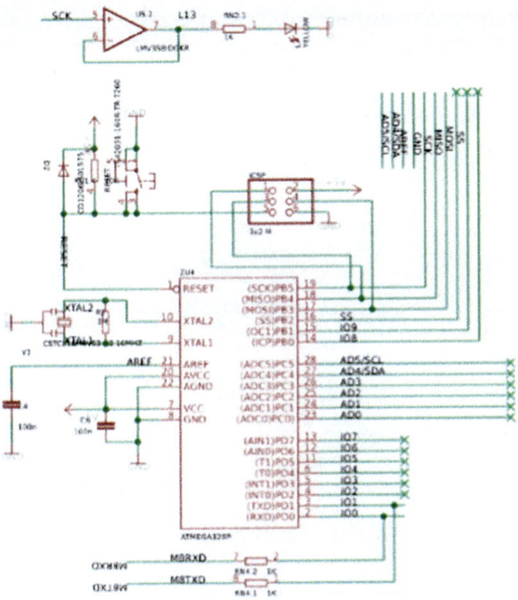

Fig. 3. ATmega328P microcontroller design

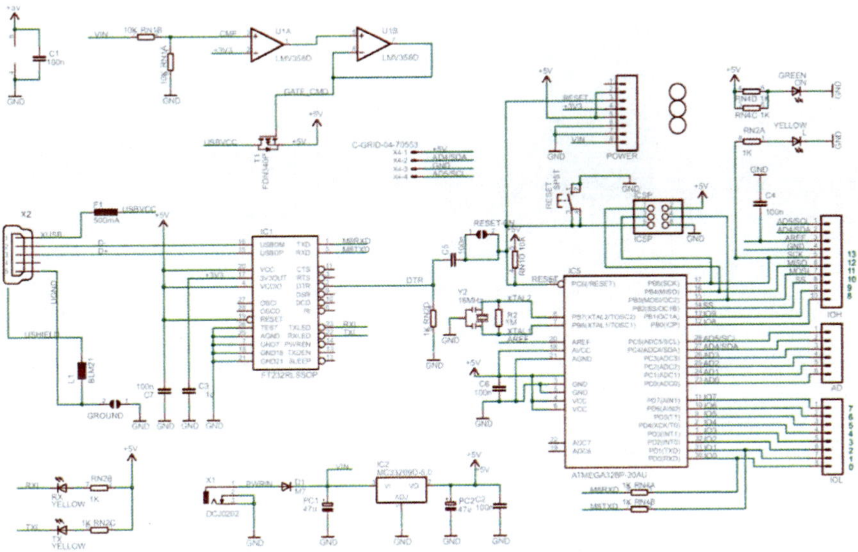

Fig. 4. OSEPP™ UNO R4 Plus design

The other components used for the monitoring device prototype development are presented below:

- ZMPT101B sensor (Fig. 5), for voltage monitoring.

Fig. 5. ZMPT101B sensor

This sensor connection to the board was made according to the diagram in the Fig. 6. The sensor is powered by breadboard because the Arduino has a single 5 V pin.

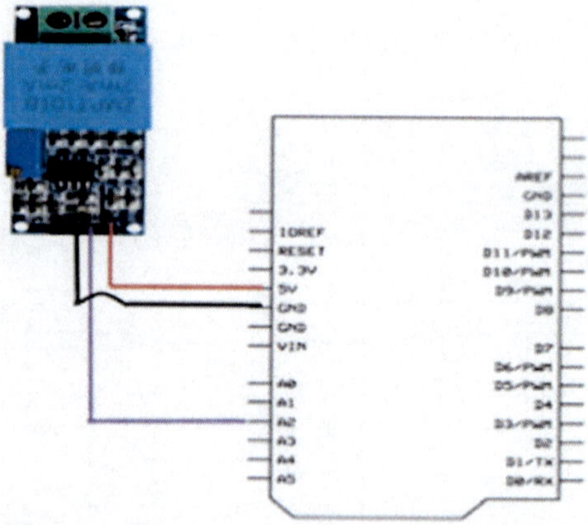

Fig. 6. ZMPT101B sensor connection diagram

- YHDC SCT013-030 sensor (Fig. 7), for electric current monitoring.

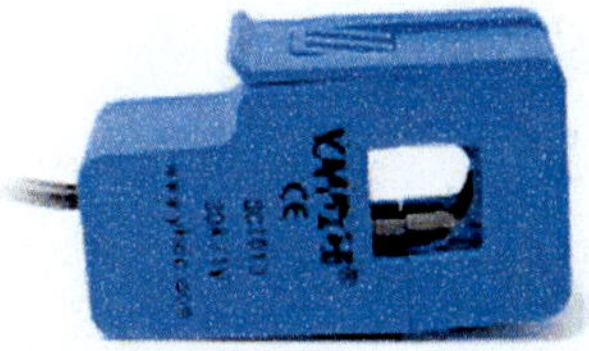

Fig. 7. YHDC SCT013-030 sensor

To connect this sensor to the Arduino it is necessary to add additional components for the construction of a voltage divider (Fig. 8).

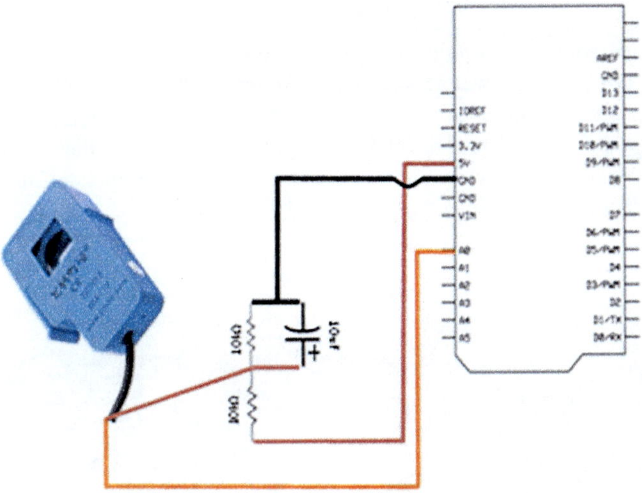

Fig. 8. YHDC SCT013-030 sensor connection diagram

- ESP8266 wireless module (Fig. 9) and Dragino LoRa de 868 MHz Shield

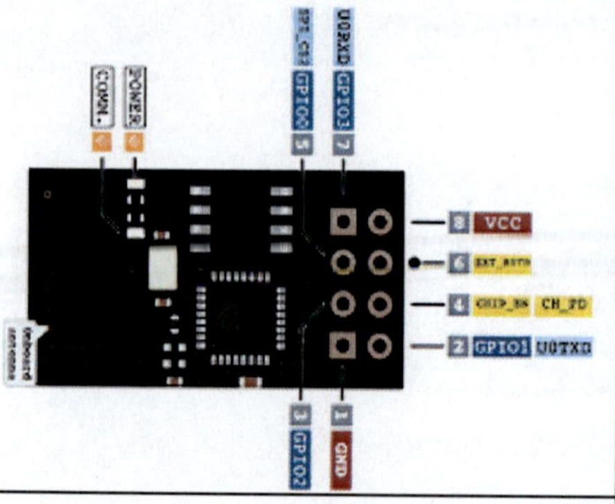	
ESP8266 wireless module	Dragino LoRa de 868 MHz Shield

Fig. 9. ESP8266 wireless module and Dragino LoRa de 868 MHz Shield

During operation this module may require a current of up to 380 mA which forces us to add an external 3.3 V power source (Fig. 10) because the 3.3 V pin from Arduino has a limitation of maximum 50 mA. The RX and TX pins of this module are also used at 3.3 V voltage and Arduino pins are at 5 V which forces us to use either a voltage divider or a 4-channel logic level converter (Fig. 11). Using this component, the connection diagram in Fig. 12 was made.

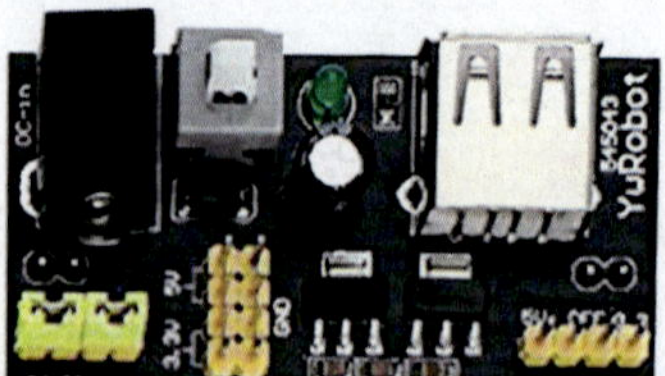

Fig. 10. Additional power source

Fig. 11. 4 channel logic level converter

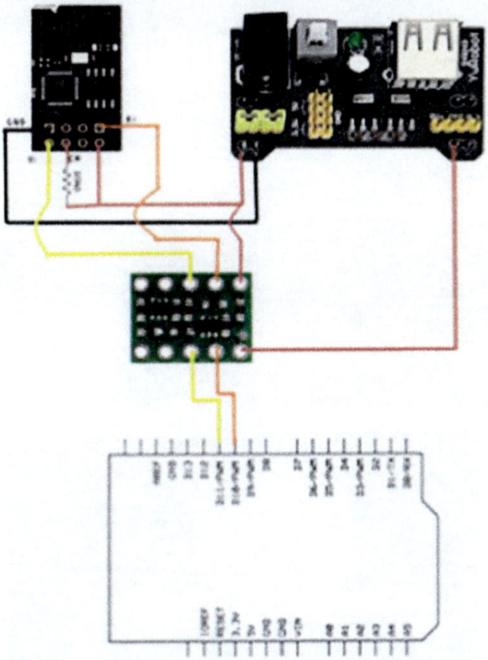

Fig. 12. ESP8266 wireless module connection diagram

3 Results

After small adjustments to the values of the resistors, capacitors for voltage dividers, for noise filtering, the prototype of the device has been finalized (Fig. 13).

The testing of the electrical quantities monitoring device was performed in real conditions (Fig. 14).

During the first test, the data was transmitted over a LoRa network using a Dragino LG01-N gateway. The LoRa network architecture is shown in the Fig. 15. As a network server we used the ThingSpeak free network server, from where we took the data through WiFi in an own developed web-based application.

The data are presented using graphs (Fig. 16, 17, 18 and 19), with the possibility of exporting data in excel and CSV formats.

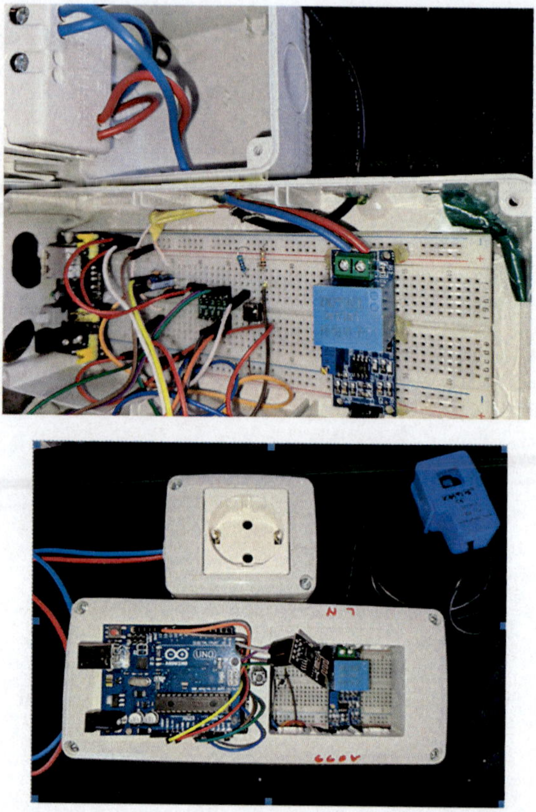

Fig. 13. Monitoring device prototype

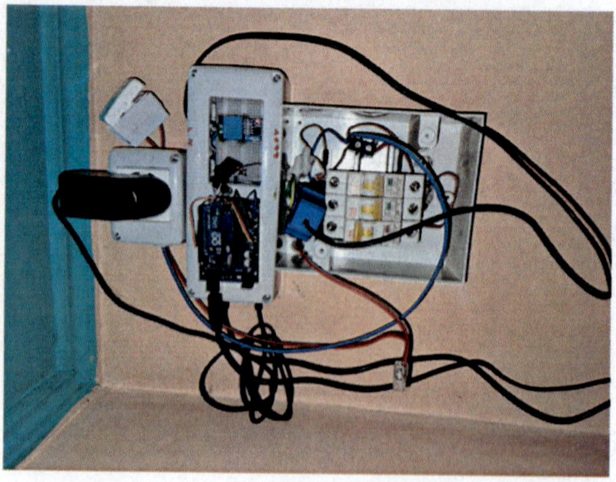

Fig. 14. Monitoring device testing in real conditions

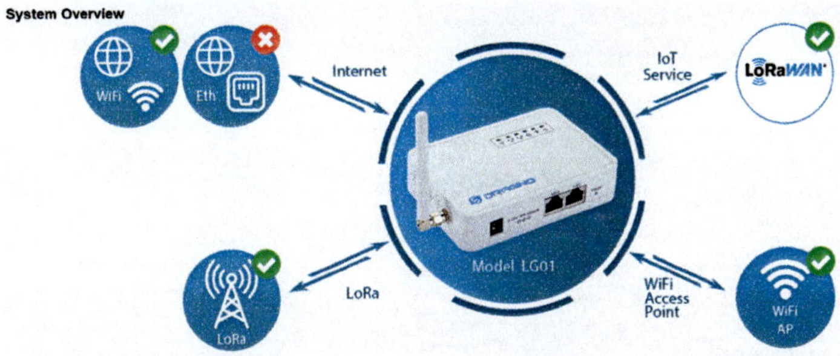

Fig. 15. LoRa network architecture using Dragino LG01-N gateway

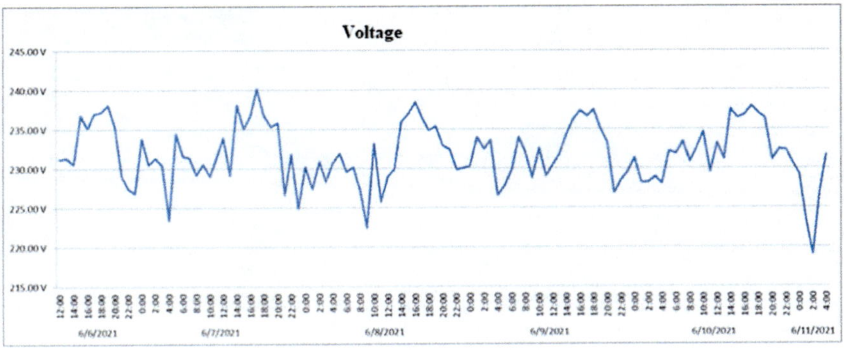

Fig. 16. Voltage monitoring

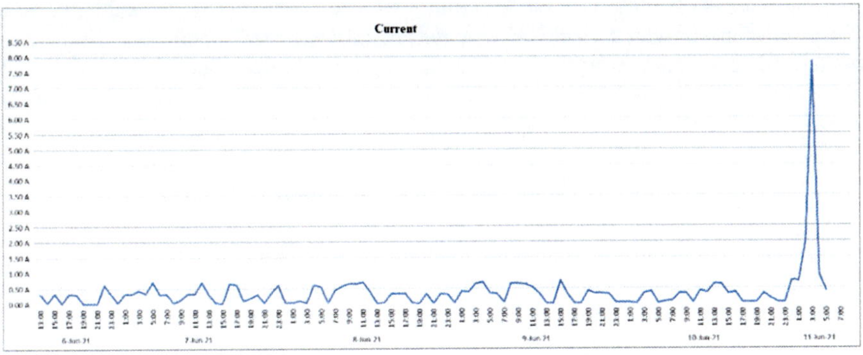

Fig. 17. Electric current monitoring

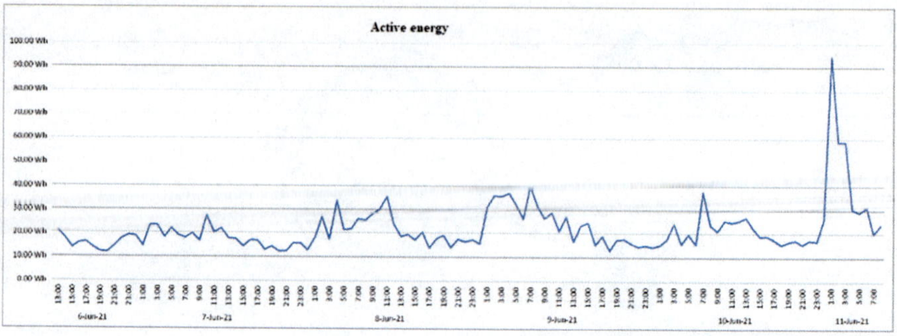

Fig. 18. Electrical active energy monitoring

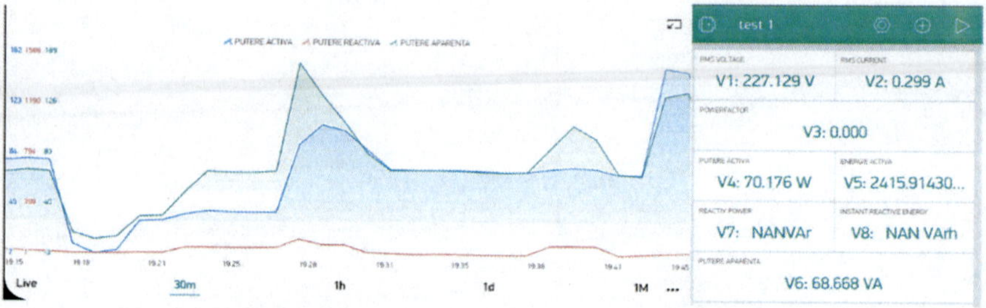

Fig. 19. Active, reactive, apparent power monitoring

4 Conclusions

The electrical energy crisis is at the beginning. We need to be more responsible for how we consume the provided electrical energy. For this we need monitoring devices, to obtain in the first stage information about each equipment how much it consumes and how. After obtaining the information we can make plans through which we can optimize these consumptions, so we don't waste energy anymore.

Through our project we focused on developing an electrical quantity monitoring device for agriculture that would meet some specific requirements, such as:

- Low-cost – Farmers do not need approved/certified instruments, for them a device that shows the consumption trend is convenient. In many cases, they can't afford to buy dozens of expensive devices on the market.
- Long-rage communication – greenhouses are often far away from people's homes. There are many situations in which data must be transmitted over long distances. LPWA technologies are very helpful here – LoRa in our case.
- Low power - using LoRa it is our duty to make these devices with low power consumption. The prototype of our monitoring device is class C.

The results of the measurements are good, they were compared with measurements made with certified devices. The developed device responds to the needs for which it was designed, it can be a real help for optimizing the electrical energy consumption for low voltage (230 V) household consumers.

Given the field of use (agriculture), in future work we are considering the extension of the device for monitoring the environmental parameters (temperature, humidity, gas concentrations, etc.).

Acknowledgment. The research has been funded by the University of Oradea, within the Grants Competition "Scientific Research of Excellence Related to Priority Areas with Capitalization through Technology Transfer: INO - TRANSFER – UO", Project No. 323/2021.

References

1. Cano, J.C., Berrios, V., Garcia, B., Toh, C.K.: Evolution of IoT: an industry perspective. IEEE Internet Things Mag. **1**(2), 12–17 (2018)
2. Raza, U., Kulkarni, P., Sooriyabandara, M.: Low power wide area networks: an overview. IEEE Commun. Surv. Tutor. **19**, 855–873 (2017)
3. Alagarsamy, G., Shanthini, J., Naveen Balaji, G.: A survey on technologies and challenges of LPWA for narrowband IoT. In: Al-Turjman, F. (ed.) Trends in Cloud-based IoT. EICC, pp. 73–84. Springer, Cham (2020). https://doi.org/10.1007/978-3-030-40037-8_5
4. Centelles, R.P., Freitag, F., Meseguer, R., Navarro, L.: Beyond the star of stars: an introduction to multihop and mesh for LoRa and LoRaWAN. IEEE Pervasive Comput. **20**(2), 63–72 (2021). https://doi.org/10.1109/MPRV.2021.3063443
5. Ntseane, L., Isong, B.: Analysis of LoRa/LoRaWAN challenges: review. In: 2019 International Multidisciplinary Information Technology and Engineering Conference (IMITEC), pp. 1–7 (2019). https://doi.org/10.1109/IMITEC45504.2019.9015864
6. Gkotsiopoulos, P., Zorbas, D., Douligeris, C.: Performance determinants in LoRa networks: a literature review. IEEE Commun. Surv. Tutor. **23**(3), 1721–1758 (2021). https://doi.org/10.1109/COMST.2021.3090409
7. Jawad, H.M., Nordin, R., Gharghan, S.K., Jawad, A.M., Ismail, M.: Energy-efficient wireless sensor networks for precision agriculture: a review. Sensors **17**, 1781 (2017). https://doi.org/10.3390/s17081781
8. Himanshu, S., Ahteshamul, H., Zainul, A.J.: Maximization of wireless sensor network lifetime using solar energy harvesting for smart agriculture monitoring. Ad Hoc Netw. **94**, 101966 (2019). https://doi.org/10.1016/j.adhoc.2019.101966, ISSN 1570-8705
9. Pokala, S.S.K., Bini, A.A.: A low cost IoT enabled device for monitoring agriculture field and smart irrigation system. In: Ranganathan, G., Chen, J., Rocha, Á. (eds.) Inventive Communication and Computational Technologies. LNNS, vol. 145, pp. 923–932. Springer, Singapore (2021). https://doi.org/10.1007/978-981-15-7345-3_78
10. Kashyap, B., Kumar, R.: Sensing methodologies in agriculture for soil moisture and nutrient monitoring. IEEE Access **9**, 14095–14121 (2021). https://doi.org/10.1109/ACCESS.2021.3052478
11. Tran, N.K.H., Nguyen, C.Q., Nguyen, P.T., Lương, V.Q.D.: A reconfigurable IoT-based monitoring and control system for small-scale agriculture. In: IOP Conference Series: Materials Science and Engineering, Volume 1088, Annual Conference on Computer Science and Engineering Technology (AC2SET), 23rd September 2020, Medan, Indonesia (2021)
12. Mandal, S., Ali, I., Saha, S.: IoT in agriculture: smart farming using MQTT protocol through cost-effective heterogeneous sensors. In: Bhattacharjee, D., Kole, D.K., Dey, N., Basu, S., Plewczynski, D. (eds.) Proceedings of International Conference on Frontiers in Computing

and Systems. AISC, vol. 1255, pp. 903–913. Springer, Singapore (2021). https://doi.org/10.1007/978-981-15-7834-2_85
13. Morais, R., Mendes, J., Silva, R., Silva, N., Sousa, J.J., Peres, E.: A versatile, low-power and low-cost IoT device for field data gathering in precision agriculture practices. Agriculture **11**, 619 (2021). https://doi.org/10.3390/agriculture11070619
14. Khan, N., Ray, R.L., Sargani, G.R., Ihtisham, M., Khayyam, M., Ismail, S.: Current progress and future prospects of agriculture technology: gateway to sustainable agriculture. Sustainability **13**, 4883 (2021). https://doi.org/10.3390/su13094883
15. Akhtar, M.N., Shaikh, A.J., Khan, A., Awais, H., Bakar, E.A., Othman, A.R.: Smart sensing with edge computing in precision agriculture for soil assessment and heavy metal monitoring: a review. Agriculture **11**, 475 (2021). https://doi.org/10.3390/agriculture11060475
16. Avram, C., Gligor, A., Avram, E.L.: New approach on optimal decision making based on formal automata models. Procedia Econ. Financ. **3**, 852–857 (2012). https://doi.org/10.1016/S2212-5671(12)00240-7
17. Sabri, Y., Siham, A.: Cloud computing in remote sensing: high performance remote sensing data processing in a big data environment. Int. J. Comput. Commun. Control **16**(6), 4236 (2021). ISSN 1841-9844

Digitization of the Characteristic Parameters of a Greenhouse in Order to Streamline Energy Consumption

Simona Dzitac[1], Adrian Cheregi[1](✉), Attila Simo[2], and Gabriela Elena Badea[1]

[1] University of Oradea, 410087 Oradea, Romania
sdzitac@uoradea.ro, andy.andain@gmail.com
[2] Politehnica University Timişoara, 300223 Timişoara, Romania
attila.simo@upt.ro

Abstract. Smart agriculture represents a management concept focused on industrial agriculture that already has the infrastructure to use advanced technologies like big data, cloud and Internet of Things (IoT) in order to supervise, automatize and analyze operations. It is known as well as precise agriculture, smart agriculture is driven by software and supervised by sensors. The importance of smart agriculture is constantly growing as it needs to use natural resources as efficiently as possible. In present paper we will consider supervising and controlling a greenhouse through sensors and software for its energy efficiency. In a first phase, the architecture of the sensor network and of the possible equipment that will control the temperature, the humidity of the greenhouse in order to make the production as efficient and the consumption of energy are established. Based on some measurements, solutions will be found to reduce the consumption of electrical energy. The conclusions present the advantages of smart agriculture, the stages of energy efficiency of a greenhouse by adding sensors and software control.

Keywords: Smart agriculture · Energy efficiency · Internet of Things

1 Introduction

Currently, technology surrounds us in many ways, mainly used to help humans to reduce labor or to optimize it. In agriculture, technology makes its way in order to increase the quality and the quantity of the crops. Since it is a fairly new term, smart agriculture [1] is still not familiar with the majority of farmers, its main reference being the usage of technologies like Internet of Things, robots, sensors, location systems and artificial intelligence on the farm [2]. Examples of such technologies used in farms are precision irrigation [3] and precise plant nutrition [4], climate management and control in greenhouses [5–7], in which this paper will cover more with sensors for soil, water, light, humidity and temperature management. All these work together with specialized software in order to supervise and control the given farm or greenhouse to bring the outmost of its crops.

S. Dzitac et al. (Eds.): ICCCC 2022, AISC 1435, pp. 135–143, 2023.
https://doi.org/10.1007/978-3-031-16684-6_11

All these technologies connected together form the Internet of Things, which is a mechanism that connects all sensors and machines resulting in a complex system that can manage a farm or greenhouse based on the data it receives. This type of system can help farmers supervise the processes on their farm and take strategic decisions remotely, either from their phone, tablet or any other mobile device.

The role of Internet-of-Things (IoT) in precision agriculture and smart greenhouses has been reinforced by recent R&D projects [8]. The utilization of IoT technologies in smart greenhouses often involves a tradeoff between the cost of agricultural production, environmental conservation, ecological degradation, and sustainability. The installation of IoT infrastructure is capital-intensive and often translates to higher energy demand that elevates the risk for climate change. The integration of IoT systems in greenhouses would be augmented by the global deployment of advanced 5G technology and Low-Earth Orbit (LEO) constellation broadband internet with low latency and high speeds. Intelligent application of agrochemicals could yield significant savings ($500/acre or more), while need-based irrigation and fertilizer application would help improve crop yields. Globally, the deployment of IoT infrastructure would yield about $500 billion of added value to the GDP by 2030. The forecasted economic benefits affirm that the applications of IoT for optimized greenhouse environment and resources management were sustainable, and any potential risks are incomparable to the long-term benefits in commercial agriculture. The review article contributes new insights on the role of IoT in agriculture 4.0, the challenges, and future prospects for developing nations, which lacked the resources to invest in precision agriculture technologies.

In [9] is presented an overview of a large knowledge based operations optimization system that is partially in operation and in part under development for implementing deep knowledge schemes as well as adapting protocols and hardware to comply with the expected wave on IOT. With the development of sensing, wireless communication, and Internet technologies, we are now living in a world that is filled with various smart things the Internet of Things. This paper introduces and prospects an emerging research area the Embedded Intelligence (EI) for agricultural operations.

The processes that are taking place on a farm using smart agriculture are data collection, diagnostics, decision making and actions. Data collection refers to the sensors installed at specific locations in the farm or greenhouse that transmit data about soil, air, light, etc. After this data is collected, it will be analyzed by the system used and conclusions will be made based on the status of the object or supervised process in which potential problems could get identified. If a problem gets identified in the previous step, either the software or a human decides what kind of actions should be taken. As the action rolls out, either by human intervention or by software, a new set of measurements are made by the sensors as the cycle starts again.

Precision Agriculture (PA) can be improved by the provision of adequate nutrients and effective control of pests and diseases. Effective analysis and utilization of IoT technologies can help in PA. IoT stipulates connectivity technology between multiple devices but limited battery energy remains as a foremost concern. This battery constraint needs a combined approach of energy-efficient and energy harvesting mechanisms for long term monitoring and data collection. [10] Outlines research questions for energy efficient wireless communication on the basis of literature evaluation. Potential research design and approach for power reduction are explored using our pilot LoRa based IoT deployment in a greenhouse. This pilot study helps in identifying challenges and limitations of IoT connectivity in greenhouse. We also discuss the possible research outlook to increase the opportunities in saving energy and efficient wireless data communication. In [11] the authors propose a cloud solution to build an Internet of Things (IoT) platform applied in a greenhouse crop production context. Real-time and historical data, as well as prediction models, can be accessed by means of representational state transfer (RESTful) Web services developed for such a purpose. Forecasting is also provided following a Greenhouse Models as a Service (GMaaS) approach. Currently, our GMaaS tool provides forecasting based on computational models developed for inside climate, crop production, and irrigation processes. Traditionally, such models are hardcoded in applications or embedded in software tools to be used as decision support systems (DSSs).

Internet of things (IoT) technology has been constantly applied in greenhouse environmental monitoring and control in recent years. The acquisition and control parameters, network protocols are different for the various purposes of the greenhouse. These factors are the keys to the abilities to effectively communicate and transfer meaningful data in IoT infrastructures. To achieve the adaptive matching of data communication between the gateway and server in a greenhouse IoT system, a data encapsulation method based on XML was designed to enable data interoperability in a distributed greenhouse IoT system. In [12] the data communication mechanism for real-time and cumulative data synchronization between the gateway and server based on JADE was tested in a specific greenhouse. The results showed that the data loss rate between the data acquisition unit and the gateway was 1.52%, and the data loss rate was 0.4% between the gateway and the server; therefore, the mechanism could be feasibly applied to the data communication for a greenhouse IoT system.

In [13] the authors propose developing a system optimally watering agricultural crops based on a wireless sensor network. This work aimed to design and develop a control system using node sensors in the crop field with data management via smartphone and a web application. The system can send notifications through LINE API for the LINE application. The developed system was implemented and tested in Makhamtia District, Suratthani Province, Thailand. The results showed the implementation to be useful in agriculture. The moisture content of the soil was maintained appropriately for vegetable growth, reducing costs and increasing agricultural productivity. Moreover, this work represents driving agriculture through digital innovation.

The main role of smart agriculture in this automated smart farming process is its high precision and 24/7 control, in order to realize savings of key resources used like water, energy, fertilizers, time spent by human resources.

2 Digitization of the Characteristic Parameters of a Greenhouse - Data Collection

Sensors are the main source for collecting data about the environment of the greenhouse, factors like the placement of the greenhouse can greatly affect the precision of measurements and surrounding buildings or other greenhouses. The measurements are of a great importance, where real opportunities exist when these measurements are applied to our work. These measurements can be classified in three types of data: instantaneous, historical and emergency data.

The first type of data, instantaneous, consists of continuous data transmission on the current status of the greenhouse like, for example, the current temperature and humidity level. Based on the data the user receives from the greenhouse using a phone or tablet, the temperature and humidity can be adjusted either by software intervention, either by its user by activating heating or cooling of the greenhouse. This type of data can be reassuring when it shows that all the systems are working properly, its value being increased by accessing it remotely.

The second type of data, historical, is a collection of instantaneous data over a period of time. This data can be broken down into recent history and year-to-year. Using the historical data by year-to-year, its user can make decisions on making improvements on the greenhouse by using more energy efficient heaters and coolers and possibly preventing failures over time. By analyzing the measurements of temperature over a long period of time, the farmer can adjust in the future the temperature based on outside climate and increase the quality and quantity of crops. Such data can be used in the long-range planning decisions, for example, negotiating fuel contractors or deciding whether heating and cooling equipment needs to be replaced.

The last type of data, emergency, is the one that requires immediate response, for example, communication loss, power outages, low or high temperature alarms. These will notice by either text message or push notification the farmer based on its severity. An emergency data can have two or more types of severity, a warning that could be represented by a yellow color, for example, temperature spikes, increase of the humidity in the greenhouse or communication loss. A more severe warning could have a red color, for example, power outage, unresponsive heating or cooling equipment. As Internet of Things advances day by day, data collected by sensors to the cloud is becoming more often used, data which is analyzed to inform farmers of environmental trends, preventive maintenance and energy savings.

Sensors covered by this paper are DHT21/AM2301 for temperature, HR202 for humidity and light sensor UV ML8511. These sensors are placed at various places in a greenhouse for measurements and data collection. Analyzing the data collected we are conducting in to determine energy consumption for heating, cooling and appliances used at a greenhouse.

3 Streamline Energy Consumption

An agreement has been made with a greenhouse owner in order to conduct measurements and to streamline energy consumption by installing the sensors with the given software in various places of the greenhouse. In the greenhouse there are various types of flowers that require controlled temperature and humidity. Currently the greenhouse is heated using conventional fuel and is cooled using two fans of 80 W each, these fans also circulate the heated air during spring in cold days. The greenhouse dimensions are 75 m by 30 m, covered by two polyethylene film plastic sheet, with space between them in order to prevent condensation and to keep warmth in the cold season. The greenhouse can be heated using three types of fuel, natural gas which is expensive nowadays, diesel which pollutes the environment and pellets which is cheaper and better than the first two. The boilers and the fans could be powered by solar panels to reduce energy consumption, even if not fully but enough to streamline it and reduce costs overall.

In order to streamline energy consumption, a 3D model of the greenhouse is created to mark the positioning of the sensors to collect data. At first, we are conducting a set of measurements and based on the measurements we integrate the software from the sensors to automate the fans and solar panels for its power source. In Fig. 1 we present the 3D model of the greenhouse and the sensors placements. The first two sensors are be placed at the two entrance pillars of the greenhouse and at the end of the greenhouse, one of the sensors being placed in the middle of it. These sensors are mainly measuring temperature and humidity, in a set period of time, with manually controlling the fans. The flowers in the greenhouse have a recommended temperature of 14 °C, the owner of the greenhouse keeping it normally at 12 °C in the cold season, during summer, the fans are usually kept on for good air circulation and to prevent rise in temperature. The recommended humidity is between 50% and 70% for normal growing of the flowers, while being below 50% is considered too try and above 80% too humid.

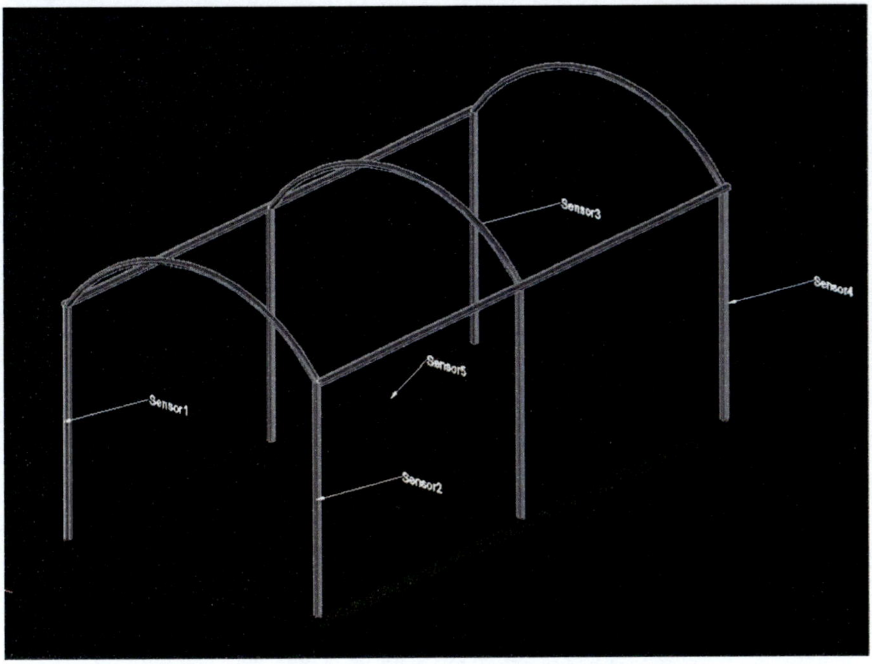

Fig. 1. 3D model greenhouse and the placement of the sensors

After the sensors are placed in the greenhouse we will conduct the first set of measurements. Over the course of a week, we are monitoring the temperature and humidity during a 24 h cycle, the dates of measurements are shown in the graph below (Figs. 2 and 3).

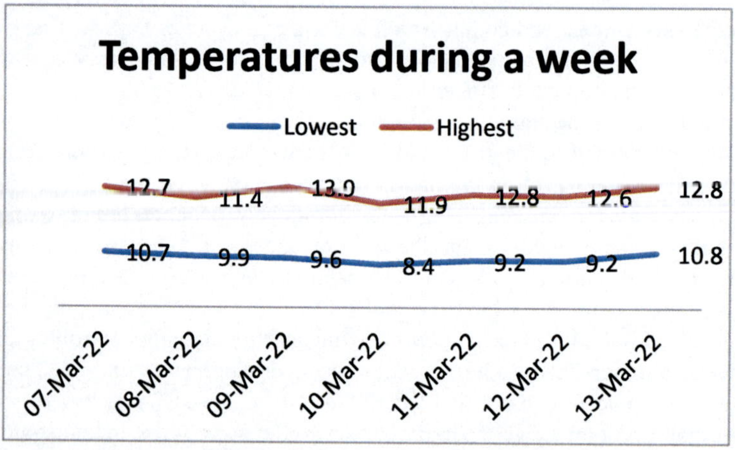

Fig. 2. Measured temperatures during a week in the greenhouse with heaters and fans on

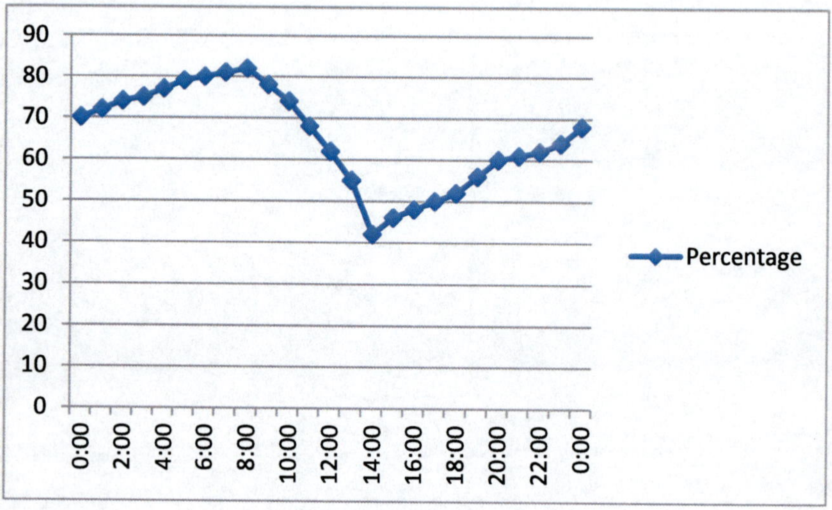

Fig. 3. Measured humidity during a day

As shown in the graphs, during the start of the day the temperature is low while during the day it can get warmer. Humidity drops as the temperature rises, fans only being used in the morning and night to circulate the heated air. The humidity is kept at around 60% to 70% by sprinkling the flowers. The measurements are made using a Raspberry Pi 4 [14] using libraries for the DHT21/AM2301 [15] and HR202 [16], while for the final implementation there will be mounted ready to use equipment so that the owner can use and replace easily without needing to change code or to remember which wires to connect.

Installing eighteen photovoltaic panels on the roof of the nearest building to the greenhouse, with a total of 8 kW total at its peak, a panel outputting 445 W each, during a sunny day, with inverter and batteries placed in a safe place in the building, it can power the boilers, fans, sprinklers and appliances of greenhouse. Automation is monitoring the power produced and if it exceeds it will inject in the grid system, as well it is powering on the fans based on the temperature and humidity measured. In the table below we present a list of the appliances used in the greenhouse with the power consumption of each including the fans and boilers (Table 1).

Table 1. Overall power consumption of the appliances in the greenhouse

No.	Appliance	Pieces	Power consumption	Total
1	Lights (LED projectors)	21	30 W	630 W
2	Boilers	3	300 W	900 W
3	Fans	2	80 W	160 W
4	Pumps	2	80 W	160 W
Total				**1850 W**

As shown in the table above, the greenhouse does not use much power overall, only when all of the appliances are active it can reach the total power consumption. In the graphs below, we present the impact that automation has on the greenhouse and how much energy has been streamlined during a day (Fig. 4).

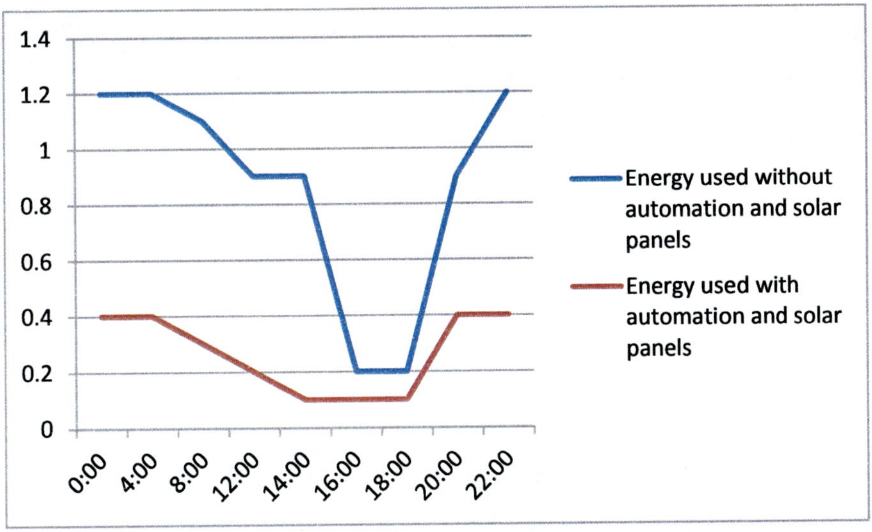

Fig. 4. Streamlined energy consumption during a day

Since the solar panels also power appliances inside the building, on a sunny day it can power both the appliances in the building and the greenhouse with little intake from the grid system.

4 Conclusions

The processes that are taking place on a farm using smart agriculture are data collection, diagnostics, decision making and actions. Data collection refers to the sensors installed at specific locations in the farm or greenhouse that transmit data about soil, air, light, etc. After this data is collected, it will be analyzed by the system used and conclusions will be made based on the status of the object or supervised process in which potential problems could get identified. If a problem gets identified in the previous step, either the software or a human decides what kind of actions should be taken. As the action rolls out, either by human intervention or by software, a new set of measurements are made by the sensors as the cycle starts again. All these technologies connected together form the Internet of Things, which is a mechanism that connects all sensors and machines resulting in a complex system that can manage a farm or greenhouse based on the data it receives. This type of system can help farmers supervise the processes on their farm and take strategic decisions remotely, either from their phone, tablet or any other mobile device.

IoTs was applied in agriculture to improve quality and reduce costs. For these reasons, we proposed these wireless sensor network through which we digitize the data from our greenhouse, using this data for command and control of other systems. We designed and implemented a system to control environmental factors in the crop fields.

To make energy consumption more efficient, we also introduce in the greenhouse electrical system producing energy through photovoltaic panels. Considerable improvement has been observed.

Acknowledgement. The research has been funded by the University of Oradea within the Grants Competition "Scientific Research of Excellence Related to Priority Areas with Capitalization through Technology Transfer: INO – TRANSFER – UO", Project no. 323/2021.

References

1. Sushanth, G., Sujatha, S.: IOT based smart agriculture system. In: 2018 International Conference on Wireless Communications, Signal Processing and Networking (WiSPNET), pp. 1–4 (2018). https://doi.org/10.1109/WiSPNET.2018.8538702
2. Naresh, M., Munaswamy, P.: Smart agriculture system using IoT technology. Int. J. Recent Technol. Eng. (IJRTE) **7**(5), 98–102 (2019). ISSN: 2277-3878
3. Abioye, E.A., et al.: Precision irrigation management using machine learning and digital farming solutions. AgriEngineering **4**, 70–103 (2022). https://doi.org/10.3390/agriengineering4010006
4. Sinha, B.B., Dhanalakshmi, R.: Recent advancements and challenges of Internet of Things in smart agriculture: a survey. Futur. Gener. Comput. Syst. **126**, 169–184 (2022)

5. Ullah, I., Fayaz, M., Aman, M., Kim, D.: An optimization scheme for IoT based smart greenhouse climate control with efficient energy consumption. Computing **104**, 1–25 (2021). https://doi.org/10.1007/s00607-021-00963-5
6. Chen, W.-H., You, F.: Semiclosed greenhouse climate control under uncertainty via machine learning and data-driven robust model predictive control. IEEE Trans. Control Syst. Technol. **30**(3), 1186–1197 (2022). https://doi.org/10.1109/TCST.2021.3094999
7. Malika, A., Faouzi, D., Benyoucef, K., Foudil, B., Mohammed Salim, H.: Comparative analysis of the management of the results of the modeling and the simulation of the evaluation of the thermal energy of the greenhouse by a fuzzy logic controller between a wet region and an arid region. Adv. Distrib. Comput. Artif. Intell. J. **10**, 77–97 (2021)
8. Maraveas, C., Piromalis, D., Arvanitis, K.G., Bartzanas, T., Loukatos, D.: Applications of IoT for optimized greenhouse environment and resources management. Comput. Electron. Agric. **198**, 106993 (2022). https://doi.org/10.1016/j.compag.2022.106993. ISSN: 0168-1699
9. Li, L., Wang, M., Zhang, M., Li, M., Sigrimis, N., Anastasiou, A.: Application of IoT technology in greenhouse management. 2012 Dallas, Texas, July 29 - August 1, 2012. 121345174. https://doi.org/10.13031/2013.42216
10. Singh, R.K., Berkvens, R., Weyn, M.: Energy efficient wireless communication for IoT enabled greenhouses. In: 2020 International Conference on COMmunication Systems & NETworkS (COMSNETS), pp. 885–887 (2020). https://doi.org/10.1109/COMSNETS48256.2020.9027392
11. Muñoz, M., Guzmán, J.L., Sánchez-Molina, J.A., Rodríguez, F., Torres, M., Berenguel, M.: A new IoT-based platform for greenhouse crop production. IEEE Internet Things J. **9**(9), 6325–6334 (2022). https://doi.org/10.1109/JIOT.2020.2996081
12. Wang, J., Chen, M., Zhou, J., Li, P.: Data communication mechanism for greenhouse environment monitoring and control: an agent-based IoT system. Inf. Process. Agric. **7**(3), 444–455 (2020). https://doi.org/10.1016/j.inpa.2019.11.002. ISSN: 2214-3173
13. Muangprathub, J., Boonnam, N., Kajornkasirat, S., Lekbangpong, N., Wanichsombat, A., Nillaor, P.: IoT and agriculture data analysis for smart farm. Comput. Electron. Agric. **156**, 467–474 (2019). https://doi.org/10.1016/j.compag.2018.12.011. ISSN: 0168-1699
14. Raspberry Pi 4, model B. https://www.raspberrypi.com/products/raspberry-pi-4-model-b/
15. Temperature and humidity sensor. https://hobbycomponents.com/sensors/840-dht21-am2301-temperature-humidity-sensor
16. Humidity sensor module. https://www.sunrom.com/p/humidity-sensor-module-hr202

Planning the Electricity Transactions Based on AI

Victorita Radulescu and Bogdan Alexandru Radulescu(✉)

University Politehnica of Bucharest, Splaiul Independentei 313, 060042 Bucharest, Romania
bogdan.radulescu97@gmail.com

Abstract. In Romania, the electricity market in 2018 and the gas supply market in 2021 have been liberalized. We are currently facing a global energy crisis with an unprecedented rise in the price of oil, electricity, and gas. The correct substantiation of the portfolio framework contracts concluded between producers, suppliers, and energy consumers is an essential condition from a financial point of view regarding transactions. They must provide a benefit to the producers, within reasonable limits, so that the consumer price is not artificially increased. This process becomes optimal if a forecast based on artificial intelligence-AI is used, assuming the recently recorded consumption is mediated annually. In the case of the EU countries, it must reflect both the domestic consumption and international trade to be achieved. The forecast must be made in the form of load curves with hourly levels for each day of the selected period. The adopted solution uses a hybrid model, optimized statistically, with neuronal networks, built based on selected preliminary data. The modeling used in the daily task schedules is presented, using self-organizing processes with Kohonen type neural networks. AI and Fuzzy logic is used in estimating energy consumption for a different scenario of average daily and monthly consumption. The results obtained by forecasting are compared with the values recorded in the months selected as a model. The errors obtained are in the range of 2.8–5.1%, validating the elaborated model.

Keywords: Artificial intelligence · Energy forecasting · Neural networks

1 Introduction

In the current conditions regarding the energy crisis and the price liberalization for electricity and gas in Romania, the trading system on the energy market represents one of the main forms of pricing in the wholesale market. The National Energy Regulatory Authority - ANRE, for more efficient control of the transactions on the energy market, has implemented differentiated framework contracts for the redistribution of energy and the establishment of different tariffs by types of consumers, [1]. These come as a complement to the portfolio framework contracts and allow energy suppliers to purchase additional quantities of electricity from different producers. The purchase is made at the price of the framework contract, but not exceeding the limit of 20% of the quantity specified in the contract. In this way, companies that supply at a better price can conclude new delivery contracts with other new, interested consumers. The redistribution market also

S. Dzitac et al. (Eds.): ICCCC 2022, AISC 1435, pp. 144–155, 2023.
https://doi.org/10.1007/978-3-031-16684-6_12

provides the possibility to organize transactions between suppliers, at prices agreed by both parties, to redistribute the energy left unconsumed by a supplier to those who have higher consumption, [2].

In recent decades, global warming has become a major issue, [3]. The use of renewable sources in new power plants instead of thermoelectric power plants becomes a major goal for the whole world. In many areas where new power plants have been built, the problem of redistribution systems in existing grids arises. Accelerating the transition to a low-carbon economy is essential and urgent, so the energy sector has become a global challenge. The use of artificial intelligence AI-based methods for planning energy transactions is proving to be useful, especially in the case of the small or medium energy market, such as Romania. A set of 9 principles related to AI to accelerate the energy transition from solid fuels to renewable sources is widely implemented, covering all areas: Governance, Planning, Enabling, Sustainability, Automation, Design, Incentives, Data, and Education. The first three principles relate to the policy of each state, the next three to the design and execution part, and the last three to the possibility of their implementation, [4, 5].

To fulfill the commitments made in the 2015 Paris Agreement, the transition needs to be accelerated and thus the energy sector increasingly digitized. International energy systems are transforming, needing the potential of AI to accelerate this energetic transition, [6, 7]. Decarburization of the energy sector must be the starting point of the whole system, [8]. As the number of electric vehicles-EVs increases and the cost of storage batteries decreases, it is estimated that the share of global demand required will increase by 60% from 2019 to 2050. If worldwide in 2020, about 12 million electric vehicles (including batteries, plug-ins, and hybrids) were recorded, 1.5 TW produced from renewable, and about 11 GW stored in batteries, the estimates for 2050 are that there will be 836 million electric vehicles, from renewable energy produced about 12 TW, and the storage in batteries will be about 1.3 TW, [9].

In this context, the correct basis for the portfolio of framework contracts is based on a medium-term forecast (one year) of electricity consumption. It is a fundamental requirement for the financial activity of electricity suppliers at the national level but is also necessary for cross-border energy exchanges.

This paper presents an original method of constructing predictions using mathematical statistics combined with AI models, based on artificial neural networks with Kohenen-type self-organization.

2 Analysis and Selection of Primary Data for Forecasting

The analysis of the primary data for the realization of the forecast is made starting from the history of electricity consumption in the last period, based on the realized and estimated production and the achievement of a necessary correlation. The basic data contains information about the amount of electricity produced from various sources, hydro, wind, coal, or solar, and the amount of energy required for internal consumption. Consumption graphs, respectively daily curves are made on hourly levels, [10]. They are necessary to determine the total detailed consumption by supply voltages and consumption categories. The database is created for consumptions recorded in 2019 and 2020, aiming to make a forecast for 2021. It was compared with the consumption recorded last year. In this way may be tested the quality of the model, and its efficiency. For example, Fig. 1 shows the hourly graphs for two days, January 9 and 10 in 2019 and 2020. The significance of the colors is blue-hydro, green-wind, red-nuclear, yellow-hydrocarbons, light brown-biomass, dark brown-solar, and black-coil. They were considered significant, being days in the middle of the week.

There is a difference between the consumptions recorded in the same period but in different years, due to the Covid-19 pandemic, when many companies, for example, hotel and restaurant networks, were closed for certain periods. There are differences in consumption between the days of the week, between summer and winter, etc.

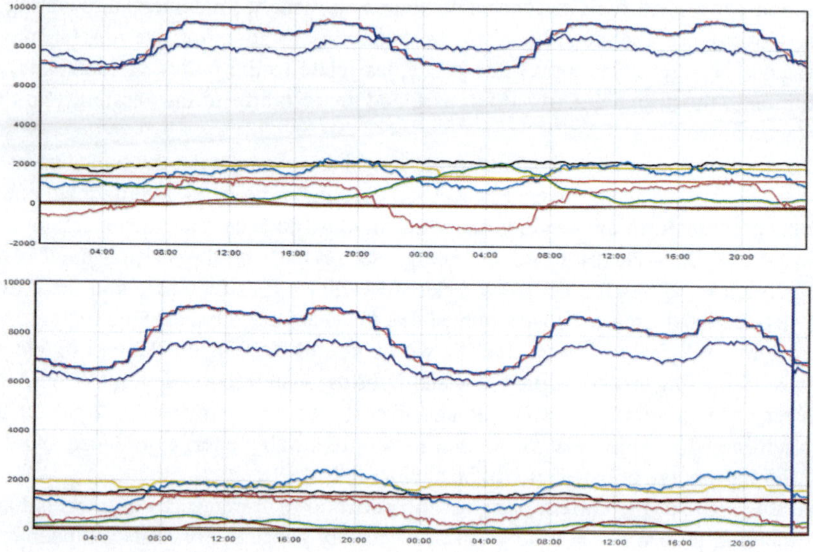

Fig. 1. Electric energy production on 9 and 10 January 2019 and 2020 in Romania.

In the interval mentioned in the graphs, in 2019 there was an average negative balance of consumption of 1011 MW, and in 2020 a positive balance of 1186 MW. The values of the main macroeconomic indicators are chosen as correlation data, for example, the consumer price index for 1 MW produced, the exchange rate, the average wage per economy, etc. The correlation analysis for the selected database showed that there is a very low level of dependence between electricity consumption and some of the mentioned macroeconomic parameters. Consequently, in this new method proposed for forecasting, these parameters of correlation are chosen and used only in estimating the daily load curves reported for the days of the week.

3 Proposed Algorithm

3.1 Applied Method

The standard days recommended by ANRE until a few years ago cannot be taken into account, because in the current conditions related to the electricity consumption during the pandemic these values are no longer significant. The balance of consumption by categories of consumers has changed recently, as many people have worked, and still work from home online, so household consumption has increased significantly. Many companies have reduced their production, and others have stopped working for certain periods.

At the level of each year, however, there is a day of maximum and a day of minimum electricity consumption. Thus, in 2019 the maximum consumption was recorded on January 21 and the minimum on August 4. In 2020, the maximum consumption was on January 14 and the minimum on August 9. It was observed that the average and maximum daily recorded graphs have a similar variation in these two years. With few exceptions, the variation curves are practically parallel. Based on this observation, the medium-term forecast is treated using the average daily consumption.

Modeling starts from the recorded primary data. Because energy demand varies both weekly and seasonally, four different types of the standard day are defined: Type 1-Monday and Friday, Type 2-Tuesday, Wednesday, and Thursday, Type 3 - Saturday, and Type 4-Sunday and other holidays. The lowest consumption is recorded on Sundays and public holidays, followed by Saturday, then Monday and Friday, and with the highest consumption on Tuesday, Wednesday, and Thursday.

As an additional observation, the temperatures and precipitations influence energy consumption, so they are considered in the forecasts. For example, January was selected as the primary data for 3 years, 2019, 2020, and 2021. In 2021, the average air temperature was higher than normal for this period, compared with previous years, across the country. The national average temperature was $T_{av} = -0.1$ °C, with 2.0 °C above the average multi-annual value of January. The minimum value $T_{min} = -23.0$ °C was recorded on January 17. The average of the minimum values was $T_{min, av} = -13.6$ °C. The maximum temperature $T_{max} = 17.0$ °C was recorded on January 23, with the average maximum value $T_{max, av} = 12.8$ °C. Total rainfall was also higher than normal in most parts of the country. The primary databases also contain information on the quantities of precipitation recorded. Based on them, synthesis data are established regarding the energy produced, the energy consumed, and the energy available to be exported to the neighboring countries, etc. Table 1 presents some of the synthesis data for these 3 years.

Most medium-term forecasting techniques must consider some macroeconomic parameters such as the Consumer Price Index - CPI, Average Wage - AW, and Foreign Exchange Rate - FER. These values are predicted by estimating their influence on energy consumption. The most significant correlations were recorded for peak consumption. Table 2 shows the values obtained for these indicators. Negative values indicate an inverse dynamic of the evolution of the maximum load. Values are determined on the third Wednesday of each month. With these indicators as independent variables, the medium-term forecast is made.

Table 1. Synthesis data for energy production.

	2019		2020		2021	
	E	P	E	P	E	P
	GWh	MWh/h	GWh	MWh/h	GWh	MWh/h
Total production	5479.61	7366.07	5349.84	7190.65	5514.33	7411.7
Variation from previous year %				−2.4		3.1
Energy exchange with neighbors	292.47	393.1	212.95	286.19	78.59	105.6
Variation from previous year %				−27.2		−63.1
Internal utilization	5777.08	7758.18	5563.77	7476.84	5592.92	7517.4
Variation from previous year %				−3.6		0.5
Maximum power in the third-wednesday		9254		9025		9367
Variation from previous year %				−2.5		3.8
Peak hourly power		9515		9207		9708
Variation from previous year %				−3.2		5.5
Average power corresponding to maximum daily internal use		8396		8201		8506
Variation from previous year %				−2.3		3.7
Minimum hourly power		5305		5239		5027
Variation from previous year %				−1.2		−4.0
Average power corresponding to minimum internal use		5995		5884		5677
Variation from previous year %				−1.9		−3.5

Table 2. Correlation coefficients.

Day	CPI	AW	FER
Monday	0.1137	0.00167	0.0068
Tuesday	0.082	0.0607	0.0293
Wednesday	−0.242	−0.1604	0.0968
Thursday	−0.6893	−0.3204	−0.3018
Friday	−0.7448	−0.5308	−0.5602
Saturday	−0.5902	−0.4221	−0.5498
Sunday	−0.3212	−0.3586	−0.4608

3.2 Neural Networks with Self-organization of Kohonen-Type

The medium-term forecasting model of energy consumption is suitable for this neural network, [11]. The network consists of two layers of neurons. The input layer has neurons with a number equal to the vectors' size applied to the network input (input data) and a layer of output neurons with several neurons greater, or equal to the number of classes in which it is desired to classify the input vectors. The Kohonen network can be of several types, rectangular - in which the neurons in the output layer are located in the nodes of a flat rectangular network, linear with the output neurons arranged in a line, and circular network with the output neurons arranged circularly. This last method was used in this paper.

This network has the advantage that it learns, without being indicated the correct answer for a model presented at the entrance. It has the advantage of discovering characteristic parameters of the input data, without using known or desired values at the output. The information about these characteristic features is created in the learning process and is stored in the weights of the subsequent connections, in the form of so-called prototypes, which take into account the macroeconomic parameters. At any time, the network outputs describe the relationship between the current input and the prototypes stored in the network.

Each network contains 4 class neurons corresponding to each of the 4 Types of days and another 50 neurons, of which 48 correspond to the hourly energy consumption for two days and the rest for macroeconomic indicators, Fig. 2. The input zone consists of neurons 1–24 representing the load curve for the 24-h values for day Z, year A-1, and neuron 25, [12], which represents the value of the macroeconomic indicator for the same day, respectively neurons 26–49 for the load curve with the 24-h levels for day Z of year A, and neuron 50 the macroeconomic value for the same day. The weights of the connections between each class of neurons and the input neurons describe the load curve associated with that standard day.

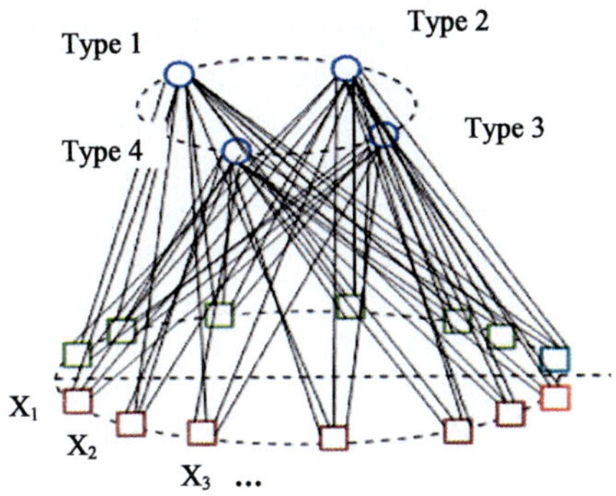

Fig. 2. Neural network with the self-organization of Kohonen-type.

The associated standard algorithm performs a vector quantization by repeatedly adapting the prototypes corresponding to each type of class. The batch training strategy uses a Fuzzy algorithm. For each model m and each neuron-class i, a learning rate $\eta_{im}(t)$ is calculated, corresponding to a training cycle t, using a Gaussian distribution:

$$\eta_{im}(t) = \left[\sum_{k=1}^{4} \left(\frac{\|X_m - W_i(t)\|}{\|X_m - W_k(t)\|} \right)^{\frac{2}{q-1}} \right]^{\frac{q_0-1}{T_{\max}}} \tag{1}$$

where X_m - is the vector of the coordinating model m, W_i - the vector of significances associated with the prototype of the neuron-class i, and r, q, q_0, and T_{max} - are coordinating parameters. In a training cycle for the Kohonen network, learning rates associated with each model m are calculated, [13]. The high value of parameter $\eta_{im}(t)$ reflects a model that will be used to adapt the prototype class neuron i. If the value of this parameter is small, it will be used to adapt a neuron-class i from the neuron-class $i + 24$. At the end of the training cycle, the learning rates thus calculated are used to adapt the prototypes of each neuron class in a single step with the relation:

$$W_i(t+1) = W_i(t) + \frac{\sum_{m=1}^{M} \eta_{im}(t)(X_m - W_i(t))}{\sum_{m=1}^{M} \eta_{im}(t)} \tag{2}$$

Here M represents the total number of training models. The process is interrupted when no significant changes occur in two successive training cycles.

4 AI and Fuzzy Logic in the Forecast of Energy Consumption

The starting point is considered the forecast of electricity consumption for 2020. The result obtained is used in the forecast for 2021 and finally, the result is compared with the consumption recorded for 2021. The average daily consumption is estimated for each week of the year, starting from the 4 Types of days. For the training stage for each month L, the last week of the month L − 1 and the first week of the month L + 1 are also considered, to include the influence of the macroeconomic factors from one month to another on electricity consumption. For the forecast, 3 associated scenarios are adopted for the evolution trend, the minimum, average, and maximum. For estimating the average daily and monthly consumption, parabolic, trigonometric and mixed approximations were performed. The calculations highlight the variant that leads to the best results. The construction of the task graphs was performed with Kohenen-type neural networks.

As a general structure, multiple modules operate together to read the dataset, train, test, and finally, save the model. PyTorch offers two very useful classes for reading the data, Dataset and Data-loader. Dataset is a class whose main purpose is to return a result via the _getitem_ method. Data-loader provides the training module with a way to iterate through the dataset, as well as to build multiple variants and batches of custom sizes, [14].

The training and testing modules are very similar, each of them being composed of a loop that iterates through the batches provided by the Data-loader, which is executed for N years. Inside the loop, the classes and examples are extracted from the batch, the examples are run through the model, and based on the results either execute the backward step and optimizer step or update the statistics, based on whether they are trained or tested. The model is saved and executed once every M year, via an easy mechanism provided by PyTorch which makes it extremely easy to save and restore the model significance, [15, 16].

The first step in solving the task is to preprocess the dataset and prepare it to be fed into the training module. The dataset consists of approximately 57,000 data, with at least 20 graphs or frames with 1920 × 1080 resolutions. This operation should be performed only once, as preprocessing of approximately 135 GB of data is expensive and time-consuming, [17]. To do this, the preprocessing has been saved and the results are stored in a separate location. Fortunately, the Dataset is reduced to a size of only 1.1 GB, which means it will provide a perfectly reasonable performance impact, representing less time spent reading the input and more time for training the network, [18]. To perform the downscaling, an OpenCV capture was realized, each graph was read, and finally, the result was obtained. An example is presented in Listing 1.

Listing 1: rgb array_method: Converting to an rgb array

```
capture = cv. GraphCapture (GraphPath )
capture. open (videoPa th )
images = [ ]
while True:
  ret , img = capture. read ( )
  if ret :
    images. append ( img )
  else:
    break
capture . release ( )
cv2 . destroy All Windows ( )
return images
```

The next step is to build the PyTorch Dataset and Data-loader classes. The dataset has only one essential method called _getitem_. In this case, it receives an index and returns an array of RGB frames. Because the data set is quite small, about 1.1 GB, it is technically possible to keep it in memory. Because the training is done on the computer, all available video RAM for the model parameters will be required. Only a list of file names will be stored in RAM and will be loaded as a CUDA tensor into VRAM with almost the same performance. The only downside is that for each year, the results will have to be processed again. However, the time required to do so is negligible. At this stage, all the necessary preprocessing transformations have been applied, standardized, and transformed. The method is illustrated in Listing 2. When the Dataset is built, two steps are required, one for training and one for testing.

Listing 2: Dataset getitem method

```
filename = self.f i l e s [idx]
video = video_to_rgb_array (file name )
if self . transform :
   video = [self. transform (x) for x in video ]
video = torch.stack ( video ).to (device) # Trans form to tensor and move to CUDA
sample = { 'frames':video , 'class ': i n t (file name [ -11: -8] ) , 'name': file name }
return sample
```

Both Data-loader processes are just basic ones of the default PyTorch class, with special parameters, namely setting the Shuffle to True and Batch_size to 10. The Shuffle term randomizes the indexes by which the data-loader iterates, i.e. it passes through the dataset in a random order every year. Batch_size represents how many samples are extracted in a single iteration step.

5 Obtained Results

For each year analyzed, the datasets used in the self-organization process were constructed for each day of the week. The curves chosen from the extended period were later regrouped for the 4 Types of days. For each day of the week and each week, 7 corresponding load charts were selected: one for the current week and the remaining 6 for the previous 3 weeks, and 3 for the following weeks. Figure 3 illustrates the load curves obtained for the 3 years using the macroeconomic indicators. There are some differences between the days of Type 2 and those of Type 4. Delays of 1–2 h are obtained for the evening peak.

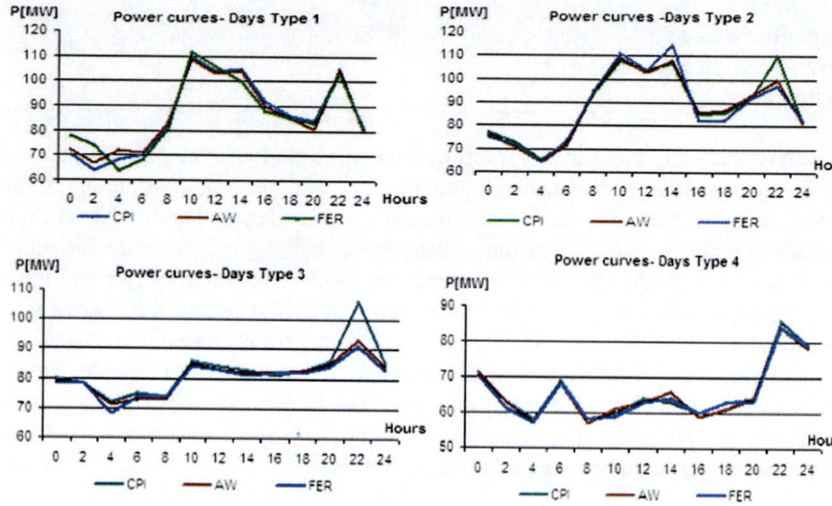

Fig. 3. Power curves for 2021 for all 4 day-types, based on macroeconomic indicators.

A large number of attempts were made which led to the selection of a maximum of 5 classes. In most cases, however, there are 2–3 more common classes, one of which is the dominant one. In the final form, the Kohonen network uses a network with 24 neurons corresponding to the time levels and 3 output neurons corresponding to the 3 distinct individualized classes during the tests performed, minimum, average, and maximum. From the 3 graphs corresponding to the class-neurons, the one associated with the dominant class is retained. Table 3 shows the number of weeks considered unrepresentative for the days of the week, out of the total number of weeks in the 3 years analyzed.

Table 3. Unrepresentative weeks.

Graph number	Mon	Tue	Wed	Thu	Fri	Sat	Sun
1	15	19	16	13	19	17	12
2	14	10	9	14	14	13	17
3	16	15	18	11	12	17	15
4	11	14	11	15	17	14	15
5	11	19	16	19	14	13	10
6	16	14	14	15	15	13	16
7	19	14	18	15	11	15	18

The quadratic percentage errors-QPE were calculated with a relation of the form:

$$QPE = \left(\frac{P_{real}(h) - P_{pr}(h)}{P_{real}(h)} \cdot 100\right)^2 \tag{3}$$

Here P_{real} and P_{pr} represent the real and the forecasted consumptions at hour h. The graphs in Fig. 3 are transformed into curves for absolute quantities and represent the actual forecast for 2021. For each day of the year, the predicted value of the average energy consumption is applied to the reported curve, generating a curve of absolute quantities. Using different calculation assumptions, energy consumption forecasts (minimum, average, maximum) were constructed. To evaluate the results of this forecasting method, the results were compared with the values recorded for 2021. The results of this comparison are presented in Table 4, for an average scenario.

Table 4 illustrates the good behavior of the adopted model, but highlights the significant sensitivity of the selected scenario. There are also significant differences between years due to the pandemic situation compared to an average year in the previous period, when electricity consumption was differently distributed.

Table 4. Annual forecast errors.

Errors (%)	2019	2020	2021	Standard day	Type 1	Type 2	Type 3	Type 4
Absolute maximum	4.2%	5.4%	5.2%	**CPI**	2.8%	4.1%	4.6%	3.4%
Absolute minimum	3.2%	3.3%	2.9%	**AW**	4.5%	4.3%	4.8%	4.6%
Absolute average	1.8%	3.8%	3.2%	**FER**	3.7%	3.9%	5.1%	4.3%

6 Conclusions

The worst results are obtained when the forecast model uses AW as an independent macroeconomic variable. For the days of Type 3 and Type 4, the differences are higher between estimation and achieved consumption, compared to the results obtained for Type 1 and Type 2 working days. There are also significant variations between years due to the pandemic. As industrial and hotel activities resumed, consumption approached a normal pre-pandemic year.

A lower sensitivity was found about the shape of the curves of variation of monthly and daily energy consumption using the proposed model, representing a realistic approach and which ensures a diversification in the approach of working hypotheses. For these completely atypical years, the results obtained are considered to be very good, the errors being between 2.8–5.1%. The analysis can be resumed when all industrial activities return to normal, possibly considering other macroeconomic parameters or other correction mechanisms.

References

1. ANRE. https://www.anre.ro/ro/. Accessed 21 Nov 2021
2. https://www.opcom.ro/pp/home.php. Accessed 01 Sep 2022
3. https://ec.europa.eu/clima/change/consequences_en. Accessed 21 Sep 2021
4. https://www.transelectrica.ro/ro/web/tel/home. Accessed 01 Oct 2022
5. https://www.weforum.org/agenda/2021/09/this-is-how-ai-will-accelerate-the-energy-transition/. Accessed 12 Oct 2021
6. Kr Jha, S., Bilalovic, J., Tha, A., Patel, N.: Renewable energy: present research and future scope of AI. Renew. Sustain. Energy Rev. **77**, 297–317 (2017)
7. Tanver, A., Zhang, D., Huang, C., et al.: Artificial intelligence in sustainable energy industry: status Quo, challenges and opportunities. J. Clean. Prod. **289**, 125834 (2021)
8. Manea, D.: Flue gas desulphurization plants in Rovinari and Craiova II power plants. Energetica **4**, 270–282 (2021)
9. Radulescu, V.: Research and solutions to minimize frontal area overheating of hydro generator stator with vertical axis. MDPI Energy **14**(5), 1243 (2021)
10. Radulescu, V.: Prediction of the atmospheric boundary layer in wind power plants by proper selection of the main air parameters. In: ImanEE Conference, May 22–24 Chisinau, Moldova, vol. 564 (2019)
11. Kalogirou, A.: Applications of artificial neural-networks for energy systems. Appl. Energy **67**(1–2), 17–35 (2000)

12. Yang, T., Chen, Y., Sze, V.: Designing energy-efficient convolutional neural networks using energy-aware pruning. In: IEEE CVPR Conference, pp. 5687–5695 (2017)
13. Chen, E., Tsao, K., Bezdek, C.: Fuzzy Kohonen clustering networks. Pattern Recogn. **27**(5), 757–764 (1994)
14. Radulescu, V.: Critical analysis of faults in operation of energy systems using fuzzy logic. In: Dzitac, I., Dzitac, S., Filip, F.G., Kacprzyk, J., Manolescu, M.-J., Oros, H. (eds.) ICCCC 2020. AISC, vol. 1243, pp. 234–244. Springer, Cham (2021). https://doi.org/10.1007/978-3-030-53651-0_20
15. Wehrens, R., Kruisselbrink, J.: Flexible self-organizing maps in Kohonen 3.0. J. Stat. Softw. **87**(7), 1–18 (2018)
16. Radulescu, B.A., Radulescu, V.: Modeling 3D convolution architecture for actions recognition. In: ASME ISPS Conference-Virtual, 2–3 June (2021)
17. Zhang, Y., Shi, Q.: An intelligent transaction model for energy block-chain based on diversity of subjects. Alex. Eng. J. **60**(1), 749–756 (2021)
18. Liu, Y.X., Zhou, F.: Electricity market transaction model design combining block-chain and machine learning. In: ICCECE Conference (2021)

Integrated Solutions and Computerized Human Gesture Control

Bogdan Alexandru Radulescu(✉)

University Politehnica of Bucharest, Splaiul Independentei 313, 060042 Bucharest, Romania
bogdan.radulescu97@gmail.com

Abstract. Recognition of human actions is a topical issue due to its applicability in different domains such as security, medicine, behavioral analysis, and education. The development of the computer processing power, as well as the quality of image capture equipment, has increased the accessibility of larger and more accurate data sets. This permanently improves the performance of the developed programs. Convolutional neural networks are the starting point for identifying facial analysis and gesture recognition. By their nature, they can recognize ideal Spatio-temporal features for RGB image processing. The proposed model with 3D Kernel is useful in certain situations where the gestures are very different from each other and must be recognized by a system that reacts quickly and with high processing speed (emergencies or critical situations). The main objective of this paper is to build a reliable and easy-to-implement human gesture recognition classifier. It must be able to easily classify a large number of images so that it can operate in real-time, processing at least 30 frames per second. This is possible by using other hardware components of a shelf computer connected to a normal CCTV-Closed Circuit TV infrastructure. The selection of the gesture class must be not related to the environment, person, background, appearance, or viewing angle. The decision should be related only to the person to whom the gesture is attributed and to the context in which he reacts.

Keywords: Artificial neural networks · Image reconstruction · Kernels

1 Introduction

In recent decades, the developments of biometric applications for facial recognition and gesture classification have become increasingly used in daily life. They have become applications associated with the terms security, smart cities, AI, etc. Facial recognition is essential in identifying those who steal from shops or ATMs, which cause road accidents, destroy property, in schools, etc., [1, 2]. This includes an overview of different techniques used in computer vision for classifying actions, [3]. Biometric applications are essential in creating identity cards or passports.

Until recently, the most commonly used type of data protection was considered password. With the development of information technology and security algorithms, many systems use more biometric factors such as iris, facial features, fingerprint, or voice, as an additional measure of identity recognition. These ID systems are very

S. Dzitac et al. (Eds.): ICCCC 2022, AISC 1435, pp. 156–166, 2023.
https://doi.org/10.1007/978-3-031-16684-6_13

attractive because they are easy to use, [4, 5]. The facial recognition system as ID is already used in airport check-in, access to sports stadiums, and concerts, where the traffic is very high. If the processing time is too long, the system is no longer useful.

Gesture analysis has become essential in medicine, in behavioral analysis of small children for the early detection of neuron-motor diseases, or adults in the diagnosis of mental illness, [6]. Biometric factors thus make it possible to identify people according to their physiological or behavioral characteristics, [7]. Identifying specific gestures allows a faster reaction in appreciating:

- The effects of verbal or visual communication: drowsiness, boredom, joy;
- Emotional states: nervousness, irritation, curiosity, interest;
- Determining defining gestures in the supervision of young children: punishment, beatings or encouragement, appreciation, tenderness;
- Caring for sick people: ignorance, rejection, compassion;
- Most importantly, in defense, the recognition of gestures before violence.

Many researchers have focused on establishing more robust and accurate computational and modeling algorithms for these applications while trying to increase the speed of information processing and accuracy of the results, [8, 9]. If a model is not well structured, it can lead to some errors in identifying the human face made up of different structures and features, or it can take too long to process, [10]. The major advantage of this system, if well structured, is that it does not require human intervention to function, [11].

Due to the large volume of data generated and the rapid progress in AI-techniques new data processing models are used, such as GPUs-Graphics Processing Units, CPUs-Central Processing Units, and FPGA-Field Programmable Gate Arrays. These are computational components required in this type of application, [12, 13]. GPUs now have computing cores a few orders of magnitude larger than traditional processors and allow greater capacity to perform parallel calculations, [14].

In contrast, FPGAs have a more flexible hardware configuration, and better performance than computing efficiency GPUs, but require longer processing time. Currently, most authors use in such applications a form of recurrent temporal neural networks-TCN, with convolutional layers, [15, 16]. This paper proposes a method that combines both possibilities of image processing, through a method that ensures both the increase of computing capacity but also the reduction of the time of processing.

2 Methods Used in Gesture Recognition

2.1 RGB-Based Classifiers

These classifiers use neural networks that either treat each frame as an image by creating a database or use multi-frame sequences to extract specific features. There are several approaches to design architecture dedicated to gesture recognition, here mentioned the direct classification method and the convolutional networks. In the present paper, the second method was developed.

A. Direct Classification Method
This method usually does not consider the temporal aspect when detecting an action. It is based on methods of classifying the closest neighbors and aims to minimize the distance between a sequence of images observed in training and those existing in the test dataset, [17]. It is easy to use, but when faced with a large dataset, the computation becomes long and expensive. Supplementary, some new methods to reduce the dimension of the analyzed database are often used. As an example, this method is used for learning to play tennis, to play piano, or to exercise in medicine for an operation. It is very useful in the first lessons, to learn the movements correctly.

B. Convolutional Networks
This method is mainly used to extract features for certain movements. The 2D and 3D convolution layers are used, many of them integrated into recurring networks. The estimation of the position can be used in CNN-convolutional Neural Network architecture, [18]. The temporal aspect is solved either by mediating predictions over time or by using 3D convolutions. Characteristics of the final layers are now easily implemented. To extract the features, a computationally efficient descriptor Center-Symmetric Motion Local Ternary Pattern-CS-Mltp is used to describe RGB sequences, both spatially and temporally. To represent the features, a time-matching of the split encoded pyramid is proposed, with a rare-time coding approach.

2.2 Skeleton-Based Classifiers

These classifiers have become more commonly used in recent years, given the improved performance of the specialized monitoring equipment and the transmission of skeleton information, such as Microsoft Kinect. They have led to the development of specialized large datasets implementation, such as NTU RGB+D-Nanyang Technological University RGB and Depth information. Skeleton classifier is easier to process than video and makes it possible to train deeper networks.

A. Spatial and Temporal Representations
This method uses the spatial and temporal performance of recurring layers, such as LSTM-Long Short-Term Memory. Many solutions rearrange the 25 joints into 5 × 5 matrixes which can be used either in convolutions or in a graph-based approach. An LSTM Global Context-Aware Attention network based on two LSTM layers and a classifier can also be used. The first LSTM encodes the skeleton sequence and generates a contextual representation of the action. This representation is introduced in the second layer, which selects the most important joints for representing an action in each frame, [19]. The model based on LSTM and CNN fusion has the advantage that LSTMs are

optimal for using temporal information, while CNN work better when exploring localized spatial information. It is a simple but very efficient method.

B. Graph-Based Representations

Temporal connections are maintained by considering each frame belonging to a sub-graph and each vertex from one graph being connected to the similar one, from the next graph. Good results were obtained even with NTU RGB + D dataset. Other solutions use neural networks and build a specialized ST-GCN Spatio-Temporal Graph Convolutional Network, capable of a better generalization.

RGB methods have the advantage of not requiring a very specialized data set. They only need a sequence of frames for classification and can run more easily on the available equipment and infrastructure. Skeleton-based datasets are lightweight but more difficult to achieve.

The solution proposed in this paper uses purely RGB methods made in neural networks, without taking into account the information in infrared, depth, or skeleton.

3 Adopted Method

Neural networks can learn connections between examples in the same class. With a parallel structure, they become computationally efficient for processing large data sets, which are often needed in classification problems.

In this proposed method two optimization methods are used, the cross-entropy loss and the adaptive moment-Adam Optimizer. The cross-entropy loss measures the performance of a model whose output is a probability distribution, being a perfect candidate for classification. The network output is a tensor of dimension C- the number of classes. Each element represents the trust of the model, respectively the certainty that an example belongs to a specific class.

The goal is to update the weights so that for each input, the probability of the correct class is as close as possible to 1, while the probability of other classes is as low as possible. This means that the model is not only rewarded for the correct prediction of one class (most likely-correct class), but also for discriminating against other classes.

Adam estimation-Adaptive Moment Estimation combines the advantages of two popular methods, AdaGrad-Adaptive Gradient algorithm, and RMS Prop-Root Mean Squared Propagation. It is suitable for large datasets and problems that require a large number of parameters, but also for non-stationary data, such as those needed in action recognition. It is difficult to cover in a dataset, large variations of non-essential parameters such as environment, other actors, other actions, lighting conditions, viewing angles, etc.

Through training, the Adam model functions well when processing new data. To solve the problem of action recognition because the examples of datasets consist of images or image sequences, the most intuitive step is to use the convolutional architecture, [20]. The advantage is of extracting spatial features from a localized region, configurable by kernel size and other hyper-parameters.

CNN is the method used here, being the most efficient solution in image classification. If inverse propagation is used to learn the coefficients of the convolution core directly from the images of handwritten numbers, this approach becomes a computerized image.

The n-D convolution layer is essentially a sliding dot product with a kernel size (x_1, x_2..., x_n) applied for an input (y_1, y_2..., y_n). Figure 1 illustrates a 2-D convolutional kernel applied over a 5 × 5 input.

Most problems can be solved using up to three dimensions, as in this study. Hidden layers consist of convolution operation, an activation function, and a pooling layer.

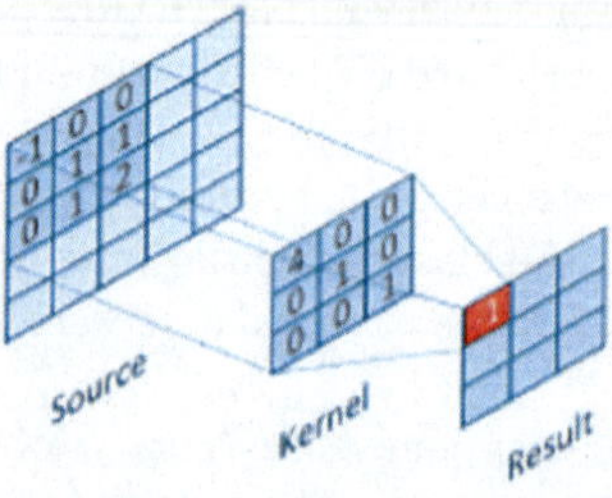

Fig. 1. Example of a convolutional kernel structure.

The activation functions in hidden layers are nonlinear and define the node output. The model uses Leaky ReLU-Leaky Rectified Linear functions, characterized by:

$$f(x) = x^+ = max(x, 0) \tag{1}$$

ReLU functions could have the problem of "dead" neurons, in this type of application. It requires a high gradient which can update the weights, but sometimes the training has no effect. Working only with the ReLU functions, it was noticed that no matter how long the training was for the tested solution, the results do not become conclusive. That's why the Leaky ReLU function was selected in this application. It allows a better preparation of the deep neural networks, exactly the type of architectures used here. The gradient of the insignificant neurons becomes small, diminishing its effect. The difference between ReLU and Leaky ReLU is illustrated in Fig. 2. The Leaky ReLU variant allows through successive training higher learning rate, that's why it was preferred:

$$f(x) = \begin{cases} x \quad if \ x \geq 0 \\ \gamma \cdot x \ if \ x < 0 \ \gamma \cong 1 \cdot e^{-2} \end{cases} \tag{2}$$

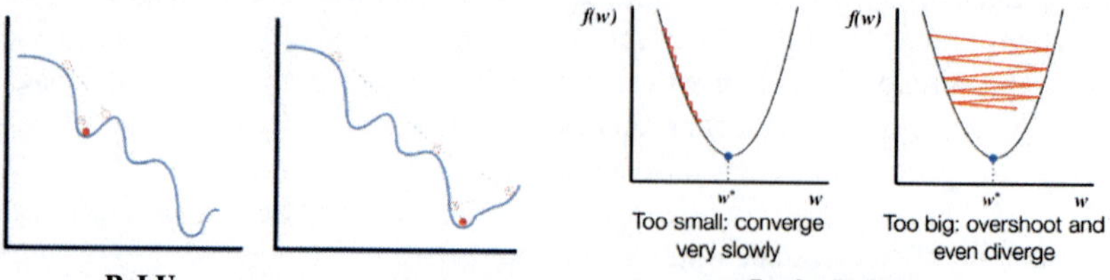

Fig. 2. Comparison between ReLU and Leaky ReLU with AdaGrad.

In Leaky ReLU the neuron always remains active. After each hidden layer consisting of convolutions and activations, a pooling layer is added to reduce the number of the model parameters and make the calculation more efficient.

When using CNNs, there are two options for choosing the pooling layer: Max Pool and Average Pool. Max Pool picks the highest value of all those present in the sliding window, while Average Pool computes their average. Figure 3 shows a down-sampling window with input from 4×4 to 2×2, using Max Pool and Average Pool. Both require a short time of computation, but the Average Pool uses more information from the previous layers, so that's why it will be used in subsequent modeling. Batch normalization was used after each convolution layer to improve learning speed.

Fig. 3. Example of down-sampling windows with Max Pool and Average Pool.

Such a hidden layer is shown in Fig. 4. Next, between Convolution and Average Pool, will be used the Batch Normalization and Leaky ReLU with Dropout.

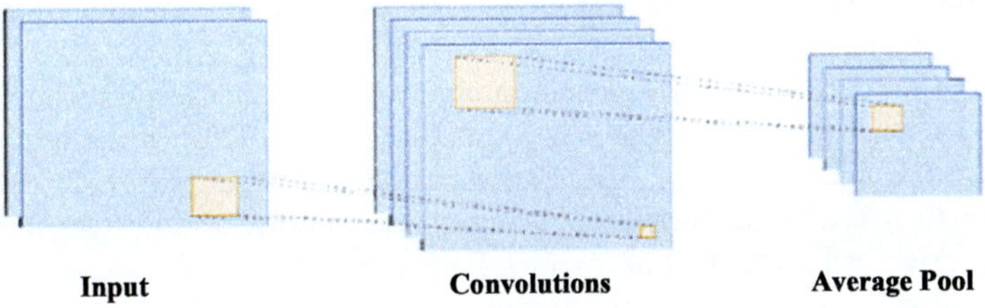

Fig. 4. Example of a typical convolution unit used in the proposed architectures.

Convolutional layers usually decrease the dimension of the output. If the input has the dimension (x; y), with a square nucleus (k, k), the output will be $(x - k + 1; y - k + 1)$. In practice, padding is introduced in all architectures that keep the shape of the input. The decrease is made by pooling.

After applying multiple convolution units, all architectures end with two or three fully connected layers. These turn the features learned in the first part of the network into class predictions. All fully connected layers function in a one-dimensional tensor with an input and an output. The input has the size Number of Channels $\times$ Depth $\times$ Width $\times$ Height, where for RGB images the Number of channels is initially 3 but may increase after passing through a convolution. Depth is initially the number of frames in the input but may decrease as it passes through the network. Each hidden layer has Batch Normalization, Activation, or Dropout Functions, where applicable. After these layers, there is an Average Pooling layer. The layers are fully connected, transforming the characteristics learned through convolutions into class predictions.

4 Used Architectures and Models in the Proposed Model

In deep neural networks, as is the case of the gesture recognition model, dropout is used to reduce over-adaptation. At the dropout layer, only some neurons are active during forwarding and backward propagation. This ensures that in the training phase, the neurons do not develop codependency and instead learn to better generalize types of movements. Each neuron has a probability "p" of being active. The model is trained on the Train partition and evaluated on the Test partition. It checks how the model works when it encounters movements it had never seen. The obtained results were very good in certain categories of movements. In this case, reducing over-adjustment is essential to improve its performance by adding Dropout layers.

4.1 Inputs and Outputs

Two types of preprocessing, scaling and normalization were used to classify human gestures. The NTU RGB + D method is used in the dataset. The selected examples have at least twenty frames, but the total number may vary. The dataset is HD (1920 × 1080), which will be downscaled to 192 × 108. Keeping enough information to classify videos reduces the training time and several features required for proper classification, by significantly reducing H and W sizes. Normalization will be applied in the range [−1, 1]. The advantage is to keep the characteristics observed by the convolution layers in approximately the same range. The performance of the model is improved by the Normalization operation. The average is 0.5 and the standard deviation is also 0.5, across all three input channels.

The input is transformed by a general architecture into a class prediction output. Input data (Depth, Channels, Height, Width) of (20, 3, 1080, 1920) with values between 0 to 255 becomes an input of (20, 3, 108, 192) with values in the range [−1, 1]. After each convolution, there is a Batch Norm layer, which acts as a normalization layer. It prevents the weights from being too high or too low. Each BatchNorm is followed by a Leaky ReLU activation layer, followed by a drop-out that acts as a regulator to prevent over-fitting. Once every few units one or more dimensions are reduced by half, preventing the model to have too many parameters. As the channels increase, the input would become larger. Reducing the number of parameters also slightly reduces over-adaptation.

4.2 Model of 3D-Convolutions

This method works with a small window of no more than three frames, corresponding to the context in which each frame is observed. It manages to learn three-dimensional features that help determine a specific action in a class of actions.

The model takes as its input the entire video, which is essentially a series of frames, and focuses on three dimensions, thus building channels. Structurally, it is composed of 3D convolution layers, with 3D Pooling, Batch Normalization, and Dropout. It works better than 2D convolutions because it differentiates between actions that have similar motions, in different orders. Architecture is composed of stacked convolution units that reduce at least one size by half. In the last part, there are three completely connected layers. Before reaching them, the input is 4-dimensional, C, D, H, and W. The fully connected layers wait for a one-dimensional input, so the result of the last convolution layer must be stacked in one dimension. Fully connected layers act as an intermediary between the features learned by the network. During the three fully connected layers, the output size changes from C _ D _ H _W to 1024, from 1024 to 256, and from 256 to 60.

5 Obtained Results

The 3D-Convolutions model has the advantage of learning Spatio-temporal features. It takes time for training and testing, but the disadvantage is offset by the precision it offers in gesture recognition. The first layer has a fixed input size, and the last layer has a fixed output size. The network has several classes necessary to learn to recognize. The middle layer can have as many inputs and outputs, but experiments show that the final number of parameters must be less than 10,000; values like 1024 and 256 are good intermediate steps.

For testing, named **Samples, 40 people**, aged between 10 and 65 were tested. The RGB + D NTU dataset used for human gesture recognition increased to **120 Classes (recognized gestures)**. The comparison is illustrated in Table 1. The realized model is complex, responding well to different challenges. It contains actions and classes of actions that overlap in the same example (as writing and drawing), actions and interactions with other people (playing, catching an object), or even multiple actions performed by different actors in the same example.

Table 1. Comparison between different datasets.

Datasets	Samples	Classes	Subjects	Views	Sensor	Modalities
MSR-action 3D	567	20	10	1	N/A	D+3D Joints
RGBD-HuDaAct	1189	13	30	1	Kinect V1	RGB+D
3D action Pairs	360	12	10	1	Kinect V1	RGB+D+3DJoints
Multi view 3D event	3815	8	8	3	Kinect V1	RGB+D+3DJoints
Online RGB+D act	336	7	24	1	Kinect V1	RGB+D+3DJoints
UWA3D multi view	900	30	10	3	Kinect V1	RGB+D+3DJoints
Office activity	1180	20	10	3	Kinect V1	RGB+D +3DJoints
UWA3D multi view	1075	30	10	5	Kinect V1	RGB+D+3DJoints
NTU RGB+D	**56880**	**60**	**40**	**80**	**Kinect V1**	**RGB+D+3DJoints**

The best performances obtained according to their accuracy are Class 59, (Going towards), Class 27 (Jumping up), Class 24 (Jumping forward), and Class 43 (Falling on the ground). All these actions have Accuracy and Recall of over 90%, which means that the model can be used successfully to recognize these actions. They have distinct Spatio-temporal characteristics, which the classifier can easily recognize.

However, a classifier is not enough for the selected dataset. Even for NTU RGB + D model with RGB, the data set has been reduced to 60 classes. Many similar actions have been abandoned, such as putting an accessory, writing and drawing, etc. and the remaining actions were divided into three groups: Daily Actions, Medical Disorders, and Two-Person Interactions. An overview of these classes is presented in Table 2.

To avoid too large databases, several convolutional layers were stacked without shrinking too quickly. The first layer focuses more on lower-level features, so higher resolution images. The statistics were calculated for the NTU RGB + D dataset, which consists of 37800 Cross-View and 40200 Cross-Subject, test videos for the 60 motion classes, at a resolution of 108 × 192 and 20 frames/subject.

Table 2. Actions and labels contained in the NTU RGB + D dataset.

Daily actions			
A1: drink water	A2: eat meal	A3:brush teeth	A4: put on shoe
…A9: stand up	A10: put on jacket	A11: take off jacket	A12: writing
A13: put on glasses	A14: take off glasses	A15: check - phone	A16: taking a selfie
…A37: phone call	A38: salute	A39: clapping	A40: shake head
Medical conditions			
A41: cough	A42: chest pain	A43: back pain	..A49: falling down
Mutual Actions			
A50:kicking	A51: pushing	..A59: hugging	A60: pat on back

The accuracy of the method was calculated. For the Cross-View Prediction, the best 5 values are between 91.17–98.73% and for Cross-Subject 88.44–99.54% and the worst 5 values for the Cross-View Prediction are between 19.06–28.12% and for Cross-Subject 20.52–27.79%. For the Recall classes, the best 5 values for the Cross-View Precision are between 94.77–99.2%, and for Cross-Subject Precision 96.27–98.50%. The worst 5 values for the Cross-View Precision are 14.24–21.89% and for Cross-Subject 15.21–28.07%.

The first test was done using the fusion of features between RGB and Depth. The second estimates the position based on the depth and fusion between the RGB subsections associated with a feature selection mechanism. The third one uses a combination of RGB features and dynamic images built from a segmented sequence of frames.

6 Conclusions

An architectural model has been developed for human gesture recognition, by implementing a method of reducing its size, to reduce the computational time. This paper aimed to build a model that can be run on CCTV infrastructures, which could be achieved by using NTU RGB + D with RGB, by splitting the dataset, into training and test partitions,

Cross-View and Cross-Subject. For each sample, was computed the prediction y_{pred} and incremented the matrix $[y_{pred}][y]$, where y is the basic truth.

The database was built on 40 subjects (Samples) aged between 10 and 65 years, on 120 types of movements for gesture recognition (Classes). From them, 60 movements were selected which were divided into three categories Daily Actions, Medical Disorders, and Two-Person Interactions. So the matrix dimension is 60×60.

An important aspect in selecting the input data is the location of the cameras, the angle, and the distance from the subject so that the model learns to correctly classify an action. In the first stage, the dataset contains three cameras that capture horizontal views of the same action, located at the same height, but at different horizontal angles, −45°, 0° and +45°. In the second stage, the distance between the cameras and the subject was modified, and in the third stage, the height of the cameras was modified but adapted to the height of the subjects. Each subject acted twice each movement, to the left and right camera.

It has been noticed that some classes of actions are more closely related to a certain type of subject. For example, horizontal jumps (A20) and upward jumps (A21) are more common among children, better defined, while medical conditions are more common in adults and the elderly cough, bending to identify spinal disorders). For this reason, good variation in the subject class is essential for increasing applicability in real-world scenarios. The results for certain types of movements have an accuracy of over 90–95%, which proved the advantage of the presented solution.

References

1. Karovaliya, M., Karedia, S., et al.: Enhanced security for ATM machine with OTP and Facial recognition features. Procedia Comput. Sci. **45**, 390–396 (2015)
2. Andrejevic, M., Selwyn, N.: Facial recognition technology in schools: critical questions and concerns. Learn. Media Technol. **45**, 115–128 (2020)
3. Poppe, R.: A survey on vision-based human action recognition. Image Vis. Comput. **28**(6), 976–990 (2021)
4. Mazura, J., Juluru, K., et al.: Facial recognition Software success rates for the identification of 3D surface reconstructed facial images: implications for patient privacy and security. J. Digit. Imagining **25**, 347–351 (2012)
5. Xu, F.Z., Zhang, Y.: Facial recognition check-in services at hotels. J. Hosp. Mark. Manag. **30**, 373–393 (2020)
6. Hurst, A.: Facial recognition software in clinical dysmorphology. Curr. Opin. Pediatr. **30**, 701–706 (2018)
7. Crook, T., Larrabee, G.: Changes in facial recognition memory across the adult life span. J. Gerontol. **47**, 138–141 (1992)
8. Suter, D., Wang, L.: Learning and matching of dynamic shape manifolds for human action recognition. IEEE TIP **16**, 1646–1661 (2007)
9. Sullivan, J., Carlsson, S.: Recognizing and tracking human action, Numerical Analysis and Computing Science. In: ECCV, Royal Institute of Technology, vol. 1 (2002)
10. Radulescu, B.A., Radulescu, V.: Model of human actions recognition based on 2D Kernel. In: ASME Conference ISPS (2021)
11. Shahroudy, A., Liu, J., Ng, T-T., Wang, G.: NTU RGB+D: a large scale dataset for 3D human activity analysis. In: IEEE Conference CVPR (2016)

12. Radulescu, B.A., Radulescu, V.: Modeling 3D convolution architecture for action recognition. In: ASME 30th Conference ISPS (2021)
13. Agarwal, R., Jain, R., Regunathan, R., Pavan Kumar, C.S.: Automatic attendance system using face recognition technique. In: Kulkarni, A.J., Satapathy, S.C., Kang, T., Kashan, A.H. (eds.) Proceedings of the 2nd International Conference on Data Engineering and Communication Technology. AISC, vol. 828, pp. 525–533. Springer, Singapore (2019) https://doi.org/10.1007/978-981-13-1610-4_53
14. Max-pooling/pooling. https://www.machinecurve.com/index.php/2020/01/30/what-are-max-pooling-average-pooling-global-max-pooling-and-global-average-pooling/. Accessed 05 June 2020
15. Radulescu, B.A., Florea, A.M., Radulescu, V.: TCN Units, solution in recognition of human activities. In: ASME Conference ISPS (2021)
16. Feichtenhofer, C., Pinz, A., Zisserman, A.: Convolutional two-stream network fusion for video action recognition. In: IEEE CVPR (2016)
17. Pang, Y., Liu, Z., Yu, N.: A new nonlinear feature extraction method for face recognition. Neurocomputing **69**, 949–953 (2006)
18. Pichao Wang, P., Li, W., Gao, Z., Tang, C., Ogunbona, P.: Depth pooling based large-scale 3-D action recognition with convolutional neural networks. IEEE Trans. Multimed. **7**(1), 90–99 (2021)
19. Huang, Z.H., Li, W.J., Shang, J., Wang, J., Zhang, T.: Non-uniform patch based face recognition via 2D-DWT. Image Vis. Comput. **37**(17), 12–19 (2015)
20. Lima, A., Zen, H., Nankaku, Y., et al.: On the use of kernel PCA for feature extraction in speech recognition. IEICE Trans. Inf. Syst. **87**, 2802–2811 (2004)

Research on the Storage of Integrated Pipe Corridor Monitoring Data Based on Blockchain Technology

Chunxiang Sun(✉), Dan Chang, Daqing Gong, and Mincong Tang

School of Economics and Management, Beijing Jiaotong University, No. 3 Shangyuancun, Haidian District, Beijing, China
{20120610,dchang,dqgong}@bjtu.edu.cn

Abstract. As an emerging urban underground life line in China, the integrated pipeline corridor has received extensive attention from many scholars, and the operation and maintenance management about the integrated pipeline corridor is gradually developing towards intelligent management with the maturity of digital technology. This paper proposes a monitoring data storage scheme based on the alliance chain to further realize the intelligent management of the integrated pipe corridor. In this paper, the data storage federated chain model is constructed by using various monitoring hosts of the integrated pipeline corridor as pre-selected nodes of the federated chain. In addition, the paper also adopts a dual-chain storage model with "on-chain" storage of key data information and "off-chain" storage of complete monitoring data to avoid the problem of communication jams among nodes caused by too much data on the federated chain. Finally, the proposed storage scheme is analyzed and discussed in terms of security performance, and the advantages of the model are demonstrated. The storage scheme proposed in this paper can safely and reliably store the monitoring data in the integrated pipe corridor, which provides a theoretical reference for further construction of an intelligent platform for integrated pipe corridor management, and to a certain extent promotes the development of intelligent management of integrated pipe corridors.

Keywords: Integrated pipe gallery · Monitoring data · Blockchain technology · Storage · Alliance chain

1 Introduction

As the number of people in urban centers around the world continues to grow, urban land resources are becoming increasingly scarce, at this time the comprehensive pipeline corridor, which can make full use of urban underground space and reasonably plan various municipal pipelines, has come into being, alleviating the problem of scarce urban land resources to a certain extent; on the other hand, the comprehensive pipeline corridor can also effectively solve many urban problems such as repeated road excavation and environmental degradation caused by the laying of traditional pipelines, and It is a key

S. Dzitac et al. (Eds.): ICCCC 2022, AISC 1435, pp. 167–181, 2023.
https://doi.org/10.1007/978-3-031-16684-6_14

project and "lifeline" of municipal infrastructure to ensure the sustainable development of cities [1]. Due to the above-mentioned advantages of integrated pipe corridors, the construction of underground integrated pipe corridors has been vigorously promoted in China, and with the completion of integrated pipe corridors across the country, the study of the operation and maintenance management of integrated pipe corridors after completion has also become a major research focus. At the same time, with the maturity of digital technology (such as ITO, BIM, GIS, etc.), the intelligent management of integrated pipe corridors has also become a major research hotspot, and has received the key attention of experts and scholars [2–7].

Comprehensive pipe corridor in the operation process will produce many real-time data as well as non-real-time data [7]. The real-time data is the monitoring data, mainly refers to the environment in the corridor, equipment operation status, etc., these data has an extremely important role in the operation and maintenance management of the integrated pipe corridor. The monitoring data directly reflects the real-time situation in the corridor and allows for timely warning of abnormal data; in addition, in the event of a corridor failure, staff can also trace the failure factor based on historical monitoring data. At present, the monitoring data storage mechanism of the integrated corridor is as follows: firstly, the monitoring data is generated by various sensors and camera monitoring equipment deployed in the corridor; then the sensors upload the collected data (such as temperature, humidity and wind speed in the corridor) to various monitoring hosts; finally, the monitoring hosts upload their respective monitoring data to the general database server. Due to the huge amount of monitoring data and the limited storage space of the various monitoring hosts, most of the monitoring hosts only store their monitoring data in their respective nodes before consolidating the data in the database server. However, this mechanism is prone to problems such as malicious forgery, tampering and attacks on the stored data. In addition, when all data is consolidated in the database server, there are also problems such as malicious attacks and illegal tampering of third-party databases.

The current research on the intelligent operation and maintenance management of integrated pipe corridors, most scholars use BIM, GIS and IoT technologies to build and study the platform of integrated pipe corridor operation and maintenance management [2–6], which provides a theoretical basis for the further development of integrated pipe corridors; however, only a few studies have been conducted to further explore the collection and storage of monitoring data in pipe corridors, but they are only limited to the macro theoretical studies [7–9]. In other fields, research on the storage of monitoring data has been conducted, such as optimizing storage technology [10], centralizing the storage of associated data [11], and using cloud computing management platforms [12] to achieve efficient access to massive amounts of data. In terms of data security, some scholars have also proposed more secure storage strategies, such as "triangular data storage strategy" to ensure data consistency through dual cloud and local backup [13], and further optimization of encryption algorithms [14]. In previous data storage research, some scholars focus on the fast storage of massive data and ignore the security of massive data, while others focus their research on data security but fail to ensure efficient access to data.

Blockchain technology is a decentralized, de-trusted distributed database with high reliability, tamper-evident and time-traceable features. Various nodes on the chain can achieve peer-to-peer transactions, coordination and collaboration based on decentralized credit without trusting each other [15]. Blockchain technology incorporates several technologies such as cryptography, P2P file transfer mechanism and consensus mechanism, which solves the problems of poor reliability, low security and high cost that exist in the current centralized model. The block consists of two parts: the block head and the block body. Among them, the block header is mainly responsible for the link between individual blocks, while the block body is responsible for storing data information. When a block is formed, the system automatically generates and timestamps it, as shown in Fig. 1.

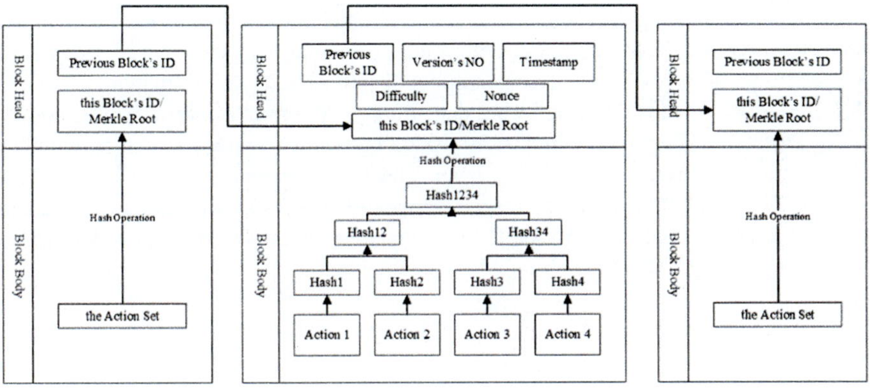

Fig. 1. Blockchain data structure

Blockchains can be divided into public, private and federated chains [16] depending on the usage authority and application scenarios. A federated chain is a blockchain in which the consensus process is controlled by pre-selected nodes, and each node can join and exit only after authorization. Usually, each node in a federated chain has a corresponding entity organization, and each federated chain participant shares the right to write to the overall network, not controlled by a single center, which promotes the cooperation of each participant and guarantees fairness among all parties. In fact, before blockchain technology was proposed, traditional distributed storage technology could solve the problem of memory capacity to a certain extent, but it could not guarantee the consistency, anti-forgery and anti-tampering of stored data. With its advantages of tamper-proof and traceability, blockchain technology has been widely used in various research fields. The literature [17–23] applied blockchain technology to data storage in agriculture, the power industry, healthcare, e-commerce and supply chain finance, respectively. The rapid development of blockchain technology has provided new solutions for the collection of massive data as well as storage security.

This paper proposes the idea of applying blockchain technology to the storage of monitoring data in integrated pipeline corridors based on the characteristics of integrated pipeline corridors. Considering the problems of slow response time and excessive redundancy of data on the chain when storing massive data, this paper proposes a "double chain" storage model - storing complete monitoring data under the chain and storing transaction history on the chain, which saves space on the chain and enables efficient sharing of monitoring data among monitoring nodes.

2 Analysis of the Monitoring Data Storage Architecture of the Integrated Pipe Corridor

The internal structure of the integrated corridor is complex, so there is a relatively large variety of monitoring data involved in the corridor. To build a clearer and more reasonable storage model for monitoring data in the integrated corridor, this paper first analyses the monitoring data storage architecture of the integrated corridor. Referring to the system architecture for the safe operation and maintenance management of integrated pipe corridors proposed by the Hefei Institute of Public Safety of Tsinghua University [24], this paper divides the monitoring data storage system into five layers, as shown in Fig. 2: data service layer, wired transmission layer, data aggregation layer, wireless transmission layer and data collection layer. The 3 data layers are responsible for data collection, integration and storage respectively, while the 2 transmission layers are the bridge to guarantee data transmission. 3 data layers are the main structure in the whole monitoring data system.

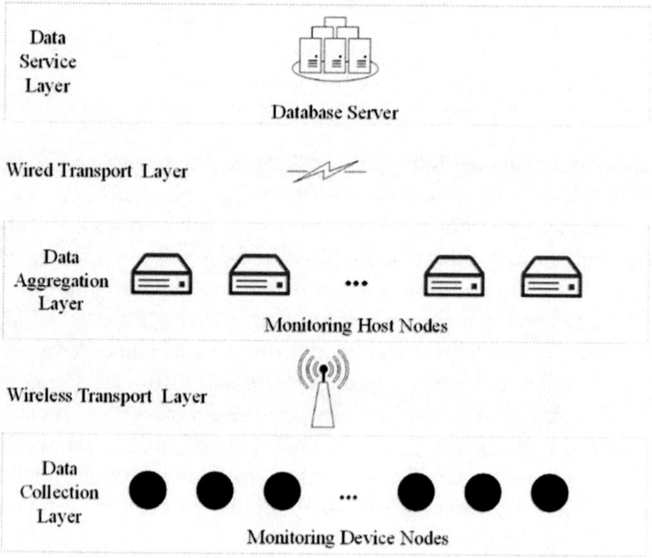

Fig. 2. Integrated corridor monitoring data storage architecture

- **Data Collection Layer**: This layer consists of various types of monitoring equipment nodes. These nodes are the base entities of the monitoring data storage architecture and are responsible for the monitoring and data collection of the environment within the integrated corridor.
- **Data Aggregation Layer**: This layer consists of a limited number of monitoring hosts and is the central link in the overall data storage system. The nodes in this layer are responsible for collecting and integrating the data collected by the collection devices in the collection layer and storing them together. Due to the complex environment in the corridor, monitoring data is also divided into different categories according to the object of detection: environment and equipment monitoring, fire monitoring, security monitoring, communication monitoring, structural monitoring and pipeline monitoring.

  - Environment and equipment monitoring. The monitoring objects are mainly the environment and the status of the equipment in the gallery, with monitoring equipment such as temperature and humidity sensors, gas detectors in the gallery, fiber optic thermometers, temperature sensitive cables and other equipment.
  - Fire monitoring. Fire detection is mainly fire-related monitoring of the gallery, such as monitoring of temperature-sensitive detectors, smoke detectors, alarm devices, fire extinguishing devices and other equipment monitoring in the gallery.
  - Security monitoring. This type of monitoring mainly includes in-porch cameras, personnel locating and access control monitoring.
  - Communications monitoring. Communications monitoring refers to the monitoring of communications-related equipment, such as walkie-talkies, fixed-line telephones and emergency communications equipment monitoring.
  - Structural monitoring. This type of monitoring is carried out on the gallery body and includes stress monitoring, gallery settlement monitoring and gallery deformation monitoring.
  - Pipeline monitoring. Pipeline monitoring refers to the monitoring of all types of pipelines entering the corridor, such as the monitoring of the flow of gases or liquids in the pipeline, the monitoring of the pressure to which the pipeline is subjected, and the monitoring of the temperature and corrosion of the pipeline.

- **Data service layer**. The data service layer, also called the data center, consists of data servers, which are the control center of the entire monitoring data storage architecture, responsible for the initialization of the entire data storage system, have the highest authority and are not controlled by third parties.

From the above data storage architecture analysis, we can see that the internal structure of the integrated corridor is relatively complex and there is a wide range of monitoring data, so it is necessary to classify and store all types of data to achieve efficient storage of monitoring data. The monitoring data storage architecture of the integrated corridor is relatively simple and hierarchical, with many devices responsible for monitoring data collection but a limited number of upper-level monitoring hosts. Therefore, this paper proposes the use of blockchain alliance chain technology for the storage of monitoring data. In addition, in order to solve the problem of long time consuming consensus between blocks in the process of storing a large amount of data, this paper adopts a "double chain" storage scheme, in which various types of complete monitoring data are stored under the chain, and key descriptive information of monitoring data (such as monitoring device information, monitoring time, data type, data file, etc.) is recorded on the chain. The complete data stored under the chain needs to be encrypted by calling the hash function before being archived to prevent malicious tampering or attacks on the monitoring data.

3 Integrated Pipe Corridor Monitoring Data Storage Model

3.1 Framework Construction for Monitoring Data Storage Nodes in Integrated Pipe Corridors

In this paper, the 3 data layers are divided into 2 types of nodes according to the monitoring data storage architecture: **Pre-selected node**: it is the main node in the federated chain storage system, which consists of the data server of the data service layer and the monitoring host of the data collection layer together. The data server of the data service layer has the highest authority in the entire storage system and is responsible for the initialization of the entire storage system. All nodes on the federated chain need to be verified by this node to have a legitimate identity. However, as this node has no subordinate monitoring device nodes, it does not participate in the competition for the blockchain record rights. Each consensus process only requires the backup of data blocks uploaded by other pre-selected nodes. The monitoring host of the data collection layer, on the other hand, is an ordinary pre-selected node on the storage federation chain, which needs to participate in the consensus, and each node needs to participate in the competition to obtain the right to upload data blocks to the chain. (2) **Side nodes**: They consist of monitoring devices of the data collection layer. They are managed by their respective monitoring device hosts and do not need to participate in the consensus algorithm of the whole chain in the process of data uploading. The monitoring devices are more flexible in uploading and downloading and only need to be verified by the data server.

In this case, the limited number of nodes avoids frequent participation in consensus and verification by the nodes on the federated chain, which greatly reduces the network overhead of the whole federated chain. The framework of monitoring data storage nodes in the federated chain is shown in Fig. 3. There are seven pre-selected nodes, of which six are monitoring hosts by type, and the last one is the data server responsible for the initialization of the whole system; the number of edge nodes is relatively large, but they are all managed by their respective monitoring hosts, and they only need to upload the monitored data to their respective monitoring hosts.

3.2 Integrated Corridor Monitoring Data Storage Mechanism

The main symbols that may be used in the data storage process and their meanings are shown in Table 1.

The data storage federated chain system is shown in Fig. 4. The entire data federated chain consists of the six monitoring host nodes mentioned above as well as one data

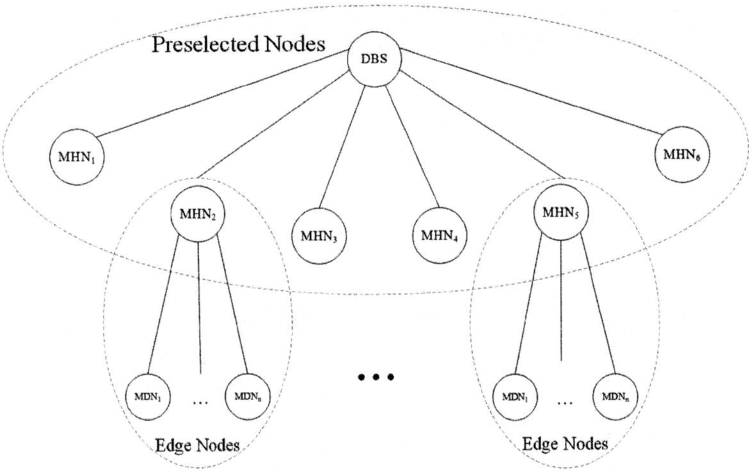

Fig. 3. Data storage nodes framework

Table 1. Key symbols and their meanings

Symbol	Implication
MDN_i	The ith monitoring device node
MHN_i	The ith monitoring host node
PK_{Hi}, SK_{Hi}	Public key, private key of the ith MHN
PK_{Di}, SK_{Di}	Public key, private key of the ith MDN
$E_{PK_R}(g)$	Encrypting messages with the public key of entity R
$Sig_{SK_i}(g)$	Sign the message with the private key of entity i
$x\|y$	Message x connects to message y
$hash(g)$	Taking hashes of data files
$x \to y$	Entity x sends a message to entity y

server node, with each monitoring host node openly competing, agreeing, verifying, and storing data throughout the storage process to form the block chain. Depending on the role of each type of node in the system, there are the following entities: monitoring device nodes, monitoring host nodes, and data server nodes.

- **Monitoring Device Nodes (MDNs)**: MDNs are a class of nodes that generate monitoring data and are managed by a specific MHN, the monitoring data generated by these nodes is a true reflection of the environment in the gallery and there is no false data, MDNs regularly upload monitoring data to their monitoring host.
- **Monitoring Host Nodes (MHNs)**: MHNs are a class of nodes that manage monitoring data and are responsible for the management of MDNs within their management scope

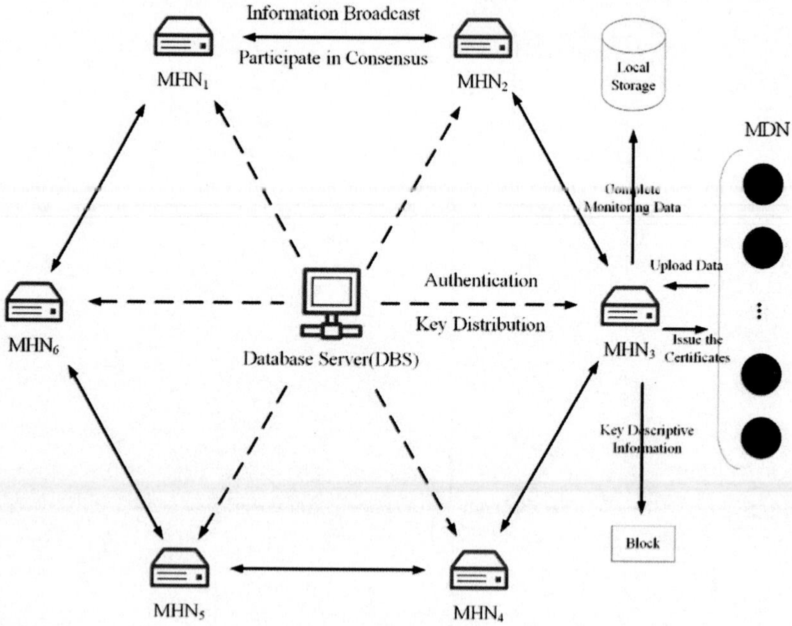

Fig. 4. System model

and the collection and storage of monitoring data generated by MDNs. They have their own storage devices and can verify the legitimacy and authenticity of their subordinate MDNs, and legitimate data can be saved to the storage devices. In addition, these nodes are also the core of the whole system to achieve the "dual-chain" storage of the federated chain. These nodes need to store the hash value of the recorded monitoring data "on-chain" and compare it with the monitoring data stored "off-chain" to achieve the integrity verification and tamper-evidence of the monitoring data.
- **Database Server (DBS)**: DBS is an important node in the federated blockchain that is responsible for the initialization of the entire storage system and does not compete for records between federated nodes, but has the highest authority over the entire storage system.

System Initialization

The data storage system in this paper is initialized using the Hash Message Authentication Code (HMAC) technique [19]. The DBS first authenticates the six MHNs as pre-selected nodes in the data storage federated chain and distributes the key pair PK_{Hi}, SK_{Hi}. After that, each MHN in the federated chain authenticates the MDNs within its scope, and the successfully authenticated MDNs become the licensed monitoring nodes within the scope of the MHN, and obtain their attribute set type and address information loc and the key pair PK_{Di}, SK_{Di}. This process can be expressed as $PK_{Di}, SK_{Di}, type, loc$, with which the MHN can authenticate, encrypt and sign the MDN.

Data Storage and the Dual Chain Storage

The data storage of the integrated corridor and the dual-chain storage scheme refer to the high-speed railway monitoring data storage scheme proposed by Zhang et al. [19]. The individual preselected nodes on the monitoring data storage federated chain have equal status. After each consensus mechanism, each participating node keeps a copy of the blockchain to ensure the security of the data on the chain. The specific dual-chain storage model of the integrated pipeline corridor monitoring data alliance chain is shown in Fig. 5.

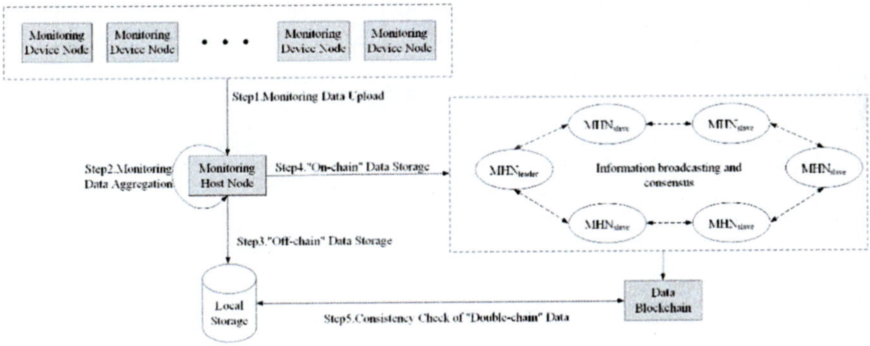

Fig. 5. Monitoring data storage model

Monitoring Data Upload

After the system initialization, the authenticated MDNs copy the latest data block index from the MHN they belong to, which indicates the specific storage location of each monitored data. When MDN_i sends a request for information to MHN_i, a pseudo-random identity (PID) is generated and then the data m_{i1} is digitally signed $Sig_{SK_i}(m_{i1})$, thus ensuring the authenticity and legitimacy of the sending entity. In addition, since the monitoring device works non-stop, to avoid duplicate uploads of data, MDNs use blockchain technology to generate a timestamp for each data upload, represented by timestamp. MDN_i encrypts the data m_{i1} with MHN_i's public key PK_{Hi}. Once MDN_i completes the authentication, it will send the final data d_i^1 to MHN_i, and the whole process can be expressed as:

$$MDN_i \rightarrow MHN_i \tag{1}$$

$$d_i^1 = \left\{v, PID, E_{PK_{Hi}}(m_{i1}), timestamp, type, loc\right\} \tag{2}$$

where, $v = HMAC_{SK_{Di}}\left(PID \parallel E_{PK_{Hi}}(m)\right)$.

$HMAC_{SK_{Di}}(\bullet)$ denotes the digital signature of the message authentication code of the data with SK_{Di} of MDN_i. d_i^1 denotes the data first uploaded by MDN_i to MHN_i.

Monitoring Data Aggregation

This process produces two types of data: complete monitoring data stored "off-chain" and key information on monitoring data stored "on-chain". Firstly, MHN_i verifies the timestamp in the MDN_i upload data to avoid replay attacks; then MHN_i uses MDN_i's PK_{Di} to authenticate the received HMAC to ensure the legitimacy of the MDN_i; after successful verification, MHN_i starts to receive legitimate monitoring data from MDNs. MHN receives a certain memory size of monitoring data and then performs an aggregation operation.

The aggregation process is as follows: MHN_i first packages and timestamps the monitoring data collected in the cache to obtain the monitoring data file; afterwards, the data file is encrypted using PK_{Hi}. The encryption operation ensures the legitimacy of the data file on the one hand, and the hash value obtained will be stored in the distributed ledger of the federated chain as part of the key data "on-chain" on the other hand.

$$MHN_i \rightarrow MHN_i \tag{3}$$

$$d = \{(d_1, d_2, \cdots, d_n)\, \mathrm{P}\; timestamp\} \tag{4}$$

$$hash = hash(d) \tag{5}$$

$$D_1 = \{d, hash\}, D_2 = \{PK_{Hi}, timestamp, hash\} \tag{6}$$

where, d_n denotes all monitoring data cached on MHN_i; d denotes the aggregated data file; D_1 denotes the data content to be stored "off-chain"; D_2 denotes the data content to be stored "on-chain".

"Off-Chain" Data Storage

To prevent the waste of storage resources and data redundancy caused by all monitoring data on the chain, the proposed monitoring data storage federated chain proposes to store only the key information of monitoring data on the federated chain for each node to share, while the complete monitoring data is stored on the storage devices of each MHN "off-chain". Before the monitoring data is stored "off-chain", a hash function is invoked to encrypt the pre-stored data files, and the hash value obtained is used to mark the complete monitoring data for subsequent verification of the integrity and consistency of the monitoring data.

"On-Chain" Data Storage

The data block is formed and linked to the data storage federated chain through a consensus process between MHNs. The MHN_i sends D_2 to the other MHNs and competes for the right to record the data block, the node with the right to record is also called the master node, MHN_{leader}, and the other MHNs are called slave nodes. At this point, the MHN_{leader} is granted write access to D_2 and the current data block, after which the MHN_{leader} has the authority to assign data D_2 to the slave nodes. Once consensus is reached, the MHN_{leader} can write the data to the blockchain, and other MHNs can obtain the latest copy of the data blockchain through the MHN_{leader}, thus enabling all MHNs in the federated chain system to keep a unique and reliable copy of the data block information. In addition, the slave nodes have the authority to check whether the MHN_{leader} is legitimate, and if the master node is abnormal, a new master node can be re-competed by the slave nodes through the protocol. The data block formation process can be represented as follows:

$$MHN_i \to MHN_{(1,\cdots,i-1,i+1,\cdots,n)} \tag{7}$$

$$(D_2, auth) \in MHN_{leader} \tag{8}$$

$$DB_i = DB_{i-1}\,\mathrm{P}\,db_d \tag{9}$$

where, *auth* denotes data block recording rights; DB_i denotes the currently formed block chain; db_d denotes the data block formed by data D_2.

Finally, to ensure that the complete monitoring data stored under the chain has not been tampered with, this paper proposes a double hash value consistency verification method for the data stored under the chain. If the hash values are equal, the data has not been tampered with, but if the hash values are different, the downloaded monitoring data has been tampered with after the chain has been uploaded. The authenticity of the monitoring data stored under the chain can be judged according to the double hash value consistency verification method.

Consensus Algorithms

The consensus algorithm is a way for each node in the blockchain network to compete for the right to record. Common consensus algorithms are: proof of workload (PoW), proof of interest (PoS), delegated proof of interest (DPoS), practical Byzantine fault tolerance (PBFT) and so on. In this paper, considering the characteristics of the integrated corridor, we propose to adopt the Byzantine (PBFT) algorithm consensus mechanism for the block consensus of the federated chain of monitoring data storage in the corridor. The consensus nodes of the PBFT algorithm are the six monitoring host nodes in the integrated corridor to keep the publication and storage of monitoring data and to guarantee the privacy, immutability and integrity of the data. The PBFT algorithm is suitable for asynchronous Byzantine fault tolerance. The main conclusion is that $n \geq 3f + 1$, where n denotes the total number of nodes involved in consensus in the system and f denotes the number of failed nodes allowed to occur in the system during the consensus process. That is, in a system with a total number of n consensus nodes, the consensus result is considered acceptable if there are no more than f faulty nodes. In this storage model $n = 6$, then $f \leq \frac{n-1}{3} = \frac{5}{3}$.The specific consensus process is shown in Fig. 6, where $MHN_i(i = 1, 2, \cdots, 6)$ denotes the monitoring host node and $MDN_{ij}(j = 0, 1, 2, \cdots, n)$ in denotes the monitoring device nodes within the range of MHN_i [19].

The MDN_{ij} sends the collected monitoring data to its monitoring host node MHN_i, which stores the data on local disks and then competes for storage privileges among

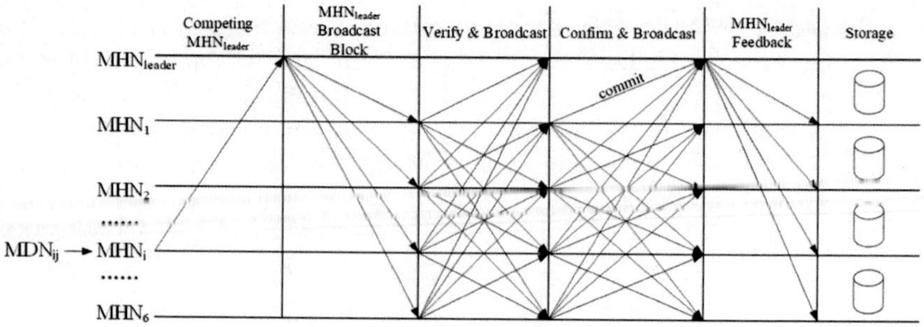

Fig. 6. Data storage consortium chain consensus process

all nodes. The node that succeeds in the competition is the only master node in the consensus process, and is known as MHN_{leader}, which has the authority to write data to the blockchain and is responsible for writing new blocks to the blockchain.

- Master node competition phase. In this phase, the six nodes in the federated chain (minus the seventh database server node) compete fairly for data logging privileges according to certain rules, and the successful node is recorded as MHN_{leader}.
- Master node broadcasts the block phase. In this phase the master MHN_{leader} broadcasts the locally stored new data block, the hash of the new data block and its digital signature to its slave nodes for verification.
- Slave nodes validate the blocks phase. In this phase, the received data blocks and digital signatures are verified by the slave nodes in the chain. Each node then broadcasts the validation result with its own digital signature to the whole network.
- Slave nodes compare and confirm phase. In this phase each node aggregates the transaction results and hash digests sent by other nodes. According to PBFT, if a slave node has more than $2f = \frac{10}{3}$ identical results to its own after verification, it broadcasts a commit message to the whole network.
- Master node feedback phase. In this phase, the master node collates all messages sent by the slave nodes and if it receives $2f + 1 = \frac{13}{3} \approx 4.33$ commit message, it submits the new blocks and the audit results of each slave node to the local blockchain and state database. The verified new data block is broadcast to the slave nodes with the master node's digital signature. From then on, the data blocks are stored in the federated chain in chronological order.

If the master node finds at least $f + 1 = 2.67$ (i.e. 3) slave nodes with abnormal verification results during the aggregation process, it means that the master node's record authority needs to be reconfirmed and the master node's hash value and digital signature need to be verified again. If the number of slave nodes with abnormal validation results is still not less than 3, the master node will need to be re-determined and another consensus calculation will be performed until the requirements are met.

4 Security Analysis of Monitoring Data Storage Models for Integrated Pipe Corridors

To demonstrate the advantages of the storage model proposed in this paper and its feasibility for the storage of monitoring data in integrated pipeline corridors, the security of the storage model is demonstrated. The features and advantages of the blockchain and the "dual-chain" solution will be discussed.

Data Tamper-Proof. The monitoring data in the integrated corridor can truly reflect the status information of the environment and equipment in the corridor and is not easily tampered with. Blockchain technology's own data tamper-proof nature protects against malicious attacks by external nodes. In addition, if an insider tries to evade responsibility by altering monitoring data, the "double-chain" mechanism will also identify the maliciously altered data. Therefore, the storage model can effectively ensure that the data cannot be tampered with and is safe and viable.

Anti-single Point of Attack. This feature is mainly for the third-party centralized storage which is vulnerable to single point of attack. This model stores all types of monitoring data in their respective node hosts, and stores the key information of the data in a distributed manner on each node of the blockchain, without any third-party restrictions. If a node's data is corrupted by a malicious attack, the entire storage system will not be significantly affected.

Data Traceability. The monitoring device nodes are encrypted and time-stamped during the data upload process; the monitoring host nodes are also digitally signed and time-stamped with a layer of the monitoring host during the data upload process. This storage mechanism ensures the legitimacy of each node and the reliability of the data source.

Collective Maintainability. The storage data on the storage federated chain is held and maintained by all nodes, even if individual nodes are damaged and lose data, other nodes on the federated chain can still transmit the data again through broadcasting, and the storage system can still operate normally, which is more robust.

Monitoring Data Sharing. The key information of monitoring data is stored in each node on the federated chain, and the node in need can submit a data viewing application to the target node through the key information, and once the verification is passed, the node can extract the corresponding data, thus realizing data sharing among monitoring nodes and solving the problem of data silos in the current monitoring data storage.

This section discusses and analyses the applicability of the storage model proposed in this paper in the context of monitoring data storage in integrated pipeline corridors. The model solves the problem that the current monitoring data storage is vulnerable to malicious attacks and tampering through the federated chain storage technology; and then avoids system lag caused by too much data on the chain by introducing the "double chain" mechanism.

5 Conclusion

Due to the complex environment in the corridor, the integrated pipeline corridor requires real-time monitoring of the corridor environment and equipment, from the corridor environment to equipment monitoring to corridor pipeline monitoring, which generates a large amount of monitoring data every day. However, although the current storage mechanism for monitoring data in integrated pipeline corridors has solved the problem of storing large amounts of data through distributed storage and cloud storage, the security of the data needs to be further strengthened. In this paper, blockchain technology is introduced based on the current data storage technology of the pipeline corridor, and a union chain monitoring data storage model is proposed to store monitoring data in a more secure and efficient way. It achieves the security, traceability and anti-tampering of monitoring data in the pipeline corridor, solving the difficulties of the previous storage technology such as centralized data storage, intractability and susceptibility to single-point attacks, laying the foundation for further promoting the development of the combination of blockchain technology and comprehensive pipeline corridor.

Acknowledgement. This work is supported by Beijing Logistics Informatics Research Base.

References

1. Qian, Q.: Transforming urban development by building integrated urban underground corridors. Tunn. Constr. **37**(6), 647 (2017)
2. Lee, P.C., Wang, Y., Lo, T.P., et al.: An integrated system framework of building information modelling and geographical information system for utility tunnel maintenance management. Tunn. Undergr. Space Technol. **79**, 263–273 (2018)
3. Yin, X., Liu, H., Chen, Y., et al.: A BIM-based lframework for operation and maintenance of utility tunnels. Tunn. Undergr. Space Technol. **97**, 103252 (2020)
4. Sampaio, A.Z., Gomes, A.M.: BIM interoperability analyses in structure design. CivilEng **2**(1), 174–192 (2021)
5. Zhou, G., Hu, W., Xiong, J.: Research and application of BIM+GIS-based urban underground integrated pipe corridor operation and maintenance management platform architecture. Intell. Build. Smart City **2018**(01), 64–68+74 (2018)
6. Ding, X.: Research on the Application of "BIM+GIS+IOT" Technology for the Operation and Maintenance of Urban Underground Integrated Pipe Corridors. Hebei University of Economics and Business (2021)
7. Yang, J.: Research on the development of intelligent management system for urban underground corridor based on cloud computing (2017)
8. Bai, Y., Yu, M., He, Y., et al.: Development and application of optical fiber security monitoring information management system for pipe gallery. Chin J. Undergr. Space Eng. **15**(S1), 391–399 (2019)
9. O-Yang, K., Kang, X., Zhang, J., et al.: Research on blockchain-driven data storage mechanisms in integrated management corridors. China High-Tech J. **2020**(16), 3 (2020)
10. Wang, D., Hua, P.: Data storage optimization design of large bridge health monitoring system. J. China Foreign Highw. **39**(05), 176–178 (2019)
11. Wang, L., Zhao, T., Zhang, Y., et al.: Hadoop-based wind power monitoring big data storage optimization and parallel query method. Electr. Meas. Instrum. **55**(11), 1–6 (2018)

12. Li, X., Sun, S.: Research on data storage technology in marine environment monitoring system. Ship Sci. Technol. **42**(08), 136–138 (2020)
13. Cao, H., Wang, L.: Research on secure storage strategies for massive digital content resources. Sci.-Technol. Publ. **2016**(09), 72–76 (2016)
14. Zhao, J.: Cloud-Based Data Storage Security Technologies and Applications. Nanjing University of Posts and Telecommunications, Nanjing (2019)
15. Yuan, Y., Wang, F.: Current situation and prospects of blockchain technology development. Acta Automatics Sinica **42**(4), 481–494 (2016)
16. Department of Informatization and Software Services, Ministry of Industry and Information Technology, China Blockchain Technology and Industry Development Forum. White Paper on China's Blockchain Technology and Application Development (2016). (2016-10-18) [2017-03-10]
17. Sheng, J., Mo, J., Yu, B., et al.: How does blockchain technology empower agricultural collaborative innovation and development: functional characteristics, efficiency enhancement mechanism and management mechanism. Chin. Rural Econ. **2021**(12), 22–43 (2021)
18. Ji, B., Chang, L., Zhu, L., et al.: Anti-tampering method and verification mechanism design of power data for private key leakage of blockchain system nodes. Electr. Power Autom. Equip. **41**(12), 87–94 (2021)
19. Zhang, L., Jiang, T., Jiang, F., et al.: Blockchain-based security storage solution for high-speed railway monitoring data. Comput. Eng. Des. **41**(4), 6 (2020)
20. Tsung-Ting, K., Hugo, Z.R., Lucila, O.M.: Comparison of blockchain platforms: a systematic review and healthcare examples. J. Am. Med. Inform. Assoc. **2019**(5), 5 (2019)
21. Ren, J., Li, J., Lin, H., et al.: Task offloading strategy with emergency handling and blockchain security in SDN-empowered and fog-assisted healthcare IoT. Tsinghua Sci. Technol. **27**(4), 760–776 (2022)
22. Zhu, X., Wang, D.: Research on blockchain application for e-commerce, finance and energy. IOP Conf. Ser. Earth Environ. Sci. **252**, 042126 (2019)
23. Sun, R., He, D., Su, H.: Application research of blockchain technology based on evolutionary game in supply chain finance. Chin. J. Manag. Sci. 1–18 (2022)
24. Integrated Corridor Safety O&M Management System. http://www.tsinghuahf.edu.cn/in-fo/?cid=104106

Review of Trends in Manufacturing Systems Based on Industry 4.0: The Opportunities

Gastón Lefranc(✉)

Pontificia Universidad Católica de Valparaiso, Valparaíso, Chile
gaston.lefranc@pucv.cl

Abstract. The paper presents a review of current trends in Manufacturing Systems that could give developing countries opportunities to improve their economy and create jobs. This would allow creating or improving companies that would have greater efficiency, flexibility, productivity and process security. These companies would produce better quality products and more profitable production. Industry 4.0 criteria are used to accelerate the automation of manufacturing, giving rise to a completely new business model, which facilitates product diversification. There are some opportunities that would allow the integration of new technologies, improve production and achieve more profits.

Keywords: Flexible manufacturing systems · Industry 4.0 · Artificial intelligence · IIoT · Automation

1 Introduction

Automation in the world of manufacturing has been mainstreaming for more than 10 years. One of the trends observed is the incorporation of new advances in technology in manufacturing companies, allowing them to be more competitive in the market. These technologies, along with new strategies, are resulting in faster automation and better production. This has made automation more accessible, changing the way companies operate. For example, automation is easy to achieve using robotic solutions that are simple to implement and use; which is complemented by simple manufacturing process management (MPM) systems and the use of human-robot collaboration. This allows companies to reduce costs, streamline workflows and get good results.

Flexible Manufacturing Systems (FMS) are used to produce a wide variety of products that are increasingly in demand by the market both in volume and speed. This implies increasing costs and slowing down production processes by incorporating automation and artificial intelligence, FMS makes it possible to speed up and improve production, even more so if it is combined with Industry 4.0 technology.

FMS is a set of manufacturing, assembly, storage and quality control cells. Cell automation is a set of machines that work as a system, consisting of processing workstations, automated material handling, and material storage. There are computers that

G. Lefranc—IEEE Senior Member.

S. Dzitac et al. (Eds.): ICCCC 2022, AISC 1435, pp. 182–192, 2023.
https://doi.org/10.1007/978-3-031-16684-6_15

manage each cell and these are connected in a hierarchical communication network to an FMS computer [1].

FMS uses software to enable machines to identify and distinguish between different styles of processed parts, make quick changes to operating instructions, and reconfigure the physical cell. The manufacturing cells of an FMS have the flexibility to use different types of numerically controlled (CNC) machines that handle machining tools, such as drills, lathes, etc. Also, you have the flexibility to route the movement of materials, parts, and workpieces from one machine to the next for each stage of production.

In the evolution of the automation of manufacturing processes, the concept of Industry 4.0 is reached, which includes the exchange of information wirelessly within industries. This criterion uses new technologies in production plants that include: Internet of Things IOT, Industrial Internet of Things (IIoT), Smart Manufacturing, Cloud Computing, Cognitive Computing, Artificial Intelligence, Cyber-Physical Systems (CPS). This means that all devices and technology are wirelessly connected and use actuators, sensors and monitors to visualize all processes, allowing a solid overview of operations in a company. In recent years, FMS has improved automation by using cobots for routing and machine process steps [2].

Manufacturing companies must enhance FMS with digital transformation technologies such as AI, generative design, and 3D printing.3D printing represents an example of such a transformation. With digital design and production of the parts and final product combined in many cases, 3D printing demonstrates true flexibility, even though it costs around 10 times more.

This paper presents a review of current trends in Manufacturing Systems that would provide developing countries with opportunities to improve their economy and create jobs. This would allow creating or improving companies that would have greater efficiency, flexibility, productivity and security of the processes in a company. These companies would produce better quality products and more profitable production. Industry 4.0 criteria are used to accelerate the automation of manufacturing, giving rise to a completely new business model, which would facilitate product diversification. There are some opportunities that would allow the integration of new technologies, improve production and achieve more profits.

2 Industry 4.0 Concept on Manufacturing Processes

Manufacturing companies should enhance FMS with digital transformation technologies like AI, generative design, and 3D printing. 3D printing represents an example of such a transformation. With digital design and the production of parts and the final product combined in many cases, 3D printing demonstrates true flexibility, although it is harder to understand the impact of the concept of industry 4.0 on manufacturing processes, it is important to understand the evolution. From Industry 1.0 to 4.0 [3].

In Industry 1.0, mechanical production facilities begin to be used in the 18th century, so that steam engines allow manufacturing.

Industry 2.0 begins at the beginning of the 20th century with the development of machines that work with electrical energy, simplifying mass production.

Industry 3.0 begins with the incorporation of electronic devices and systems in the 20th century. This creates machines with software and the use of processors.

Industry 4.0 is said to start in 1990 when the Internet and telecommunications transformed the way we connect and exchange information. This technology merged the physical world, the virtual world, and cyber-physical systems (CPS).

3 The Flexible Manufacturing Cell

A flexible manufacturing system (FMS) is a production method that is designed to easily adapt to changes in the type and quantity of the product being manufactured. Machines and computerized systems can be configured to manufacture a variety of parts and handle changing levels of production [4].

The general architecture of the flexible manufacturing cell (FMC) is Fig. 1.

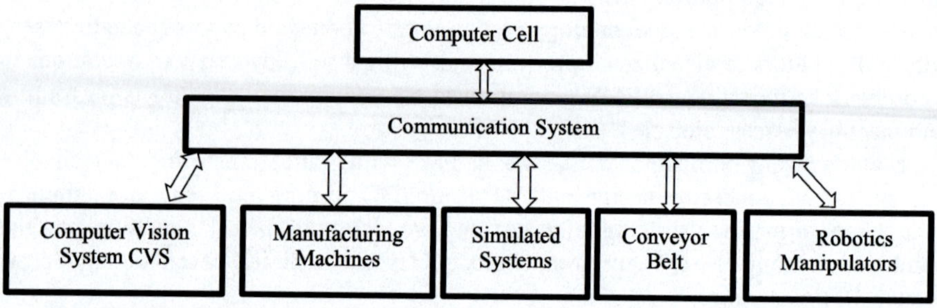

Fig. 1. General architecture of the flexible manufacturing cell FMC

Typically, an FMC is comprised of a cell computer, a vision system, a robotic manipulator, a conveyor, and a simulator system.

The information from the FMC is handled by a cell computer, which has a hierarchical structure: for example, the movement of the robotic manipulator RM and the pick and place of its effector is a function of the vision system CVS and the movement of the conveyor belt (Fig. 2) [5, 6].

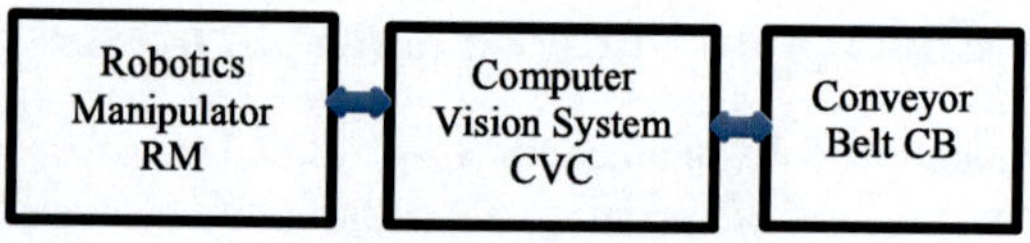

Fig. 2. FMC hierarchical structure

In Fig. 3 is shown the real FMC that use a Kuka robotic manipulator, a conveyor belt and pieces made using 3D printer.

In Fig. 4 is the same FMC simulated using CopeliaSim [7].

They both work alone and can also work in coordination, that is, if in the real FMC the task is to pick and place tools or parts, the simulated FMC is performing the same task. On the other hand, if the simulated FMC is to pick and place tools or parts, the real FMC does the same job.

Fig. 3. Real FMC [Kuka]

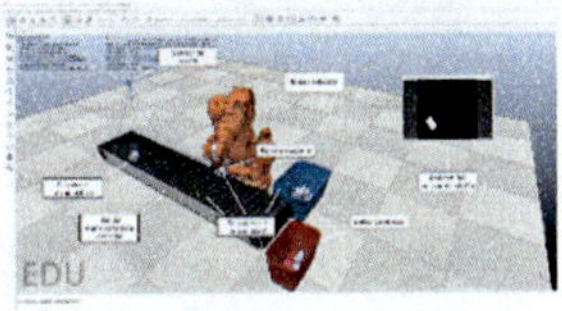

Fig. 4. FMC Simulation [CoppeliaSim] [7]

In addition, we can carry out hybrid work, that is, you can replace a real system with a simulated system. In the example, we can replace the real CVS using the CVS from the simulation. This idea has been used before using mobile robot control using software on a computer.

The images (Fig. 5) show that the X axis of the FMCF is a mirror with respect to that of the simulator. It is possible to send and/or receive the pick or place coordinates, if the proper calibration is done, since the robot's movement depends on its own reference system.

(a) Cartesian axes of real FMC [Kuka authorization]

(b) Cartesian axes of FMC simulated [CoppeliaSim] [7]

Fig. 5. Cartesian axes of the physical and virtual FMC.

A hybrid solution combines a virtual part with another physical part. For example, a CVS machine vision system is in a simulator [5] and sends the grip positions to both the real Kuka manipulator and the virtual manipulator; both controllers travel the same scheduled routes. These paths are created for the robot to traverse and track the part, along with its movement, from the starting point [8, 9] (Fig. 6).

This configuration works robustly in case of failures, since there have been no problems in the evaluations, and a correct follow-up of the built part is achieved, even if there are parts that are not fully supported on the web and/or in some orientation. Of the piece. Obtaining the gripping coordinates of the object is done with the BOF vector [10].

4 Industry 4.0 Automation

To improve the automation of FCM, some concepts of Industry 4.0 are used, mainly the exchange of information wirelessly within the Cell, MQTT an Internet of Things protocol

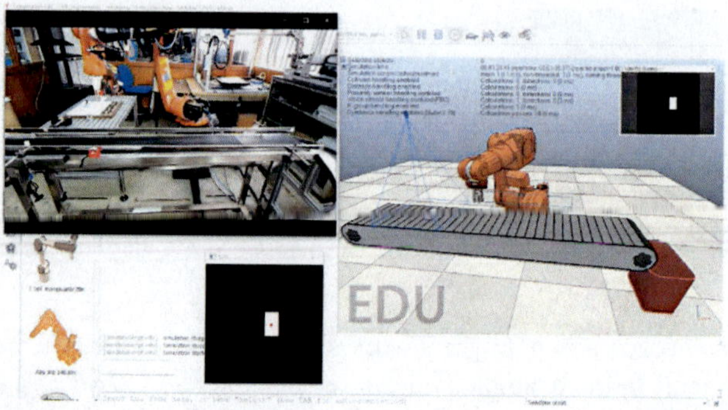

Fig. 6. Monitoring of a manufacturing part by artificial vision [Kuka authorization] and simulated using [CoppeliaSim] [7]

IOT and Cyber-Physical Systems (CPS) creating 3D parts. This criterion complements the uses of Cloud Computing and Artificial Intelligence.

This means that all devices and technology are wirelessly connected and use actuators, sensors and monitors to visualize all processes, allowing a robust overview of cell operations.

The new FCM architecture is (Fig. 7):

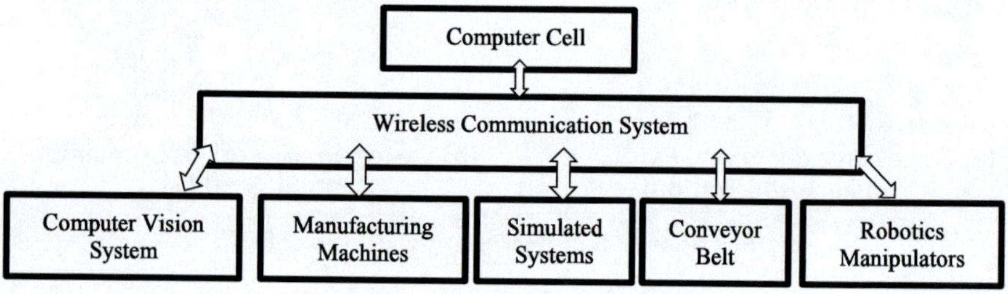

Fig. 7. New architecture of the FCM

The wireless communication system is based on the Standard for IoT Messaging, the MQTT protocol [11], which needs an intermediary to manage all the information and devices that are connected to the network, that is why the intermediary was used as in Mosquitto [12]. This agent is perfect as it is lightweight, secure, suitable for use on a variety of devices, and is open source. This type is fine for this purpose, due to its potential use and low cost, specifically a Raspberry Pi 4 Model B. Setting up a Mosquitto runner inside a Raspberry is easy.

5 Trends on Manufacturing

The following trends are considered to be already influencing manufacturing now or will in the near future.

Industry 4.0 describes the current trend in manufacturing towards "smart factories", Industry 4.0 is the fourth industrial revolution. Many of the following trends contribute to and make Industry 4.0 possible.

The widespread adoption and understanding of the latest information and communication technologies (ICTs) and manufacturing technology, such as IoT, the cloud, advanced computers, robotics, and the human workforce, work together to drive Industry 4.0. Companies looking to stay relevant, competitive and thrive in the marketplace must take advantage of the technological advances that have brought about the fourth revolution in manufacturing [3].

5.1 Automation for All

Automation is now more accessible to manufacturers changing the way companies operate. "Automation is for all" and is the next step in the industry.

Automation is possible with easy-to-use robotic solutions, easy-to-use manufacturing process management (MPM) systems, and with human-robot collaboration. With this automation, enterprises and companies can reducing costs, optimizing workflows and increasing their bottom line.

5.2 Supply Chain Technology

Effective supply chain management is vital for manufacturers looking to save costs while delivering product to customers when and how they want it. Procurement is affected when manufacturing is high mix and low volume production. New technology affects inventory, assembly, logistics, transportation, and sales. For example, adding sensors or barcodes to items allows companies to scan and track parts throughout the process, helping companies quickly see where improvements can be made.

5.3 The Internet of Things (IoT)

The Internet of Things (IoT) makes it possible to connect and to monitor the operations of various intelligent sensors and actuators. This connectivity provides valuable data that enables them to change, optimize and improve every stage of the manufacturing process. This connectivity is in the cloud powered by the Internet, IoT. Companies are also improving security, saving money, speeding up manufacturing and even creating new products with IoT capabilities.

Forbes projects investment, globally, of more than $267 billion in IoT by 2022, with 50% focused on manufacturing, transportation, and logistics.

5.4 Autonomous Mobile Robots (AMR) Drons.

Since 2018, robotics shipments were up 41% from 2018, especially in non-automotive workplaces. The main trend are autonomous mobile robots (AMR) and drons. Waypoint's easy-to-use AMRs are reckoned to be the future of manufacturing and are reinventing the way workers operate within the industry. The Vector and MAV3K AMRs can navigate on their own, can be configured with Dispatcher, the Waypoint robot software is intuitive allowing it to be used with different fixtures and can be customized to suit most applications. Mobile manipulators.

They are manipulators mounted on an AMR mobile robot, with automatic navigation. With this mobility, they can be moved to automate other manufacturing areas, while optimizing the use of these manipulators.

5.5 Collaborative Robots (Cobots)

Manufacturing trends are collaborative work between robots and people, resulting in faster, safer work. Cobots are a valuable tool that help improve efficiency in the workplace. AMRs are well suited to this collaborative work, as multiple non-value-added jobs, such as moving heavy goods, can be performed while people focus on skilled labor.

A use of cobots in a logistics environment is for order picking. In traditional order fulfillment roles, the physical act of walking to pick up the product accounts for 50% of the entire process, consuming valuable time and money.

Automated picking is achieved when a robot is used to go to the warehouse and pick up a product, leaving people to do the work of delivery. A human r picks between 60 and 80 orders per hour, an automated system can pick up to 300 in the same period of time, according to Westernacher Consulting.

5.6 Workforce Attraction, Retention, and Training

Companies want to have trained and qualified employees. There is a labor shortage in the industry, and Deloitte has projected that almost half of the 4.6 million manufacturing jobs that will be needed in the next 10 years will go unfilled. To have such workers, they must be trained so that they can easily master automation tools such as robots and collaborative work with cobots. This optimizes processes and reduces repetitive and tiring tasks.

5.7 Connectivity

The new manufacturing trends have the ability to connect universally, with technologies that allow connecting with any device or system. Waypoint's omnidirectional AMRs, for example, can be adapted to any manufacturing configuration, as the device itself and its operating technology can support a wide variety of operations. That is, they can be tailored to your operations, without the cost of a fully customized solution.

5.8 Additive Manufacturing

This is the term is better known as 3D printing. Through the use of computer-aided design (CAD) software, custom parts and products can be created layer by layer for your customers. Additive manufacturing can also create models, prototypes, moulds, wax castings or components of final finished products.

Additive manufacturing is a huge growth trend in the industry, and an important tool to save time and money. New selective laser sintering (SLS) technology for 3D printing metals enables companies to fabricate metal products by eliminating tooling costs and dramatically shortening product development time.

5.9 Enterprise Resource Planning ERP

Enterprise resource planning (ERP) has been used in manufacturing for many years. Only now are there more cloud-based options that are easier to deploy and easier for small businesses to use. ERP systems facilitate automation in different areas of operations, obtaining information necessary to supervise the entire manufacturing.

5.10 Artificial Intelligence and Machine Learning

Companies have access to more data than ever, the artificial intelligence (AI) and the machine learning allow them to make the most of that information. AI does not mean walking, talking robots, but rather refers to a computer system's ability to recognize trends and draw logical conclusions that can help manufacturers make data-driven decisions. AI and machine learning can help improve many aspects of a manufacturing operation, such as: inventory management; supply chain visibility; reduction of storage costs; asset tracking; forecast accuracy; reduction of transportation costs; and more predictive maintenance.

5.11 Predictive Maintenance

Predictive maintenance is possible by IoT, AI and machine learning that helps manufacturers avoid downtime by detecting problems before they arise.

This means saving maintenance costs such as downtime, extending the life of the machines, and extending the life of the machines used in the manufacturing operation. McKinsey & Company projects that predictive maintenance programs can reduce maintenance costs by approximately 20% and reduce unplanned machine outages by 50%.

6 Opportunities for Developing Countries

Manufacturing Systems could be used in developing countries giving opportunities to improve their economy and create jobs. This would make it possible to create new companies or improve current ones with industry 4.0, resulting in companies with greater efficiency, flexibility, productivity and process safety. This is an opportunity for third world countries.

Another opportunity is to try to attract some reshoring companies, as explained below.

The relocation or the return of companies from China to the US and possibly to Europe provides job opportunities in those countries, and it is because of the shortage of products, because of the pandemic, logistics and also the business because the cost of hand of work in China has increased and no longer gives so much advantage. There are indications that some of these companies would establish themselves in stable and secure developing countries.

The shortage of products in the United States would have become a constant. For this reason, in June 2021, a specialized work team is created in the US to deal with the shortage crisis, from video game consoles to furniture, clothing, food and even cars. This is due to the companies' dependence on producing parts or all of the product in other latitudes. Some companies want to bring factories from the country. This phenomenon is called "reshoring" to the USA or in Mexico, Colombia or Chile, being an economic opportunity, it represents for those countries [13, 14].

With globalization, many manufacturing companies move to China, Vietnam or India, where wages are much lower, and then import and sell them in the large North American market. This is called "offshoring". This led to the US economy losing an alarming number of 2 to 2.4 million jobs lost between 1999 and 2011 due to increased imports from China.

Harry Moser says: "Wages in China have been growing 10–15% a year for most of the last 20 years," says Moser, "so even before covid, before the trade war, there are jobs in cheaper places like Vietnam, Cambodia or even Mexico. This means that around 15% are leaving China"[15].

The US president, Joe Biden, seeks to encourage reshoring to solve problems in the production chains. To encourage relocation to the US, by the end of 2020 companies announced that 160,649 jobs would return to the country or be created as a result of foreign direct investment.

During 2021, some of the largest companies in the US announced multi-million-dollar investments to boost domestic production. During 2021, some of the largest companies in the US announced multi-million-dollar investments to boost domestic production, creating a new magnet factory and a battery factory for electric cars.

Toyota North America announced that it has broken ground on a battery plant in North Carolina that would begin production in 2025 and employ more than 1,800 people.

In the same way, Intel, the company that manufactures microprocessors, will build the world's largest microchip factory in New Albany, Ohio and two new plants in Arizona, which would have an investment of close to US$20,000 million [16].

7 Conclusion

A review of current trends in Manufacturing Systems that could be used in developing countries giving opportunities to improve their economy and create jobs has been presented. This would allow creating or improving companies that would have greater efficiency, flexibility, productivity and security of the processes in a company.

If there are companies that have FMS Systems, it is possible to convert the systems with industry 4.0 criteria, into a wireless communications system. These companies would produce better quality products and more profitable production. Industry 4.0 criteria are used to accelerate the automation of manufacturing, giving rise to a completely new business model, which facilitates product diversification.

The trends in manufacturing that are observed can be summarized as:

Automation is accessible to all manufacturers easy to use;

Improvements in supply chain technology affect inventory, assembly, logistics, transportation, and sales;

The Internet of Things (IoT) makes it possible to connect and monitor the operations of various intelligent sensors and actuators. This connectivity allows them to change, optimize and improve every stage of the manufacturing process;

Robotics improves the manufacturing process with autonomous mobile robots (AMR), robot manipulator mobile robots, with automatic navigation, optimizing the use of these manipulators; the use of Collaborative Robots (Cobots) between robots and people, resulting in faster and safer work. Additionally, cobots are used in a logistics environment to pick product orders from the warehouse;

Another trend is for companies to train workers in automation tools such as robots and collaborative work with cobots to optimize processes by reducing repetitive and tiring tasks;

The use of Additive Manufacturing, known as 3D printing, with computer-aided design (CAD) software, allows obtaining parts, models, prototypes, molds, wax castings or components of final finished products;

Enterprise resource planning (ERP) is used in making cloud-based options that are easier for small businesses to implement and use, making automation easier;

Artificial intelligence (AI) and machine learning help improve inventory management; supply chain visibility; storage cost reduction; asset tracking; Forecast accuracy; reduction of transportation costs; and more predictive maintenance.

Predictive maintenance is made possible by IoT, AI, and machine learning that help manufacturers avoid downtime by detecting issues before they arise.

Manufacturing Systems could be used in developed countries giving opportunities to improve their economy and create jobs. This would make it possible to create new companies or improve current ones with industry 4.0, resulting in companies with greater efficiency, flexibility, productivity and process safety. This is an opportunity for developing countries. The other opportunity is to try to attract some reshoring companies to our countries.

References

1. Lefranc, G.: Some trends in robotics and automation. In: 2021 IEEE CHILEAN Conference on Electrical, Electronics Engineering, Information and Communication Technologies (CHILECON), pp. 1–8 (2021). https://doi.org/10.1109/CHILECON54041.2021.9703003
2. Sullivan, M.: Flexible manufacturing systems combined with industry 4.0 can further progress toward mass customization (luxresearchinc.com) (2020)
3. Oztemel, E., Gursev, S.: Literature review of Industry 4.0 and related technologies. J. Intell. Manuf. **31**(1), 127–182 (2018). https://doi.org/10.1007/s10845-018-1433-8
4. Leighton, F., Osorio, R., Lefranc, G.: Modelling, implementation and application of a flexible manufacturing cell. Int. J. Commun. Comput. Control 1 (2011)
5. Osorio-Comparán, et al.: Recognition and simulation of a vision system for a KUKA KR-5 industrial manipulator. In: 2021 IEEE CHILEAN Conference on Electrical, Electronics Engineering, Information and Communication Technologies (CHILECON), pp. 1–6 (2021). https://doi.org/10.1109/CHILECON54041.2021.9702984

6. Osorio-Comparan, R., et al.: Vision system for a Kuka KR-5 industrial manipulator. 2019 IEEE CHILEAN Conference on Electrical, Electronics Engineering, Information and Communication Technologies (CHILECON), pp. 1–7 (2019). https://doi.org/10.1109/CHILECON47746.2019.8988027
7. CoppeliaSim Release 4.2.0 is out, 6 April 2021. https://www.coppeliarobotics.com
8. Peña-Cabrera, M., Amezquita, L., Lomas-Barrié, V., Osorio, R., López, I.J., Lefranc, G.: Artificial vision in embedded system for rigid parts recognition. In: 2019 IEEE CHILEAN Conference on Electrical, Electronics Engineering, Information and Communication Technologies (CHILECON), pp. 1–6 (2019). https://doi.org/10.1109/CHILECON47746.2019.8988010
9. López-Juárez, I., Castelán, M., Castro-Martínez, F.J., Peña-Cabrera, M., Osorio-Comparan, R.: Using object's contour, form and depth to embed recognition capability into industrial robots. J. Appl. Res. Technol. **11**, 5–17 (2013)
10. López-Juárez, I., Rios-Cabrera, R., Peña-Cabrera, M., Méndez, G.M., Osorio, R.: Fast object recognition for grasping tasks using industrial robots. Computación y Sistemas **16**(4), 421–432 (2012). ISSN 1405–5546
11. R MQTT: The Standard for IoT Messaging. https://mqtt.org/
12. Mosquito Version 2.0.13 released, Eclipse Mosquitto. https://mosquitto.org/
13. Filip, F.G.: DSS—a class of evolving information systems. In: Dzemyda, G., Bernatavičienė, J., Kacprzyk, J. (eds.) Data Science: New Issues, Challenges and Applications. Studies in Computational Intelligence, vol. 869. Springer, Cham (2020). https://doi.org/10.1007/978-3-030-39250-5_14
14. Abuchaibe, R.: Reshoring: why US companies want to return to manufacturing in the country. BBC News World, 27 January 2022
15. Moser, H.: Company studies cases of Reshoring, pdf slides. https://reshorenow.org/content/companies_reshoring/Cases9_26_16.pdf. Accessed 31 Dec 2021
16. Moser, H.: Company list of manufacturing companies company list thru, 31 January 2021. (reshorenow.org)

Soft Computing Including Fuzzy System Approach

Non-parametric Rank Correlation Trained Single-Hidden Layer Feedforward Neural Networks for Medical Data

Smaranda Belciug[1](✉) and Renato Constantin Ivanescu[2]

[1] Department of Computer Science, Faculty of Sciences, University of Craiova, 200585 Craiova, Romania
sbelciug@inf.ucv.ro

[2] Department of Computers and Information Technology, Faculty of Automation, Computer and Electronics, University of Craiova, 200585 Craiova, Romania
constantin.ivanescu@edu.ucv.ro

Abstract. Massive amounts of medical data led to a faster, more accurate diagnosis and course of treatment for cancer. Machine learning (ML) algorithms have been developed increasingly in order to process medical data. By embedding knowledge from data into the algorithm's architecture we customize MLs to provide personalized diagnosis and personalized treatment plan. The goal of this study is to statistically analyze non-parametric correlation coefficients for initializing the input-hidden weight matrix of three single hidden layer feedforward neural networks (SLFNs). A SLFN is able to approximate a continuous function given a training dataset as input for a desired output. Using different non-parametric correlation coefficients, ρ, τ, and γ, we developed three novel SLFNs that have their input hidden layer weight matrix initialized using the above-mentioned correlation coefficients computed between the input and the output. The matrix that contains the hidden output weights is computed in just one step, using the Moore-Penrose pseudoinverse matrix. A statistical analysis is performed for assessing the models' results when applied on three medical datasets that regard differentiating lung, breast, and liver cancer. The statistical benchmarking showed that the ρ-SLFN is superior to the other two. All three models were overall competitive to other state-of-the-art algorithms.

Keywords: Spearman correlation · Kendall tau correlation · Goodman Kruskal Gamma · Somer's D · Single hidden feedforward network · Statistical assessment

1 Introduction

Establishing a fast and accurate diagnoses is not an easy task in the light of the huge amount of medical data that is produced by the healthcare system. A viable solution is using intelligent systems to assist doctors and transform the medical act into an automatic one. The involvement of artificial intelligence in the healthcare sector led to an overall decline of 29% in cancer deaths between 1991 and 2017, [1]. Artificial neural networks

S. Dzitac et al. (Eds.): ICCCC 2022, AISC 1435, pp. 195–207, 2023.
https://doi.org/10.1007/978-3-031-16684-6_16

(NNs) are known for their adaptive learning and nonlinear mapping properties, therefore are ubiquitous when it comes to automated medical diagnosis.

Different approaches have been developed to speed the NNs' convergence. One way to achieve this convergence is to try initializing the synaptic weights the best possible way. Kathirvalayakumar and Subavathi used Cauchy's inequality merged with sensitivity analysis to establish the initial weights in a NN, [2]. Yam and Chow used two different approaches for synaptic weights' initialization: the first was based on the Cauchy inequality and linear algebra, [3], while the second was based on a geometrical technique, [4]. Belciug used logistic regression to initialize the synaptic weights for a two-class decision problem, [5], and extended it to multiple classes by comparing a cascade approach versus a parallel one, [6]. Hsiao et al. designed a hybridization between the backpropagation algorithm and the partial-least square method, [7].

Mishkin and Matas proposed a Layer-sequential unit-variance method that preinitialized weights for each convolution or inner-product layer with orthonormal matrices for a deep neural network, [8]. A proper random initialization together with an increasing schedule for the momentum parameter was proposed as training technique for deep and recurrent neural networks, [9].

Another idea would be to compute the correlations between the input and the output and assign them as the values of the synaptic weights between the input and the hidden layer, so that we could quantify the goodness-of-fit between them. Non-parametric correlation is used in machine learning to state in quantitative terms the strength and direction of the relationship between variables, since the value of one variable is related to the value of the other variable in such a manner that the hazard is excluded. This idea is based on the subjective Bayesian paradigm, by considering the synaptic weights as posterior probabilities which are computed using the priors (specific information about an object through the correlation of its attributes and the class label). Through this process we can build competent predictive algorithms because the synaptic weights will explain certain amount of variance in the outcome. The scope of this paper was two-fold: (a) at first, to develop three novel SLFNs which use three non-parametric coefficients to initialize the input-hidden weights, and secondly, (b) to perform a statistical comparison between these rank correlation approaches. We are interested in the relationship between variables that are related through a non-linear function, hence, we have included in our study the following approaches: Spearman, Kendall Tau, Goodman-Kruskal's Gamma, and Somers' D correlations. We have used each one of these methods to initialize the weights in different single-hidden layer feedforward neural networks (SLFNs), and then benchmarked the resulting novel algorithms on three different cancer datasets. The training process is performed in a single step using the Moore-Penrose pseudo-inverse matrix.

The experiments were performed on datasets that used different diagnosing techniques: *i)* through ribonucleic acid (RNA) sequencing expression data for differentiating breast and lung cancer, and *ii*) through the METAVIR F score for differentiating hepatic cancer. We used multiple datasets with different types of attributes (categorical and numerical) to observe the model's behavior in each situation. A thorough statistical analysis was performed to determine which model is more robust and performs better.

The paper is organized in the following manner: Sect. 2 briefly presents the dataset used in the study and some state-of-the-art techniques and their obtained results.

Section 3 outlines the rank correlation approaches and the novel obtained SLFN algorithms. The paper continues with Sect. 4 that presents the design of the experiments and the results which we have obtained by applying the novel SLFNs. A thorough statistical analysis compares the rank correlation methods, demonstrating the advantages and disadvantages of each novel SLFN. In Sect. 5 we discuss a statistical comparison between the new SLFNs and other high performance state-of-the-art neural networks. The paper ends with the final conclusions in Sect. 6.

2 Materials

2.1 RNA and METAVIR F Datasets

The novel SLFNs have been tested for comparison on two medical datasets that contain RNA gene expression regarding breast, and lung cancer. Both datasets are publicly available (mldata.org). The first data dataset is *Breast cancer Duke (BD)*. There are 68 observations with 7129 attributes each. It has two decision classes and is divided as such: 45 cases with positive estrogen receptor, and 41 cases with negative estrogen receptor. The dataset is balanced, [10].

The second data set is *Lung cancer Michigan (LM)*. It contains RNA gene expressions for 86 cases with 7139 features divided in two classes. The dataset is unbalanced. The first class contains 67 malignant tumors (67 stage I and 19 stage III), and the second class contains 10 non-neoplastic lung tissue, [11].

The third dataset is METAVIR F *(FIB)*. The dataset contains 722 with 25 attributes and five decision classes that represent the liver fibrosis stages. FIB dataset does not contain RNA sequences, the attributes representing the liver stiffness measurements, clinical, biological, epidemiological, and anthropological, [12]. This dataset is not balanced either. We have five classes with the following distribution of data: 30 samples in the first class, 226 in the second, 165 in the third, 88 in the fourth, and 213 in the fifth.

For all datasets we have used the 10-fold cross-validation in order to split them into training and testing.

2.2 State-of-the-Art Machine Learning Algorithm Applied on These Datasets

Different ML techniques have been applied on three above mentioned datasets. In what follows, we are going to briefly present their results reported in literature. Shevade and Keerthi preprocessed the BD dataset to standardize it before applying a novel training algorithm, [12]. The method was based on the Gauss-Seidel algorithm, which is generally used for the unconstrained optimization problem. In this study the authors used it for the sparse logistic regression problem. In the preprocessing step, the dataset was arranged as a matrix with N columns and m rows, every row being standardized having mean zero and standard deviation one. The authors reported 81.9% accuracy on the BD dataset. Belciug and Gorunescu used an adaptive single-hidden layer feedforward neural network, [13]. The dataset was not preprocessed, and the authors reported 83.31% accuracy. Czajkowski et al. proposed an evolutionary algorithm for induction of top scoring pair decision trees on the LM dataset. They reported 95.87% accuracy, [14]. An adaptive SLFN which used the Goodman-Kruskal gamma rank correlation together with a filtering module based on the corresponding p-value obtained 92.84% accuracy, [15].

Stoean et al. selected the most important features from the FIB dataset using a hill-climbing algorithm followed by a cooperative co-evolutionary classifier. In this study, the authors reported testing accuracies between 58.89% and 62.11%, [16]. Another approach that merged a genetic algorithm together with a support vector machine obtained accuracies between 61.5% and 64.07%, [17]. Other approaches regarded: a tandem feature selection and an evolutionary-driven neural network which obtained 65.21%, [18]; another evolutionary strategy-based decision system which obtained 62.11%, [19]; a 3-layer and a 4-layer perceptron obtained 54% mean accuracy, [20]; a radial basis function which obtained 57% accuracy, [21]; a cascaded logistic regression trained SLFN and a parallel logistic regression trained SLFN which obtained 52%, and 80.87% respectively, [6].

3 Materials

In general, the hidden neurons of a SLFN are randomly generated. In this study, we want to statistically compare four different alternatives to this initialization. Instead of randomly generating the hidden units, we make them problem dependent and estimate them using different statistical rank correlations between the features and class labels. Because the relationship between them is nonlinear, we decided to use the Spearman, Kendal Tau, Somers' D, and Goodman Kruskal Gamma ranks, so that we can embed knowledge into the network's architecture.

Let us denote the number of cases in the training set with N. Each case contains p features, $\mathbf{x}_k = (x_1^k, x_2^k, \ldots, x_p^k)$, where $x_i^k, i = 1, 2, \ldots, p$ is the i-th attribute of $\mathbf{x}_k$, and $y_j, j = 1, 2, \ldots, q$ is the decision class Ω_j, where q is the number of decision classes. From a probabilistic point of view, we can presume that for each object $\mathbf{x}_k$, $k = 1, 2, \ldots, N$, the values x_i^k that correspond to the attributes A_i, are governed by a random variable denoted X_i. Continuing our presumption, we can say that for each object $\mathbf{x}_k$, the corresponding class labels y_j are governed by a categorical random variable Y. Thus, we can use the above mentioned non-parametric ranks to quantify the non-linear relationship between the two random variables.

The Spearman rank correlation ρ is the nonparametric version of the Pearson coefficient. Different from Pearson, Spearman is not as sensitive when it comes to outliers, because it uses ranks instead of values. We can compute ρ using the following formula:

$$\rho(X, Y) = 1 - \frac{6\sum d_i^2}{N\left(N^2 - 1\right)}, \tag{1}$$

where d_i is the difference between variable rankings, [22, 23]. This formula can be used only if there are no tied ranks. If we encounter tied ranks we use the following formula:

$$\rho(X, Y) = \frac{\frac{1}{N}\sum_{i=1}^{N}\left(R(X_i) - \overline{R(X)}\right)\cdot\left(R(Y_i) - \overline{R(Y)}\right)}{\sqrt{\left(\frac{1}{N}\sum_{i=1}^{N}\left(R(X_i) - \overline{R(X)}\right)^2\right)\cdot\left(\frac{1}{N}\sum_{i=1}^{N}\left(R(Y_i) - \overline{R(Y)}\right)^2\right)}} \tag{2}$$

where $R(X_i)$ and $R(Y_i)$ are the ranks, and $\overline{R(X)}$ and $\overline{R(Y)}$ are the mean ranks.

Kendall Tau τ is another non-parametric alternative of the Pearson's correlation. We mentioned that if we have many tied ranks, we can use Spearman's formula (2). A better way would be to use the Kendal Tau coefficient if the sample size is small. To compute τ we use the following formula:

$$(X,Y) = \frac{C - D}{\sqrt{(n_0 - n_1)(n_0 - n_2)}} \tag{3}$$

where C represents the concordant pairs, and D represents the discordant pairs, $n_0 = \frac{N(N-1)}{2}$ is the binomial coefficient [28], $n_1 = \sum_i \frac{t_i(t_i-1)}{2}$, where t_i is the number of tied values of the ith sample of ties for the first variable, whereas $n_2 = \sum_j \frac{u_j(u_j-1)}{2}$, where u_j is the number of tied values for the jth sample of ties for the second variable [24–26].

Somers' D rank rank correlation can be defined through Kendall Tau rank using the following equation:

$$D(Y,X) = \frac{\tau(X,Y)}{\tau(X,X)} \tag{4}$$

Because in our case the X random variable has a continuous probability distribution, then $\tau(X,X)$ equals 1, hence Somers' D has the same value Kendall's Tau. Thus, from our statistical comparison we have removed Somers D based SLFN [27, 28].

The Gamma statistic γ, also known as the Goodman Kruskal Gamma rank correlation, is generally preferred when we are dealing with many tied ranks or the data contains outliers. The returned values are the same as for the rest of the ranks. When computing Gamma, the variables are treated symmetrically, meaning that one does not have to hypothesize which variable is dependent and which is independent [29]. We obtain γ using the following formula:

$$\gamma(X,Y) = \frac{C - D}{C + D} \tag{5}$$

where C represents the concordant pairs, and D represents the discordant pairs.

3.1 Non-parametric SLFNs

The hidden nodes in the three proposed novel parametric SLFNs (ρ, τ, γ- SLFNs) are initiated considering the natural connection between the features and the decision classes, technically adapting the NNs to the knowledge embedded in the data, and not randomly generating them. Practically, the hidden neurons are transformed from being independent to be being dependent.

Let us denote $\tilde{N}$ the number of hidden nodes in the network. $\tilde{N}$ is flexible, ranging from 2 to 30, its value being chosen depending on the dataset used. The choice of $\tilde{N}$ was of heuristic nature, even if in literature there are multiple ways to set it. For instance, a SLFN that is activated through a nonpolynomial activation function can approximate any continuous function, if $\tilde{N}$ is large enough, [30]. Other theories state that a preliminarily given number $\tilde{N}$ can approximate univariate and multivariate functions, [31, 32].

Let w_{ih}, $i = 1, 2, \ldots, p$, $h = 1, 2, \ldots\tilde{N}$ be the synaptic weights that connect the input to the hidden layer. W is the input hidden matrix. For each novel non-parametric SLFN we initialized the weights by computing the corresponding rank correlation (ρ, τ, γ) between each attribute and the decision class. We considered the strength of the association, as well as the direction. The hyperbolic tangent played the role of the non-linear activation function, $f(u) = 1.7159 \cdot \tanh\left(\frac{2u}{3}\right)$, due to its fast convergence, [3]. M is the hidden layer output matrix, and $M^+ = \left(M^T \cdot M\right)^{-1} \cdot M^T$ the Moore-Penrose pseudo-inverse matrix, where M^T is M's transpose. β is the hidden-output matrix, and o the output.

The algorithm is as follows:

1. Using the *one-hot-encoding* rule for categorical data, given a decision class $\Omega_j, j = 1, 2, \ldots, q$, transform the output label y_j into an array, e.g. $y_1 \sim (0, 0, \ldots, 1), y_2 \sim (0, 0, \ldots, 1, 0), \ldots, y_q \sim (1, 0, \ldots, 0)$.
2. For each hidden neuron $h_j, j = 1, 2, \ldots \tilde{N}$, compute the synaptic weights w_{ij} using one of the following formulas (depending on which SLFN we are training on): for the ρ-SLFN, $w_{ij} = \rho\left(x_i^k, y_k\right)$, for the τ-SLFN, $w_{ij} = \tau\left(x_i^k, y_k\right)$, and for the γ-SLFN: $w_{ij} = \gamma\left(x_i^k, y_k\right)$, $i = 1, 2, \ldots, p$, and $k = 1, 2, \ldots, q$.
3. Compute the hidden layer output matrix M, using the corresponding dataset.
4. Compute $\beta = M^+ \cdot o$, the matrix containing the hidden-output weights.

The resulting matrixes W and β are going to be used in the testing phase. The three SLFNs have been implemented in Python and ran on 2.7 GHz Intel Core i5, 8 GB (RAM).

4 Results

The goal of this study was to assess the performances of three novel SLFNs that are based on three nonparametric correlation coefficients. We decide not to preprocess the data, so all the SLFNs have been trained and tested on raw unprocessed datasets. We have used for all the algorithms the 10-fold cross validation method for validation purposes. All algorithms have been run independently for 130 times (e.g. 130 runs for each complete cross-validation cycle). The statistical power (two-tailed type of null hypothesis with default statistical power goal $P \geq 95\%$ and type I error $\alpha = 0.05$) was achieved through this number of computer runs. When designing the experiments, we need to compute the statistical power to obtain the sample size estimation. This process allows us to balance accuracy and computational cost, a sample size too low would imply a loss in performance, on the other hand a too large sample size would lead to an increase in computational and time costs. Each algorithm's performance is computed as the average accuracy (ACA) obtained on the 130 complete cross-validation cycles. We have computed the standard deviation (SD) of the obtained ACAs, and the 95% confidence interval (95% CI). The SD provides crucial information regarding the model's stability and robustness. Recall that we have tested the algorithms on three different datasets, and the SD may fluctuate from one dataset to another, having a smaller or a larger size, meaning that the algorithms do not have *omnibus robustness*.

Besides these descriptive statistical parameters, we have computed the Precision-Recall curves (PR-AUC). The area under the ROC curve (AUC) was ruled out of the statistical discussion, since we are dealing with unbalanced data, hence the results might have led to a modification of the false positive rate, that is used for computing the AUC, [33].

To objectively compare the three algorithms, we have performed a throughout statistical analysis that involved: the verification of the performance samples' distribution (*Kolmogorov-Smirnov & Liliefors* and *Shapiro Wilk W* tests), the verification of variances (*Levene's* and *Brown-Forsythe* tests), and one way ANOVA followed by *Tukey*'s post-hoc test.

The ρ, τ, γ-SLFNs have been compared with following state-of-the-art neural networks: extreme (ELM), radial basis functions (RBF), and a 3 multi-layer perceptron (3-MLP). The chosen parameters for the experiment are: SLFN 30 hidden neurons, ELM 12 hidden neurons, RBF 17 hidden nodes, MLP 300 hidden nodes, learning rate 0.25, and momentum 0.6.

5 Results

For an objective comparison, all the algorithms were run in the same conditions. In Table 1, we present the results obtained on the three datasets. Each experiment is described in terms of the ACA averaged over 130 computer runs, SD, 95% CI, and PR AUC.

Table 1. Performance indicators for the ρ, τ, γ-SFLNs

Dataset	Algorithm	ACA (%)	SD	95% CI	PR AUC
BD	ρ-SLFN	65.32	9.31	(61.47, 65.16)	0.602
	τ-SLFN	51.94	5.55	(47.73, 54.15)	0.498
	γ-SLFN	57.88	12.51	(55.40, 60.37)	0.523
LM	ρ-SLFN	90.90	5.28	(89.85, 91.94)	0.888
	τ-SLFN	81.90	11.25	(79.66, 84.13)	0.796
	γ-SLFN	85	6.59	(83.69, 86.30)	0.810
FIB	ρ-SLFN	52.16	10.73	(50.03, 54.29)	0.478
	τ-SLFN	57.16	9.42	(55.29, 59.03)	0.512
	γ-SLFN	39.53	22.61	(35.04, 44.02)	0.239

From Table 1 we can see that on the BD dataset, the best performing algorithm was the ρ – SLFN which obtained 63.32%, surpassing the τ and γ-SLFNs that obtained 51.94%, and 57.88% respectively. An interesting fact is that even if the hierarchy in terms of accuracy is ρ followed by γ and τ- SLFNs, the SDs show that the most stable and robust model is τ – SLFN, having the smallest SD value, 5.55, followed by ρ-SLFN with an SD of 9.31, and γ-SLFN with 12.51. The PR AUC show that the first two models perform poorly, whereas the third is a failure. On the LM dataset, the performance hierarchy remains the same, ρ-SLFN leading with 90.90%, followed by γ-SLFN with 85%, and τ-SLFN with 81.90%. There exists a difference in terms of the SD, this time the best two performing models being more stable and robust, SD 5.28 and 6.59, whereas the latter is unstable, SD 11.25. The PR AUC demonstrates that the models perform good. On the FIB dataset the hierarchy changes, the best performing algorithm being τ-SLFN leading with a 57.16% ACA, followed by ρ-SLFN with a 52.16% ACA, and γ-SLFN with a 39.52% ACA. All the SDs have high values ranging from 9.42 to 22.61. In terms of PR AUC the first two models performed fairly, whereas the last was a failure. As an overall conclusion, we can state that the results on these three datasets prove that the ρ-SLFN performs the best, two out of three times, whereas the τ -SLFN performed the poorest.

The benchmarking process continues with the *data screening* process, more specifically with applying the *Kolmogorov-Smirnov & Lilliefors* test and the *Shapiro-Wilk W* test for establishing the samples distributions. Table 2 present the results.

Table 2. Testing the normality of the ρ, τ, γ-SFLNs performances

Dataset	Algorithm	Kolmogorov-Smirnov		Shapiro-Wilk W	
		K-S max D	Lilliefors *p*	S-W W	*p*-level
BD	ρ-SLFN	0.230	<0.01	0.873	0.000
	τ-SLFN	0.139	<0.01	0.932	0.000
	γ-SLFN	0.181	<0.01	0.927	0.000
LM	ρ-SLFN	0.270	<0.01	0.720	0.000
	τ-SLFN	0.305	<0.01	0.752	0.000
	γ-SLFN	0.194	<0.01	0.829	0.000
FIB	ρ-SLFN	52.16	<0.01	0.907	0.000
	τ-SLFN	57.16	<0.01	0.958	0.003
	γ-SLFN	39.53	<0.01	0.936	0.000

Table 2 shows us that no matter the algorithm, nor the dataset used, the samples are not governed by the Normal distribution, *p*-level <0.05. To bypass this, we shall use the *Central limit theorem* that states that if the sample size is large enough (over 30) the sample distribution is approximately Gaussian, [34]. Thus, we can proceed further with our statistical analysis.

We have compared the 3 SLFNs with state-of-the-art NNs. All the algorithms have been run under the same conditions. The results are depicted in Table 3.

From Table 3 we can see that the performances of the six competitors are fairly equal when it comes to the LM dataset. An interesting fact is observed on the BD and FIB datasets. ELM and RBF behave the same, but all the other models do not. ρ, γ -SLFNs perform better than τ-SLFN and 3-MLP on the BD dataset, and exactly the opposite on the FIB dataset, result that is consisting with the '*no-free-lunch*' theorem. There is no

Table 3. Average accuracies of different NNs and ρ, τ, γ-SFLNs

Algorithm	Dataset testing accuracies		
	BD	LM	FIB
ρ-SLFN	63.32	90.90	52.16
τ-SLFN	51.94	81.90	57.10
γ-SLFN	57.88	85	39.53
ELM	69.07	87.54	55.35
RBF	68.09	80.95	56.51
3-MLP	51.38	89.91	60.80

algorithm that performs the best in every situation. It is up to us to choose the correct one considering the priceless statistical analysis.

The data screening process continued with the equality of variances verification.

After applying Levene and Brown-Forsythe tests we obtained the following results:

- On the BD dataset the variables concerning ρ-SLFN vs. τ-SLFN, ρ-SLFN vs. γ-SLFN, τ-SLFN vs. γ-SLFN, ρ-SLFN vs. RBF, τ-SLFN vs. RBF, γ-SLFN vs. RBF, ρ -SLFN vs. ELM, τ -SLFN vs. ELM, have equal variances (p-level > 0.05), the rest do not;
- On the LM dataset: the variables τ-SLFN vs. RBF, τ-SLFN vs. 3-MLP, and γ -SLFN vs. 3-MLP have equal variances, the rest do not;
- For the FIB dataset: only variables concerning ρ -SLFN vs. τ-SLFN have equal variances, the other do not.

Models that have equal variances behave similarly. To bypass the inconvenience given by these results, we took the same number of observations (130 computer runs for each model. Let us see the results of the one-way ANOVA in Table 5 in terms of sum of squares (SS), degrees of freedom (df), mean squares (MS), F-value, and p-level (contrast quadratic polynomial), [35, 36].

Table 4. One-way ANOVA results

Dataset	SS	DF	MS	F-value	p-level
BD	29553	5	5911	36.36	0.000
LM	8495	5	1699	23.37	0.000
FIB	27600	5	5520	44.54	0.000

Table 4 reveals that there are significant differences between the models' testing performances (p-level < 0.05). The only two questions that remain to be answered are: which models perform differently, and whether they perform differently on all datasets?

To find out the result we have performed Tukey's post-hoc test. Table 5 depicts the results.

Table 5. Tukey post-hoc test

BD	ρ-SLFN	τ-SLFN	γ-SLFN	ELM	RBF	3-MLP
ρ-SLFN		**0.000**	**0.047**	**0.034**	0.080	**0.000**
τ-SLFN	**0.000**		**0.008**	**0.000**	**0.000**	0.999
γ-SLFN	**0.047**	**0.008**		**0.000**	**0.000**	**0.002**
ELM	**0.034**	**0.000**	**0.000**		0.999	**0.000**
RBF	0.080	**0.000**	**0.000**	0.999		**0.000**
3-MLP	**0.000**	0.999	**0.002**	**0.000**	**0.000**	
LM	ρ-SLFN	τ-SLFN	γ-SLFN	ELM	RBF	3-MLP
ρ-SLFN		**0.000**	**0.000**	0.059	**0.000**	0.964
τ-SLFN	**0.000**		0.104	**0.000**	0.970	**0.000**
γ-SLFN	**0.000**	0.104		0.283	**0.010**	**0.000**
ELM	0.059	**0.000**	0.283		**0.000**	0.359
RBF	0.000	0.970	**0.010**	**0.000**		**0.000**
3-MLP	0.964	**0.000**	**0.000**	0.359	**0.000**	
FIB	ρ-SLFN	τ-SLFN	γ-SLFN	ELM	RBF	3-MLP
ρ-SLFN		**0.0186**	**0.000**	0.064	0.329	**0.000**
τ-SLFN	**0.018**		**0.000**	0.998	0.858	0.189
γ-SLFN	**0.000**	**0.000**		**0.000**	**0.000**	**0.000**
ELM	0.064	0.998	**0.000**		0.977	0.069
RBF	0.329	0.858	**0.000**	0.977		**0.007**
3-MLP	**0.000**	0.189	**0.000**	0.069	**0.007**	

Table 5 reveals:

- On the BD dataset there are significant difference between ρ-SLFN and γ -SLFN, -SLFN and τ-SLFN, ρ -SLFN and MLP, τ -SLFN and ρ -SLFN, τ -SLFN and γ-SLFN, τ -SLFN and ELM, τ -SLFN and RBF, γ-SLFN and ρ -SLFN, γ-SLFN and τ -SLFN, γ-SLFN and ELM, γ -SLFN and RBF, γ -SLFN and 3-MLP.
- On the LM dataset there are significant differences between -SLFN and τ-SLFN, ρ -SLFN and γ-SLFN, ρ -SLFN and RBF, τ -SLFN and γ -SLFN, τ-SLFN and ELM, τ-SLFN and MLP, γ -SLFN and ρ -SLFN, γ -SLFN and RBF, γ -SLFN and 3-MLP.
- On the FIB dataset there are significant differences between: ρ -SLFN and τ -SLFN, ρ -SLFN and MLP, τ-SLFN and ρ -SLFN, τ -SLFN and γ-SLFN, τ-SLFN and ρ -SLFN, τ -SLFN and γ-SLFN.

The obtained results demonstrate that overall, the best performing SLFN is the ρ-SLFN that overtook the others on the BD and LM dataset and occupied the second place on the FIB dataset. The second place in the hierarchy is occupied by γ-SLFN, followed by τ– SLFN. Nevertheless, even if γ-SLFN came in second, its performance is very low (almost 10% difference) when compared to ρ– SLFN for the FIB dataset. Again, this is an exceptionally good example for the *no free lunch theorem.* In terms of comparison to other state-of-the-art NNs, the statistical analysis revealed that ρ-SLFN performs the same as RBF on the BD dataset, even if on this dataset, RBF led the race by 5%; it performs the same as MLP and ELM on the LM dataset, and the same as ELM and RBF on the FIB dataset. Regarding γ-SLFN, the post-hoc tests revealed that it performs the same as ELM on the LM dataset, whereas on the other datasets there exist significant differences. τ – SLFN performs the same as 3-MLP on the BD dataset, the same as RBF on the LM dataset, and exactly the same as all the classical NNs on the FIB dataset.

6 Conclusions

The overwhelming amount of data that is produced on a daily basis by the healthcare system puts pressure on data scientists all over the world to develop new ML algorithms that can deal with the famous two curses: *the curse of sparsity* and *the curse of dimensionality*. The diagnosis, prognosis, and treatment are currently custom-made so that they consider the particular feature of each individual through gene expression array or clinical and bio-chemical parameters.

This paper deals with the statistical comparison of non-parametric correlation coefficients that are used for initializing the input-hidden weight matrix of three ρ, τ, γ-SLFNs. The three new models are used for differentiating cancer from unbalanced, and unprocessed databases that contain either DNA arrays or clinical and bio-chemical features. The objective statistical benchmarking process in terms of ACA, PR AUC, SD, Kolmogorov – Smirnov & Lilliefors test, Shapiro-Wilk W test, Levene test, Brown Forysthe, test, one-way ANOVA followed by Tukey post-hoc tests established the following hierarchy: ρ, γ, τ. All models are comparable to other state-of-the-art NNs in terms of performance and robustness.

References

1. Siegel, R.L., Miller, K.D., Jemal, D.V.M.A.: Cancer statistics. CA Cancer J. Clin. https://doi.org/10.3322/caac.21590 (2020)
2. Kathirvalayakumar, Y., Subavathi, S.J.: A new weight initialization method using Cauchy's inequality based on sensitivity analysis. J. Intell. Learn. Syst. Appl. **3**, 242–248 (2011)
3. Yam, J.Y.F., Chow, T.W.S.: A weight initialization method for improving training speed in feedforward neural networks. Neurocomputing **30**, 219–232 (2000)
4. Yam, J.Y.F., Chow, T.W.S.: Feedforward networks training speed enhancement by optimal initialization of the synaptic coefficients. IEEE Trans. Neural Networks **12**(2), 430–444 (2001)
5. Belciug, S.: Logistic regression paradigm for training a single-hidden layer feedforward neural network. J. Biomed. Inf. **102**, 103373 (2020). https://doi.org/10.1016/j.jbi.2019.103373
6. Belciug, S.: Parallel versus cascaded logistic regression trained single-hidden feedforward neural network for medical data. Exp. Sys. App. **170**, 114538 (2021)

7. Hsiao, T.C.R., Lin, C.W., Chiang, H.K.: Partial least squares algorithm for weights initialization of backpropagation network. Neurocomputing **50**, 237–247 (2003)
8. Mishkin, D., Matas, J.: All you need is a good init. In: ICLR (2016). https://arxiv.org/abs/1511.06422v7
9. Sutskever, I., Vinyals, O, Le, Q.: Sequence to sequence learning with neural networks. arXiv:1409.3215 (2016)
10. West, M., et al.: Predicting the clinical status of human breast cancer by using gene expression profiles. In: Proceedings of the National Academy of Sciences USA, vol. 98, no. 20, pp. 11463–11467 (2001)
11. Beer, D., Kardia, S., Huang, C.C., et al.: Gene-expression profiles predict survival of patients with lung adenocarcinoma. Nat. Med. **8**(8), 816–823 (2002)
12. Shevade, S.K., Keerthi, S.S.: A simple and efficient algorithm for gene selection using sparse logistic regression. Bioinformatics **19**(17), 2246–2253 (2003)
13. Belciug, S., Gorunescu, F.: Learning a single-hidden layer feedforward neural network using rank correlation-based strategy with application to high dimensional gene expression and proteomic spectra datasets in cancer detection. J. Biomed. Inf. **83**, 159–166 (2018)
14. Czajkowski, M., Kretowski, M.: Global top-scoring pair decision tree for gene expression data analysis. In: Krawiec, K., Moraglio, A., Hu, T., Etaner-Uyar, A.Ş, Hu, B. (eds.) EuroGP 2013. LNCS, vol. 7831, pp. 229–240. Springer, Heidelberg (2013). https://doi.org/10.1007/978-3-642-37207-0_20
15. Belciug, S., Gorunescu, F.: Error-correction learning for artificial neural networks using the Bayesian paradigm. Application to automated medical diagnosis. J. Biomed. Inf. **52**, 329–337 (2014)
16. Stoean, C., Stoean, R., Lupsor, M., Stefanescu, H., Badea, R.: Feature selection for cooperative coevolutionary classifier in liver fibrosis diagnostic. Comput. Biol. Med. **41**, 238–246 (2011)
17. Stoean, R., Stoean, C., Lupsor, M., Stefanescu, H., Badea, R.: Evolutionary-driven support vector machines for determing the degree of liver fibrosis in chronic hepatitis C. Art Intel. Med. **51**(1), 53–65 (2011)
18. Gorunescu, F., Belciug, S., Gorunescu, M., Badea, R.: Intelligent decision-making for liver stadialization based on tandem feature selection and evolutionary-driven neural network. Exp. Syst. App. **39**(17), 12824–12832 (2012). https://doi.org/10.1016/j.eswa.2012.05.011
19. Gorunescu, F., Belciug, S.: Evolutionary strategy to develop learning based decision systems. Application to breast cancer and liver fibrosis stadialization. J. Biomed. Inf. **49**, 112–118, (2014)
20. Belciug, S., Lupsor, M.: A multi-layer based procedure for detecting liver fibrosis. Ann. Univ. Craiova, Math. Comput. Sci. Ser. **36**(1), 64–70 (2009)
21. Gorunescu, F., Belciug, S., Gorunescu, M., Lupsor, M., Badea, R., Stefanescu, H.: Radial basis function network based diagnosis for liver fibrosis estimation. In: Proceedings of the 2nd International Conference on E-health and Bioengineering, pp. 209–212 (2009)
22. Spearman, C.: The proof and measurement of association between two things. Am. J. Psychol. **15**(1), 72–101 (1904)
23. Glasser, G.J., Winter, R.F.: Critical values of the coefficient of rank correlation for testing hypothesis of independence. Biometrika **48**, 444–448 (1961)
24. Kendall, M.: A new measure of rank correlation. Biometrika **30**(1–2), 81–89 (1938)
25. Kendall, M.: Rank Correlation Methods, 2nd ed. APA (1955)
26. Kight, W.R.: A computer method for calculating Kendall's Tau with ungrouped data. J. Am. Stat. Assoc. **61**, 314(1), 436–439 (1996)
27. Somers, R.H.: A new asymmetric measure of association for ordinal variables. Am. Sociol. Rev. **27**(6), 2090408 (1962)
28. Newson, R.: Parameter behind "nonparametric" statistics: Kendall's tau, Somers'D, and median differences. Stata J. **2**(1), 45–64 (2002)

29. Goodman, L., Kruskal, W.H.: Measures of association for cross classifications. J. Am. Stat. Assoc. **49**(268), 732–764 (1954)
30. Leshno, M., Lin, V.Y., Pinkus, A., Schocken, S.: Multilayer feedforward networks with a nonpolynomial activation function can approximate any function. Neural Netw. **6**(6), 861–867 (1993)
31. Gulyev, N.J., Ismailov, V.E.: A single hidden layer feedforward network with only one neuron in the hidden layer can approximate any univariate function. Neural Comput. **28**(7), 1289–1304 (2016)
32. Ismailov, V.E.: On the approximation by neural networks with bounded number of neurons in hidden layers. J. Math. Anal. Appl. **417**(2), 963–969 (2014)
33. Belciug, S.: Artificial Intelligence in Cancer: Diagnostic to Tailored Treatment. Elsevier, San Diego (2020)
34. Altman, D.G.: Practical Statistics for Medical Research. Chapman and Hall, New York (1991)
35. Wallenstein, S.: Some statistical methods useful in circulation research. Circ. Res. **47**(1) (1980)
36. Demsar, J.: Statistical comparisons of classifiers over multiple datasets. J. Mach. Learn. Res. **7**, 1–30 (2006)

Nonstandard Fuzzy Sets: A General View

Sorin Nădăban(✉) and Dan Deac

Department of Mathematics and Computer Science, Faculty of Exact Sciences, Aurel Vlaicu University of Arad, Elena Drăgoi 2, 310330 Arad, Romania
snadaban@gmail.com, dan.deac@uav.ro

Abstract. Nonstandard fuzzy sets are extensions, generalizations of fuzzy sets introduced by Zadeh. They represent a research domain of great interest due to the multiple applications in ambiguity situations and in problems with incomplete information. We make a systematic review of these types of fuzzy sets in order to provide a framework for new research in this field, to strengthen the available theoretical results, to establish the relations among them as well as their various applications. Finally, we introduce the concept of vector fuzzy set and give different representation of Pythagorean fuzzy sets.

Keywords: Fuzzy sets · Intuitionistic fuzzy sets · Interval-valued fuzzy sets · Pytagorean fuzzy sets

1 Introduction

After L.A. Zadeh introduced the concept of fuzzy set [32], many researchers have introduced other types of fuzzy sets in order to deliver membership functions to model better ambiguity situations and to be helped in problems of approximate reasoning, computing with word, decision making etc.

The starting point in this approach is paper [4], in which the authors make a study of the various extensions of fuzzy sets. We intend to continue and develop this study in order to provide a framework for new research in this field, to strengthen the available theoretical results, to establish the relations among fuzzy sets as well as their various applications.

From now on we denote by X the universe.

2 Types of Fuzzy Sets

2.1 Type-1 Fuzzy Sets

In 1965, in his famous paper [32], L.A. Zadeh introduced the concept of fuzzy set (or type-1 fuzzy set). A fuzzy set (FS) is a mapping $A : X \to [0, 1]$. The value $A(x)$ represents the membership degree of the element x to the fuzzy sets A. We will denote by $\mathcal{F}(X)$ the family of all fuzzy sets in X. IF A, B are fuzzy

S. Dzitac et al. (Eds.): ICCCC 2022, AISC 1435, pp. 208–218, 2023.
https://doi.org/10.1007/978-3-031-16684-6_17

sets in X their union $A \vee B$, their intersection $A \wedge B$ and the complement of A, denoted $\mathcal{C}(A)$, are fuzzy sets in X defined by:

$$(A \vee B)(x) = \max\{A(x), B(x)\}, (\forall)x \in X;$$

$$(A \wedge B)(x) = \min\{A(x), B(x)\}, (\forall)x \in X;$$

$$\mathcal{C}(A) = 1 - A(x), (\forall)x \in X.$$

Fuzzy numbers (FNs) are the most used fuzzy sets. A fuzzy set $A : \mathbb{R} \rightarrow [0,1]$ is called fuzzy number if: (a) A is convex; (b) A is upper semicontinuous; (c) A is normal, i.e. $(\exists)t_0 \in \mathbb{R}\ A(t_0) = 1$. Among the various types of fuzzy numbers the most common are:

(1) *Triangular FNs* which have membership function

$$A(t) = \begin{cases} 0 & \text{if} \quad t < a \\ \frac{t-a}{b-a} & \text{if } a \le t < b \\ \frac{c-t}{c-b} & \text{if } b \le t < c \\ 0 & \text{if} \quad t > c \end{cases}, \text{where } a \le b \le c\,,$$

and they are denoted by $\tilde{A} = (a, b, c)$.

(2) *Trapezoidal FNs* defined by membership function

$$A(t) = \begin{cases} 0 & \text{if} \quad t < a \\ \frac{t-a}{b-a} & \text{if } a \le t \le b \\ 1 & \text{if } b < t < c \\ \frac{d-t}{d-c} & \text{if } c \le t \le d \\ 0 & \text{if} \quad t > d \end{cases}, \text{where } a \le b \le c \le d\,,$$

and expressed as $\tilde{A} = (a, b, c, d)$.

2.2 L-Fuzzy Sets

In 1967, J. Goguen [11], remarked that we can replace the truth values $[0,1]$ with a lattice L, so he defined L-fuzzy sets as mappings $A : X \rightarrow L$. The family of all L-fuzzy sets in X will be denoted $\mathcal{L}(X)$.

2.3 Type-n Fuzzy Sets

In 1971, L.A. Zadeh [33] introduced the concept of type-2 fuzzy set. In this case the values of the membership function are allowed to be fuzzy subsets of the interval $[0,1]$. Thus a type-2 fuzzy set is a mapping $A : A \rightarrow \mathcal{F}([0,1])$ or a type-2 fuzzy set can seen as a function $A : X \rightarrow [0,1]^{[0,1]}$.

Later on, in 1975, L.A. Zadeh [34] defined, in a recursive way, type-n fuzzy sets: "a fuzzy set is of type n, n = 2,3, . . . , if its membership function ranges over fuzzy sets of type $n-1$".

In paper [34], L.A. Zadeh considered a particular case of type-2 fuzzy sets, the one in which $A(x)$ is a closed interval in $[0,1]$. Thus we obtained to the notion of interval-valued fuzzy set (IVFS). Let us denote by $\mathcal{D}([0,1])$ the family of all closed intervals of $[0,1]$. An IVFS is a mapping $A : X \to \mathcal{D}([0,1])$. We mention that these IVFSs were also presented and studied by R. Sambuc [22] with the name of ϕ-fuzzy sets.

As $\mathcal{D}([0,1])$ is a bounded lattice with the operations

$$[a,b] \cup [c,d] = [\max(a,c), \max(b,d)], [a,b] \cap [c,d] = [\min(a,c), \min(b,d)],$$

it results that IVFSs are a particular case of L-fuzzy sets.

Another particular case of type-2 fuzzy sets are set-valued fuzzy sets(SVFS) defined by I. Grattan-Guinness [12] in 1976. In this case the value of the membership are subsets of $[0,1]$. Thus a SVFS is a mapping $A : X \to 2^{[0,1]} \setminus \emptyset$.

2.4 Intuitionistic Fuzzy Sets

In 1983, K.T. Atanassov [1] introduced the notion of intuitionistic fuzzy set (IFS). The paper was written in Bulgarian. In 1986 he presented in English [2] his ideas. An IFS is given by

$$A = \{< x, \mu_A(x), \nu_A(x) >: x \in X\},$$

where $\mu_A, \nu_A : X \to [0,1]$ such that $0 \leq \mu_A(x) + \nu_A(x) \leq 1$ and represent the degree of membership and the degree of non-membership of an element x to A. The mapping $\pi_A(x) = 1 - \mu_A(x) - \nu_A(x)$ represents the hesitance grade. It is obvious that FSs are a particular case of IFSs (in the situation $\mu_A(x) + \nu_A(x) = 1, (\forall) x \in X$).

IFSs were also generalized (see [18], [15]).

2.5 Fuzzy Multisets

R.R. Yager [27] introduced in 1986 the idea of fuzzy multiset (MS). A FM is a mapping $A : X \to [0,1]^n$, where $n \in \mathbb{N}^*$. In fact $A = (A_1, A_2, \cdots, A_n)$, where each A_i is a FS ($i = \overline{1,n}$). We note that FMs are a particular case of SVFSs.

2.6 Interval Valued Intuitionistic Fuzzy Sets

In 1989, K.T. Atanassov and G. Gargov [3] introduced the concept of interval valued intuitionistic fuzzy sets (IVIFS). An IVIFS in X is $A = \{< x, \mu_A(x), \nu_A(x) > : x \in X\}$, where

$$\mu_A, \nu_A : X \to \mathcal{D}([0,1]) \ : \ 0 \leq \sup_{x \in X} \mu_A(x) + \sup_{x \in X} \nu_A(x) \leq 1.$$

A can be written $A = ([\mu_1(x), \mu_2(x)], [\nu_1(x), \nu_2(x)])$. If $\mu_1(x) = \mu_2(x)$ and $\nu_1(x) = \nu_2(x)$, for all $x \in X$, A became IFS. Thus IFSs are a particular case of IVIFSs. As an IVIFS is a mapping $A : X \to \mathcal{D}([0,1]) \times \mathcal{D}([0,1])$ and $\mathcal{D}([0,1]) \times \mathcal{D}([0,1])$ is a bounded lattice it result that IVIFSs are a particular case of L-fuzzy sets.

2.7 Fuzzy Rough Sets

To study situations of uncertainty we can use rough sets, firstly introduced by Z. Pawlak [19] in 1982.

Let X be the universe and R be an equivalence relation on X. We denote by X/R the family of all equivalence classes and by $[x]$ the equivalence class of x. A definable set is a set which is a union of $\emptyset$ and elements from X/R. We will denote by $Def(X, R)$ the family of all definable sets.

Let A be a subset of X which cannot be accurately described. We will introduce a pair $(\underline{A}, \overline{A})$ such that $\underline{A} \subset \overline{A}$ and $\underline{A}, \overline{A} \in Def(X, R)$ which approximates the set A, i.e. $\underline{A} \subseteq A \subseteq \overline{A}$. The sets $\underline{A}, \overline{A}$ will be called the lower approximation and the upper approximation and they are defined by:

$$\underline{A} = \{x \in X \ : \ [x] \subset A\} \ , \ \overline{A} = \{x \in X \ : \ [x] \cap A \neq \emptyset\}.$$

Starting from the concept of rough set, in 1990, Dubois and Prade [6] introduced the notion of fuzzy rough set. However, some preliminary notions are necessary.

Triangular norms (shortly t-norms) were introduced by A.K. Menger [17] and then studied by B. Schweizer and A. Sklar [23] and by many others. They are important in fuzzy logic because they generalize conjunction. A t-norm is a binary operation $* : [0,1] \times [0,1] \to [0,1]$ which is commutative, associative, increasing and $a * 1 = a, (\forall) a \in [0,1]$. Closely related to t-norms are triangular conorms (shortly t-conorms) which retains the main properties of disjunction. A t-conorm is a binary operation $\circ : [0,1] \times [0,1] \to [0,1]$ which is commutative, associative, increasing and $a \circ 0 = a, (\forall) a \in [0,1]$.

A negator is a decreasing mapping $N : [0,1] \to [0,1]$ such that:

$$N(0) = 1, \ N(1) = 0 \ , \ N(N(x)) = x, (\forall) x \in [0,1].$$

The standard negator is $N(x) = 1 - x$.

An implicator is a mapping $I : [0,1] \times [0,1] \to [0,1]$ decreasing in its first component, increasing in its second component and $I(0,0) = 1, I(1,a) = a, (\forall) a \in [0,1]$.

A t-norm $*$ and a negator N induce an implicator: $I(x,y) = N(x * N(y))$. A t-norm $*$ and the standard negator induce the standard implicator: $I(x,y) = 1 - x * (1-y)$.

We note that the most important t-norm is the minimum t-norm $a \wedge b = \min\{a, b\}$, with the corresponding t-conorm $a \vee b = \max\{a, b\}$ and the corresponding standard implicator $I(x,y) = 1 - \min\{x, 1-y\} = \max\{1-x, y\}$.

A mapping $R : X \times X \to [0,1]$ is called fuzzy relation. A fuzzy relation is called fuzzy equivalence relation if

1. $R(x,x) = 1, (\forall) x \in X$;
2. $R(x,y) = R(y,x), (\forall) x, y \in X$;
3. $R(x,y) * R(y,z) \leq R(x,z), (\forall) x, y, z \in X$.

Let now A be a fuzzy set in X and R be a fuzzy equivalence relation. We define the fuzzy sets $\underline{A}, \overline{A}$ by:

$$\underline{A} = \inf x \in XI(R(x,y), A(x)),\ \overline{A} = \sup_{x\in X} R(x,y) * A(x).$$

In the particular case where $*$ is the minimum t norm and I is the corresponding implicator, we will obtain that

$$\underline{A} = \inf_{x\in X} \max\{1 - R(x,y), A(x)\},\ \overline{A} = \sup_{x\in X} \min\{R(x,y), A(x)\}.$$

2.8 Bipolar Fuzzy Sets

By a bipolar fuzzy sets (BFS), as Zhang [36] introduced this notion in 1994, we understand a mapping $A : X \to [0,1] \times [-1,0]$. Usually we write $A(x) = (A^+(x), A^-(x))$, where $A^+ : X \to [0,1]$ and $A^- : X \to [-1,0]$.

In 2012, the concept of bipolar fuzzy rough set [30] is introduced.

2.9 Shadowed Sets

In 1998, W. Pedrycz [20] introduced the notion of shadowed set. This is a mapping $A : X \to \{0, 1, [0,1]\}$. When an element certainly belongs to A we associate him the value 1, if an element does not belong to A we associate him the value 0 and when uncertainty exists we put $A(x) = [0,1]$.

2.10 Neutrosophic Sets

F. Smarandache [25] proposes in 1999 the concept of neutrosophic set. A neutrosophic set in X is defined as

$$A = \{< x, T_A(x), I_A(x), F_A(x) >: x \in X\}$$

where $T_A(x), I_A(x), F_A(x)$ are subsets of $]0^-, 1^+[$ and represent the truth-membership function, indeterminacy-membership function and falsity-membership function such that

$$0^- \leq \sup_{x\in X} T_A(x) + \sup_{x\in X} I_A(x) + \sup_{x\in X} F_A(x) \leq 3^+,$$

where

$$a^- = \{a - \epsilon\ :\ \epsilon \in \mathbb{R}^*, \epsilon \text{ is infinitesimal}\}$$
$$b^+ = \{b + \epsilon\ :\ \epsilon \in \mathbb{R}^*, \epsilon \text{ is infinitesimal}\}.$$

For applications we can consider that $T_A(x), I_A(x), F_A(x)$ are subsets of $[0,1]$.

2.11 Complex Fuzzy Sets

In 2002, D. Ramot et al. [21] defined complex fuzzy sets. In this case the values of membership function are extended to the unit circle in the complex plane. Thus, by a complex fuzzy set we understand a mapping

$$A : X \to D(0,1)\ ,\ A(x) = r(x)e^{i\omega(x)},$$

where r is magnitude ($r(x) \in [0,1]$) and $\omega(x)$ is the phase ($\omega(x) \in \mathbb{R}$). A good review of complex fuzzy sets is paper [31].

In 2021, the notion of complex bipolar fuzzy set [13] is studied.

2.12 Hesitant Fuzzy Sets

V. Torra and Y. Narukawa [26] introduced the concept of hesitant fuzzy set (HFS). They defined HFSs as functions defined on X with values subsets of $[0,1]$. We remark that this definition coincides with Grattan-Guinness's definition for SVFSs, but V. Torra and Y. Narukawa proposed new definitions for union and intersection.

2.13 Pythagorean Fuzzy Sets

Pythagorean fuzzy sets were proposed by R.R. Yager in 2013 (see [28]). A Pythagorean fuzzy set is characterized by the functions $\mu, \nu : X \to [0,1]$ which give us the degree of membership and degree of non-membership, respectively and $0 \leq (\mu(x))^2 + (\nu(x))^2 \leq 1, (\forall)x \in X$. The function $\pi : X \to [0,1]$ defined by $\pi(x) = \sqrt{1 - [(\mu(x))^2 + (\nu(x))^2]}$ is called the degree of indeterminacy.

R.R. Yager [29] introduced q-rung orthopair fuzzy sets (qROFs). A qROFs can be defined

$$A = \{< x, \mu_A(x), \nu_A(x) >:\ x \in X\},$$

where $\mu_A, \nu_A : X \to [0,1]$ such that $0 \leq (\mu_A(x))^q + (\nu_A(x))^q \leq 1$, for some $q \leq 1$, and represent the degree of membership and the degree of non-membership of an element x to A.

For $q = 1$ we obtain IFSs, for $q = 2$ we obtain PFSs and for $q = 3$ we obtain Fermatean fuzzy sets [24].

A $(3,2)$-fuzzy set (see [14]) is defined as

$$A = \{< x, \mu_A(x), \nu_A(x) >:\ x \in X\},$$

where $\mu_A, \nu_A : X \to [0,1]$ such that $0 \leq (\mu_A(x))^3 + (\nu_A(x))^2 \leq 1$.

3 Some Remarks on Pythagorean Fuzzy Sets

In this section we will give other representations of Pythagorean fuzzy sets.

Definition 1. *A vector fuzzy set (VFS) on the universe X ia a mapping $A : X \to U_+(\mathbb{R}^3)$, where $U_+(\mathbb{R}^3)$ is the family of all unit vectors in $\mathbb{R}^3$ (vectors of length 1) such that $u_1 \geq 0, u_2 \geq 0, u_3 \geq 0$, for $u = (u_1, u_2, u_3)$.*

Such an approach could be generalized, replacing $\mathbb{R}^3$ with a separable Hilbert space H, which leads us to think of a connection between fuzzy logic and quantum logic, where each state of a physical system is described by unitary vectors $u \in H$.

If we denote $A(x) = u(x) = (u_1(x), u_2(x), u_3(x))$, as $u(x)$ is an unitary vector and $u_1, u_2, u_3 : X \to [0,1]$ satisfy $u_1^2(x) + u_2^2(x) + u_3^2(x) = 1, (\forall)x \in X$, we will have that u_1, u_2, u_3 define a Pythagorean fuzzy set. More precisely, $u_1 = \mu$ represents the degree of membership, $u_2 = \nu$ represents the degree of non-membership and $u_3(x) = \pi(x) = \sqrt{1 - (\mu^2(x) + \nu^2(x))}$ represents the degree of indeterminacy.

If $A(x) = u(x)$ and the vector $u(x)$ is expressed as OM, then M can be represented by Cartesian coordinates (u_1, u_2, u_3) or by spherical coordinates. In order to obtain the spherical coordinates of M we denote by θ the angle between OM and positive z axis. Let P be the projection of M in the plan $x0y$ and P_1 be the projection of P to $0x$. We denote by φ the angle between OP and $0x$. We have that $\theta \in [0, \pi/2]$ and $\varphi \in [0, \pi/2]$ (Fig. 1).

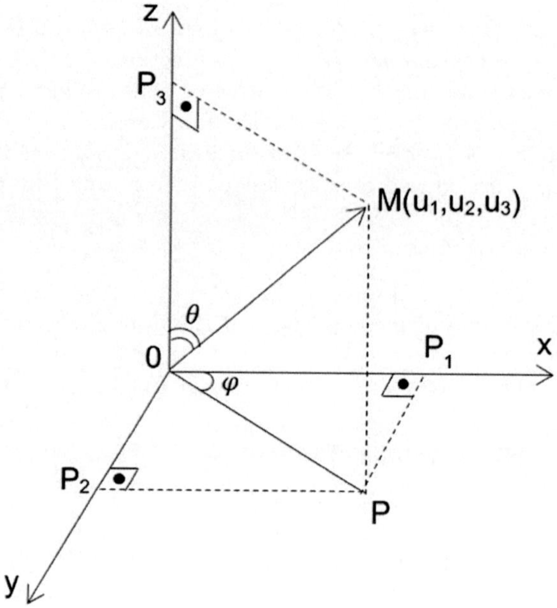

Fig. 1. Spherical representation of a vector fuzzy set

Thus $u_3 = \cos\theta$; $OP = \sin\theta$; $u_1 = \sin\theta\cos\varphi$; $u_2 = \sin\theta\sin\varphi$. Hence

$$A(x) = (\sin\theta(x)\cos\varphi(x), \sin\theta(x)\sin\varphi(x), \cos\theta(x)).$$

The vector fuzzy set A can be seen as a mapping $A : X \to [0, \pi/2] \times [0, \pi/2]$, $A = (\theta, \varphi)$.

If we denote $\sin\theta(x) = r(x) \in [0, 1]$ we have that

$$A(x) = (r(x)\cos\varphi(x), r(x)\sin\varphi(x), \sqrt{1 - r^2(x)}).$$

So, the vector fuzzy set A can be seen as a mapping $A : X \to [0, 1] \times [0, \pi/2]$, $A = (r, \varphi)$.

Finally, we note that the Pythagorean membership grades

$$(\mu(x), \nu(x)) = (r(x)\cos\varphi(x), r(x)\sin\varphi(x))$$

can be seen as the real part and imaginary part of the complex number

$$z(x) = r(x)\cos\varphi(x) + ir(x)\sin\varphi(x) = r(x)e^{i\varphi(x)}.$$

Definition 2. *1. The complement of the vector fuzzy set* $A = (u_1, u_2, u_3)$ *is defined as* $\overline{A} = (u_2, u_1, u_3)$.

2. Let $A = (u_1, u_2, u_3)$, $B = (v_1, v_2, v_3)$ *be two VFS. We say that* $A \subseteq B$ *if* $u_1(x) \le v_1(x), u_2(x) \ge v_2(x), (\forall)x \in X$.

Remark. $A = (u_1, u_2, u_3)$, $B = (v_1, v_2, v_3)$ be two VFS. As $\overline{A} = (u_2, u_1, u_3)$ and $\overline{B} = (v_2, v_1, v_3)$, if $A \subseteq B$, then $\overline{B} \subseteq \overline{A}$.

Definition 3. *Let* $A = (u_1, u_2, u_3)$, $B = (v_1, v_2, v_3)$ *be two VFS.*

1. *We define* $A \cap B = (g_1, g_2, g_3)$, *where* $g_1 = \min\{u_1, v_1\}, g_2 = \max\{u_2, v_2\}, g_3 = \sqrt{1 - (g_1^2 + g_2^2)}$.
2. *We define* $A \cup B = (h_1, h_2, h_3)$, *where* $h_1 = \max\{u_1, v_1\}, h_2 = \min\{u_2, v_2\}, h_3 = \sqrt{1 - (h_1^2 + h_2^2)}$.

Remark. The previous definition is correct if we show that $g_1^2 + g_2^2 \le 1$ and $h_1^2 + h_2^2 \le 1$. Indeed,

$$\begin{aligned}
g_2^2 &= \max\{u_2^2, v_2^2\} = \max\{\sin^2\theta_1(x)\sin^2\varphi_1(x),\ \sin^2\theta_2(x)\sin^2\varphi_2(x)\} \\
&= \max\{\sin^2\theta_1(x)(1 - \cos^2\varphi_1(x)), \sin^2\theta_2(x)(1 - \cos^2\varphi_2(x))\} \\
&= \max\{\sin^2\theta_1(x) - u_1^2, \sin^2\theta_2(x) - v_1^2\} \\
&\le \max\{\sin^2\theta_1(x) - \min\{u_1^2, v_1^2\}, \sin^2\theta_2(x) - \min\{u_1^2, v_1^2\}\} \\
&\le \max\{1 - \min\{u_1^2, v_1^2\}, 1 - \min\{u_1^2, v_1^2\}\} = 1 - \min\{u_1^2, v_1^2\} = 1 - g_1^2.
\end{aligned}$$

Similarly it is shown that $h_1^2 + h_2^2 \le 1$.

4 Some Recent Applications

The special types of fuzzy sets that we have presented allow us to work in conditions of uncertainty, in situations of ambiguity and to solve problems with incomplete information. They can be applied in fuzzy image processing, fuzzy reasoning etc.

Numerous applications are found in the problems of Multi-Criteria Decision Making (MCDM). In these problems decision makers generally use real numbers, but in many situations that come from the real world, in evaluating alternatives, in establishing the importance of various criteria, the use of real numbers is not enough. By using some types of fuzzy sets, decision makers can build models in situations of uncertainty or incomplete information.

Base on q-rung orthopair fuzzy numbers, in paper [16], the authors present new approaches of MCDM problems using q-rung orthopair fuzzy generalized Maclaurin symmetric mean operators.

Pythagorean fuzzy sets are applied in [5] in selecting of the appropriate project delivery system.

H. Garg [7–9] discussed MCDM problems under Pythagorean fuzzy environment by proposing some new generalized aggregation operators.

In paper [37] a new approach is presented in MCDM problems based on interval-valued Pythagorean fuzzy linguistic sets.

To see various applications of complex fuzzy sets we refer to the paper [31].

Some applications of (3, 2)-fuzzy sest can be found in paper [14].

A new decision-making approach, based on Fermatean fuzzy sets can be found in [10]. In this paper the authors also make an important study on MCDM applications in construction industry.

For applications of bipolar fuzzy metric spaces we refer to [35].

5 Conclusions

In this paper we have investigated various types of fuzzy sets. We believe that this study can be a foundation for new research in the field and that it will be useful to researchers. A section has been dedicated to Pythagorean fuzzy sets. Here we introduced the concept of vector fuzzy set we also gave various representations of Pythagorean fuzzy sets. Finally we presented some recent applications of nonstandard fuzzy sets.

References

1. Atanassov, K.T.: Intuitionistic fuzzy sets, VII ITKR's Session, Sofia, 1697-1684 (in Bulgarian) (1983)
2. Atanassov, K.T.: Intuitionistic fuzzy sets. Fuzzy Sets Syst. **20**, 87–96 (1986)
3. Atanassov, K.T., Gargov, G.: Interval valued intuitionistic fuzzy sets. Fuzzy Sets Syst. **31**(3), 343–349 (1989)

4. Bustince, H., et al.: A historical account of types of fuzzy sets and their relationships. IEEE Trans. Fuzzy Syst. **24**, 179–194 (2016)
5. Cao, Y.C., Li, H.M., Su, L.M.: Decision-making for project delivery system with related-indicators based on Pythagorean fuzzy weighted Muirhead mean operator. Information **11**, 451 (2021)
6. Dubois, D., Prade, H.: Rough fuzzy sets and fuzzy rough sets. Int. J. Gen. Syst. **17**, 91–209 (1990)
7. Garg, H.: A novel accuracy function under interval-valued Pythagorean fuzzy environment for solving multicriteria decision making problem. J Intell Fuzzy Syst. **31**(1), 529–540 (2016)
8. Garg, H.: A new generalized Pythagorean fuzzy information aggregation using Einstein operations and its application to decision making. Int J Intell Syst. **31**(9), 886–920 (2016)
9. Garg, H.: Generalized pythagorean fuzzy geometric aggregation operators using Einstein t-norm and t-conorm for multicriteria decision-making process. Int J Intell Syst. **32**(6), 597–630 (2017)
10. Keshavarz-Ghorabaee, M., Amiri, M., Hashemi-Tabatabaei, M., Zavadskas, E.K., Kaklauskas, A.: A New Decision-Making Approach Based on Fermatean Fuzzy Sets and WASPAS for Green Construction Supplier Evaluation. Mathematics **8**, 2202 (2020)
11. Goguen, J.: L-fuzzy sets. J. Math. Anal. Appl. **18**, 145–174 (1967)
12. Grattan-Guinness, I.: Fuzzy membership mapped onto interval and many-valued quantities. Zeitschrift fur mathematische Logik und Grundladen der Mathematik **22**, 149–160 (1976)
13. Gulistan, M., Yaqoob, N., Elmoasry, A., Alebraheem, J.: Complex bipolar fuzzy sets: an application in a transport's company. J Intell Fuzzy Syst **40**(3), 3981–3997 (2021)
14. Ibrahim H.Z.; Al-Shami T.M.; Elbarbary O.G.: (3, 2)-Fuzzy sets and their applications to topology and optimal choices, computational intelligence and neuroscience, Article ID 1272266 (2021)
15. Liu, H.C.: Liu's generalized intuitionistic fuzzy sets. J. Educ. Meas. Stat. **18**, 1–14 (2010)
16. Liu, P.D., Wang, Y.M.: Multiple attribute decision making based on q-rung orthopair fuzzy generalized Maclaurin symmetric mean operators. Inform. Sci. **518**, 181–210 (2020)
17. Menger, K.: Statistical metrics. Proc. Nat. Acad. of Sci. U.S.A. **28**, 535–537 (1942)
18. Mondal, T.K., Samanta, S.K.: Generalized intuitionistic fuzzy sets. J. Fuzzy Math. **10**, 839–862 (2002)
19. Pawlak, Z.: Rough sets. Int. J. Comput. Inf. Sci. **11**(5), 341–356 (1982)
20. Pedrycz, W.: Shadowed sets: representing and processing fuzzy sets. IEEE Trans. Syst. Man Cybern. part B **28**, 103–109 (1998)
21. Ramot, D., Milo, R., Friedman, M., Kandel, A.: Complex fuzzy sets. IEEE Trans. Fuzzy Syst. **10**(2), 171–186 (2002)
22. Sambuc, R.: Function ϕ-Flous. Application a l'aide au Diagnostic en Pathologie Thyroidienne, These de Doctorat en Medicine, University of Marseille (1975)
23. Schweizer, A., Sklar, A.: Statistical metric spaces. Pac. J. Math. **10**, 314–334 (1960)
24. Senapati, T., Yager, R.R.: Fermatean fuzzy sets. J. Ambient Intell. Humanized Comput. **11**(2), 663–674 (2019). https://doi.org/10.1007/s12652-019-01377-0
25. Smarandache, F.: A Unifying Field in Logics. Neutrosophy: Neutrosophic Probability, Set and Logic, American Research Press, Rehoboth (1999)

26. Torra, V., Narukawa, Y.: On hesitant fuzzy sets and decision. In: IEEE International Conference on Fuzzy Systems, pp. 1378–1382 (2009)
27. Yager, R.R.: On the theory of bags. Int. J. Gen. Syst. **13**, 23–37 (1986)
28. Yager, R.R.: Pythagorean fuzzy subsets. In: Joint IFSA World Congress and NAFIPS Annual Meeting (IFSA/NAFIPS), pp. 57–61. Edmonton, AB (2013)
29. Yager, R.R.: Generalized orthopair fuzzy sets. IEEE Trans. Fuzzy Syst. **25**(5), 1222–1230 (2017)
30. Yang, H.-L., Li, S.-G., Guo, Z.-L., Ma, C.-H.: Transformation of bipolar fuzzy rough set models. Knowl. Based Syst. **27**, 60–68 (2012)
31. Yazdanbakhsh, O., Dick, S.: A systematic review of complex fuzzy sets and logic. Fuzzy Sets Syst. **338**, 1–22 (2018)
32. Zadeh, L.A.: Fuzzy Sets. Inf. Control **8**, 338–353 (1965)
33. Zadeh, L.A.: Quantitative fuzzy semantics. Inf. Sci. **3**, 159–176 (1971)
34. Zadeh, L.A.: The concept of a linguistic variable and its application to approximate reasoning-1. Inf. Sci. **8**, 199–249 (1975)
35. Zararsız, Z., Riaz, M.: Bipolar fuzzy metric spaces with application. Comput. Appl. Math. **41**(1), 1–19 (2022). https://doi.org/10.1007/s40314-021-01754-6
36. Zhang, W.R.: Bipolar fuzzy sets and relations: a computational framework for cognitive modeling and multiagent decision analysis. In: Proceedings of IEEE Conference, pp. 305–309 (1994)
37. Zhou, Y., Xu, Y., Xu, W.H., Wang, J., Yang, G.M.: A novel multiple attribute group decision-making approach based on interval-valued Pythagorean fuzzy linguistic sets. IEEE Access **8**, 176797–176817 (2020)

Extension-Principle-Based Approach to Least Square Fuzzy Linear Regression

Bogdana Stanojević[1(✉)] and Milan Stanojević[2]

[1] Mathematical Institute of the Serbian Academy of Sciences and Arts, Kneza Mihaila 36, 11000 Belgrade, Serbia
bgdnpop@mi.sanu.ac.rs

[2] Faculty of Organizational Sciences, University of Belgrade, Jove Ilića 154, 11000 Belgrade, Serbia
milans@fon.bg.ac.rs

Abstract. The regression method is widely used in predictive analysis. Its role is to derive an analytic estimation of the outputs expected for given inputs based on observed input-output data. The objective function that is optimized within the regression model is generally the representation of the approximation error comparing to the observed data. Nowadays, the uncertainty is commonly taken into consideration when modeling real systems, and vectorizing information is an important aspect of addressing big data in computer science. Consequently, finding pertinent fuzzy regression models is of great importance within mathematical modeling. In this paper we report our findings related to the full use of the extension principle in solving the optimization model comprised in a least square fuzzy linear regression methodology. We propose a solution approach based on mathematical programming to estimate the fuzzy outputs of the observed fuzzy data; and group our experiments in two categories with respect to the crispness of the observed input data. The first category uses crisp input data and is considered to better explain the advantage of using the extension principle within the solution approach; while the second category, having fuzzy both input and output observed data, is included to prove the relevance of the new approach compared to methodologies from the recent literature.

Keywords: Fuzzy linear regression · Extension principle · Least square

1 Introduction

The regression method is widely used in predictive analysis. Its role is to derive an analytic estimation of the outputs expected for given inputs based on observed input-output data. The objective function that is optimized within the regression model is generally the representation of the approximation error comparing to the observed data. Nowadays, the uncertainty is commonly taken into consideration when modeling real systems, and vectorizing information is an important aspect of addressing big data in computer science. Consequently, finding

S. Dzitac et al. (Eds.): ICCCC 2022, AISC 1435, pp. 219–228, 2023.
https://doi.org/10.1007/978-3-031-16684-6_18

pertinent fuzzy regression models is of great importance within mathematical modeling.

Recently, Kashani et al. [5] introduced a fuzzy penalized regression model with variable selection. They addressed problems with a fuzzy output variable and crisp input variables; and proposed optimal iterative solutions to them. They formulated a variable selection criterion for relevant input variables; and developed a unified approach for multicollinearity and variable selection.

He et al. [3] proposed a model to a fuzzy-in fuzzy-out system. They used a Random Weight Network to develop a model with inputs and outputs described by triangular fuzzy numbers. Their approach is based on the minimization of the integrated squared error between those α-cuts that correspond to the predicted and targeted fuzzy outputs.

Zeng et al. [17] proposed a new way to evaluate the distance between two triangular fuzzy numbers, and used the least absolute deviation method with the new distance definition to model the fuzzy regression.

Chukhrova and Johannssen [2] provided a comprehensive systematic review on fuzzy regression methodology. A wide range of papers and practical applications were analyzed, structured and categorized. They offered a critical discussion of the most relevant methods and approaches, suggested several directions for further researches; and included 455 items in the references list.

In this paper we report our findings related to the full use of the extension principle in solving the optimization model comprised in a least square fuzzy linear regression methodology. We propose a solution approach based on mathematical programming to estimate the fuzzy outputs of the observed fuzzy data. Our goal is to predict the fuzzy outputs as accurate as possible, namely in a complete accordance to the extension principle, even thought that such prediction does not provide the best values for the performance measures when comparing to other approaches.

We group our experiments in two categories with respect to the crispness of the observed input data. The first category uses crisp input data and is considered to better explain the advantage of using the extension principle within the solution approach; while the second category, having fuzzy both input and output observed data, is included to prove the relevance of the new approach compared to methodologies from the recent literature.

The main contribution of this paper is two-fold: (i) we develop a novel approach to fuzzy least square linear regression fully based on the extension principle and applicable to crisp-in–fuzzy-out systems; (ii) using a Monte Carlo simulation procedure we extend the applicability of the proposed approach to fuzzy-in–fuzzy-out systems.

The paper is organized as follows: in Sect. 2 we include the basic notation and summarize the needed terminology. We introduce our approach in Sect. 3, and report our numerical results in Sect. 4. Our final conclusion and some directions for future research are presented in Sect. 5.

2 Notation and Terminology

2.1 Fuzzy Sets and Fuzzy Numbers

Zadeh [15] introduced the main concepts of fuzzy theory. A fuzzy set $\widetilde{A}$ over the universe X is a collection of pairs $\left(x, \mu_{\widetilde{A}}(x)\right)$, with $x \in X$ and $\mu_{\widetilde{A}}(x) \in [0,1]$. The function $\mu_{\widetilde{A}} : X \to [0,1]$ is the membership function describing the fuzzy set $\widetilde{A}$. The value $\mu_{\widetilde{A}}(x)$ is the membership degree of x in $\widetilde{A}$.

The support of a fuzzy set $\widetilde{A}$ is the set of values with non-zero membership degree, i.e.

$$\text{Supp}\left(\widetilde{A}\right) = \left\{x \in X | \mu_{\widetilde{A}}(x) > 0\right\}. \tag{1}$$

The α-cut of a fuzzy set $\widetilde{A}$ is denoted by $\left[\widetilde{A}\right]_\alpha$ and it is defined as the set of values with a membership degree greater or equal to α, $0 < \alpha \leq 1$, i.e.

$$\left[\widetilde{A}\right]_\alpha = \left\{x \in X | \mu_{\widetilde{A}}(x) \geq \alpha\right\}. \tag{2}$$

A fuzzy set $\widetilde{A}$ of the universe R of real numbers is called fuzzy number if and only if: (i) it is fuzzy normal and fuzzy convex; (ii) its membership function $\mu_{\widetilde{A}}$ is upper semi-continuous; and (iii) its support $\left\{x \in R | \mu_{\widetilde{A}}(x) > 0\right\}$ is bounded.

Through the paper we use triangular fuzzy numbers. Generally, a triangular fuzzy number $\widetilde{A}$ is expressed by a triple (a_1, a_2, a_3), such that $a_1 \leq a_2 \leq a_3$. The interval (a_1, a_3) is the support of $\widetilde{A}$. The α-cut of a triangular fuzzy number $\widetilde{A} = (a_1, a_2, a_3)$ is the interval

$$\left[\widetilde{A}\right]_\alpha = \left[(1-\alpha)\, a_1 + \alpha a_2, \alpha a_2 + (1-\alpha)\, a_3\right]. \tag{3}$$

2.2 The Extension Principle

Zadeh [16] introduced the extension principle and applied it to develop the fuzzy arithmetic on fuzzy quantities.

According to the extension principle, the fuzzy set $\widetilde{B}$ over the universe Y that is the result of evaluating the function f at the fuzzy sets $\widetilde{A}_1, \widetilde{A}_2, \ldots, \widetilde{A}_r$ over their universes $X_1, X_2, \ldots, X_r$ is defined through its membership function

$$\mu_{\widetilde{B}}(y) = \begin{cases} \sup\limits_{(x_1,\ldots,x_r) \in f^{-1}(y)} \left(\min\left\{\mu_{\widetilde{A}_1}(x_1), \ldots, \mu_{\widetilde{A}_r}(x_r)\right\}\right), & f^{-1}(y) \neq \emptyset, \\ 0, & \text{otherwise.} \end{cases} \tag{4}$$

Stanojević et al. [11] provided a general view on the use of extension principle within mathematical programming. Their perspective concerning the use of the extension principle in all stages of an optimization process is adopted here in the context of fuzzy regression modeling. Previous papers [12, 14] that addressed fuzzy regression via the extension principle only partly involved it in their models.

2.3 Least Square Method for Linear Regression

The linear regression is used to model the relationship between a scalar output and one or more explanatory (input) variables. Such relationships is modeled by a linear function whose unknown parameters are estimated from the observed data.

Given n observed input k-tuples $(x_{i1}, x_{i2}, \ldots, x_{ik})_{i=1,\ldots,n}$ and n output scalars $(y_i)_{i=1,\ldots,n}$ one searches for the coefficients $a_0, a_1, \ldots, a_k$ of the linear function $f(x_1, x_2, \ldots, x_k) = a_0 + a_1x_1 + a_2x_2 + \cdots + a_kx_k$ able to give a good approximation $f(x_{i1}, x_{i2}, \ldots, x_{ik})$ to y_i, for each $i = 1, \ldots, n$. We call

$$\widehat{y}_i = f(x_{i1}, x_{i2}, \ldots, x_{ik}) \tag{5}$$

the i-th estimated output.

In the least square linear regression the coefficients $a_0, a_1, \ldots, a_k$ are determined such that the sum of the squared Euclidean distances between y_i and $f(x_{i1}, x_{i2}, \ldots, x_{ik})$ is minimal. The well known formula for computing the coefficients is

$$(a_0, a_1, \ldots, a_k)^T = \left(X^TX\right)^{-1} X^TY, \tag{6}$$

where

$$X = \begin{pmatrix} 1 & x_{11} & x_{12} & \ldots & x_{1k} \\ 1 & x_{21} & x_{22} & \ldots & x_{2k} \\ \vdots & \vdots & \vdots & \vdots & \vdots \\ 1 & x_{n1} & x_{n2} & \ldots & x_{nk} \end{pmatrix}, \; Y = \begin{pmatrix} y_1 \\ y_2 \\ \vdots \\ y_n \end{pmatrix}. \tag{7}$$

Linear regression models may also be formulated in other ways than using the least square method. For instance they can rely on minimizing the "lack of fit" in some other norm (as with least absolute deviations regression), or by minimizing a penalized version of the least squares cost function as in RIDGE regression (L_2-norm penalty) and LASSO (L_1-norm penalty).

2.4 Least Square Method for Fuzzy Linear Regression

Generalizing the least square linear regression to the fuzzy environment, let us consider that there are given n k-tuples of fuzzy numbers $(\widetilde{x}_{i1}, \widetilde{x}_{i2}, \ldots, \widetilde{x}_{ik})_{i=1,\ldots,n}$ as inputs, and n fuzzy numbers $(\widetilde{y}_i)_{i=1,\ldots,n}$ as outputs. In this case one searches for the fuzzy coefficients $\widetilde{a}_0, \widetilde{a}_1, \ldots, \widetilde{a}_k$ of the fuzzified linear function

$$f(\widetilde{x}_1, \widetilde{x}_2, \ldots, \widetilde{x}_k) = \widetilde{a}_0 + \widetilde{a}_1\widetilde{x}_1 + \widetilde{a}_2\widetilde{x}_2 + \cdots + \widetilde{a}_k\widetilde{x}_k$$

able to give a good approximation $f(\widetilde{x}_{i1}, \widetilde{x}_{i2}, \ldots, \widetilde{x}_{ik})$ to $\widetilde{y}_i$, for each $i = 1, \ldots, n$.

By analogy, the fuzzy least square linear regression method should determine the fuzzy coefficients $\widetilde{a}_0, \widetilde{a}_1, \ldots, \widetilde{a}_k$ such that the sum of the squared distances between the fuzzy numbers $\widetilde{y}_i$ and $f(\widetilde{x}_{i1}, \widetilde{x}_{i2}, \ldots, \widetilde{x}_{ik})$ is minimal. We call $\widehat{\widetilde{y}}_i = f(\widetilde{x}_{i1}, \widetilde{x}_{i2}, \ldots, \widetilde{x}_{ik})$ the i-th estimated fuzzy output.

With respect to different ways to evaluate the distance between two fuzzy numbers, different fuzzy least square linear regression models can be formulated. In the next section we introduce the fuzzy least square linear regression model, and solve it in full accordance with the extension principle.

2.5 Performance Measures

In this section we briefly present the similarity, distance and credibility measures that we used for comparing the quality of our results to the results from the literature. We recall these performance measures as they were described in [1].

The similarity measure is based on the concept of area. It is computed as

$$S = \sum_{i=1}^{n} \left(1 - \left(\int_{\Omega_i} \left|\mu_{\widetilde{y}_i}(x) - \mu_{\widehat{\widetilde{y}}_i}(x)\right| dx\right) / \left(\int_{\Omega_i} \mu_{\widetilde{y}_i}(x)\, dx + \int_{\Omega_i} \mu_{\widehat{\widetilde{y}}_i}(x)\, dx\right)\right),$$

where $\mu_{\widetilde{y}_i}$ is the membership function of the observed output $\widetilde{y}_i$, $\mu_{\widehat{\widetilde{y}}_i}$ is the membership function of the estimated output $\widehat{\widetilde{y}}_i$, and Ω_i is the union of the intervals over which these membership functions have strictly positive values.

The distance measure is based on a certain definition of the distance between two fuzzy numbers. It is computed as

$$D = \frac{1}{4} \sum_{i=1}^{n} \left(\left|(\widetilde{y}_i)^L - \left(\widehat{\widetilde{y}}_i\right)^L\right| + \left|(\widetilde{y}_i)^U - \left(\widehat{\widetilde{y}}_i\right)^U\right| + 2\left|(\widetilde{y}_i)^C - \left(\widehat{\widetilde{y}}_i\right)^C\right|\right),$$

where $\left((\widetilde{y}_i)^L, (\widetilde{y}_i)^C, (\widetilde{y}_i)^U\right)$ is the triple representing the triangular fuzzy number $\widetilde{y}_i$, and $\left(\left(\widehat{\widetilde{y}}_i\right)^L, \left(\widehat{\widetilde{y}}_i\right)^C, \left(\widehat{\widetilde{y}}_i\right)^U\right)$ is the triple representing the triangular fuzzy number $\widehat{\widetilde{y}}_i$.

The credibility measure is computed as

$$C = \sum_{i=1}^{n} \left(\max\left(\mu_{\widetilde{y}_i} \sqcap \mu_{\widehat{\widetilde{y}}_i}\right)\right) / \left(\left|\left(\widehat{\widetilde{y}}_i\right)^U - \left(\widehat{\widetilde{y}}_i\right)^L\right|\right),$$

where $\mu_{\widetilde{y}_i} \sqcap \mu_{\widehat{\widetilde{y}}_i}$ represent the intersection of the surfaces under the graphs of the functions $\mu_{\widetilde{y}_i}$ and $\mu_{\widehat{\widetilde{y}}_i}$.

3 Our Approach

In this section we formulate an algorithm to determine the predicted fuzzy outputs with respect to observed crisp inputs and triangular fuzzy outputs using a fuzzy least square linear regression model based on the extension principle.

Our goal is to define predicted outputs that are in full compliance to the extension principle applied to the classic least square linear regression model.

Given n k-tuples of real numbers $(x_{i1}, x_{i2}, \ldots, x_{ik})_{i=1,\ldots,n}$ as inputs, we construct the matrix Xas in (7). Further on, given n triangular fuzzy numbers $(\widetilde{y}_i)_{i=1,\ldots,n}$ as outputs, for each index $h = 1, \ldots, n$ and for any fixed $\alpha \in [0,1]$ we solve the optimization model

$$\begin{array}{ll} \min a_0 + a_1 x_{h1} + a_2 x_{h2} + \cdots + a_k x_{hk} & \\ \text{s.t.} & \\ \quad (a_0, a_1, \ldots, a_k)^T = \left(X^T X\right)^{-1} X^T \left(y_1, y_2, \ldots, y_n\right)^T, & \\ \quad (1-\alpha)\left(\widetilde{y}_i\right)^L + \alpha \left(\widetilde{y}_i\right)^C \leq y_i \leq \alpha \left(\widetilde{y}_i\right)^C + (1-\alpha)\left(\widetilde{y}_i\right)^U, & i = 1, \ldots, n, \\ \quad a_j \text{ free variable}, & j = 0, \ldots, k, \end{array} \tag{8}$$

to derive the left endpoint of the α-cut of the predicted output $\widehat{\widetilde{y}}_h$. In a similar way, solving the optimization model

$$\begin{array}{ll} \max a_0 + a_1 x_{h1} + a_2 x_{h2} + \cdots + a_k x_{hk} & \\ \text{s.t.} & \\ \quad (a_0, a_1, \ldots, a_k)^T = \left(X^T X\right)^{-1} X^T \left(y_1, y_2, \ldots, y_n\right)^T, & \\ \quad (1-\alpha)\left(\widetilde{y}_i\right)^L + \alpha \left(\widetilde{y}_i\right)^C \leq y_i \leq \alpha \left(\widetilde{y}_i\right)^C + (1-\alpha)\left(\widetilde{y}_i\right)^U, & i = 1, \ldots, n, \\ \quad a_j \text{ free variable}, & j = 0, \ldots, k, \end{array} \tag{9}$$

we derive the right endpoint of the α-cut of the predicted output $\widehat{\widetilde{y}}_h$.

Due to the linearity of the models (8) and (9), we can obtain the full description of the predicted fuzzy output $\widehat{\widetilde{y}}_h$ by solving both models for $\alpha = 0$ to determine its support, and any of them for $\alpha = 1$ to determine its value with maximal amplitude.

Trying to apply a similar idea to determine the predicted fuzzy outputs with respect to observed fuzzy inputs and outputs two issues arise: (i) the predicted outputs are not triangular fuzzy numbers anymore, and (ii) the generalized models (8) and (9) are non-linear. Because of these issues the definition of the membership function given by the extension principle (4) is difficult to be directly used in solving the fuzzy linear regression problem. Having in mind that the well known Monte Carlo method is able to provide numerical results in a wide variety of problems by random sampling certain parameters, we propose the following solution procedure for the fuzzy linear regression with observed fuzzy inputs and outputs: (i) generate random crisp values in the range of the α-cuts for the observed input and output fuzzy numbers; (ii) use them for calculating the regression's coefficients as in (6) and the predicted crisp outputs as in (5); and (iii) empirically construct the fuzzy predicted outputs for the original fuzzy linear regression problem. In the next section this procedure will be applied to solve a numerical example involving observed fuzzy inputs and outputs.

4 Computation Results

4.1 The Case of Crisp Observed Inputs and Fuzzy observed Outputs

In this section we present detailed results obtained by our approach applied to the numerical example given in Tanaka et al. [13] and compared to the results given in [4,6–9] and [13]. In [10] the same example is addressed but the results were not reported in detail, thus we did not include them in the comparison.

Table 1. Numerical data for the example presented in 4.1: x – crisp input, $\widetilde{y}$ – fuzzy output, $\widehat{\widetilde{y}}$ – fuzzy output predicted by our approach, estimated error as difference of the membership functions between the observed and estimated fuzzy numbers, and the values of the performance measures of the results obtained by our approach

Observed values		Estimated values	Performance measures		
x_i	$\widetilde{y}_i = (l_i, c_i, r_i)$	$\widehat{\widetilde{y}}_i = \left(\widehat{l}_i, \widehat{c}_i, \widehat{r}_i\right)$	Credibility	Distance	Similarity
1	$(6.2, 8, 9.8)$	$(3.62, 6.7, 9.7)$	0.20	1.34	1.25
2	$(4, 6.4, 8.8)$	$(6.15, 8.37, 10.59)$	0.12	1.97	0.33
3	$(6.9, 9.5, 12.1)$	$(7.72, 10.08, 12.44)$	0.17	0.58	0.78
4	$(10.9, 13.5, 16.1)$	$(9.29, 11.79, 14.29)$	0.13	1.71	1.11
5	$(10.6, 13, 15.4)$	$(10.14, 13.5, 16.86)$	0.19	0.73	0.83

Table 1 contains the given data, our estimated outputs and the performance measures of our results. Table 2 reports the comparison between our results and the results reported in the literature.

Table 2. Comparison results for the example presented in 4.1 – using three ways to measure the difference between the observed and estimated fuzzy numbers. The results marked with a * are better than the results obtained by the new proposed method.

Method		Similarity	Distance	Credibility
Savic & Pedrycz [9]		4.87*	7.85	0.87*
Kim & Bishu [6]		4.07	6.05*	0.78
Tanaka et al. [13]	min	4.46*	7.30	0.82*
	max	2.57	7.39	0.57
	conj.	3.40	9.12	0.43
Chen & Hsueh [7]		4.25	5.60*	0.81
Nasrabadi & Nasrabadi [8]		4.10	5.93*	0.79
Hojati et al. [4]		3.79	5.6*	0.83*
Our approach		4.30	6.33	0.81

4.2 The Case of Fuzzy Observed Inputs and Outputs

Our second numerical example is taken from [14]. The observed values are given in Table 3. The observed and estimated outputs are shown in Fig. 1. In this figure one can see the observed outputs (in blue), estimated outputs obtained by Wu [14] (in black) and the estimated outputs empirically obtained by our Monte Carlo simulation based approach (in green). Since the results of our Monte Carlo simulation of the extension principle do not fit within the bounds of the estimated outputs derived by Wu's method [14], we may conclude that Wu's method does not fully comply to the extension principle even though it is based on it.

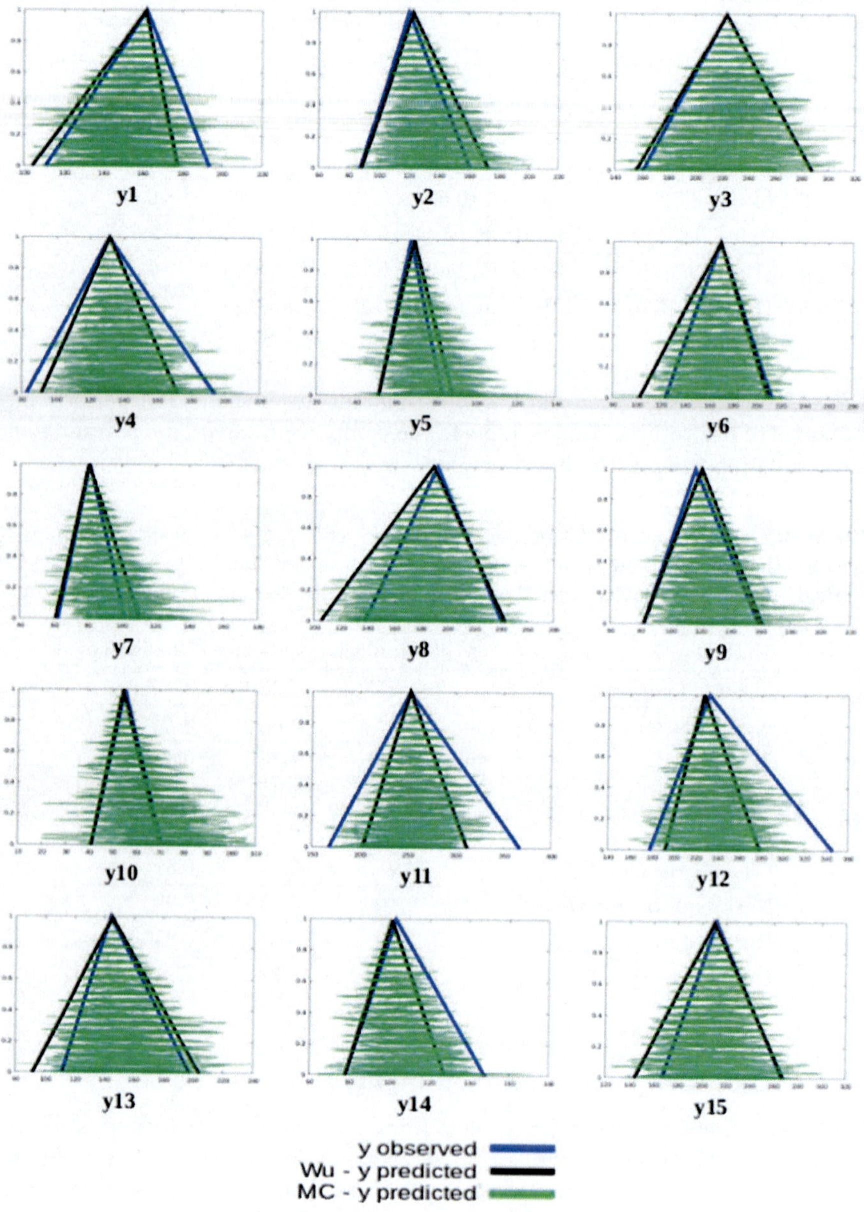

Fig. 1. Observed and estimated outputs for the example presented in 4.2.

Table 3. Observed inputs and outputs for the example presented in 4.2.

i	$\widetilde{x}_{i1} = (l_i, c_i, r_i)$	$\widetilde{x}_{i2} = (l_i, c_i, r_i)$	$\widetilde{y}_i = (l_i, c_i, r_i)$
1	(151, 274, 322)	(1432, 2450, 3461)	(111, 162, 194)
2	(101, 180, 291)	(2448, 3254, 4463)	(88, 120, 161)
3	(221, 375, 539)	(2592, 3802, 5116)	(161, 223, 288)
4	(128, 205, 313)	(1414, 2838, 3252)	(83, 131, 194)
5	(62, 86, 112)	(1024, 2347, 3766)	(51, 67, 83)
6	(132, 265, 362)	(2163, 3782, 5091)	(124, 169, 213)
7	(66, 98, 152)	(1687, 3008, 4325)	(62, 81, 102)
8	(151, 330, 463)	(1524, 2450, 3864)	(138, 192, 241)
9	(115, 195, 291)	(1216, 2137, 3161)	(82, 116, 159)
10	(35, 53, 71)	(1432, 2560, 3782)	(41, 55, 71)
11	(307, 430, 584)	(2592, 4020, 5562)	(168, 252, 367)
12	(284, 372, 498)	(2792, 4427, 6163)	(178, 232, 346)
13	(121, 236, 370)	(1734, 2660, 4094)	(111, 144, 198)
14	(103, 157, 211)	(1426, 2088, 3312)	(78, 103, 148)
15	(216, 370, 516)	(1785, 2605, 4042)	(167, 212, 267)

5 Conclusions and Future Works

In the present study we reported our findings related to the full use of the extension principle in solving the optimization model comprised in a least square fuzzy linear regression methodology. We proposed a solution approach based on mathematical programming to predict the fuzzy outputs of the observed fuzzy data. Our main goal was to predict the fuzzy outputs as accurate as possible, i.e. in a complete accordance with the extension principle. Providing the best values for the performance measures when comparing to other approaches was not essentially targeted.

We grouped our experiments in two categories with respect to the crispness of the observed input data. The first category used crisp input data and was considered to better explain the advantage of using the extension principle within the solution approach; while the second category, having both input and output observed data expressed by fuzzy numbers, was included to prove the relevance of the new approach compared to methodologies from the recent literature.

Our future research will focus on non-linear regression models for crisp in – crisp out systems that can be adapted with respect to the extension principle to fuzzy in – fuzzy out systems.

Acknowledgments. This work was supported by the Serbian Ministry of Edu-cation, Science and Technological Development through Mathematical Instituteof the Serbian Academy of Sciences and Arts and Faculty of Organizational Sciences of the University of Belgrade.

References

1. Chen, L.H., Nien, S.H.: A new approach to formulate fuzzy regression models. Appl. Soft Comput. **86**, 105915 (2020)
2. Chukhrova, N., Johannssen, A.: Fuzzy regression analysis: systematic review and bibliography. Appl. Soft Comput. **84**, 105708 (2019)
3. He, Y., Wang, X., Huang, J.Z.: Fuzzy nonlinear regression analysis using a random weight network. Inf. Sci. **364–365**, 222–240 (2016)
4. Hojati, M., Bector, C.R., Smimou, K.: A simple method for computation of fuzzy linear regression. Eur. J. Oper. Res. **166**(1), 172–184 (2005). Metaheuristics and Worst-Case Guarantee Algorithms: Relations, Provable Properties and Applications
5. Kashani, M., Arashi, M., Rabiei, M.R., D'Urso, P., De Giovanni, L.: A fuzzy penalized regression model with variable selection. Expert Syst. Appl. **175**, 114696 (2021)
6. Kim, B., Bishu, R.R.: Evaluation of fuzzy linear regression models by comparing membership functions. Fuzzy Sets Syst. **100**(1), 343–352 (1998)
7. Chen, L.H., Hsueh, C.C.: A mathematical programming method for formulating a fuzzy regression model based on distance criterion. IEEE Trans. Syst. Man Cybern. B Cybern. **37**(3), 705–12 (2007)
8. Nasrabadi, M.M., Nasrabadi, E.: A mathematical-programming approach to fuzzy linear regression analysis. Appl. Math. Comput. **155**(3), 873–881 (2004)
9. Savic, D.A., Pedrycz, W.: Evaluation of fuzzy linear regression models. Fuzzy Sets Syst. **39**(1), 51–63 (1991)
10. Skrabanek, P., Marek, J., Pozdilkova, A.: Boscovich fuzzy regression line. Mathematics **9**(6), 685 (2021)
11. Stanojevic, B., Stanojevic, M., Nadaban, S.: Reinstatement of the extension principle in approaching mathematical programming with fuzzy numbers. Mathematics **9**(11), 1272 (2021)
12. Tanaka, H., Uejima, S., Asai, K.: Linear regression analysis with fuzzy model. IEEE Trans. Syst. Man Cybern. **12**, 903–907 (1982)
13. Tanaka, H., Hayashi, I., Watada, J.: Possibilistic linear regression analysis for fuzzy data. Eur. J. Oper. Res. **40**(3), 389–396 (1989)
14. Hsien-Chung, W.: Linear regression analysis for fuzzy input and output data using the extension principle. Comput. Math. Appl. **45**(12), 1849–1859 (2003)
15. Zadeh, L.A.: Fuzzy sets. Inf. Control **8**(3), 338–353 (1965)
16. Zadeh, L.A.: The concept of a linguistic variable and its application to approximate reasoning i. Inf. Sci. **8**(3), 199–249 (1975)
17. Zeng, W., Feng, Q., Li, J.: Fuzzy least absolute linear regression. Appl. Soft Comput. **52**, 1009–1019 (2017)

Fuzzy Logic Applied in Hierarchical Secondary Regulation Voltage-Reactive Power

Victorita Radulescu(✉)

University Politehnica of Bucharest, Splaiul Independentei 313, 060042 Bucharest, Romania
victorita.radulescu@upb.ro

Abstract. In the current conditions of internal energy market liberalization and permanent international trades, new solutions of hierarchical regulation between voltage and reactive power for the electricity supply systems became necessary. Electricity transmission and distribution companies in Romania specifically have approaches to this regulation. Another problem is the power lines overloading in the distribution nodes, especially in the transition zones from one company to another. Regulation between voltage and reactive power is useful in normal operation conditions but especially in the emergency mode when it is necessary to supply many consumers, in a short time. The focus centers on secondary regulation based on Fuzzy Logic for the optimal management of the complex problem of redistribution and the hierarchical secondary regulation in overloaded nodes. This adjustment allows the changes of the generator's voltage in the control zone, necessary when a variation appears in the pilot node, a consequence of modification of the reactive power consumption. The Fuzzy model is tested in the Galati County, for a monofilament diagram with a developed Power Test line for the power station from the Smardan node, considered a national strategic point. Finally, some numerical results and remarks are mentioned.

Keywords: Fuzzy control · Reactive power control · Voltage control

1 Introduction

In the current context of the energy crisis throughout Europe and beyond, significant transformations appear in the operating, regulation, and control systems of the electricity distribution, [1]. Liberalization of the electricity market, from the production and continuing with its transport and distribution until the final consumer, raised new problems of competitive level between the companies involved in this process. The integration of the Romanian national electricity distribution network in the European market highlighted some issues mainly related to the territorial distribution and overloading of some nodes and lines from the system. The newly liberalized electricity market led the consumers from some geographical areas to be attracted by the local suppliers, in any conditions, [2]. They set a price that may be, or not, correlated with the price of other suppliers, from other areas. Under the implemented European legislation, the system operators are obliged to ensure non-discriminatory access to infrastructure for all participants in the energy

S. Dzitac et al. (Eds.): ICCCC 2022, AISC 1435, pp. 229–238, 2023.
https://doi.org/10.1007/978-3-031-16684-6_19

market, [3]. This concludes with bilateral contracts between the customer-consumer and supplier. The system of transmission and distribution guarantees the transport of the necessary electricity flow, [4].

On the other hand, due to global warming, there is the problem of minimizing the operation of thermal power plants on coal, oil, and gas and replacing them with power plants based on renewable sources; in Romania mainly wind and solar, [5]. EU member states and beyond signed the Paris Climate Agreement, which has set the goal of maintaining global warming at a maximum of 2oC, [6]. Even so, nowadays there is an acceleration of the ice melting, at a rate about 4 times faster than it was 20 years ago.

In Romania, in the last decade, several wind farms have been built, whose production ensures about 16–24% of the internal electricity production and photovoltaic power plants, with about 3–5% of it. They are added to the hydroelectric power plants whose capacity provides over 30% of domestic production so that approximately 47–60% of total electricity produced in Romania is provided from renewable resources.

Many of the newly built wind and photovoltaic power plants have been realized in less populated areas, where the distribution networks of electric lines are reduced. New problems have arisen related to the transmission lines, undersized compared to the current capacity of electricity production, and improperly distributed in areas with high wind or photovoltaic potential. To solve this problem, the modernization and rehabilitation of the existing electric lines, where the installations of protection and automation play a special role, has started. The control must be provided to ensure the transit of electric energy between any two points of the transmission lines. The voltage control and reactive power management become the top technical objective, especially for the overloaded nodes, [7].

This paper proposes an algorithm for the hierarchical organization of automatic voltage regulation systems, emphasizing the secondary regulation between voltage and reactive power. It starts from the available transport capacity between two nodes by applying the Fuzzy logic in the secondary regulation, [8]. The algorithm offers a method of selecting the problems that occur, and proposes a control solution capable of avoiding the overloading in the distribution nodes, [9].

2 Material and Methods

2.1 Estimation of Capacity and Appeared Problems in Electricity Transmission Lines

Due to the development of new capacities of electricity production based on renewable resources, some physical limitations of the existing transmission lines have appeared in many areas, which do not allow voltage regulation during the operation. Figure 1 shows the distribution of the Romanian energy production in January 2022; with blue-hydro, green-wind, red-nuclear, and light brown-solar. Together, they represent over 60% of the entire electricity production, [10].

Before 2007, the Dobrogea was considered an area with low economic potential and with a reduced population, so the electricity lines are undersized. Due to the wind farms realized here, currently, there is a significant amount of produced electricity that must be transported and delivered to the national energy system.

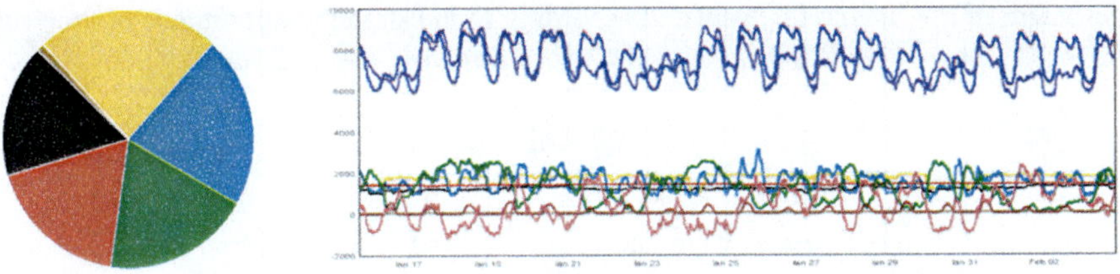

Fig. 1. Electric energy production in January 2022 in Romania.

Starting in 2021, in Romania, based on an international agreement, several thermal power plants-TPPs have been closed, such as Borzesti and Mintia. Now, differences between the distribution lines from different geographical zones and variations in supplying them appear often.

The Romanian electric transmission system is illustrated in Fig. 2-a, where LEA-represents the airlines and LES-underground power lines. New zones with overloaded electric lines appear by developing new power plants, and by closing TPPs areas with under-solicited nodes are recorded, [11]. Figure 2-b shows the available capacity of the current electrical transmission lines. In the Dobrogea area with wind farms and Banat with photovoltaic power plants, even high electric energy producer zones are less equipped with power lines in terms of transport capacity and the possibility of delivering it in optimal conditions.

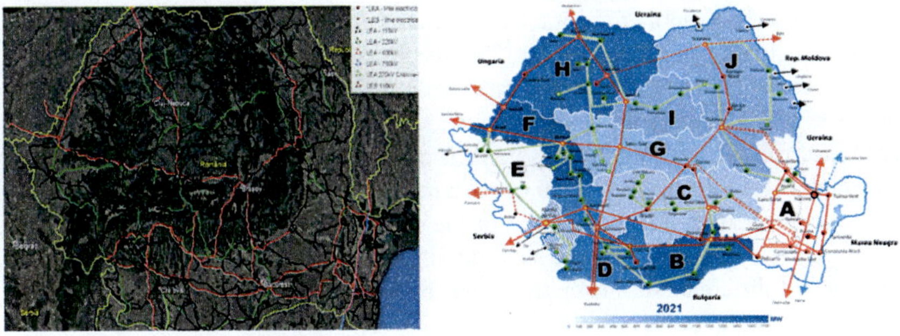

Fig. 2. Distribution of electric lines and capacity of transport.

In other areas, due to the large number of transactions recorded simultaneously, the transmission network can sometimes be congested. The operating system should have solutions to eliminate them.

2.2 Hierarchical Structure of the Regulation Voltage - Reactive Power

Shutting down an operating node or a power line may affect their safety. This is expressed by the Performance Indicator-PI, which is an index of the severity in functioning, where $f_j(x)$ - is the function of the active/reactive power transactions in the grid line j, p_j - the coefficient of the line/node, evaluated according to the strategic importance of the grid, and m – an exponent with different values for improving the range of PI values.

$$PI = \sum_{j} p_j [f_j(x)]^m \quad (1)$$

The regulation between voltage and reactive power is required both in normal-safe operation and in emergency mode. Normally, voltage regulation provides consumers with electricity at the contracted quality parameters, in the most economically efficient conditions for the supplier. If a fault occurs, the voltage control must make a selection to ensure the operating conditions for as many consumers as possible. There are several approaches to solving this problem, specific to each supplying company. Some of them use a hierarchical structure with two or three levels of voltage-reactive power regulation. However, the most complex solution includes a primary, secondary, and tertiary adjustment.

Primary voltage regulation is performed based on local voltage regulation equipment, with synchronous generators, and synchronous or static compensators, when a disturbance is recorded, [12]. The control actions are based on local measures and intend to bring the voltage from the respective node to the normal working value. Any distribution company makes this adjustment, which is minimal in the case of a fault.

Regarding the secondary regulation, there are several options, chosen by each company. Some of them use secondary regulation as a static voltage-reactive power optimization problem, treated as an open loop. In this case, the permanent regime with minimal power losses is first identified. The second solution involves an online closed-loop, being faster and more efficient. Regardless of the adopted solution, the principle of secondary regulation consists in controlling the voltage in an area of the electrical lines, called the control zone. Using the control methods in this area, the voltage in nodes is maintained within the predetermined limits and the reactive power circulation, in lines interconnected with the neighboring areas, is preserved.

Tertiary regulation acts as global coordination, for maintaining the circulation of the reactive power between different areas of the system. It ensures the decrease of losses from the entire system and increases its stability. Tertiary adjustment involves real-time monitoring of different areas from the operating system.

3 Fuzzy Logic in Secondary Voltage-Reactive Power Regulation

The trend of using Fuzzy logic in controlling industrial processes, including energy, has been developed in recent decades. Fuzzy Logic allows the modeling of individual possibilities of some industrial components and their inclusion in a semantic concept. A fuzzy controller can perform the same functions as a conventional one. It manages the complex control space by reading the information received from the sensors, [13]. Fuzzy logic allows the application of a predetermined mathematical model and transmits appropriate signals to the control equipment, [14].

Voltage control is generally done online in a closed loop and assumes that the interaction with neighboring areas is negligible. It starts from the idea of creating a pilot node, in which the voltage evolution is significant for the voltage variation in the rest of the consuming nodes. Voltage measurements are made only in these nodes. The secondary control makes a record for all these sources, called the regulating nodes.

If X is the set of control nodes and A is the regulation/control range, the fuzzy set consists of the pairs $A = \{x,\ \mu_A(x)|x \in X\}$, where $\mu_A(x)$ is the membership function. Its value indicates the voltage in the respective node for secondary regulation. Membership functions can be different but interconnected. The association of linguistic terms for a set of numerical values must be well correlated with the possibilities of local regulation of voltage and reactive power. A basic tool is represented by the general principle of determining the fuzzy sets, as a result of regulation in consumer nodes, [15, 16]. This represents the extension process of the MIN-MAX rule or the principle of trust made in a heuristic manner.

The fuzzy controller differs from the conventional one in the management of the complex control space. It is capable of approximating linear or nonlinear functions with any level of accuracy, and assumes the existence of the blocks mentioned in Fig. 3, where:

- A - Sensors that measure the instantaneous values of the process parameters;
- B - Fuzzy block of the quantities considered from the regulation process;
- C - Logical calculation block;
- D - Block of storage the fuzzy parameters.
- E – The un-fuzzy block;
- F - Transducers that transmit the adjustment signal to the adjustment equipment;
- G - Regulation process.

The fuzzy block makes the transition from the field of numerical quantifications specific to technical and mathematical systems to the field of logical qualitative assessments, specific to human thinking. The un-fuzzy block realizes the inverse transformation, from the numerical quantification of the logically qualitative assessments making them accessible to the technical systems.

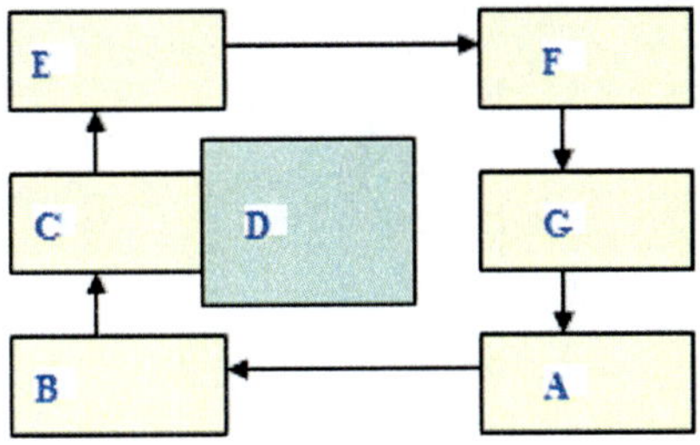

Fig. 3. Structure of the Fuzzy controller.

The model represents a Fuzzy secondary controller with a single channel. The principle of regulation consists in modifying the voltage of the generators, considered execution elements in the control area. The voltage of the pilot node is considered the input parameter, which changes when a modification appears in the reactive power consumption for that area, [17, 18]. The model of the secondary voltage-reactive power regulation is characterized by a set of rules: R1 and R4- NG-Not Good, R5-LG-Less Good, R3-G-Good, R2- TG-Too good.

R1 - Voltage of the pilot node is low;
- Voltage imposed on the generators participating in the regulation is high;
R2 - Voltage of the pilot node is high;
- Voltage imposed on the generators participating in the regulation is high;
R3 - Voltage of the pilot node is normal;
- Voltage imposed on the generators participating in the regulation is normal;
R4 - Voltage of the pilot node is low;
- Voltage imposed on the generators participating in the regulation is low;
R5 - Voltage of the pilot node is high;
- Voltage imposed on the generators participating in the regulation is low.

The creation of a Fuzzy controller involves the encoding of the control function into a set of rules that operate on fuzzy elements, obtained by fusing the domain of definition, corresponding to the voltage of the pilot node. The result of applying this set of rules will be found in another fuzzy set corresponding to the action of the element that performs the adjustment, respectively the voltage imposed on the generators.

4 Simulation of Secondary Voltage-Reactive Power Regulation

The simulations were made on a Test Network realized with 13 nodes; its scheme is presented in Fig. 4. Based on the situation recorded in Romania in the last decades in the distribution system, there are zones where the existing transport electric lines have physical limitations, with overloaded nodes. The Test Network is considered with a single adjustment, with no interconnection lines with other zones, modeled on the pattern of the electrical network Case Study, the Smardan node, from Galati County.

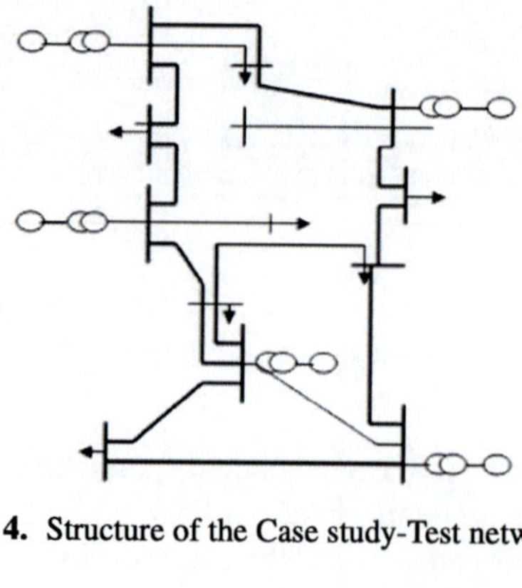

Fig. 4. Structure of the Case study-Test network.

The method of eliminating the load of nodes by redistribution was used to optimize the regulation. The pilot node selected for this network is node 10, based on the representative degree of the voltage evolution recorded here. It connects the electricity lines from the Dobrogea area with Moldova.

There are situations when the network congestion must be rapidly eliminated through operational measures, to reallocate available transfer capacity to other electric lines from the network, or through a secondary regulation between voltage-reactive power adjustments.

Power variation at the generator node will be compensated by a proportional variation at the equilibrium node, while the power at the rest of the generators remains unchanged. PI appears as a variation of the main voltage or power, compared to their maximum allowable values. The power variation capacity on each line-P_{ik}, as a function of variation of the generated power-P_g, is determined from the active power:

$$P_{ik} = B_{ik} U_{n,i} U_{n,k} (D_i - D_k) \tag{2}$$

Variation of the active power circulation on lines can be estimated according to the variation of the generated power:

$$a_{ik,g} = \frac{dP_{ik}}{dP_g} \tag{3}$$

By considering:

$$a_{ik,g} = B_{ik} U_{n,i} U_{n,k} \left(\frac{dD_j}{dP_g} - \frac{dD_k}{dP_g} \right) \tag{4}$$

The project that includes the Smardan node ensures the increase of the National Energy System –NES capacity by transporting the energy produced in the wind power plants from Dobrogea and its transmission in the deficit area, south Moldova and Baragan. These changes and adjustments also support an efficient interconnection of the electricity networks between Romania and the Republic of Moldova.

In Galati County, the power station from the Smardan node becomes a strategic point through which the power injection is made in the 110 kV network to supply consumers in the Galati area and south Moldova. This project contributes to the development of the European electricity transmission infrastructure. At the European level, it ensures the consolidation of the transport corridor along the Black Sea coast - Ukraine, Romania, Bulgaria, and the rest of Europe – the Black Sea Corridor.

The included simulations increase the reactive power consumed in nodes with already low voltage levels. The voltage regulation sources available in the analyzed network are the generators from nodes 1, 2, and 3. The generator from node 4, after modeling the increase of the reactive power, was charged to the maximum. In the second stage of simulation, it will be treated as a consumer node, without participation in voltage regulation. This is the case that happened in July 2019 when the Smardan Node Station was overloaded.

For similar situations, fast coordinated adjustment situations had to be found. An increase in reactive power consumption was simulated from 70 MVAr to 90 MVAr in node 9, which now will have a low voltage level. When the power disturbance occurs, the voltage in the pilot node decreases from 210.62 kV to 206.39 kV, and in node 9 where the power increases, the voltage decreases from 211.99 kV to 209.38 kV. The degrees of belonging of the pilot node voltage after the reactive power increase are:

$$\mu_{low}(U_p) = 0.3951;$$
$$\mu_{normal}(U_p) = 0.6049; \quad (5)$$
$$\mu_{high}(U_p) = 0$$

Finally, the rules applied considering the recorded situations in the node are:

R1: *If* Voltage of the pilot node is low *Then* Voltage imposed on the generators participating in the regulation should be high

$$\mu_{U_{g-high}}(R1) = 0.3951 \quad (6)$$

R2: *If* Voltage of the pilot node is normal *Then* Voltage imposed on the generators must be normal.

$$\mu_{U_{g-normal}}(R2) = 0.6049 \quad (7)$$

Based on the extension principle, the domain for the secondary regulation of voltage-reactive power is obtained by reuniting the two domains from these rules.

$$\mu_{U_g} = \mu_{U_{g-high}}(0.3951) \cup \mu_{U_{g-normal}}(0.6049) \quad (8)$$

Figure 5 illustrates the action of the secondary Fuzzy regulator in case of increasing the reactive power consumed in node 9 by 20 MVAr (from 70 MVAr to 90 MVAr). With green is Generator 1, with orange-Generator 2, and with blue-Generator 3.

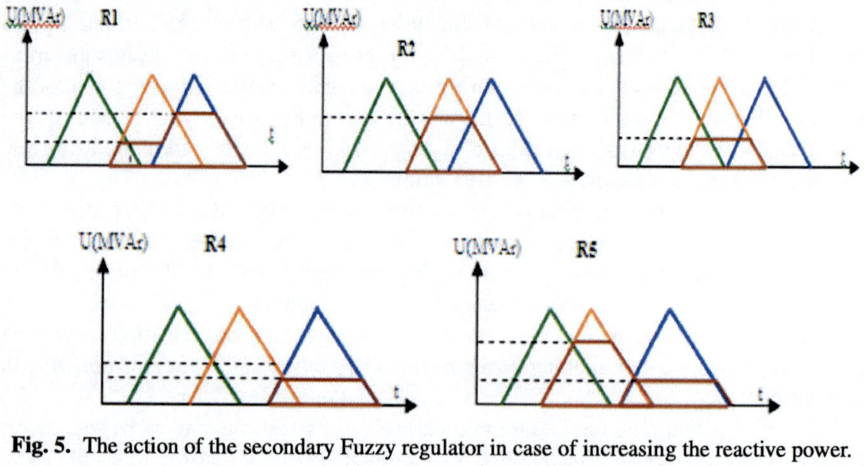

Fig. 5. The action of the secondary Fuzzy regulator in case of increasing the reactive power.

The domain is un-fuzzy by the method of the center of gravity of the domain to which it belongs. New values of voltages imposed on the generators that participate in secondary voltage-reactive power regulation are next obtained. For this case, Table 1 presents the coefficients for voltage in the pilot node, respectively the voltage imposed at the terminals of the generators participating in the adjustment.

The active power losses on the entire power system for the three mentioned regimes are presented in Table 2. The losses of active power on the entire analyzed electric lines are smaller in the case of the secondary voltage-reactive power regulation using Fuzzy logic, compared to the case without secondary regulation.

Table 1. Values of coefficients from Fuzzy function.

	α	β	γ	δ
Pilot node	0.973	0.985	1.015	1.034
Generator 1	0.964	0.983	1.023	1.041
Generator 2	0.963	0.982	1.024	1.042
Generator 3	0.984	0.992	1.011	1.021

Table 2. Power losses in the energy system during the test.

Functioning regime	ΔP [MW]
Initial	25.88
$\Delta Q_9 = 20$ MVAr without secondary fuzzy regulator	26.47
$\Delta Q_9 = 20$ MVAr with secondary fuzzy regulator	26.12

By using the secondary Fuzzy regulator, the voltage of the pilot node increased to 207.64 kV, and in node 9 the voltage increased to 210.87 kV.

5 Conclusions

In Romania, the electricity distribution network was built in the 1960s. In the last decades, several new wind and photovoltaic power plants were realized and need to be connected to the existing power lines distribution system. In this situation, sometimes overloaded nodes appear, especially in areas with a reduced number of electric lines, or with low population. In these conditions, there is either the problem of redistribution on other lines or of making adjustments that allow safe operation. The secondary voltage-reactive power regulator based on Fuzzy logic keeps the voltage in the pilot node close to the set point and assures the voltage values in the rest consuming nodes to act within the predetermined limits. A major advantage of this regulator is the simplicity of the operations necessary for its parameterization. Depending on the selection of the coefficients α, β, γ, and δ from the fuzzy functions, the conditions imposed on the voltage of the generators participating in the regulation will be more restrictive or more relaxed. These coefficients allow the selection of the generators that can have an important contribution to the secondary regulation.

This solution was tested for a monofilament diagram and the developed Power Test line from Smardan node, Galati County. By this hierarchical secondary regulation between voltage and reactive power, an improvement of about 4% in current functioning in this node was obtained, reducing the overloading.

References

1. LNCS. http://www.springer.com/lncs. Accessed 21 Nov 2016
2. https://ec.europa.eu/clima/change/consequences_en. Accessed 21 Sep 2021
3. https://ire.ro/evenimente/mese-rotunde/managementul-congestiilor/. (in Romanian). Accessed 12 Oct 2021
4. Ponce, P., Oliviera, C., Alvarez, V., Rio-Rama, M.: The liberalization of the internal energy market in the European Union: evidence of its influence on reducing environmental pollution. MDPI Energy 13 (2020)
5. Bojnec, S., Krizaj A.: Electricity markets during the liberalization: the case of a European Union country. MDPI Energy 14 (2021)
6. Raptis, T.E., Vokas, G.A., Langouranis, P.A.: Total power quality index for electrical networks using neural networks. Energy Procedia **74**, 1499–1507 (2015)
7. Tanner, T., Phathanothai L. H.: Climate Change and Development, New York (2014)
8. Singh, S.: Energy Crisis and Climate Change Global Concerns. Wiley, Hoboken (2021)
9. Li, N., Qu, G., Dahleh, M.: Real-time decentralized voltage control in distribution networks. In: 52nd Annual Allerton Conference on Communication, Control, and Computing, pp. 582–588 (2014)
10. Corsi, S.: Voltage Control and Protection in Electrical Power Systems from System Components to Wide-Area Control, Springer, (2015)
11. Alaqeeli, R., Habibaliah, I.: Live line maintenance of transmission lines: a review. Int. J. Eng. Tech. Res. **9**(11), 1–5 (2020)
12. Piana, E.A., Bignucolo, F., Donini, A., Spezie, R.: Maintenance of a high-voltage overhead transmission line: sustainability and noise. MDPI: Sustainability (2018)
13. Feng, G.: A survey on analysis and design of model-based fuzzy control systems. IEEE Trans. Fuzzy Syst. **14**, 676–697 (2006)
14. Radulescu, V.:Critical analysis of faults in operation of energy systems using fuzzy logic. In: Advances in Intelligent Systems and Computing, pp. 234–244 (2021)
15. Raju, G.V.S., Zhou, J., Kisner, R.: Hierarchical fuzzy control. Int. J. Control **54**(5), 1201–1216 (1991)
16. Michels, K., Klawonn F., Kruse R.: Fuzzy Control: Fundamentals, Stability and Design of Fuzzy Controllers, Springer, Heidelberg
17. Feng, G.: Analysis and Synthesis of Fuzzy Control Systems: A Model-Based Approach. CRC, Taylor and Francis Group (2018)
18. Du, T.W., Zhang, K., Chen, K., Pangqi, F.X.: Reliability evaluation of high voltage direct current transmission protection system based on interval analytic hierarchy process and interval entropy method mixed weighting. ICPE Guangzhou China **7**(1), 90–99 (2021)

Modeling with Fuzzy Logic the Impact of Overhead Power Lines on the Environment

Victorita Radulescu(✉)

University Politehnica of Bucharest, Splaiul Independentei 313, 060042 Bucharest, Romania
vradul4@gmail.com

Abstract. Power plants based on renewable sources have been built in Romania in the last decade, so it was necessary to develop overhead power lines-OHL. During their realization and operation, they have an impact on the environment, flora, and fauna but also on economic activities. The induced current, the magnetic field, and the noise produced affect quite large areas in their vicinity. The effect on the environment cannot be separated from the economic aspect regarding distances required in their realization associated with the maintenance costs and the access roads. By applying Fuzzy logic, a balance is realized between these two major issues, economic and environmental. Some factors need to be quantified, including the impact on agriculture, the visual impact, the unusable occupied areas for new roads and housing constructions, also associated with population health, animal life, migratory bird flight routes, TV signals, and radio reception. A multi-criteria model based on Fuzzy inequalities is used to solve this problem. The known values for the electric and magnetic fields are compared with values obtained by numerical modeling, and if necessary, the Fuzzy parameters are corrected. The 400 kV overhead power line from Project 138-Black Sea Corridor, Smardan – Gutinas area is analyzed as a Case Study. The Fuzzy method allows a greater degree of flexibility than the classic solutions used, being easier to adapt in various other cases.

Keywords: Environmental impact · Fuzzy logic · High voltage lines

1 Introduction

In response to global warming and the need to reduce carbon emissions, a plan of measures was adopted, to close thermal power plants and build new power plants based on renewable sources - RES, in Romania mainly wind and solar. The energy produced in these areas must be transported and connected to the National Energy System-SEN, [1]. For the provision of electricity transmission and distribution service, the National Electricity Transmission Company-Transelectrica SA plans to develop the electricity transmission networks – RET. The current situation and the forecast evolution of consumption, production capacity, and electricity exchanges shall be taken into account. Every two years, it draws up a Development Plan for the next 10 years, which must be approved by the National Energy Regulatory Authority-ANRE. RET development planning mainly pursues the following objectives:

S. Dzitac et al. (Eds.): ICCCC 2022, AISC 1435, pp. 239–249, 2023.
https://doi.org/10.1007/978-3-031-16684-6_20

- Safe operation and transmission of electricity at an appropriate quality level;
- Development of RET to be properly sized for the transmission of electricity estimated to be produced, consumed, imported, transited, and exported;
- Proper functioning of the internal energy market;
- Increasing the interconnection capacity of energy networks;
- Minimizing investment costs when choosing RET development solutions;
- Sustainability by integrating energy from RES into the network and by transporting the energy generated to the main consumption centers;
- Ensuring non-discriminatory access of consumers to the public network, under the conditions provided by rules in force;
- Compliance with all environmental regulations, public health, and local and regional economic development regulations.

Romania is part of two energy systems at the European level – Continental Central-East and Continental South-East. It must thus meet the requirements of both distribution areas, [2]. The development of RET adapted to the European requirements must ensure the increase of cross-border capacity (MW), social and economic well-being (€), security of electricity supply (MWh), integration of RES (MWh), reduction of CO_2 emissions (kt) and electricity losses (€), system resilience security (++/−), and flexibility (++/−), taking into account the total cost of the project (€), length (km), and social impact. Environmental issues and impact assessments need to be addressed separately.

The proposed Fuzzy method achieves more flexible coordination of all these parameters, especially those of the environment which is much more difficult to quantify economically. For the Case Study, the Smardan - Gutinas Network from Project 138-Black Sea Corridor of 400 kV-LEA suspended power lines are analyzed using two routes, with different lengths and different intersection points with the communication lines, and a different number of critical points.

2 Materials and Methods

2.1 Characteristics of the Overhead Power Lines

In December 2018, Directive Energy set a target for the EU to produce at least 32% of final gross consumption by 2030 from RES. In Romania, the energy market has been liberalized since January 1, 2018, which has led many entrepreneurs to invest in wind or photovoltaic power plants. At that time, Green certificates could be traded either on the centralized green certificate market (spot or forward) or on the electricity market, sustained by the State aid scheme. The RET-electric transport network is considered of national strategic interest, so most of it is owned by the state, Fig. 1. Its planning, development, and modernization aim to maintain in conditions of economic efficiency the quality of the transport system and services, taking into account:

- The evolution of consumption;
- The emergence of new production groups based on RES;
- Withdrawal from the operation of some production capacities, mainly thermal power plants;

- The evolution of the demand for cross-border electricity exchanges, especially in the current conditions of the energy crisis;
- Physical and moral wear and tear of transport equipment;
- Changes in the dominant power flows in the network.

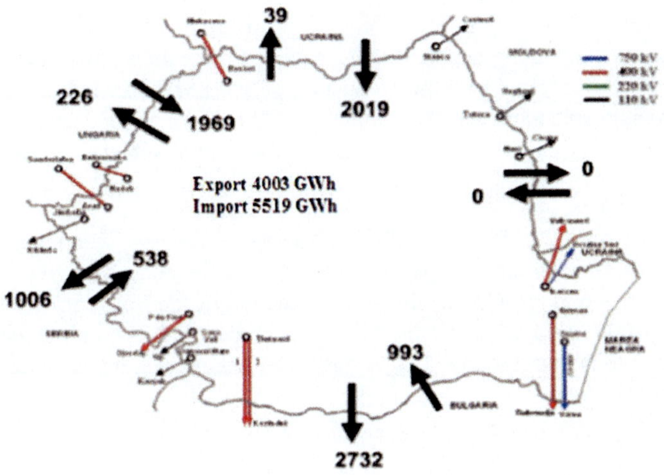

Fig. 1. Import-export-transit of energy through the suspended power lines in 2021.

The development of the network involves the completion of primary stages of analysis:

- Forecast of energy consumption and active and reactive power level on characteristic levels of the load curve, including its peak and the loading gap between winter and summer, in territorial profile and for each station separately;
- Electricity import/export/transit forecast;
- Estimating the availability of production capacities, considering the programs of scrapping, rehabilitation, and installation programs of new groups;
- Analysis of operating regimes in the reference period;
- Calculation of reliability indicators for nodes;
- Establishing the optimal technical-economic solutions.

The power installed in SEN at the end of 2021 was about 24,600 MW, which decreased to about 20,700 MW after closing two thermal power plants. The RET system includes overhead power lines-OHL with rated voltage of 400 kV, 220 kV, and 110 kV and buried lines. Only the airlines that cause the most disturbances are considered in this analysis. The total length of the national RET is 8890.87 km, of which the interconnection lines are 489.04 km long. The lines and power stations that make up the national transport system were built, mostly in 1960–1980, at the technological level of that period, so it is obvious the need to modernize them. Romania is in an extensive refurbishment program, already completed in a proportion of about 56%, respectively 45 stations out of 81 existing, [3].

Analysis of the loading degree of the equipment is performed on the reference regimes from the selected period, respectively winter season 2020–2021 and the summer season of 2021. It was also considered the unavailability of some transformation units due to their withdrawal from the operation, [4]. Calculations regarding the working regime consider the consumptions in the power stations read at the characteristic consumption level WEP-Winter Evening Peak, respectively SMP-Summer Morning Peak. The analysis of the loading degree of RET equipment for the summer of 2021 is performed on an SMP regime with the production of 2790 MW in wind farms-WPP, and 700 MW in photovoltaic plants-PVP, with an export balance of 1000 MW. The network corresponding to the calculation scheme has the following main characteristics:

Summer Season

In the summer of 2020, the OHL 400 kV Smardan – Isaccea - Circuit 2 was out of operation for refurbishment but put into operation at the end of the year. As a Case Study from the summer of 2021, the 400 kV section of the Smardan - Gutinas OHL of about 680 MW was analyzed. Line's loading compared to its nominal power for the analyzed regime is under-loaded $P_{nat} = 450$ MW, being used only in a proportion of about 85% of the total.

Winter Season

Analysis of the equipment's loading for winter 2020–2021 is performed for the WEP regime. The production is 2840 MW in WPP, with an export balance of 800 MW, for a network corresponding to a calculation scheme of about 697 MW. The line's loading compared to the nominal power in the analyzed regime is charged only in a proportion of about 82% of the total.

In terms of power consumption, performance indicators are defined by:

- Average Interruption Time – AIT

$$AIT = 8760 \times 60 \times NE/EC \ \left[\text{min/year}\right] \quad (1)$$

  where: NE - is undelivered energy due to transmission service interruptions [MWh/year], and EC - annual net energy consumption for the electricity system, excluding own technological consumption [MWh/year].
- Severity indicator - SI, which is a transport service performance parameter that estimates the average duration of a transmission service interruption:

$$SI = AIT/NI \ \left[\text{min/interruption}\right] \quad (2)$$

  Here NI is the number of incidents in the RET, associated with the interruptions in the supply of the consumers per year.
- System minutes Indicator-SM is a transport service performance parameter that estimates the average annual downtime through the peak of annual consumption:

$$SM = \frac{NE[MWh/year] \times 60}{PV[MW]} \ \left[\text{min/system}\right] \quad (3)$$

In Table 1 these parameters are summarized for the period 2012–2021. By national and EU laws, the operation of electricity transmission networks is only allowed if they comply with environmental protection and water management rules, [5].

Table 1. Performance indicators for the section analyzed as a case study.

Year	2012	2013	2014	2015	2016	2017	2018	2019	2020	2021
AIT[min/year]	1.54	0.98	0.85	0.48	1.98	2.17	1.83	1.24	1.11	1.42
IS [min/inter.]	0.05	0.03	0.03	0.03	0.2	0.18	0.07	0.05	0.03	0.04
MS [min sys.]	0.76	0.32	0.59	0.28	1.52	1.98	0.81	0.68	0.71	0.67

2.2 Environmental Impact Assessment

Effects of Power Lines on the Ground

Overhead power lines do not cause changes in the properties of the soil on which they are implemented, except in the event of accidental spills of oil or other chemicals that may lead to soil or groundwater pollution. Their impact mainly relates to the land used and the agricultural impact.

The land is occupied by the organization of the construction sites, with OHL routes, location of foundations, stations, sometimes even warehouses, or the generation of waste in the form of metal, ceramic material, glass, plastics, oil, rubble, concrete, or electrical insulation. Restrictions are imposed on the plants that can grow in their vicinity due to the footprint of the poles. If the power lines cross agricultural lands, the production is affected, due to the systematic deforestation that must be done and the impossibility of land irrigation in their vicinity.

Impact of Power Lines on Some Objectives

Airlines can interfere with different objectives, depending on their type: conductors or non-conductors of electricity, stationary or mobile. The OHL may cross or be located near electrified railways, roads with different classes of importance as highways, European, county, and municipal roads, and may intersect other power lines, watercourses, buildings, and warehouses, overhead pipelines for gas, oil, or water supply, underground pipes, or telecommunication lines. All these crossings or intersections are regulated by international norms and legislation that ensure the protection of the environment and population of the area.

Impact of Power Airlines on Living Organisms

There is a magnetic field around AC power lines. It affects the life of living organisms because it pulses with the frequency of the network. However, the biological impact depends on several factors. The most important are: the electric field on the ground, the induced current, the magnetic field, the risk of accidents, and the fragmentation of habitats. The electric field at the ground surface causes currents and voltages induced in humans and animals. It affects wildlife habitats as an obstacle to the flight of birds and especially migratory birds through the induced magnetic field.

The phenomena of direct perception of the electric field create only a feeling of discomfort in the short term, but in the long term, it can produce dizziness, a perception like a breeze, and stings in the case of intense fields. For the safety of living organisms, a width of protection must be established, equal to the distance between the extreme phases of the OHL, respectively 2.5 m left-right of the line. They can also present a potential risk of collision with aircraft, risk of falling if they are near roads, railways, watercourses, buildings, and last but not least fire hazards due to damage to the insulation or accidental touching of conductors by different objects.

The presence of OHL can lead to air pollution by emissions from heating systems, accumulator batteries, sulfur hexafluoride due to equipment leaks, ozone generation, and nitrogen oxides by high voltage Corona effect, CO_X, SO_X, gas emissions NO_X, combustion, Volatile Organic Compounds-VOC, dust in the atmosphere.

Visual and Audible Impact of Overhead Power Lines

The presence of OHL raises concerns about their visual and audible effects. The power lines induce a visual impact by damaging the landscape and are considered proportional to the nominal voltage of the electromagnetic installation located in the external environment and to the height of the voltage poles. The most polluting by their extension are considered the high and very high voltage lines.

The noise produced by the operation and vibration of OHL must also be considered. Sound and light effects of the Corona phenomenon on very high voltage lines, noise produced by the operation of equipment and means of transport often leads to disruption of social activities, including changes in the structure of the population.

Impact on Radio and TV Receptions

Radio and TV receiver interference is also a consequence of Corona discharges. These disturbances occur in a high-frequency spectrum, in the range of 0.5–1.6 MHz for radio reception and the range of 24–126 MHz for TV. The disturbing effect is assessed by the ratio between the intensity of the disturbing field and the quality of the reception. The level of disturbance depends on the electric field at the surface of the conductors, the condition of the conductors, the weather conditions, air humidity, and the distance between the lines and the radio and TV reception system.

2.3 Coefficients for Quantifying Future and Uncertain Events

Future and uncertain events with an impact on the environment are the result of pollution risks whose probabilities of occurrence may or may not change. Risk quantification is based on a classification system used in Fuzzy logic, taking into account the probability of the event occurring, the severity of the impact, and the length of time the environment is affected, Table 2.

The magnitude of the action and its intensity are considered when assessing the severity of the impact, Table 3. The risk is multiplied by the value of the probability of occurrence of the event and the value of the severity of the impact. This sets out some areas that need special attention.

Table 2. Quantification factors.

Probability of occurrence of the event	Value	Severity of the impact of the event	Value
High	2.84	Major	3.04
Medium	1.72	Average	2.11
Low	0.77	Light	0.83

Table 3. Quantification of environmental risks for a high voltage line.

Event	Value of the probability of occurrence	Gravity	Risk
Breakage of the line and its fall to the ground	1	1	1
Fall of high voltage conductors on low or medium voltage line	1	2	2
Breakage of conductors and their fall on roads, railways, etc.	1	2	2

3 Fuzzy Logic Applied in the Analysis of the Impact of Air Lines on the Environment

The mentioned criteria are often difficult to quantify economically because they depend on some concrete conditions, but also many uncertainties. The first category includes the height of the poles - H, area of the foundation - S_f, area of the rectangle on which the pillars are placed LxH, and the intensity of electric and magnetic fields, (which may vary over time) if they cross agricultural land (also variable over time). Other important parameters that can also vary are the line's integration in the power supply system, power losses, and the degree of imbalance of the network. All are considered Fuzzy sets and the consumers' consumption of Fuzzy tasks. The Fuzzy multi-criteria algorithm-FMCDM is used for this type of problem. Requires the following steps:

Step 1: Define the evaluation criteria.
Step 2: Select the appropriate linguistic parameters.
Step 3: Choosing the weights for the structure of the criterion, correlated with the opinion of the decision-makers; through an evaluation of each sub-criterion.
Step 4: Build the Fuzzy decision matrix.
Step 5: Realized the normalized Fuzzy decision matrix and the weighted normalized fuzzy decision matrix.
Step 6: Determine ideal positive/negative value for the chosen solution-FPIS/FNIS.
Step 7: Determine the deviation of each alternative studied from FPIS and FNIS and determine the approximation coefficient of each alternative.
Step 8: Determine the distance of each alternative to FPIS and FNIS.
Step 9: Compute the approximation coefficient of each alternative.

Step 10: Determine the order of ranking of all alternatives, established according to approximation, depending on the approximation coefficients.

The **Selected Criteria**: C1-Costs; C2-Productivity; C3-Electricity production in the area, C4-Environmentally friendly; C5-Air, soil, and water pollution; C6-Global warming; C7 use of resources (C7–1-RES, C7–2-Fossils); C8-Sustainability.

All these criteria are evaluated in terms of benefit, with different degrees of importance for the decision-maker, Table 4. Fuzzy language terms and fuzzy numbers are determined based on the criteria of importance and evaluation of the strategy, [6].

Table 4. Fuzzy language terms and fuzzy values.

Importance	Fuzzy numbers	Criteria
Very Important/Very Good (VI/VG)	0.83–1.00	C1, C4
Important (I)	0.675–0.83	C3
Quite Important (QI)	0.5–0.675	C2
Normal/Medium (N/M)	0.4–0.5–0.6	C5
Medium Unimportant (MUI)	0.25–0.4	C7–1, C6
Unimportant (UI)	0.15–0.25	C7–2
No Importance (NI)	0.0–0.15	C8

The assessment of the impact of power lines on the environment cannot be studied separately from the cost of design, installation, operation, and maintenance. Consequently, the impact needs to be only part of the global model. The decision group in our case has 4 elements, determined by the electric nodes and is based on 2 distinct situations regarding the impact:

- The optimal solution for the realization of power lines – with minimal impact
- Modification of the route and operation of the power system to ensure a compromise between cost and the ecological balance with all the mentioned aspects.

Fuzzy numbers and their weight are associated with a certain decision-maker factor k concerning an alternative $i = 1, .., m$ for a criterion $j = 1, .., 8$ in this case. Each situation has two possible models:

- A model in which different alternatives are ordered separately according to their importance and impact on the environment and their cost so that they are minimal. The average is weighted according to their maximum values. In the end, the optimal solution is chosen, admitting some tolerances allowed in terms of cost;
- A model in which all criteria are admitted simultaneously.

The first method is adopted in the analyzed Case Study. For the optimal solution of the lines' location, the Fuzzy model is described by the following inequalities, [7]:

$$\sum_{i=1}^{m} I < I_{adm}; \quad H_k < H_{adm} \tag{4}$$

$$E_k < E_{adm}; \quad S < S_{adm} \tag{5}$$

where the notations are: I - investments, $k = 1, .., n_{cr}$, n_{cr} - number of critical points, E - the intensity of the electric field, H - intensity of the magnetic field, S - visual impact area, and *adm* - maximum permissible value. In addition to the selected criteria, the influence of lands, whether cultivated or not, the presence of localities, the influence of the existence of the transport lines, roads, railways, riverbeds, etc. can also be taken into account. However, they have a lower share in the optimal cost estimation.

By multiplying the weights of the evaluation criteria by the standardized fuzzy decision matrix, the weighted matrix is obtained. At this point, each selected criterion is important and must be treated carefully. The correspondent fuzzy numbers used in assessing the criteria for the power lines and assessing their impact on the environment are presented in Table 5.

Table 5. Fuzzy language terms and fuzzy numbers.

Criteria	Electric lines				Environmental impact			
C1	1.00	0.96	0.88	0.91	0.25	0.36	0.4	0.29
C2	0.52	0.64	0.59	0.66	0.11	0.14	0.18	0.02
C3	0.7	0.88	0.84	0.8	0.14	0.21	0.18	0.19
C4	0.56	0.49	0.44	0.58	0.9	0.92	0.99	1
C5	0.52	0.64	0.68	0.74	0.88	0.95	0.94	0.9
C6	0.68	0.71	0.72	0.68	0.67	0.73	0.72	0.77
C7–1/C7–2	0.11/0.12	0.06/0.15	0.08/0.21	0.17/0.18	0.11/0.89	0.12/0.93	0.08/0.88	0.1/0.92
C8	0.48	0.53	0.6	0.47	0.66	0.68	0.63	0.59

The optimization model is based on inequalities (4), and (5). The known values for the electric and magnetic fields are compared with the values obtained by Fuzzy logic. The functions for the fuzzy optimal values for the electric and magnetic field intensity are illustrated in Fig. 2-a, and the ideal values for the cost are described in Fig. 2-b. The determined solution must satisfy the mentioned inequalities and meet the Min-Max conditions. The cost variation must be modified to reduce the voltage and therefore the impact of the electromagnetic field.

Case Study: The numerical model was tested for the Smardan-Gutinas 400 kV power line in two variants. The Variant-V1 is shorter, but has more points of intersection with the communication lines and is closer to them. The Variant-V2 is longer, so it is more expensive but has fewer critical points. The electric and magnetic field intensities were

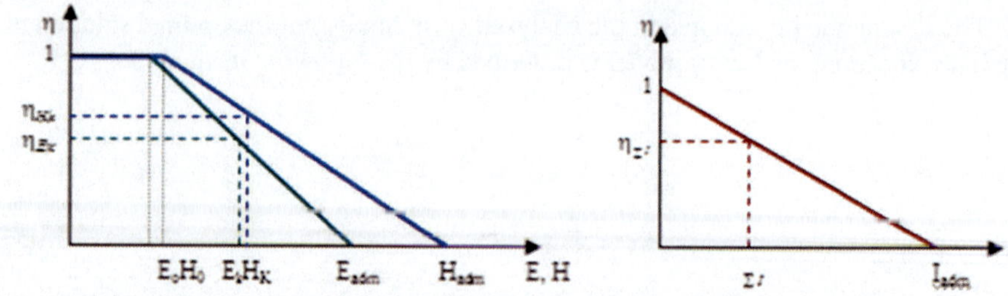

Fig. 2. Optimal Fuzzy values for electric and magnetic fields and the cost.

calculated for the same values for $E_0 = 0.6$ kV/m, $E_{adm} = 4$ kV/m, $H_0 = 1$ A/m, and $H_{adm} = 10$ A/m, used in the calculation of Fuzzy functions.

In V1 the power lines cross the communications lines twice, so d = 0 m, they reach close to them 4 times, at d = 15 m two times, and at d = 20 m two times; d[m] – the distance between the electric lines and the transport routes. In V2 the power lines cross the communication paths only once d = 0 and are close to them twice, once at d = 20 m and once at d = 25 m. The results obtained are presented in Table 6.

The Fuzzy MCDM method is applied to power lines by assessing the relationship between their costs and environmental conditions. The power of strategies based on each criterion can always be adapted. An optimal strategy can be found, obtained by mutual evaluation of all strategies. If environmental factors are essential, V2 is better. If a cost-environment balance is preferred, select V2.

Table 6. Obtained results.

Variant	L [km]	Costs [mil €]	D [m]	E [kV/m]	H [A/m]	η_E	η_H	η_l	η_{Emed}	η_{Hmed}
V1	200	72	0	2.51	6.82	0.32	0.67	0.72	0.64	0.71
			15	1.68	4.78	0.68	0.81			
			20	1.28	3.46	0.89	0.89			
V2	320	128	0	2.51	6.82	0.85	0.67	0.46	0.87	0.73
			20	1.24	3.91	0.87	0.79			
			25	1.12	2.82	0.89	0.87			

4 Conclusions

This study allows the extension of the Fuzzy method to other criteria and allows the determination of the optimal operating conditions. Environmental protection is a complex phenomenon, which includes many domains and energy variables. This solution can be used to determine energy policy strategies, modeled on the current international objectives. The use of Fuzzy logic for establishing the routes of power lines and determining an optimal cost has the advantage that it allows changing the weight of different criteria depending on the determined route and importance of the network. Other criteria can be entered at any time. An acceptable approach to the optimal balance between environmental protection and the energy transport sector minimizes transport costs and improves its sustainability. This reduces the cost of energy, contributes to a sustainable environment, and can prevent climate change to some extent. However, the correct analysis and decision-making are managed by each company.

References

1. Romanian RET Development Plan 2018–2027, Transelectrica 2018. Accessed 23 Nov 2021
2. Project Black Sea Corridor https://eepublicdownloads.entsoe.eu/clean-documents/tyndp-documents/TYNDP%202016/projects/P0138.pdf. Accessed 08 Dec 2021
3. Biasotto, L., Kindel, A.: Power lines and impacts on biodiversity: a systematic review. Environ. Impact Assess. Rev. **71**, 110–119 (2018). https://doi.org/10.1016/j.eiar.2018.04.010
4. Radulescu, V.: Critical analysis of faults in operation of energy systems using fuzzy logic. In: Advances in Intelligent Systems and Computing, pp. 234–244 (2021)
5. Gonos, J., et al.: Environmental impact analysis of electric power lines. In: IEEE International Conference on Environment and Electrical Engineering, pp. 1–5 (2018)
6. Gerlak, A.K., Weston, J., et al.: Climate risk management and the electricity sector. Clim. Risk Manag. **19**, 12–22 (2018)
7. Moghimi, R., Anvari, A.: An integrated fuzzy MCDM approach and analysis to evaluate the financial performance of Iranian companies. J. Adv. Tech. **71**, 685–698 (2014)

Fuzzy Fractal: An Information Entropy View

Qianli Zhou[1], Chenhui Qiang[1,2], and Yong Deng[1,3,4,5](✉)

[1] Institute of Fundamental and Frontier Science, University of Electronic Science and Technology of China, Chengdu 610054, China
dengentropy@uestc.edu.cn

[2] Yingcai Honors College, University of Electronic Science and Technology of China, Chengdu 610054, China

[3] School of Education Shaanxi Normal University, Xi'an 710062, China

[4] School of Knowledge Science, Japan Advanced Institute of Science and Technology, Nomi, Ishikawa 923-1211, Japan

[5] Department of Management, Technology, and Economics, ETH Zurich, 8093 Zurich, Switzerland

Abstract. We model intuitionistic fuzzy sets (IFS) by Dempster-Shafer theory (DST), and use fractal-based belief (FB) entropy to propose a new intuitionistic fuzzy measure. Combining the Shannon entropy measure of fuzzy sets (FS) and the proposed IFS measure, we explore the fractal features of FS and IFS from the perspective of information entropy. Besides, we give the fuzzy information fractal dimension according to the growth rate of normalized information entropy of the joint distributions.

Keywords: Fuzzy set · Intuitionistic fuzzy set · Information entropy · Dempster-shafer theory · Intuitionistic fuzzy belief entropy · Fuzzy information fractal dimension

1 Introduction

Fuzziness, as a unique property of fuzzy sets (FS), can model imprecise information from the perspective of membership. As an extension of the crisp set, FS allow soft relationships between elements and sets, which is more intuitive in describing human behavior. Hence, it is widely used in group decision-making [10,16], fuzzy reasoning [11,18], and medical diagnosis [23,33]. The expert system is a usual way to determine memberships. In order to realize reasonable decision-making, some non-standard FS such as intuitionistic fuzzy sets (IFS) [13], Pythagorean fuzzy sets (PFS) [14,27], Z numbers [30], D numbers [8], and random permutation sets [9] are proposed.

Modeling fuzziness is a key issue in decision-making. De Luca and Termini [7] measure fuzziness from the view of discord of probabilities. Hung and Yang

Supported by National Natural Science Foundation of China (Grant No. 61973332), JSPS Invitational Fellowships for Research in Japan (Short-term).

S. Dzitac et al. (Eds.): ICCCC 2022, AISC 1435, pp. 250–259, 2023.
https://doi.org/10.1007/978-3-031-16684-6_21

[15] utilize Rényi entropy to extend it to IFS, but it generated counter-intuitive results because probabilities can not handle the relationship between membership and hesitation directly. Kosko [17] proposes a fuzzy measure from the perspective of distance, Szmidt and Kacprzyk [21] extend it to IFS. Yager [25,26,28] models fuzziness and proposed fuzzy measures based on Dempster-Shafer theory (DST) to apply them to decision-making. Based on Gini index, Xie *et al.* [24] proposed the information quality of IFS. In addition, there are some heuristic methods for measuring and modeling fuzziness [1,4], such as hybrid entropy, higher order entropy, etc. In the previous methods, most of them are proposed for discrete fuzzy numbers, and their connection with continuous fuzzy numbers does not be discussed.

Fractal dimension as a usual index to measure the complexity of nonlinear systems also has been applied in the information system. The information dimension (ID) proposed by Rényi [20] represent the increasing rate of normalized entropy of the refinement random variable. Ashkenazy [2] utilizes generalized ID to measure the complexity of time series. Wen *et al.* [22] identify influences of nodes in complex networks based on ID. Zhou *et al.* [32] calculate the ID of Galton board and Pascal's triangle. For imprecise information, Gao *et al.* [12] propose the information volume fractal dimension of mass function. Fuzzy fractal dimension proposed by Pedrycz and Bargiela [19], and Castillo and Melin [6] utilize it to model non-linear dynamic systems.

In this paper, we first give a measure of IFS based on belief entropy under framework of DST. Then, the IDs of FS and IFS are proposed, which provides a new idea for fuzzy information measures. Section 2 introduces the common fuzzy measures and their requirements. Section 3 proposes a new IFS measure based on belief entropy. In Sect. 4, we discuss the fuzzy fractal in an information entropy view and calculate its dimensions. Section 5 summarizes the paper and speculates application fields.

2 Preliminaries

2.1 Fuzzy Sets

Classical Fuzzy Sets [29]. For the universe of discourse X with objects $\{x_1, \cdots, x_n\}$, the fuzzy set A_{F} is defined on x_i: $A_{\mathrm{F}} = \{\langle x_i, \mu(x_i)|x_i \in X\rangle\}$, where $\mu(x_i)$ is a real value in interval $[0, 1]$ and describes the membership degree of object x_i to domain X. When the number of targets is infinite, the membership function can be written as a continuous function based on the continuous variable x.

Intuitionistic Fuzzy Sets [3]. In classical fuzzy sets, the negation of the membership degree of x_i is $v(x_i) = 1 - \mu(x_i)$, which represents non-membership degree. However, in the expert system, the membership degree is determined subjectively by experts. IFS utilizes hesitation degree to simulate the experts' determination more reasonably. For the universe of discourse X with objects

$\{x_1, \cdots, x_n\}$, the IFS A_{IF} is defined on x_i: $A_{\mathrm{IF}} = \{\langle x_i, \mu(x_i), v(x_i) | x_i \in X\rangle\}$, where membership $\mu(x_i)$ and non-membership $v(x_i)$ satisfy $0 \leq \mu(x_i) + v(x_i) \leq 1$. For each IFS, hesitation $\pi(x_i) = 1 - \mu(x_i) - v(x_i)$ represent the part can not be distinguished into membership and non-membership.

2.2 Existing Fuzzy Measures

Classical Fuzzy Measures. De Luca and Termini [7] regard the objects in X as 2-dimensional random variables, and used the average Shannon entropy of them to measure fuzziness of X. For a fuzzy set A_{F}, its information entropy is

$$H_{\mathrm{DT}}(\mu_{\mathrm{A}}) = -\mu_A \log \mu_A - (1 - \mu_A) \log(1 - \mu_A). \tag{1}$$

And the fuzziness of X equals $S_{\mathrm{DT}}(A_{\mathrm{F}}) = \sum_{x_i \in X} H(\mu_A(x_i))/K$, where K is the normalization coefficient.

Kosko [17] utilize the distance of fuzzy sets to propose a measure

$$H_{\mathrm{Ks}}(A_{\mathrm{F}}) = l^p(A_{\mathrm{F}}, A_{\mathrm{F}}^{\mathrm{near}})/l^p(A_{\mathrm{F}}, A_{\mathrm{F}}^{\mathrm{far}}), \tag{2}$$

where $l^p(A_{\mathrm{F}}, B_{\mathrm{F}})$ is the distance between fuzzy sets

$$l^p(A_{\mathrm{F}}, B_{\mathrm{F}}) = \left[\sum_{i=1}^{n} |\mu(x_i) - \mu(y_i)|^p\right]^{1/p}. \tag{3}$$

Besides, there are many fuzziness measures. Based on the above classical methods, the necessary 4 properties are given.

Minimality : When A_{F} is a crisp set ($\forall x_i \in X$, $\mu_A(x_i) \in \{0, 1\}$), fuzziness reaches the minimum.
Maximality : When all memberships in A_{F} equal 0.5, i.e. $\forall x_i \in X$, $\mu_A(x_i) = 0.5$, fuzziness reaches the maximum.
Resolution : When A_{F} is a sharpened version of B_{F} (if $\mu_B(x_i) \leq 0.5$, $\mu_A(x_i) \leq \mu_B(x_i)$; if $\mu_B(x_i) \geq 0.5$, $\mu_(x_i) \leq \mu_A(x_i)$.), the fuzziness of A_{F} is less than B_{F}.
Symmetry : $A_{\mathrm{F}}^{\mathrm{c}}$ is the complementary of A_{F} and they satisfy $\mu(A_{\mathrm{F}}) = 1 - \mu(A_{\mathrm{F}}^{\mathrm{c}})$. They have same fuzziness.

After verification, S_{DT} and H_{Ks} satisfy all requirements.

Intuitionistic Fuzzy Measures. According to the classical fuzzy measure requirements, Szmidt and Kacprzyk extend them to IFS [21].

Minimality : When A_{IF} is a crisp set ($\forall x_i \in X$, $\mu_A(x_i) \in \{0, 1\}$), fuzziness reaches the minimum.
Maximality : If $\forall x_i \in A_{\mathrm{IF}}$, $\mu_A(x_i) = v_A(x_i)$, the fuzziness reaches the maximum.

Resolution : When A_{IF} is a sharpened version of B_{IF} (if $\mu_B(x_i) \leq v_B(x_i)$, $\mu_A(x_i) \leq \mu_B(x_i)$ and $v_A(x_i) \geq v_B(x_i)$; $\mu_A(x_i) \geq \mu_B(x_i)$ and $v_A(x_i) \leq v_B(x_i)$, the fuzziness of A_{IF} is less than B_{IF}.

Symmetry : The complementary satisfies $\mu(A_{\text{IF}}) = v(A^c_{\text{IF}})$ and $v(A_{\text{IF}}) = \mu(A^c_{\text{IF}})$, and they have same fuzziness.

Though the requirements do not interpret the physical meaning of hesitation, they emphasized the specialty of intuitionistic fuzzy measure to an extent. They proposed a new IFS entropy based on cardinality and the above requirements. For an IFS with domain X, Szmidt and Kacprzyk's method is

$$H_{\text{SK}}(A_{\text{IF}}) = \frac{1}{n}\sum_{x_i \in X} \frac{\max \text{Count}(A_{\text{IF}} \cap A^c_{\text{IF}})}{\max \text{Count}(A_{\text{IF}} \cup A^c_{\text{IF}})}, \tag{4}$$

where

$$A_{\text{IF}} \cap A^c_{\text{IF}} = \langle \min(\mu_A(x_i), \mu_{A^c}(x_i)), \max(v_A(x_i), v_{A^c}(x_i))\rangle,$$
$$A_{\text{IF}} \cup A^c_{\text{IF}} = \langle \max(\mu_A(x_i), \mu_{A^c}(x_i)), \min(v_A(x_i), v_{A^c}(x_i))\rangle,$$

and

$$\max \text{Count}(A_{\text{IF}}) = \mu_A + \pi_A.$$

In addition, Hung and Yang [15] extend the idea of De Luca and Termini to IFS, they propose an IFS measure based on Rényi entropy,

$$H^{\alpha}_{\text{HY}}(A_{\text{IF}}) = \frac{1}{1-\alpha}\log(\mu^{\alpha}_A + v^{\alpha}_A + \pi^{\alpha}_A), \tag{5}$$

where $\alpha > 0$, and $H^1_{\text{HY}}(A_{\text{IF}}) = -\mu_A \log \mu_A - v_A \log v_A - \pi_A \log \pi_A$.

Burillo and Bustince [5] propose several fuzzy measures based on the hesitation, and they can be summarized as

$$H_{\text{BB}}(A_{\text{IF}}) = \sum_{x_i \in X} \pi_A \cdot f(\pi_A). \tag{6}$$

These three methods can handle most of the intuitionistic fuzzy measurement problems, we qualitatively analyze them in the following sections and give a new intuitionistic fuzzy measure based on belief entropy.

3 Measure IFS in a Dempster-Shafer Theory View

3.1 Discuss Previous IFS Measures

A reasonable description of the relationship between membership, non-membership, and hesitation is the premise of proposing an intuitive fuzzy measure. We discuss it from 2 perspectives. First, for the generation of hesitation, because experts cannot give a definite membership and non-membership, we use hesitation to express the unreliable part. In other words, hesitation is potential membership or non-membership. Therefore, if an object is regarded as a random variable, μ, v, and π are not mutually exclusive, so using information entropy may generate counter-intuitive results. For example, given 2 IFSs $A_{\text{IF1}} = \langle 1, 0 \rangle$ and $A_{\text{IF2}} = \langle 0, 0 \rangle$, the results of Hung and Yang's methods are same. However, A_{IF1} is a crisp set, which does not have fuzziness, and A_{IF2} has no idea whether the object belongs to the domain, which must exist fuzziness. Second, consider the composition of fuzziness. For the classical fuzzy set, the fuzziness comes from the discord of μ and v. When they are equal, the conflict is the largest, so the fuzziness is the largest. In IFS, however, there is not only conflict between μ and v, but also ignorance about how to assign π to μ and v. Therefore, the fuzziness composition of IFS should have two parts. Szmidt and Kacprzyk's method cannot distinguish the cases that π are different with equal μ and v. Fuzziness reaches the maximum in their method when μ and v are equal, so it ignores the existence of π. Conversely, Burillo and Bustince's method can be viewed as a function of π, which ignores the case that μ and v take different values with fixed π. Based on the above, the previous methods have some counter-intuitive measurements.

3.2 Modeling IFS From a Dempster-Shafer View

Dempster-Shafer theory (DST) is proposed based on interval probability and defines mass functions distributed on a power set of a frame. Given an n-element closed set Ω, its power set is $2^{\Omega} = \{\{\varnothing\}, \{\omega_1\}, \cdots, \{\omega_n\}, \{\omega_1\omega_2\}, \cdots, \{\omega_1 \cdots \omega_n\}\}$. The mass functions satisfy $m(F) \in [0, 1]$, $\sum_{F \subseteq \Omega} m(F) = 1$, and $m(\varnothing) = 0$ and compose the basic probability assignment (BPA), which is the basic information unit in DST. As an extension of probability distribution, BPA allows belief to exist on multi-element subsets, which as the underlying support degree of the elements. Since the elements in Ω are mutually exclusive, the belief is usually assigned subjectively to represent the case where the exact probability distribution is unknown because of lacking information.

Yager uses the belief structure in DST to model classical fuzzy information [26], and proposed several related measures. In this paper, we take a different perspective, treating fuzzy sets and probability distributions as juxtaposed, and think that BPA and IFS are generated by adding ignorance to probability distribution and fuzzy set respectively.

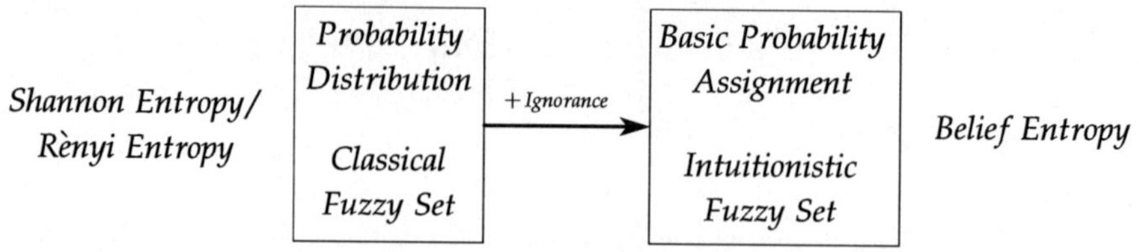

Fig. 1. Extend probability and fuzzy measures to BPA and IFS measures based on ignorance.

As shown in Fig. 1, these ignorances are generated by subjective decisions. For an intuitionistic fuzzy subset $A_{\mathrm{IF}}(x_i)$, we regard it as a 2-element frame $\{\mu, v\}$ in DST, whose BPA is $m = \{\mu_A(x_i), v_A(x_i), \pi_A(x_i)\}$. This transformation type has appeared in previous research, and is usually used to measure the distance and similarity between non-standard fuzzy sets [23,33]. As an important method to measure the total uncertainty of BPA, belief entropy can also be applied to measure the fuzziness of IFS. Though there are many belief uncertainty measures, fractal-based belief (FB) entropy is proved to have advantages in expressing total uncertainty [31].

Given a BPA m under an n-element frame Ω, the FB entropy is

$$E_{\mathrm{FB}}(m) = -\sum_{F \subseteq \Omega} \sum_{F \subseteq G} \frac{m(G)}{2^{|G|}-1} \log \sum_{F \subseteq G} \frac{m(G)}{2^{|G|}-1}, \tag{7}$$

where $|G|$ is the cardinality of G. Based on the above, we define the Intuitionstic fuzzy belief entropy.

Definition 1 (IFB entropy). *Given an intuitionistic fuzzy set $A_{\mathrm{IF}} = \langle \mu_A, v_A \rangle$, the hesitation is $\pi_A = 1 - \mu_A - v_A$, and the intuitionistic fuzzy belief (IFB) entropy is defined as*

$$H_{\mathrm{IFB}}(A_{\mathrm{IF}}) = -(\mu_A + \frac{\pi_A}{3}) \log(\mu_A + \frac{\pi_A}{3}) - (v_A + \frac{\pi_A}{3}) \log(v_A + \frac{\pi_A}{3}) - (\frac{\pi_A}{3}) \log(\frac{\pi_A}{3}). \tag{8}$$

Remark 1. Verification IFB entropy by the properties in Sect. 2.2, only the *Maximality* is not satisfied. When IFB entropy reaches maximum, $\forall x_i \in A_{\mathrm{IF}}$ satisfies $\pi_A(x_i) = 1$. □

In order to show the advantages of IFB entropy comprehensively, we calculate the fuzziness of a dynamic IFS by 4 methods respectively in Example 1.

Example 1. Given an intuitionistic fuzzy set $A_{\mathrm{IF}} = \langle \mu_A, v_A \rangle$, the hesitation is $\pi_A = 1 - \mu_A - v_A$. When μ_A and v_A change on interval $[0, 1]$ and satisfy $0 \leq \mu_A + v_A \leq 1$. The fuzziness change trends of TFB entropy and methods in Eqs. 4, 5, and 6 are shown in Fig. 2.

In this section, we discuss the insufficiency of previous intuitionistic fuzzy measures and give the intuitionistic fuzzy belief entropy by modeling IFS with DST. There are many methods to measure the fuzziness of a single object, but for infinite numbers of objects (continuous fuzzy sets), how to reasonably measure them has not been discussed carefully. In the next section, we try to analyze the problem by fractals.

4 Fuzzy Fractal: In an Information Entropy View

4.1 Information Dimension

Fractal dimension, as a scale-free measure, can measure the complexity of non-linear systems in nature. As a statistical law, fractal features can be broadly

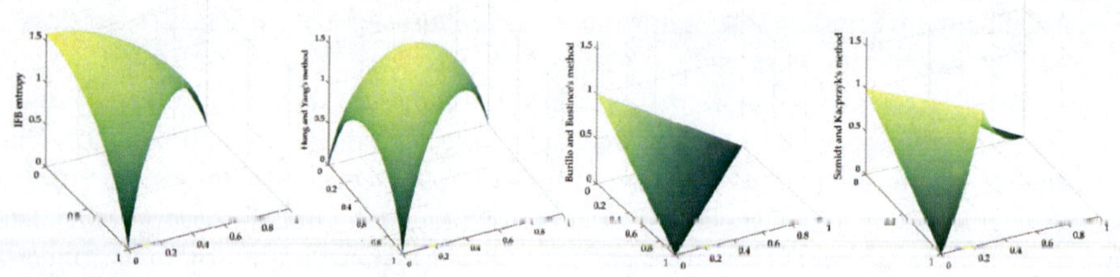

Fig. 2. By comparison, it can be found that the proposed method is more intuitive than the other 3 methods. Although in Eq. 8, only a part of the hesitation is equally distributed to the membership and the non-membership, this special assignment method allows the IFB entropy to measure discord and ignorance together.

described. For object A, measure N with scale ε leads to the result $N(\varepsilon)$. If they satisfy $N(\varepsilon) \sim \varepsilon^{-d}$, it means that object A has fractal features and the dimension is $d = \lim_{\varepsilon \to 0} \log N(\varepsilon) / \log \varepsilon$. The existence of logarithmic function in information entropy provides favorable conditions for exploring fractal features. Based on the above, Rényi proposes the information dimension (ID) [20]. Given a continuous random variable X, ID measures it by the growth rate of the discretized normalized entropy. Assuming the maximum scale is 1, the discretization operation is equivalent to using a scale of $1/s$ to measure information. According to the calculation of the fractal dimension, the information dimension is

$$d_{\mathrm{I}} = - \lim_{s \to 0} \frac{H(s)}{\log s}, \tag{9}$$

where $H(s)$ is the Shannon entropy of the coarse probability distribution under the s-dimensional random variable generated from X.

4.2 Fuzzy Information Fractal Dimension

In the fuzzy entropy proposed by De Luca and Termini (Eq. 1), the membership and non-membership of an object are regarded as a 2-dimensional random variable. For a multi-object domain, the fuzzy measure is taken by averaging the Shannon entropy. However, for domains where objects are continuous (such as Gaussian fuzzy numbers), their fuzziness cannot be directly calculated. Based on the above, we apply the information fractals to fuzzy sets. Although the previous fuzzy measures have used similar expressions, they are not related to fractals. We regard the random variables in a domain are independent, and for an n-object domain, a 2^n-dimensional joint probability distribution can be determined. We sample the continuous domain to determine the discrete fuzzy sets, and give the fuzzy information fractal dimension (FIFD) based on the normalized entropy of the joint probability distribution.

Definition 2 (FIFD). *For a continuous fuzzy number A_{F} in the domain X, the fuzzy information fractal dimension (FIFD) is*

$$d_{\mathrm{FIF}} = -\lim_{s \to \infty} \frac{\sum_{i=1}^{L/s} H_{\mathrm{DT}}(\langle A_{\mathrm{F}}(x_i) \rangle_{L/s})}{s \log 2}, \tag{10}$$

where L is the length of the domain, s is the scale of measures, L/s is the sample numbers, and $\langle A_{\mathrm{F}} \rangle_n$ is the coarse fuzzy set with n average sampling.

H_{DT} is chosen as the numerator because of the *additivity* of Shannon entropy and the importance of logarithmic function in fractal features. In [31], Zhou and Deng prove the *additivity* of FB entropy, so the intuitionistic fuzzy information fractal dimension (IFIFD) can be defined based on IFB entropy.

Definition 3 (IFIFD).
For a continuous intuitionistic fuzzy number A_{IF} in the domain X the intuitionistic fuzzy information fractal dimension (IFIFD) is

$$d_{\mathrm{IFIF}} = -\lim_{s \to \infty} \frac{\sum_{i=1}^{L/s} H_{\mathrm{IFB}}(\langle A_{\mathrm{IF}}(x_i) \rangle_{L/s})}{s \log 3}, \tag{11}$$

where the symbols have same meaning with Definition 2, and the $\log 3$ *means the maximum IFB entropy of a IF subset.*

Similarly, IFIFD can also be seen as the normalized belief entropy changes of the joint BPA distribution. Although the discrete fuzziness is calculated by sampling in Definitions 2 and 3, due to the independence of objects, can be more vividly shown as measuring continuous fuzzy numbers process with different scales s. As shown in Fig. 3, different s represents different fuzzy numbers to describe a Gaussian fuzzy number ($\sigma = 3$). With the $s \to 0$, the fuzzy number approaches the Gaussian distribution. From this point of view, the fuzzy information fractal can be more intuitive than the probability information fractal. Figure 4 shows the information fractal features of the Gaussian fuzzy distribution under different standard deviations. As the standard deviation increases, the slope of the fitting curve gradually increases, which is an intuitive result.

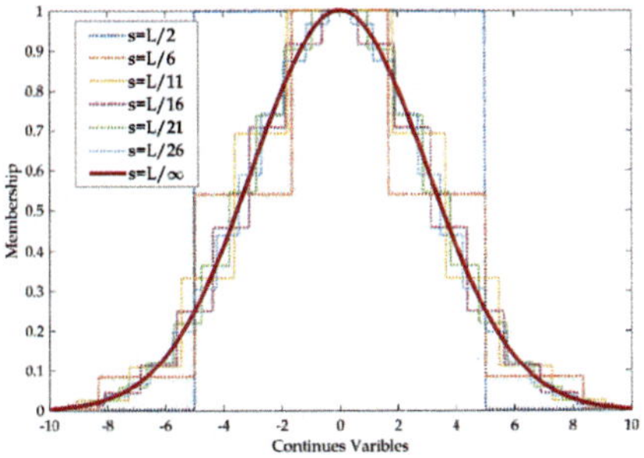

Fig. 3. Process of approximating a Gaussian fuzzy number through different scales.

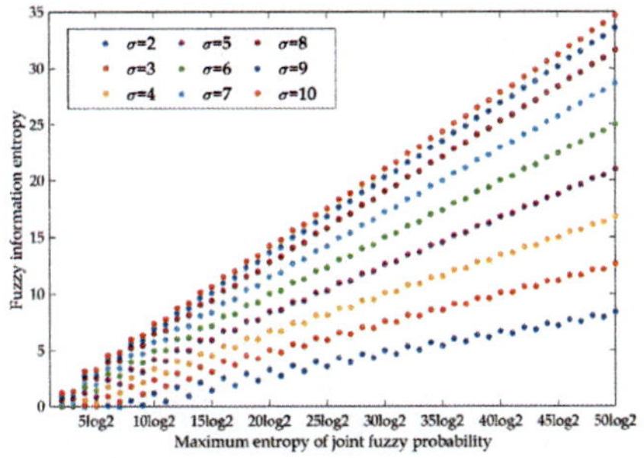

Fig. 4. Double logarithmic fitting curve of Gaussian fuzzy number fractal

5 Conclusion

In this paper, we describe the fractal features of classical fuzzy sets and intuitionistic fuzzy sets from the perspective of information entropy. We model IFS based on DST, propose the intuitionistic fuzzy belief entropy, and demonstrate its advantages over other measures. Then the probability distribution fractal is applied to fuzzy sets, and the calculation methods of fuzzy fractal information dimension and intuitionistic fuzzy fractal information dimension are defined based on joint probability distribution and BPA. From the visualization perspective, we show that fuzzy information has more intuitive fractal features than probability information. In future research, we will further explore fuzzy distributions with special dimensions, and continue to extend the fractal modeling fuzzy information to other fuzzy applications.

References

1. Al-Sharhan, S., Karray, F., Gueaieb, W., Basir, O.: Fuzzy entropy: a brief survey. In: 10th IEEE International Conference on Fuzzy Systems. (Cat. No. 01CH37297), vol. 3, pp. 1135–1139. IEEE (2001)
2. Ashkenazy, Y.: The use of generalized information dimension in measuring fractal dimension of time series. Physica A Stat. Mech. Appl. **271**(3–4), 427–447 (1999)
3. Atanassov, K.T.: Intuitionistic fuzzy sets. In Intuitionistic fuzzy sets, pp. 1–137. Springer (1999). https://doi.org/10.1007/978-3-7908-1870-3_1
4. Bhandari, D., Pal, N.R.: Some new information measures for fuzzy sets. Inf. Sci. **67**(3), 209–228 (1993)
5. Burillo, P., Bustince, H.: Entropy on intuitionistic fuzzy sets and on interval-valued fuzzy sets. Fuzzy Sets Syst. **78**(3), 305–316 (1996)
6. Castillo, O., Melin, P.: A new fuzzy fractal control approach of non-linear dynamic systems: the case of controlling the covid-19 pandemics. Chaos Solitons Fractals, **151**, 111250 (2021)
7. De Luca, A., Termini, S.: A definition of a nonprobabilistic entropy in the setting of fuzzy sets theory. Inf. Control **20**(4), 301–312 (1972)
8. Deng, X., Deng, Y.: D-AHP method with different credibility of information. Soft. Comput. **23**(2), 683–691 (2017). https://doi.org/10.1007/s00500-017-2993-9
9. Deng, Y.: Random permutation set. Int. J. Comput. Commun. Control **17**(1), 4542 (2022)
10. Dzitac, I.: The fuzzification of classical structures: a general view. Int. J. Comput. Commun. Control **10**(6), 12–28 (2015)
11. Fujita, H., Rudas, I.J., Fodor, J., Kurematsu, M., Hakura, J.: Fuzzy reasoning for medical diagnosis-based aggregation on different ontologies. In: 2012 7th IEEE International Symposium on Applied Computational Intelligence and Informatics (SACI), pp. 137–146. IEEE (2012)
12. Gao, Q., Wen, T., Deng, Y.: Information volume fractal dimension. Fractals, 2150263 (2021)
13. Garg, H.: A new generalized improved score function of interval-valued intuitionistic fuzzy sets and applications in expert systems. Appl. Soft Comput. **38**, 988–999 (2016)

14. Garg, H.: A novel correlation coefficients between pythagorean fuzzy sets and its applications to decision-making processes. Int. J. Intell. Syst. **31**(12), 1234–1252 (2016)
15. Hung, W.-L., Yang, M.-S.: Fuzzy entropy on intuitionistic fuzzy sets. Int. J. Intell. Syst. **21**(4), 443–451 (2006)
16. Kacprzyk, J., Fedrizzi, M., Nurmi, H.: Group decision making and consensus under fuzzy preferences and fuzzy majority. Fuzzy Sets Syst. **49**(1), 21–31 (1992)
17. Kosko, B.: Fuzzy entropy and conditioning. Inf. Sci. **40**(2), 165–174 (1986)
18. Nădăban, S., Dzitac, S., Dzitac, I.: Fuzzy topsis: a general view. Procedia Comput. Sci. **91**, 823–831 (2016)
19. Pedrycz, W., Bargiela, A.: Fuzzy fractal dimensions and fuzzy modeling. Inf. Sci. **153**, 199–216 (2003)
20. Rényi, A.: On the dimension and entropy of probability distributions. Acta Mathematica Acad. Scientiarum Hung. **10**(1–2), 193–215 (1959). https://doi.org/10.1007/BF02063299
21. Szmidt, E., Kacprzyk, J.: Entropy for intuitionistic fuzzy sets. Fuzzy Sets Syst. **118**(3), 467–477 (2001)
22. Wen, T., Deng, Y.: Identification of influencers in complex networks by local information dimensionality. Inf. Sci. **512**, 549–562 (2020)
23. Xiao, F.: A distance measure for intuitionistic fuzzy sets and its application to pattern classification problems. IEEE Trans. Syst. Man Cybern. Syst. **51**(6), 3980–3992 (2019)
24. Xie, D., Xiao, F., Pedrycz, W.: Information quality for intuitionistic fuzzy values with its application in decision making. Eng. Appl. Artif. Intell. **109**, 104568 (2022)
25. Yager, R.R.: On the entropy of fuzzy measures. IEEE Trans. Fuzzy Syst. **8**(4), 453–461 (2000)
26. Yager, R.R.: Uncertainty representation using fuzzy measures. IEEE Trans. Syst. Man Cybern. B (Cybern.) **32**(1), 13–20 (2002)
27. Yager, R.R.: Pythagorean membership grades in multicriteria decision making. IEEE Trans. Fuzzy Syst. **22**(4), 958–965 (2013)
28. Yager, R.R.: Uncertainty modeling using fuzzy measures. Knowl.-Based Syst. **92**, 1–8 (2016)
29. Zadeh., L.A.: Fuzzy sets. In: Fuzzy Sets, Fuzzy Logic, and Fuzzy Systems: Selected Papers by Lotfi A Zadeh, pp. 394–432. World Scientific (1996)
30. Zadeh, L.A.: A note on z-numbers. Inf. Sci. **181**(14), 2923–2932 (2011)
31. Zhou, Q., Deng, Y.: Fractal-based belief entropy. Inf. Sci. **587**, 265–282 (2022)
32. Zhou, Q., Deng, Y., Pedrycz, W.: Information dimension of galton board. Fractals **30**(04), 2250079 (2022)
33. Zhou, Q., Mo, H., Deng, Y.: A new divergence measure of pythagorean fuzzy sets based on belief function and its application in medical diagnosis. Mathematics **8**(1), 142 (2020). https://doi.org/10.3390/math8010142

IV-FEC: Information Volume-Based Fuzzy Evidential Classifier

Jixiang Deng[1] and Yong Deng[1,2,3,4](✉)

[1] Institute of Fundamental and Frontier Science, University of Electronic Science and Technology of China, 610054 Chengdu, China
dengjixiang@std.uestc.edu.cn, dengentropy@uestc.edu.cn
[2] School of Education, Shaanxi Normal University,710062 Xi'an, China
[3] School of Knowledge Science, Japan Advanced Institute of Science and Technology, 923-1211 Nomi, Ishikawa, Japan
[4] Department of Management, Technology, and Economics, ETH Zurich, 8093 Zurich, Switzerland

Abstract. Classification based on fuzzy set theory and evidence theory has attracted much attention. However, traditional fuzzy classifiers and evidential classifiers may encounter two issues when facing data with high uncertainty. For one thing, the knowledge represented by the classifier may be in conflict, making classification difficult. For the other thing, most classifiers only consider one type of uncertainty. Hence, in this paper, information volume-based fuzzy evidential classifier (IV-FEC) is proposed by considering both the fuzziness and the total uncertainty (TU) of the classification problem. IV-FEC incorporates fuzzy information volume (FIV) and evidential information volume (EIV) for conflict management under the different types of uncertainty. FIV and EIV respectively estimate the fuzziness of the Gaussian membership degree of the samples and the TU for each attribute. An FIV and EIV-based discounting technique is utilized to modify the mass functions in order to manage conflicts in evidence combination and improve classification performance. The proposed IV-FEC is applied in target classification and verified based on several experiments. The results show that IV-FEC has a better classification accuracy compared with the existing models.

Keywords: Dempster-Shafer evidence theory · Fuzzy set theory · Fuzzy evidential classifier · Information volume · Target classification

1 Introduction

Uncertainty is ubiquitous in real practice. In order to deal with the uncertain information, researchers have proposed lots of theories, such as probability theory, Dempster-Shafer evidence theory [1,14], and fuzzy set theory [8,12,17].

Supported by National Natural Science Foundation of China (Grant No. 61973332); Invitational Fellowships for Research in Japan (Short-term).

S. Dzitac et al. (Eds.): ICCCC 2022, AISC 1435, pp. 260–269, 2023.
https://doi.org/10.1007/978-3-031-16684-6_22

According to [10], uncertainty can be broadly divided into two types. One is fuzziness, which can be modeled with fuzzy set theory. The other is total uncertainty (TU), which can be handled in terms of mass functions in evidence theory. Different kinds of uncertainty measures have been presented to evaluate fuzziness [13,19] and TU [5,6]. Recently, Deng proposed a novel entropy to determine TU, which is called Deng entropy [6]. For better modeling the higher-order uncertainty in evidence theory, Deng entropy is generalized into information volume of mass function, also known as evidential information volume (EIV) and information volume (IV) [5]. EIV has a wide range of applications, including fuzzy IV [3], fractal-based IV [18], extropy-based IV [11], complex-valued IV [2], and EIV-based pattern recognition [9].

As one of the topics of uncertain information processing, classification refers to the problem of identifying the category of an object. Over the past decades, various classification algorithms have been developed based on different techniques [4,13,16]. Target classification algorithms based on fuzzy set theory, namely fuzzy classifiers, are able to process uncertain information with fuzziness. Many approaches have been systematically designed for fuzzy classification models [13,19]. Evidence theory is a powerful framework of knowledge representation and logical inference in the environment with TU. A growing number of evidence theory-based classification methods, also called evidential classifiers, have been reported [7,16].

However, traditional fuzzy classifiers and evidential classifiers have limitations in two aspects, which may be problematic when faced with highly uncertain data. (i) In scenes with uncertainty and imprecision, objects can be highly overlapping or affected by noise, which can lead to conflicts within the knowledge represented by the classifier, making classification difficult. Thus, effective conflict management should be designed for classifiers. (ii) Most traditional classifiers only take one type of uncertainty into consideration, that is, fuzzy classifiers only consider fuzziness, and evidential classifiers only consider TU. Consequently, a hybrid classifier using both techniques is needed.

To address the two issues mentioned above, this paper proposes a novel classifier, called information volume-based fuzzy evidential classifier (IV-FEC), which considers both the fuzziness and TU of the classification problem. In IV-FEC, two types of information volume, *i.e.*, fuzzy information volume (FIV) and evidential information volume (EIV), are applied to manage conflict caused by different types of uncertainty. In specific, FIV determines the fuzziness of the Gaussian membership degree of the sample, and EIV evaluates the TU and derives the risk degree for each attribute. In addition, based on FIV and EIV, a discounting technique is used to modify the mass functions so as to manage conflicts in evidence combination and improve classification performance. Moreover, several experiments are conducted to verify the proposed IV-FEC for the target classification problem. The results show that IV-FEC has a better classification accuracy compared with the existing models.

The rest of this paper is organized as follows. Section 2 briefly reviews some preliminaries. Section 3 proposes information volume-based fuzzy evidential classifier. In Sect. 4, IV-FEC is applied in target classification and tested based on several experiments. Section 5 makes a brief conclusion.

2 Preliminaries

2.1 Dempster-Shafer Evidence Theory

Frame of Discernment and Mass Function. Several conceptions of evidence theory [1,14] are as follows. Frame of discernment (FOD) is an exhaustive nonempty set of N hypotheses: $\Theta = \{\theta_1, \theta_2, \cdots, \theta_N\}$. 2^Θ denotes the power set of Θ:

$$2^\Theta = \{\emptyset, \{\theta_1\}, \{\theta_2\}, \cdots, \{\theta_N\}, \{\theta_1, \theta_2\}, ..., \{\theta_1, \theta_N\}, ..., \Theta\} \tag{1}$$

which contains all 2^N possible subsets of Θ. Mass function is a mapping function $m : 2^\Theta \rightarrow [0, 1]$, which is constrained by $\sum_{A \in 2^\Theta} m(A) = 1$ and $m(\emptyset) = 0$.

Dempster's Rule of Combination. Given two mass functions m_1 and m_2, Dempster's rule of combination is the orthogonal sum $m_1 \oplus m_2$ defined as [14]:

$$m(A) = \begin{cases} \frac{1}{1-K} \sum_{B \cap C = A} m_1(B) \cdot m_2(C) & A \in 2^\Theta, A \neq \emptyset \\ 0 & A = \emptyset \end{cases} \tag{2}$$

where K can be calculated by $K = \sum_{B \cap C = \emptyset} m_1(B) \cdot m_2(C)$.

2.2 Fuzzy Set Theory

Given a universe of discourse X, a fuzzy set A is defined as [17]:

$$A = \{\langle x, \mu(x) \rangle | x \in X\} \tag{3}$$

where $\mu(x) : X \rightarrow [0, 1]$ is the membership function of A, describing the membership degree for each element x to the fuzzy set A. The closer $\mu(x)$ is to 1, the more likely x belongs to A.

2.3 Information Volume

Evidential Information Volume (EIV). The calculating procedure for EIV is to continuously separate the mass function of focal elements with multi-sets, and then measure the uncertainty of the separated mass function based on Deng entropy [6] until convergence. Given a mass function defined on FOD Θ, its EIV can be calculated by the following steps [5]:

Step 1: Input the mass function m.
Step 2: Based on the distribution of maximum Deng entropy [6], proportionally separate the mass function $m(A)$ with $|A| > 1$. Then, measure the uncertainty of the separated mass function based on Deng entropy [6].
Step 3: Repeat Step 2 until the increment of Deng entropy is less than the allowable error ε. Then, output the uncertainty measured by Deng entropy in the last iteration.

Fuzzy Information Volume (FIV). The basic principle of FIV is just the same as EIV. According to [3], FIV has two types of expression. One is for classical fuzzy set and the other is for intuitionistic fuzzy set. This paper considers the the former expression of FIV. Let the universe discourse be X, and the fuzzy set be $\{\langle x, \mu\rangle | x \in X\}$ where μ is membership function and $\pi = 1 - \mu$ is the hesitancy degree. FIV is defined as follows:

$$FIV(\mu) = -\sum_{i=1}^{\infty} \mu_i \log \mu_i - \lim_{i\to\infty} \pi_i \log \frac{\pi_i}{C} \tag{4}$$

where C is usually assigned to be 2 for classical fuzzy sets. $\mu_1 = \mu$ and $\pi_1 = \pi = 1 - \mu$ denote the initial states of membership function and hesitancy degree. i denotes the times of separation. μ_i and π_i $(i > 1)$ are the terms derived by separating π_{i-1} based on the following rules: $\mu_i = k_1\pi_{i-1}$ and $\pi_i = k_2\pi_{i-1}$, in which $k_1 = \mu$ and $k_2 = \pi = 1 - \mu$ are the separating proportions.

3 Information Volume-Based Fuzzy Evidential Classifier

Assume that the original dataset, containing N classes, is denoted by FOD $\Theta = \{\theta_i \,|\, i = 1, 2, ..., N\}$, and the sample with K-dimensional attributes is represented by vector $\mathbf{x} = [x_j \,|\, j = 1, 2, ..., K]$. Then, the proposed IV-FEC is detailed as follows. For better understanding, the flowchart of the proposed IV-FEC is illustrated in Fig. 1.

Step 1: Input the original dataset and divide it into two subsets, namely the training set and the test set.

Step 2: Build the fuzzy discriminant model based on training set.

(i) Assume that the number of training samples in the class θ_i is M_i. Let $x_{ij}^{(l)}$ denote the j-th attribute of the l-th sample in the class θ_i. Then, the mean value and the standard deviation of the training samples in class θ_i can be calculated as:

$$\bar{x}_{ij} = \frac{1}{M_i}\sum_{l=1}^{M_i} x_{ij}^{(l)} \tag{5}$$

$$\sigma_{ij} = \sqrt{\frac{1}{M_i - 1}\sum_{l=1}^{M_i}\left(x_{ij}^{(l)} - \bar{x}_{ij}\right)^2} \tag{6}$$

(ii) Establish the fuzzy discriminant model for each attribute based on Gaussian fuzzy membership function:

$$\mu_{ij}(x) = \exp\left(-\frac{(x - \bar{x}_{ij})^2}{2{\sigma_{ij}}^2}\right) \tag{7}$$

Step 3: Determine the fuzzy membership degree of each test sample, and then measure the associated fuzziness based on FIV.

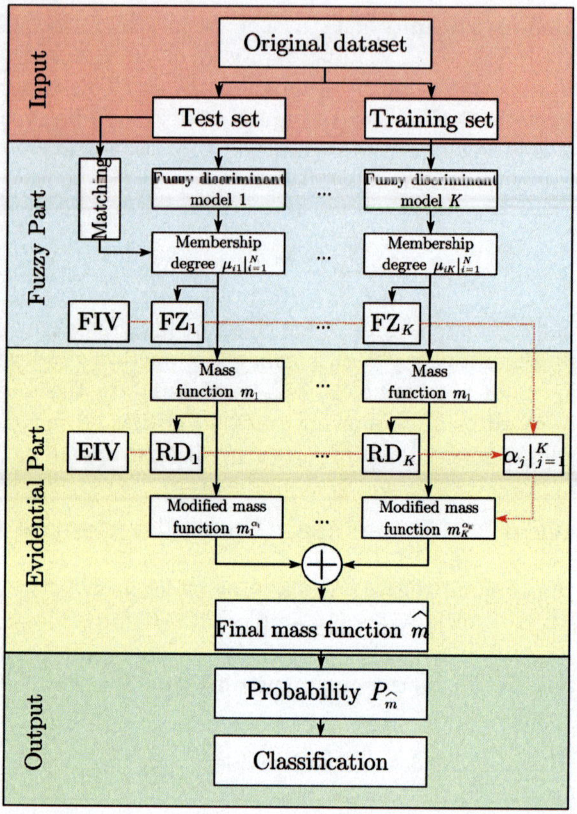

Fig. 1. The flowchart of the proposed IV-FEC

(i) For a certain test sample with K-dimensional attributes $\mathbf{x} = [x_j \,|\, j = 1, 2, \ldots, K]$, the corresponding fuzzy membership degree for the j-th attribute of this sample in class θ_i can be determined by:

$$\mu_{ij} \triangleq \mu_{ij}(x)\,|_{x=x_j} \tag{8}$$

(ii) Based on μ_{ij}, the fuzzy set representation for the j-th attribute of the test sample is that:

$$S_j = \{\langle x_j, \mu_{ij} \rangle \,|\, i = 1, 2, \ldots, N\} \quad (j = 1,\, 2,\, \ldots,\, K) \tag{9}$$

which describes the membership degree of the j-th attribute with respect to all the classes.

(iii) Measure the fuzziness for the j-th attribute of the test sample based on FIV:

$$FZ_j = \frac{1}{N} \sum_{i=1}^{N} FIV(\mu_{ij}) \quad (j = 1,\, 2,\, \ldots,\, K) \tag{10}$$

Step 4: Convert the fuzzy set into the mass function.
(i) For the j-th attribute, rank the N fuzzy membership degree $\tilde{\mu}_{ij}|_{i=1}^{N}$ in descending order $\tilde{\mu}_{rj}^{R}|_{r=1}^{N}$.
(ii) Let the corresponding classes for $\tilde{\mu}_{rj}^{R}|_{r=1}^{N}$ be denoted as $\theta_r|_{r=1}^{N}$. For the j-th attribute, convert fuzzy membership degree $\tilde{\mu}_{ij}|_{i=1}^{N}$ into the associated mass function m_j based on the following rule:

$$m_j : \begin{cases} m_j\left(\{\theta_r | r=1\}\right) = \tilde{\mu}_{1j}^{R} \\ m_j\left(\{\theta_r | r=1,2\}\right) = \tilde{\mu}_{2j}^{R} \\ \dots \\ m_j\left(\{\theta_r | r=1,2,...,N\}\right) = \tilde{\mu}_{Nj}^{R} \end{cases} \tag{11}$$

Step 5: Calculate the risk degree for each mass function based on EIV.
(i) Build the EIV-based dissimilarity matrix (EIV-DM) of K mass functions $m_j|_{j=1}^{K}$:

$$EIV\text{-}DM = \begin{bmatrix} 0 & d_{12} & \cdots & d_{1K} \\ d_{21} & 0 & & d_{2K} \\ \vdots & & \ddots & \vdots \\ d_{K1} & d_{K2} & \cdots & 0 \end{bmatrix} \tag{12}$$

where $d_{pq}\,(p, q = 1,\, 2, \dots, K)$ is the EIV-based divergence between m_p and m_q, which is defined as:

$$d_{pq} = EIV\left(\frac{m_p + m_q}{2}\right) - \frac{1}{2}EIV(m_p) - \frac{1}{2}EIV(m_q) \tag{13}$$

(ii) Based on EIV-DM, calculate the risk degree for mass function m_j:

$$RD_j = \sum_{k \neq j} d_{kj} \quad (j = 1,\, 2,\, \dots, K) \tag{14}$$

The lower RD_j is, the more reliable the mass function m_j is.

Step 6: Obtain the discounting factor α_j, and then modify the mass function.
(i) Calculate the normalized fuzziness (NFZ) and the normalized risk degree (NRD) for each attribute:

$$NFZ_j = \frac{FZ_j}{\sum_{j=1}^{K} FZ_j} \quad (j = 1,\, 2,\, \dots, K) \tag{15}$$

$$NRD_j = \frac{RD_j}{\sum_{j=1}^{K} RD_j} \quad (j = 1,\, 2,\, \dots, K) \tag{16}$$

(ii) Based on NFZ_j and NRD_j, get the discounting factor α_j:

$$\alpha_j = \frac{e^{-0.1*NFZ_j} + e^{-NRD_j}}{\max\limits_{j}\left[e^{-0.1*NFZ_j} + e^{-NRD_j}\right]} \quad (j = 1,\, 2,\, \dots, K) \tag{17}$$

(iii) Inspired by [14], we present an FIV and EIV-based discounting technique. Modify the mass function m_j by discounting factor α_j and acquire the modified mass function $m_j^{\alpha_j}$:

$$m_j^{\alpha_j}(A) = \begin{cases} \alpha_j \cdot m_j(A) & \forall A \subset \Theta \\ \alpha_j \cdot m_j(\Theta) + (1 - \alpha_j) & A = \Theta \end{cases} \tag{18}$$

Step 7: Combine the K modified mass functions, and then convert the combined mass function into probability.

(i) Combine the K modified mass functions $m_j^{\alpha_j}$ ($j = 1, 2, \ldots, K$) via Dempster's rule of combination [14]:

$$\widehat{m} = \bigoplus_{j=1}^{K} m_j^{\alpha_j} \tag{19}$$

(ii) Transform $\widehat{m}$ into probability based on Pignistic Probability Transformation (PPT) [15]:

$$P_{\widehat{m}}(\theta_i) = \sum_{\theta_i \in A \in 2^\Theta} \frac{\widehat{m}(A)}{|A|} \quad (i = 1, 2, \ldots, N) \tag{20}$$

Step 8: Classify the test sample $\mathbf{x}$ based on probability $P_{\widehat{m}}$, and then output the classification result.

$$\theta^{\mathbf{x}} \triangleq \arg\max_{1 \le i \le N} [P_{\widehat{m}}(\theta_i)] \tag{21}$$

4 Application in Target Classification

In this section, the proposed IV-FEC is applied in target classification and tested based on several experiments. The experiment descriptions are as follows.

(i) **Datasets:** Five datsets are considered in the experiments, including Iris, Seeds, Wine, Sonar (Sonar, Mines vs. Rocks), Heart (Statlog). These datasets are available in the UCI machine learning repository[1].

(ii) **Selected classifiers:** In the following experiments, IV-FEC is compared with eight selected classifiers, including classifiers in machine learning, fuzzy classifiers, and evidential classifiers. Four well-known classifiers are chosen: support vector machine with radial basis function (SVM-RBF), Bayes network (BayesNet), decision tree (C4.5), and AdaBoost with decision stump. We select two fuzzy classifiers: Zhou et al.'s Pythagorean fuzzy divergence-based method [19] and Pan et al.'s intuitionistic fuzzy similarity-based method [13]. Two evidential classifiers are considered: Deng et al.'s evidential distance-based method [7] and Xiao's belief divergence-based method [16].

(iii) **Experimental design:** The eight classifiers and the proposed IV-FEC are implemented on the five datasets based on five-fold cross-validation. The performance of the classifier is evaluated in terms of classification accuracy.

Table 1. The classification accuracy results based on five-fold cross-validation

	SVM-RBF	BayesNet	C4.5	AdaBoost	Proposed IV-FEC
Iris	0.911	0.937	0.945	**0.951**	0.949
Seeds	0.893	0.887	**0.910**	0.894	0.882
Wine	0.405	**0.983**	0.924	0.890	0.958
Sonar	0.677	0.751	0.730	**0.771**	0.749
Heart	0.829	0.819	0.772	0.816	**0.839**
Average	0.743	0.875	0.856	0.864	**0.876**
	Zhou's method	Pan's method	Deng's method	Xiao's method	Proposed IV-FEC
Iris	0.924	0.921	0.927	0.929	**0.949**
Seeds	0.870	0.876	0.874	0.871	**0.882**
Wine	0.919	0.924	0.928	0.929	**0.958**
Sonar	0.720	0.738	0.735	0.736	**0.749**
Heart	0.820	0.823	0.827	0.824	**0.839**
Average	0.851	0.857	0.858	0.858	**0.876**

The classification accuracy results of the classifiers are shown in Table 1. In general, the proposed IV-FEC and BayesNet perform about the same on most datasets, and the average accuracy of IV-FEC is the best (0.876) among the classifiers. It can be observed that, on the Heart dataset, IV-FEC has the best performance (0.839) compared with the other classifiers, and its accuracy is much higher than that of fuzzy classifiers and evidential classifiers. As for the Wine dataset, BayesNet outperforms the other classifiers with an accuracy of 0.983, which is much higher than the other classifiers. The accuracy of IV-FEC on the Wine dataset is the third highest (0.749), which is better than other fuzzy classifiers and evidential classifiers. From the table, it is clear that IV-FEC is much better than the existing fuzzy and evidential methods, especially on the Wine dataset (0.958). To summarize, the experimental results show that the performance of IV-FEC is better than the existing classifiers, further demonstrating the efficiency of conflict management based on FIV and EIV.

5 Conclusion

Designing classifiers based on fuzzy set theory and evidence theory has attracted much attention. However, when facing data with high uncertainty, traditional fuzzy classifiers and evidential classifiers may have problems in two aspects. For one thing, the knowledge represented by the classifier may be in conflict, resulting in difficulties of classification. For the other thing, most classifiers only consider

[1] http://archive.ics.uci.edu/ml/datasets.

one type of uncertainty. To solve these issues, this paper proposes information volume-based fuzzy evidential classifier (IV-FEC). The major contributions of the paper and some remarks are summarized below:

(i) By considering both the fuzziness and the total uncertainty (TU), IV-FEC is proposed, which incorporates fuzzy information volume (FIV) and evidential information volume (EIV) for conflict management under the different types of uncertainty.
(ii) Based on FIV and EIV, a discounting technique is designed to modify the mass functions for conflict management in evidence combination, thereby improving classification performance.
(iii) The proposed IV-FEC is applied in target classification and verified based on several experiments. The results show that IV-FEC has a better classification accuracy compared with the existing models.

Acknowledgements. The work is partially supported by National Natural Science Foundation of China (Grant No. 61973332); Invitational Fellowships for Research in Japan (Short-term).

References

1. Dempster, A.P.: Upper and lower probabilities induced by a multivalued mapping. Ann. Math. Stat. **38**(2), 325–339 (1967)
2. Deng, J., Deng, Y.: CIV: information volume of complex mass function. In: 2021 IEEE International Conference on Unmanned Systems (ICUS), pp. 438–442. IEEE (2021)
3. Deng, J.X., Deng, Y.: Information volume of fuzzy membership function. Int. J. Comput. Commun. Control **16**(1), 1–14 (2021)
4. Deng, J., Deng, Y., Cheong, K.H.: Combining conflicting evidence based on Pearson correlation coefficient and weighted graph. Int. J. Intell. Syst. **36**(12), 7443–7460 (2021)
5. Deng, Y.: Information volume of mass function. Int. J. Comput. Commun. Control **15**(6), 3983 (2020)
6. Deng, Y.: Uncertainty measure in evidence theory. Sci. China Inf. Sci. **63**(11), 1–19 (2020)
7. Deng, Y., Shi, W.K., Zhu, Z.F., Liu, Q.: Combining belief functions based on distance of evidence. Decis. Support Syst. **38**(3), 489–493 (2004)
8. Dzitac, I., Filip, F.G., Manolescu, M.-J.: Fuzzy logic is not fuzzy: world-renowned computer scientist Lotfi A. Zadeh. Int. J. Comput. Commun. Control **12**(6), 748–789 (2017)
9. Gao, X., Pan, L., Deng, Y.: A generalized divergence of information volume and its applications. Eng. Appl. Artif. Intell. **108**, 104584 (2022)
10. Klir, G.J., Yuan, B.: Fuzzy sets and fuzzy logic: theory and applications. Possibility Theory Versus Probab. Theory **32**(2), 207–208 (1996)
11. Liu, J., Xiao, F.: Information volume of mass function based on extropy. Soft. Comput. **26**, 2409–2418 (2021). https://doi.org/10.1007/s00500-021-06410-z
12. Nădăban, S., Dzitac, S., Dzitac, I.: Fuzzy TOPSIS: a general view. Procedia Comput. Sci. **91**, 823–831 (2016)

13. Pan, L., Deng, Y.: A novel similarity measure in intuitionistic fuzzy sets and its applications. Eng. Appl. Artif. Intell. **107**, 104512 (2022)
14. Shafer, G.: A Mathematical Theory of Evidence, vol. 7. Princeton University Press, Princeton (1976)
15. Smets, P.: Decision making in the TBM: the necessity of the pignistic transformation. Int. J. Approximate Reasoning **38**(2), 133–147 (2005)
16. Xiao, F.: Multi-sensor data fusion based on the belief divergence measure of evidences and the belief entropy. Inf. Fusion **46**, 23–32 (2019)
17. Zadeh, L.A.: Fuzzy sets. Inf. Control **8**(3), 338–353 (1965)
18. Zhou, Q., Deng, Y.: Higher order information volume of mass function. Inf. Sci. **586**, 501–513 (2022)
19. Zhou, Q., Mo, H., Deng, Y.: A new divergence measure of pythagorean fuzzy sets based on belief function and its application in medical diagnosis. Mathematics **8**(1), 142 (2020)

Boosting Data Management Using Artificial Neural Networks

Francisco Sepúlveda-Ortiz[1], Claudia Durán[1], Diego Fuentealba[2], Manuel Vargas[3], and Raúl Carrasco[4](✉)

[1] Departamento de Ingeniería Industrial, Universidad Tecnológica Metropolitana, 7800002 Santiago, Chile

[2] Departamento de Informática y Computación, Universidad Tecnológica Metropolitana, 7800002 Santiago, Chile

[3] Departamento de Ingeniería Industrial, Universidad de Santiago de Chile, 9170124 Santiago, Chile

[4] Facultad de Ingeniería y Negocios, Universidad de Las Américas, 7500975 Santiago, Chile

rcarrasco@udla.cl

Abstract. The present work promotes the study of artificial neural networks (ANN) in the management of data from logistics chains. The main aspects and advantages of applying this technology to business are reviewed. Through a literature review, the business domains, models, algorithms and benefits of ANN forecasting are determined. A case study is generated with historical data to predict with ANN the flow of cargo moving by land and sea transport during the year 2020. The results are analyzed and when compared with the company's actual data and the least squares method, they show greater efficiency when using ANN. Finally, conclusions are drawn on the advantages of incorporating this technology in a public company and the need to generate more research in this area to reduce costs, increase process efficiency and improve decision making.

Keywords: Artificial neural networks · Decision making · Data management

1 Introduction

The current pandemic has reduced the export of products to international markets due to the interruption of logistic chains worldwide, which has been generated in part by the sanitary protocols that have been implemented in the production processes that have reduced the labor work force [17]. This situation has caused inefficiency in the planning of cargo and port capacities, since there are delays in the flow of containers that are transported by sea and/or land, this uncertainty makes it difficult for port managers to make effective strategic decisions to increase competitiveness [24].

S. Dzitac et al. (Eds.): ICCCC 2022, AISC 1435, pp. 270–280, 2023.
https://doi.org/10.1007/978-3-031-16684-6_23

More flexibility is required in the flow of shipping and distribution, adding intelligent information and communication technologies that deliver real-time cargo traceability data to reduce the costs generated by inefficient scheduling of operational activities [11]. It is important to reduce the inefficiencies caused by delay time and lead time as they affect the flow of export information and the decision making of exporters, customs agencies, freight forwarders, shipping companies, land transport and shipping companies [8].

Logistics chains generate a large amount of data that lacks structure and does not always contribute to the decision making process of company managers. It is important that the flow of information be in accordance with the strategies and negotiations that are generated between the actors in the logistics chain [5,8]. Systematized and organized information systems are required to handle large amounts of data with the help of new trends in machine learning [19].

The implementation of new technologies such as Artificial Neural Networks (ANN) can help to solve the problems associated with the lack of updated predictive models and adequate instruments in chilean ports. Neural networks can be used to learn and store a large number of mapping relationships of an input-output model. It is currently widely applied in many areas such as pattern recognition, classification, data compaction and data prediction. ANN has also been applied to ports, however the applications predict the total volume and do not care to disaggregate by container type [6].

The purpose of this article is to study the potential of using ANNs and to develop a case study in the Port of Valparaiso to predict the flow of containers that will transit in the export of one of the main port operators in Chile.

2 Background

2.1 ANN Classification According to its Domain

In Table 1 a literature review is generated to identify the domains and advantages of a production or service company using ANN.

Table 1. Domains and advantages of using ANN

Domain	Description	Advantages
Logistics decision making	Development of Bat-type Backpropagation algorithm based on Chinese metallurgical factory inventory data collection [13]	Greater accuracy in inventory prediction compared to traditional models [16]
Strategic planning	Supports strategic container flow planning in major Asian ports, compares traditional forecasting models (least squares, linear regression) with multilayer neural perceptron modeling [23]	Obtains low error rate prediction with neural networks of the flow of containers with macroeconomic and operational variables [23]
Storage control	Using historical data and neural network (ANN) recurrent short and long term memory (LSTM) prediction of daily container throughput in Asian port areas [10]	Prediction with LSTM is more reliable [18], the port's space and assets are better optimized
Operations planning	Optimizes operations management and planning with traditional methods (ARIMA, SES, ARIMA+ANN) that measure errors and compare them with LSTM neural network results [21]	The more LSTM is trained, the better it forecasts container flows. LSTM delivers more accurate results than traditional methods [25]
Cost reduction	Decreases the costs of damaged products from ground transportation that are stored and distributed with Back Propagation Neural Networks (BBNN), General Regression (GR), Adaptive Fuzzy and Radial Basis (ANFIS) [14]	ANFIS delivers better results than the other three types of neurons
Environmental control	Historical air quality and meteorological data are used to forecast the level of air pollution around Busan Port using a recurrent neural network and the short-term memory model (LSTM) [12]	The highest performance of the neural network is determined and it is found that large ships generate the most air pollution [2]
Service efficiency	Evaluates measures of the operational quality of vehicles used in dairy freight transport services, using the multilayer perceptron neural network (MLP) [22]	MLP obtains high efficiency in predicting the quality of the transport service. Enables effective decision making [3] short or long term
Performance Control	To forecast container throughput at Qindao port using deep learning creates the two-dimensional short-term memory network model (BSTM). Compare BSTM with LSTM and ARIMA model [15]	BSTM delivers higher empirical predictive accuracy with multi-criteria [15]
Hybrid model prediction	It is applies the mixed prediction model combining a backpropagation neural network (BP) and the support vector machine (SVM), to calculate the quantity of containers used for export. BP forecasts with historical data and the SVM calculates the residual correction of the results obtained by the BP [6]	The combined BP and SVM prediction model shows a lower and more accurate overall relative error

It can be seen that all the domains are related to the decisions made at the strategic, business and operational levels in logistics chains to improve efficiency and effectiveness.

2.2 Business Context

The web repositories are studied and identified in Table 2 las companies developing neural networks. These actors are classified according to the industrial or service sectors that use them and the business advantages they obtain with the use of ANN are described.

By comparing the Table 2 with the Artificial Intelligence (AI) Annual Report Index rankings [20], it is observed that the countries in which the companies are installed coincide with the developed countries that invest the most in AI. It is observed that in Latin American countries private investment is low, it is worth saying that Chile allocated approximately 0.25% of what the United States invested during the period 2018–2019.

Table 2. Digital repositories

Sector	Company	Advantages	Repository
Ecommerce	Amazon	History search prediction	www.ideamotive.com
Ecommerce	PicxelDTGAN	Cost reduction	https://github.com/fxia22/PixelDTGAN
Finance	SAS Real Time Decision Manager	Fraud prevention	https://support.sas.com
Finance	Finprophet	Prediction of financial instruments	www.ideamotive.co
Logistics	FourKites	Estimated delivery time of a shipment	https://www.fourkites.com/
Health	IBM Watson	Performing medical diagnostics	https://www.ibm.com/watson
Security	ICSP Neural of Symantec	Protection of critical systems in the enterprise	https://neural.iot.symantec.com/
Transportation	Tesla	Autonomous driving	https://www.tesla.com
Transportation	Waymo	Autonomous driving	https://waymo.com

2.3 ANN in Chilean Ports

The Ministry of Science, Technology, Knowledge and Innovation (MINCIENCIA) creates a national artificial intelligence policy in 2019 with initiatives that impact society, the economy and human capital [1].

State initiatives include an AI program for port technology transformation, in which major investments in this area are planned for public ports to upgrade to a smarter generation [1]. With this innovation, they could evolve from industry 3.0 with some characteristics of intelligent ports to industry 4.0, in addition to incorporating sustainable aspects that improve competitiveness, minimize the risks involved in operations, generate quality jobs and reduce pollution [7].

Port management is beginning to implement new technologies for decision making at a national level, policies have been created focused on improving processes and automation in ports to achieve greater efficiency of key resources. Currently, companies are looking for new methods related to artificial intelligence to help in strategic planning, optimize their processes, generate better predictions and are able to process large amounts of data [4,9].

3 Materials and Methods

To determine in which areas the use of neural networks are being investigated and to determine the domains shown in the Table 1, an advanced search is generated with the keywords `TS="artificial neural networks"` the period 2014–2021 in the databases of the Web of Science Core Collection (WoS). As a result, 21,963 publications are obtained, from which those related to decision making in companies are selected.

In order to identify the main companies that develop ANNs and their characteristics, the following Table 2 is generated.

A prediction case is developed in the Port of Valparaiso with ANN, which follows the Knowledge Discovery database (KDD) process. The KDD is adapted according to the following phases and activities:

Phase 1 - Port data collection:

- The study domain is understood and the problem is characterized by reviewing the website of the Cámara Marítima y Portuaria de Chile A.G. (CAMPORT).
- Collect data from the CAMPORT web site related to the shipment process to vessels.
- Load the data observing the most relevant elements for the prediction of exported containers, according to each type.
- We analyzed 2,675,679 data during the years 2017–2019 and selected a sample of 613,594 data corresponding to containers transiting through the Port of Valparaiso.

Phase 2 - Pre-processing of export data:

- The following variables are filtered: year, customs office responsible for shipment, port of shipment, weight of package (container) and type of package (container).

Phase 3 - Transformation of export data:

- The data obtained from DIRECTEMAR are aggregated and outlier and noise are removed with KNIME software, and specialized filters are used to isolate the conditions.
- After cleaning, algorithms are used to quantify each type of container.
- The descriptive data report is analyzed.

Phase 4 -Data Mining.

- To build the sequential predictive model with neural networks, the containers are summed. In Python, the "KERAS" library is used with two input nodes, two hidden layers and one output node. The neuron is activated with "relu" with an input value greater than zero and trained with the "ADAM" algorithm that adjusts the weights and biases. The loss function used is the "Mean Squared Error".
- The neural network is evaluated.
- The prediction is made.

The phases of the KDD are shown in Fig. 1.

On the other hand, Fig. 2 shows the architecture to be used in the case study, which considers a node in charge of the data extraction, transformation and loading (ETL) process and another node in charge of processing the model.

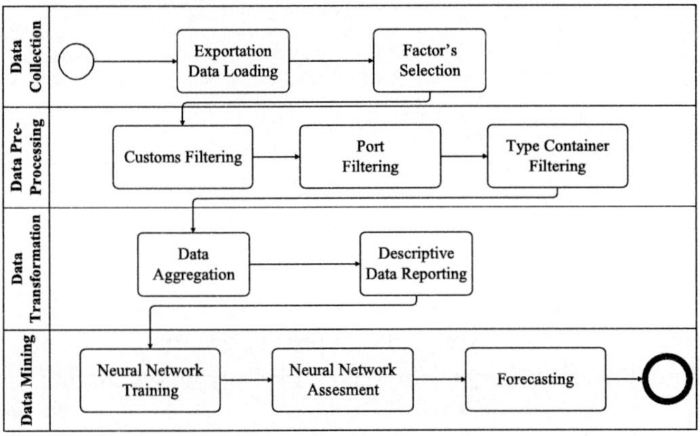

Fig. 1. Adapted KDD process for port data analytics.

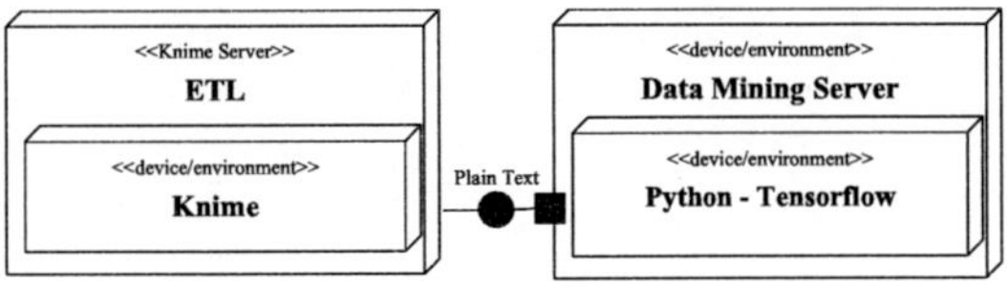

Fig. 2. Proposed architecture.

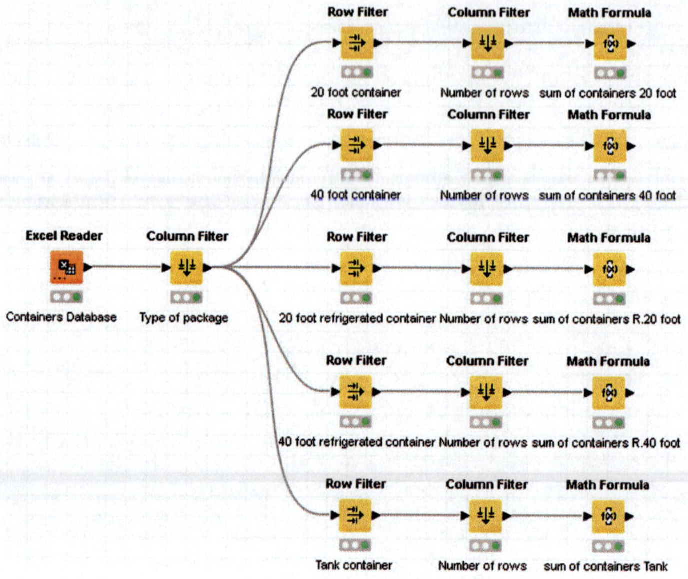

Fig. 3. ETL workflow used during 2017–2021.

4 Results

As a result of the KDD phases related to ETL, we obtain Fig. 3, the diagram shows the workflow used for each year of the 2017–2019 period for the Port of Valparaiso developed in the Knime data mining platform.

With the databases obtained in Fig. 3, phase 4 of the KDD is developed. The dense neural network with 2 hidden layers is trained on keras generated in Google Collaboratory with 1000 repetitions. As shown in Fig. 4 after 450 Epochs, the network stops learning.

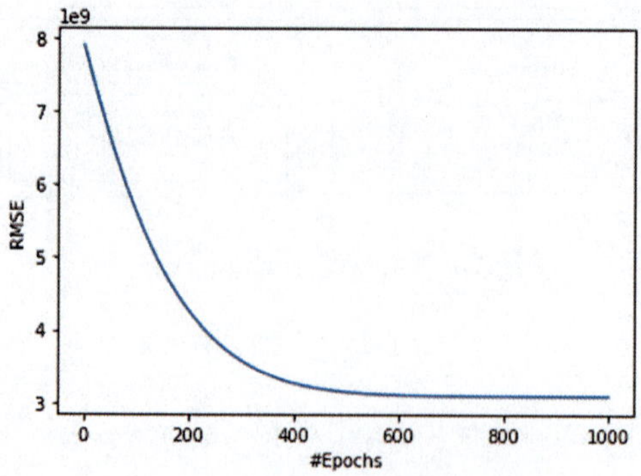

Fig. 4. Epochs vs RMSE graph.

As a result of the ANN training, the prediction for the year 2020 shown in Fig. 5. It is observed that when comparing 2018 with 2019, the transit of refrigerated and non-refrigerated containers decreases. This decrease is due to lower logistics activity caused by the lack of dock workers due to the Covid-19 variants. It is estimated that the losses due to the cessation of activities amount to 400 million dollars.

When comparing 2019 with 2020, it is noticeable that the movement of cargo increases but it is less than the years prior to the pandemic.The most affected containers during the sanitary crisis are refrigerated (reefer) of 20 ft or 40 ft. The main products that were not exported were fresh grapes, cherries, plums and fresh apples.

In relation to non-refrigerated containers, the most traded products during 2019–2020 were industrial type oil and different kinds of wines and the main countries to which they exported goods were the United States, Germany, Holland, Spain, United Kingdom and Holland [4].

To verify the accuracy of the results provided by the neural network for the year 2020, we compare the predictions made with ANN, with the least squares (LS) method and the real data obtained in the year 2020 [4]. Since CAMPORT's container movement figures are in tons [4]the data in the Fig. 5 are calculated with the following formula: Number of containers by type ($units$) $*$ maximum load capacity ($tons$)

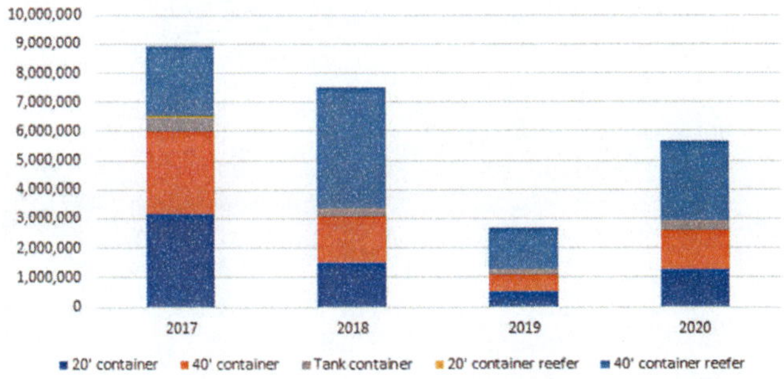

Fig. 5. Containerized cargo movement during 2017–2020

Table 3. Comparison of predictions with actual data for the year 2020 (ton).

Method	Forecast (ton)	Actual real (ton) [4]	Difference (ton)
ANN	5,709,189	5,663,878	45,311
LS	5,774,505	5,663,878	110,627

When comparing the ANN and LS methods with the real data it can be seen that the neural network had a higher accuracy.

5 Conclusion

ANN is a technology that allows creating more accurate predictions based on historical data in supply chains. In the logistics area, its use is key because it can predict future demand, obtain patterns related to operational activities, manage key resources for decision making, structure preventive maintenance, optimize shipping routes, and plan inventories, among others. In Chile its application is incipient, since 2019 the State is including in its public policy discussions artificial intelligence development initiatives for ports to improve competitiveness.

In spite of having different sources of information, the work is able to aggregate the data and predict for the year 2020 how many tons and what type of containers are used in exports. The results show the impact of the sanitary crisis generated by the port protocols at national level and by the requirements of the main trading partners. The application of ANN shows a greater accuracy of the container flow, which presents an advantage in the planning of activities due to the importance for the business of complying with the times established for each type of operation. Reducing lost time increases productivity and efficiency [8].

It is necessary to continue with the research, improve the predictions by incorporating new data such as the number of fractionated containers. Better cargo planning could reduce the number of trips by ships and trucks, reducing the costs and pollution generated by land and sea transport. In addition, data management is improved and decision making at the Port of Valparaíso is optimized.

Acknowledgements. This research aknowledge to the government of Chile, the project FONDEF ID21I10191 and Centre for Multidisciplinary Research in Telecommunication Technologies (CIMTT).

References

1. Arancibia, D., Ávila, C., Caro, M.J., Girardi, J., González, N., Guridi, J.A., Rivera, A.: Política Nacional de Inteligencia Artificial. Chile IA. Ministerio de Ciencia, Tecnología, Conocimiento e Innovación, Santiago, Chile (2022)
2. Aulinger, A., Matthias, V., Zeretzke, M., Bieser, J., Quante, M., Backes, A.: The impact of shipping emissions on air pollution in the greater North Sea region - Part 1: current emissions and concentrations. Atmos. Chem. Phys. **16**(2), 739–758 (2016)
3. Borges, A.F.S., Laurindo, F.J.B., Spínola, M.M., Gonçalves, R.F., Mattos, C.A.: The strategic use of artificial intelligence in the digital era: systematic literature review and future research directions. Int. J. Inf. Manag. **57**, 102225 (2021)
4. Cámara Marítima y Portuaria de Chile CAMPORT. Desafíos (2022)
5. Czachorowski, K.: Solesvik, Marina, Kondratenko, Yuriy: the application of blockchain technology in the maritime industry. In: Kharchenko, V., Kondratenko, Y., Kacprzyk, J. (eds.) Green IT Engineering: Social. Business and Industrial Applications, volume 171 of Studies in Systems, Decision and Control, pp. 561–577. Springer International Publishing, Cham (2019)

6. Ding, M.J., Zhang, S.Z., Zhong, H.D., Yao Hui, W., Zhang, L.B.: A prediction model of the sum of container based on combined BP neural network and SVM. J. Inf. Process. Syst. **15**(2), 305–319 (2019)
7. Duran, C., Palominos, F., Carrasco, R., Carrillo, E.: Influence of strategic interrelationships and decision-making in Chilean port networks on their degree of sustainability. Sustainability **13**(7), 3959 (2021)
8. Durán, C.A., Córdova, F.M., Palominos, F.: Method for improving critical strategic and operational success factors in a port system. Procedia Comput. Sci. **139**, 448–455 (2018)
9. Durán, C.A., Córdova, F.M., Palominos, F.: A conceptual model for a cyber-social-technological-cognitive smart medium-size port. Procedia Comput. Sci. **162**, 94–101 (2019)
10. Gao, Y., Chang, D., Fang, T., Fan, Y.: The daily container volumes prediction of storage yard in port with long short-term memory recurrent neural network. J. Adv. Transp. **2019**, 5764602 (2019)
11. Gharehgozli, A.H., Roy, D., de Koster, R.: Sea container terminals: new technologies and OR models. Marit. Econ. Logistics **18**(2), 103–140 (2016)
12. Gürbüz, F., Eski, I., Denizhan, B., Daglı, C.: Prediction of damage parameters of a 3PL company via data mining and neural networks. J. Intell. Manuf. **30**(3), 1437–1449 (2019). https://doi.org/10.1007/s10845-017-1337-z
13. Han, C., Wang, Q.: Research on commercial logistics inventory forecasting system based on neural network. Neural Comput. Appl. **33**, 691–706 (2021). https://doi.org/10.1007/s00521-020-05090-4
14. Hong, H., Jeon, H., Youn, C., Kim, H.: Incorporation of shipping activity data in recurrent neural networks and long short-term memory models to improve air quality predictions around Busan port. Atmosphere **12**(9), 1172 (2021)
15. Jiang, F., Xie, G., Wang, S.: Forecasting port container throughput with deep learning approach. In: The 5th International Conference on Computer Science and Application Engineering, CSAE 2021, pp. 1–4, October 2021. Association for Computing Machinery, New York, NY, USA (2021)
16. Lahmiri, S.: Wavelet low- and high-frequency components as features for predicting stock prices with backpropagation neural networks. J. King Saud Univ. - Comput. Inf. Sci. **26**(2), 218–227 (2014)
17. Larraguibel, J., Carrasco, R., Banguera, L., Lagos, C., Durán, C., Osorio-Valenzuela,L.: Supply chain and logistics in pandemic. In: 2021 IEEE International Conference on Automation/XXIV Congress of the Chilean Association of Automatic Control (ICA-ACCA), pp. 1–6. IEEE, Valparaíso, Chile, March 2021
18. Livieris, I.E., Kiriakidou, N., Stavroyiannis, S., Pintelas, P.: An advanced CNN-LSTM model for cryptocurrency forecasting. Electronics **10**(3), 287 (2021)
19. Mi, J.-X., Li, A.-D., Zhou, L.-F.: Review study of interpretation methods for future interpretable machine learning. IEEE Access **8**, 191969–191985 (2020)
20. Perrault, R., et al.: Artificial intelligence index 2019 annual report. Human-Centered AI Institute, Stanford University, Stanford, CA, USA, AI Index Steering Committee (2019)
21. Shankar, S., Ilavarasan, P.V., Punia, S., Singh, S.P.: Forecasting container throughput with long short-term memory networks. Ind. Manag. Data Syst. **120**(3), 425–441 (2020)
22. Swiderski, A., Jozwiak, A., Jachimowski, R.: Operational quality measures of vehicles applied for the transport services evaluation using artificial neural networks. Eksploatacja I Niezawodnosc **20**(2), 292–299 (2018)

23. Tsai, F.-M., Huang, L.J.W.: Using artificial neural networks to predict container flows between the major ports of Asia. Int. J. Prod. Res. **55**(17), 5001–5010 (2017)
24. Wang, Z., Guo, C.: Minimizing the risk of seaport operations efficiency reduction affected by vessel arrival delay. Ind. Manag. Data Syst. **118**(7), 1498–1509 (2018)
25. Zhou, J., Xu, W.: End-to-end learning of semantic role labeling using recurrent neural networks. In: ACL-IJCNLP 2015 - 53rd Annual Meeting of the Association for Computational Linguistics and the 7th International Joint Conference on Natural Language Processing of the Asian Federation of Natural Language Processing, Proceedings of the Conference, vol. 1, pp. 1127–1137 (2015)

Decision Making and Support Systems

Algorithmic Modeling to Individualize Touristic Experience in Petrila Theme Park

Ionela Samuil, Andreea Ionica(✉), and Monica Leba

University of Petrosani, 332006 Petrosani, Romania
andreeaionica@upet.ro

Abstract. The perception of tourism activity has changed a lot in recent years, as a result of the innovation of services, now accessible through the Internet and smart devices. The tourist offer is of an impressive variety and the tourist is looking for personalized services. In the light of this information, it is necessary to develop a tool for automatic planning of a tourist route based on the requirements defined by a tourist in order to increase his satisfaction. The personalized planning of the tour involves the selection, from the base of information related to the destination, of the elements that correspond to the requirements of the tourist, and the configuration of the tourist route according to the selected elements. In order to achieve this goal, in this article, the personalized planning of the tourist route in the context of the industrial tourism within the proposed Petrila Theme Park, is modeled on the basis of the knapsack problem, the discreet form. Gamification elements are also introduced in the analysis to enhance the result. Thus, in the case under study, the tourist represents the knapsack, the elements to be introduced in the knapsack are represented by the characteristics that define each type of player and the constraint is given by the time that allows the player to explore the destination by buying the ticket.

Keywords: Industrial tourism · Gamification · Knapsack problem

1 Introduction

It is a certainty that planning a trip is a complicated process, consuming energy and time [1], given that any tourist destination has several points of interest (PI) and tourists must choose the most relevant to them. [2] concluded that tourists are increasingly preferring personalized tours over standard options, with predefined, printed guides not being sufficient to meet the specific preferences of tourists. In the digital age, digital tools allow users not only access to the opinion and point of view of an expert, but also to find out opinions from other users. In this way, the content created allows the permanent updating of information in the public space, giving users the opportunity to make well-informed decisions.

[3] considers that mobile travel guides tend to become tools for planning personalized tours and propose an approach that combines artificial intelligence with metaheuristics in order to solve the "tourist trip design problem" (TTDP).

S. Dzitac et al. (Eds.): ICCCC 2022, AISC 1435, pp. 283–292, 2023.
https://doi.org/10.1007/978-3-031-16684-6_24

In this paper we explore the possibility of developing an instrument, with the help of which a tourist can configure his own tourist route in a destination, using his own preferences and constraints. To achieve this goal, we aim to maximize the satisfaction obtained from the tourist experience, approaching the problem as a knapsack problem, enhancing this experience with elements of gamification, optimization that will be implemented for points of interest in the Petrila Theme Park. The knapsack problem is a matter of combinatorial optimization: given a set of items, each with a weight and a value, the number of each item in the set that can be included in a collection is determined so that the total weight is less or equal to a certain limit and the total value to be as high as possible. It derives its name from the problem faced by someone who is forced to fill a fixed-sized knapsack with the most valuable items.

2 Background and Related Work

2.1 Algorithmic Approach in the Research Context

Algorithmic approaches are the core of the research in this field, being developed or refined Greedy algorithms with which to configure the shortest route in a destination, so that the tourist should visit all the selected places [4].

[5] explores the state of the art on Kitchenham-Charters guidelines and the PICOC method on the use of artificial intelligence tools and expert systems in product customization, with systems designed according to a set of rules being considered of utmost interest. [6] concludes that modeling the planning process as an orienteering problem is the most popular approach used to automate it. The orienteering problem is considered a combination of Traveling Salesman Problem and Knapsack Problem [7].

The knapsack problem is a combinatorial optimization problem [8] being widely studied in industry and financial management but also in tourism, not only due to its applicability and theoretical reasons. Issues such as resource allocation, investment decision making and warehouse loading can be modeled as knapsack issues [9].

[10] studied the Online Constrained Multiple-Choice Promotions Personalization Problem, the aim of this study being to optimize the promotions applicable to each type of customer that will ultimately lead to the completion of online orders, initiated while respecting the imposed budgetary constraints. [11] used the 0–1 knapsack issue to develop a system for recommending the most suitable travel routes to a tourist destination, taking into account the travel time and price constraints of tourists.

2.2 Gamification in Tourism

The dynamics of the game consist in the actions that derive from the game, namely completion, competition, cooperation, communication and creation [12]. The first four correspond to the four types of players of performers, explorers, killers and socializers [13]. Based on these realities [14], it examines the reasons that lead people to make a gamified journey and places players in six different typologies (knowledge collector, reward seeker, explorer, curious, sensation seeker, flow experimenter).

The importance of gamification in tourism, management and marketing of tourist destinations, is indisputable, having a significant contribution in obtaining a memorable experience, by immersing tourists in a simulated world [15].

This study examines the possibility of developing a tool that would help tourists in tracing personalized tourist routes, by integrating gamification elements, using the six types of players identified by [16].

In order to achieve the proposed objective, we approach the problem as a knapsack problem, with the aim of increasing the satisfaction of tourists. The tool proposed in this article aims to increase the satisfaction of the tourist after completing the route, configured in relation to the characteristics of the type of players to which it belongs, and the rank it gives during and at the end of their journey. In this way you can get a memorable tourist experience, using as a support a mobile application based on game-specific techniques. For efficient communication, it must be personalized and use interactivity tools.

The novelty of the proposed topic is supported by the lack of studies that integrate all the elements discussed in the study of existing literature at this time, not identifying any research that would take a step forward in this direction.

3 Materials and Methods

The research process was developed using 5W1HQuestions. The Simon Sinek's "golden cycle" (Fig. 1) – Why, How, What – was adapted and extended in a "processing spiral" (Fig. 2), starting with Who and adding Where and When.

Starting with the question Who will benefit, the answer comes clear: any person who plans to make a visit/trip to a destination, looking for a unique experience, using as mapping elements their own interests. The research offers the possibility to generalize the use of the proposed tool in different tourist destinations in order to increase customer satisfaction.

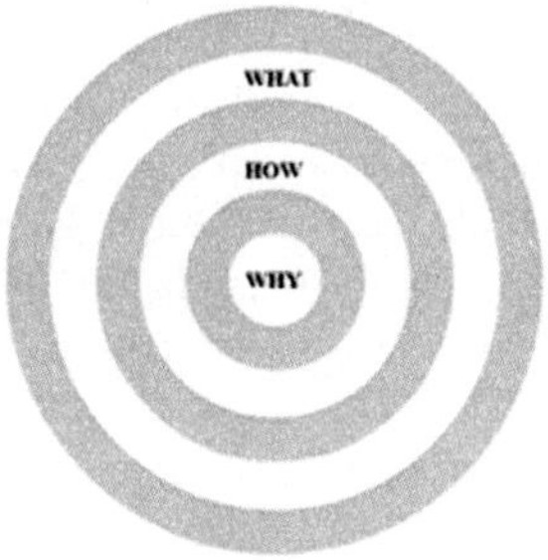

Fig. 1. Golden cycle

Fig. 2. Processing spiral

The proposed tool is useful in the development of a new tourist destination, thus appearing the answer to the question Why, and also being established the purpose of the research. How is this goal achieved? Through an algorithm based on knapsack problem, elements of gamification and classification of tourists into 6 player types [14] (Fig. 3).

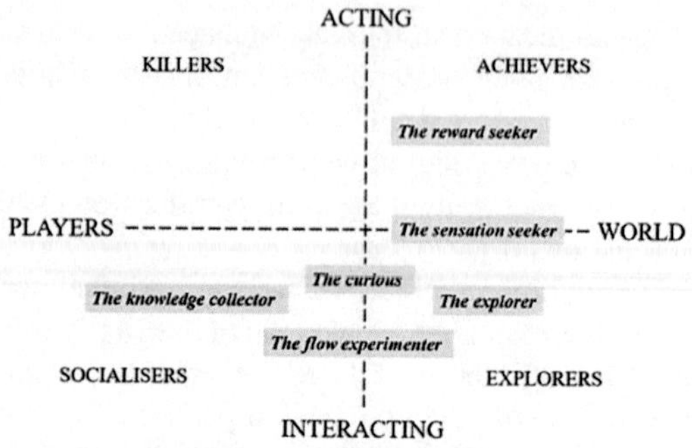

Fig. 3. Types of players

The six types of players, customized for generating tourist routes, are placed in Fig. 3, in the context of this research, in the 2D representation of the types of players specific to computer games [13]. The 2D representation is highlighting the source of players' interest, where the four extreme corners of the graph show the four typical game preferences associated with each quadrant.

Each of the six types of players is motivated by certain elements to choose a gamified journey. While the knowledge collector is motivated by the thirst for knowledge of cultures, places and facts and combines fun with culture, the reward seeker seeks rewarding and prefers interactive maps as a tool for collecting destination information in an interesting way, in real time, with the possibility of winning prizes. The explorer, eager to interact with the locals uses the interactive elements to explore different aspects of the destination, gamification fulfilling the function of guide. The curious is looking for the element of surprise, its purpose for a visit is the desire to find or see something that would make the trip interesting and memorable. While the sensation seeker is not interested in socializing, he prefers to follow a circuit generated by a system with the possibility to live different adventures and to achieve challenges while appreciating the riddles. Looking to explore a destination in an interesting way, the flow experimenter seeks memorable experiences and excursions through games that make travel interesting and allow them to fully engage in tours.

The problem is to find the optimal route to visit by each type of player in a period of time and taking into account the preferences of each. From the statement of the problem, it can be seen that we are dealing with an optimal problem in the category of Greedy algorithms. Among the best-known algorithms, the knapsack problem is best suited for solving the formulated problem.

In the research, the player represents the knapsack, the capacity of the knapsack is the time allocated to the route, the elements that are inserted in the knapsack are represented by the time element allocated to each point of interest for each type of player and the constraint is given by the time that allows the player to explore the destination by buying the ticket.

Thus, first the points of interest for each type of player are identified, then each point of interest, corresponding to a player, is assigned a value (materialized in a number of points) according to the degree of interest it arouses in the player. Also, each point of interest is dimensioned as time duration, using the same principle.

In this sense, the results of a qualitative research based on focus group are used. Each type of player is represented in the focus group and participates voluntarily. The previous stage of the research presupposes quantitative research that would allow the classification of the subjects, potential tourists in one of the types. The resulting perceptions of points of interest following the application of the focus group then allowed the researchers to identify the values of each point of interest. Table 1 shows for each point of interest (PI_j) the values for each type of player (P_i).

Table 1. PI assigned values

Pi	PI_1	PI_2	PI_3	PI_4	PI_5	...	PI_m
Vi	V_{i1}	V_{i2}	V_{i3}	V_{i4}	V_{i5}	...	V_{im}
D_i	D_{i1}	D_{i2}	D_{i3}	D_{i4}	D_{i5}	...	D_{im}

What is desired to be obtained as results? The personalized tourist routes that maximize the tourist's satisfaction. The question ***Where***, identified the research area being the former Petrila mining site, where it is proposed to develop a theme park as an alternative to revitalize the mono-industrial area [17]. The research presents the example of the Petrila Theme Park, but the algorithm can be applied to any tourist area or any developing area with tourist potential.

Further research will be done by answering the question ***When*** according to the "Strategy for the economic, social and environmental development of the Jiu Valley (2021–2030)".

The proposed algorithm takes into consideration $n = 6$ players ($P_1, \ldots, P_6$), representing the number of players and $m_i \left(i = \overline{1,\ 6}\right)$ possible points of interest, representing the destinations within the existing building inside the tourist attraction. Each destination has one or more points of interest, but there can be players with no points of interest per destination. So, having the 6 players will generate a total of $m_1, m_2, \ldots, m_6$ points of interest per player.

In order to design the selection algorithm, we need several inputs, as follows:

- $m_i\left(i = \overline{1,\ 6}\right)$ – the number of points of interest for each player;
- 6 arrays ($V_1, \ldots, V_6$) containing the values (in points) of each point of interest for each player. The dimensions of those arrays are $1 \times m_i$;
- 6 arrays ($D_1, \ldots, D_6$) containing the needed time to explore each point of interest for each player. We considered on these times the time allocated for actual visit of the point of interest and the average time needed to reach the point of interest. The dimensions of those arrays are also $1 \times m_i$;
- the time T, chosen by the player to explore tourist attraction (can be a time-based ticket bought by the player).

The output of the algorithm should be a list of points of interest chosen as exploration path for the player in order to get maximum value in the chosen time T.

For a player $P \in \{P_1, \ldots, P_6\}$:

$$\text{Path}\ (P_i) = \left\{ d_k \sum\nolimits_{k=1}^{X} D_i(d_k) \cdot r_k = T \text{ and } \frac{V_i(d_k)}{D_i(d_k)} = \max \right\} \tag{1}$$

where:

$$r_k = \begin{cases} 1 & \text{for } k = \overline{1,\ X-1} \\ (0, 1] & \text{for } k = X \end{cases} \tag{2}$$

and X = number of points of interest in the generated path.

The above algorithm allows the generation of the optimal route for each type of player depending on the time allocated to the visit and its preferences. The algorithm is designed to be a backend in a smartphone or online application that, in addition to determining the optimal route, performs other functions such as: guiding the tourist through the personalized route map, counting the points collected, taking feedback from him for updating dynamics of the times and points allocated to each point of interest per type of player and the update of the generated route in case time remains uncovered.

4 Results – The Case of the Personalized Routes Within the Theme Park

The research area is defined by a former mining perimeter that has the potential to be transformed into a theme park [17].

The results of running the proposed selection algorithm to solve the problem of generating the most suitable tourist route within the Petrila Theme Park are presented, taking into account the characteristics of the types of players analyzed, using the knapsack problem.

For example, we run an analysis for player number 3 (P_3). Thus, in the analyzed case, P_3 represents the knapsack, the elements that are inserted in the knapsack are represented by the time element (D_3) allocated to each point of interest, corresponding to this type of player, and the imposed constraint is defined by the time interval for which the ticket is purchased, allowing the player to explore the destination.

In the research area, for P_3 are identified 9 points of interest related to a number of seven existing buildings in the perimeter and with potential for use in tourism that will be presented below (Fig. 4). The former medical office, in the current structure is the Cafeteria, can serve as an information point and/or resting place, here being arranged a store with local products, various souvenirs related to the past of the perimeter (PI_1). The former administrative building, proposed to become an Event Space, offers tourists the opportunity to experience the route of a miner, with the whole process (PI_2). Here you can find information about the evolution of the area through leaflets, exhibits and video projections (PI_3). At higher floors, tourists learn creative arts from different craftsmen, with the possibility of acquiring the result of their own work (PI_4). Given that in the building next to the school mine, to be transformed into an educational space, were performed geomechanical characteristics determinations of various rocks, but also environmental analysis, here tourists can learn and experience how to perform these procedures, the space being a laboratory equipped for such determinations (PI_5). The school mine, to be transformed into a museum mine, was and remains a small-scale underground mine, becoming the central point of any tourist route, tourists being able to get an experience as close as possible to the real life of a miner in a coal mine, enhanced with augmented reality devices (PI_6). The mechanical workshop, proposed to be transformed into a mechanical workshop museum, dating from 1879, an element of heritage, offers tourists a real history lesson with the help of existing machines and equipment in the perimeter (PI_7). The extraction complex, Center shaft, which is to become Center shaft museum + education space, built in 1943, also a heritage element, offers tourists the opportunity to find out information about its history and operation (PI_8). The Skip shaft, in the proposed structure of the Park becoming a skip shaft museum, built in 1887, a heritage element as well, allows tourists to witness the extraction process, but also to experience the activity of the mechanic responsible for handling the extraction machine (PI_9).

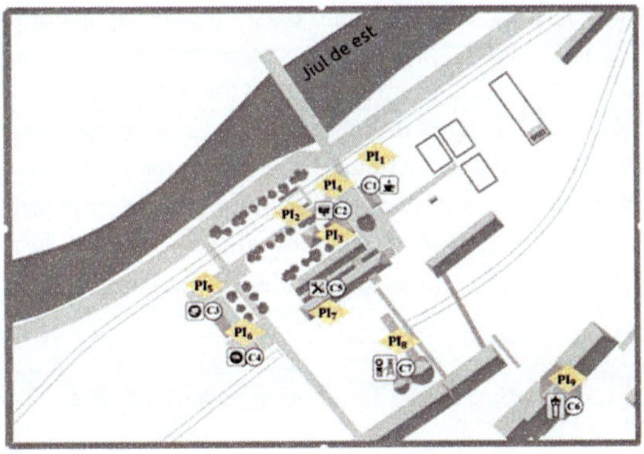

Fig. 4. Points of interest for player type 3

Thus, for $m = 9$ *and*, we have the following arrays:

D3: 10 10 5 2 10 15 20 2 20 (min)
V3: 5 10 12 4 15 20 4 1 30 (value points)

Following the running of the developed algorithm, the generated results are as follows:

$$
\begin{array}{lccccccccc}
{}^{v}/_{d}: & 0.5 & 1 & 2.4 & 2 & 1.5 & 1.33 & 0.2 & 0.5 & 1.5 \\
 & \uparrow & \uparrow & \uparrow & \uparrow & \uparrow & \uparrow & \uparrow & \uparrow & \uparrow \\
 & 7 & 6 & 1 & 2 & 3 & 5 & 9 & 8 & 4
\end{array}
$$

Given that, in the case under study, the tourist bought a ticket that gives him 60 min to explore the destination, the results show that his time does not allow him to fully explore one of the points of interest selected, so that the algorithm applies a correction coefficient. So, we get:

$$r_1 = r_2 = r_3 = r_4 = r_5 = 1;\ r_6 = 0.8 \tag{3}$$

The total reward points is computed as:

$$\begin{array}{llllll} \{d_3 & d_4 & d_5 & d_9 & d_6 & d_2 \cdot 0.8\} \\ 5 & 2 & 10 & 20 & 15 & {}^{10}/_{8} \\ 12 & +4 & +15 & +30 & +20 & +10 \cdot 0.8 = 89 \text{ value points} \end{array}$$

For a centralization of the tourists' opinions, based on which to update the algorithm data, the tool allows the continuous evaluation by tourists, of the points of interest covered, at the time of the visit. This will allow the adjustment of the value points for each point of interest and each player type for a better correlation of algorithm data to the tourists' expectations.

Based on the algorithm described there are designed the mockups for a smartphone application that has the algorithm as backend.

Upon first contact with the application, the user creates an account, defines a profile and selects a number of preferences. The user creates an initial account, but this profile will be improved as a result of interactions within the application.

After defining the profile, based on the preferences expressed by the user, the algorithm places the tourist in a player type and recommends a tourist route map (Fig. 5), the main screen highlighting the type of player, the path to follow drawn by joining the most interesting points valuable for the player and the time allotted for the exploration of the destination, purchased by the tourist, in the form of a ticket.

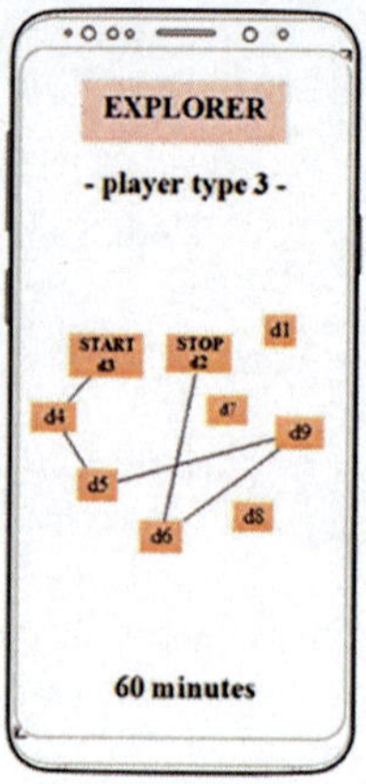

Fig. 5. Map screen

Fig. 6. PI check in screen

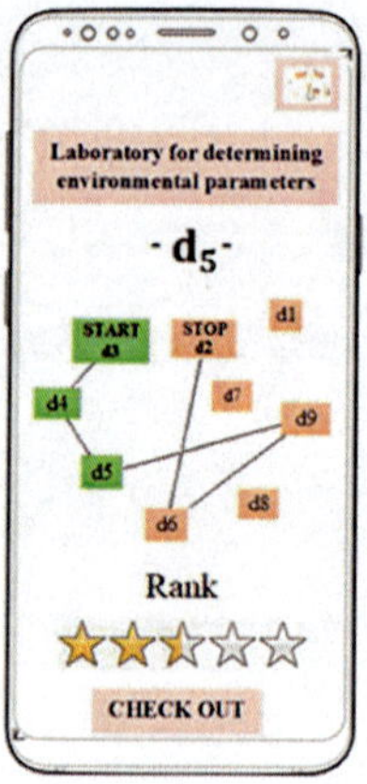

Fig. 7. PI check out screen

For a maximum gain (value of points) the user will explore the points of interest in the recommended order, trying to fit in the time allocated to each point of interest, any excess of time can lead to not achieving the entire proposed route. The exploration is performed sequentially, in Fig. 6 there is a point of interest from the route, on the device screen being highlighted the name and a photo showing the point of interest, next to the Check In button. At the end of the visit, the application screen will highlight the map with visited points of interest, highlighted in green but also a ranking section, next to the Check Out button (Fig. 7).

At any intermediate point, the application allows the user to return to the recommended route in the form of a map, using the button on the top right. After completing these steps, the application calculates and centralizes the accumulated value points but also the time elapsed from the total time purchased with the ticket by the user (Fig. 8).

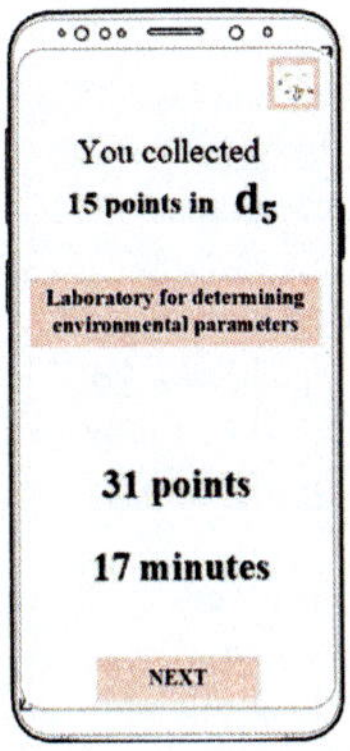

Fig. 8. Path intermediary screen

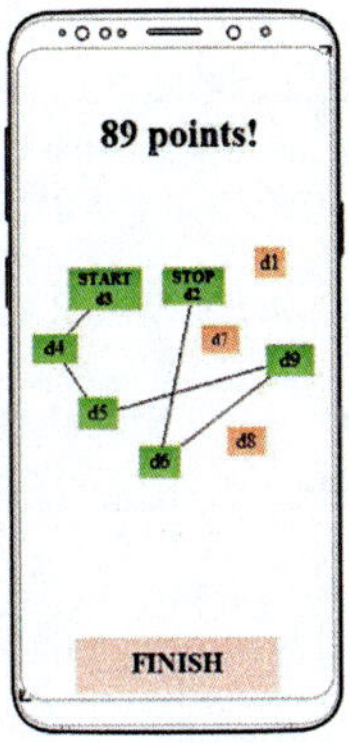

Fig. 9. Finish screen

At the end of the route, the application screen will highlight the route map in green, a sign that all the points of interest have been visited together with the total points accumulated (Fig. 9). In general, points can be used either in an in-app ranking, in the form of in-app badges, or for various in-store purchases, but also can be kept for later experience, depending on the type of player identified.

5 Conclusions

The implementation of the knapsack algorithm together with the gamification techniques for generating tourist routes provides a useful IT application for optimizing the experience and motivating tourists to visit different tourist areas.

By using the proposed tool, the aim is to create an interactive and individualized tourist experience, offering the user the possibility to configure their own routes based on two parameters: the interest expressed for different activities and the maximum time allocated for the tourist experience. In this way, for a visitor who, based on the expressed preferences, has been included in a certain type of players, a tourist route is recommended that includes the points of interest that are based on his preferences.

The application can be adapted not only to optimize tourist routes based on the selected options, but also to optimize different routes based on the previous interests of the tourist in other visited destinations. This goal can be achieved if we give up the unreasonable assumption that the tourist already knows what he wants before arriving in the park and starting the tour and we create the context for making personalized tours based on historical selections, using the tourist's previous behavior. For example, if in another destination he has chosen activities related to extreme tourism, it is reasonable to consider that he will be interested in this type of activity in the area under study.

References

1. Souffriau, W., Vansteenwegen, P., Vanden Berghe, G., Van Oudheusden, D.: The multiconstraint team orienteering problem with multiple time windows. Transp. Sci. **47**(1), 53–63 (2013)
2. Hyde, K.F., Lawson, R.: The nature of independent travel. J. Travel Res. **42**(1), 13–23 (2003)
3. Souffriau, W., Vansteenwegen, P., Vertommen, J., Vanden Berghe, G., Van Oudheusden, D.: A personalized tourist trip design algorithm for mobile tourist guides. Appl. Artif. Intell. **22**, 964–985 (2008)
4. Hashim, Z., Desa, W.L.H.M., Aziz, H.A., Zaki, N.M.Z.M.: Greedy method for solving the langkawi tourist route: a case study. J. Technol. Oper. Manag. **13** (1), 18–25 (2018)
5. Benjamin, A.M., Abdullah, A.S., Abdul-Rahman, S., Nazri1, E.M., Yahaya, H.Z.: Developing a comprehensive tour package using an improved greedy algorithm with tourist preferences. J. Sustain. Sci. Manag. **14** (4), 106–117 (2019)
6. Bruzza, M., Cabrera, A., Tupia, M.: Survey of the state of art based on PICOC about the use of artificial intelligence tools and expert systems to manage and generate tourist packages. In: International Conference on Infocom Technologies and Unmanned Systems (2017)
7. Wibowo, B., Handayani, M. K.: A genetic algorithm for generating travel itinerary recommendation with restaurant selection. In: International Conference on Industrial Engineering and Engineering Management (2018)
8. Vansteenwegen, P., Souffriau, W., Van Oudheusden, D.: The orienteering problem: a survey. Eur. J. Oper. Res. **209**(1), 1–10 (2011)
9. Nemhauser, G.L., Wolsey, L.A.: Integer and Combinatorial Optimization. Wiley, Hoboken (1988)
10. Zhao, B., Deng, C., Yang, Y., Peng, H.: Novel binary biogeography-based optimization algorithm for the knapsack problem. In: Tan, Y., Shi, Y., Ji, Z. (eds.) ICSI 2012. LNCS, vol. 7331, pp. 217–224. Springer, Heidelberg (2012). https://doi.org/10.1007/978-3-642-30976-2_26
11. Albert, J., Goldenberg, D.: E-Commerce Promotions Personalization via Online Multiple-Choice Knapsack with Uplift Modeling, Computer Science (2021). https://arxiv.org/abs/2108.13298v2
12. Kanimozhi, U., Sannasi, G., Manjula, D., Arputharaj, K.: A user preference tree based personalized route recommendation system for constraint tourism and travel. In Review (2021)
13. Zalewska-Turzyńska, M.: Gamification – an invaluable source for gaining information on the customer in the contemporary market. Handel Wewnętrzny **6**(371), 417–423 (2017)
14. Bartle, R.A.: Hearts, clubs, diamonds, spades: players who suit MUDs (1996). https://www.researchgate.net/publication/247190693
15. Shen, Y.S.: Gamification in Tourism and Hospitality: Theoretical Foundations, Player Typology, and Effectiveness, Ph D Thesis (2020)
16. Xu, F., Buhalis, D., Weber, J.: Serious games and the gamification of tourism. Tour. Manag. **60**, 244–256 (2017)
17. Ionică, A.C., Samuil, I., Leba, M., Toderas, M.: The path of Petrila mining area towards future industrial heritage tourism seen through the lenses of past and present. Sustainability **12**(23), 9922 (2020)

A Combined VIKOR-TOPSIS Method with Application in Cloud Services Provider Selection

Constanta Zoie Radulescu[1](✉), Marius Radulescu[2], and Radu Boncea[1]

[1] National Institute for Research and Development in Informatics, 8-10, Mareşal Averescu, 011455 Bucharest, Romania
radulescucz@yahoo.com, {zoie.radulescu,radu.boncea}@ici.ro

[2] "Gheorghe Mihoc-Caius Iacob" Institute of Mathematical Statistics and Applied Mathematics of the Romanian Academy, Calea 13 Septembrie, No.13, 050711 Bucharest, Romania

Abstract. In the last decades multi-criteria methods have known an important development and their area of application was significantly enlarged. Multi-criteria methods are used for making decisions that involve multiple and conflicting criteria. These methods consider a set of alternatives and a set of criteria. Each alternative is evaluated according to each criterion. As a result, a best alternative or an alternatives ranking is obtained. The selection of the appropriate multi-criteria method, for a decision, is an important problem. Two classical multi-criteria methods are TOPSIS and VIKOR, which have been proven effective through a large number of applications. Sometimes the two methods give different solutions regarding the alternatives ranking when they are starting from the same input data. In this paper we propose a new method that is a combination of these two methods. The combined VIKOR-TOPSIS method, is applied to evaluate and ranking a set of CSP (Cloud Service Provider) for selecting a best CSP, taking into account the user requirements and Quality of Service (QoS) criteria.

Keywords: Multi-criteria methods · VIKOR method · TOPSIS method · Cloud Service Provider Selection · Quality of Service criteria

1 Introduction

New multi-criteria methods have been developed, in the last decades. They were applied in various domains. Many of these methods are extensions or improvements of the existing methods [9, 11, 12].

Multi-criteria methods are used for making decisions that involve multiple and conflicting criteria. The multi-criteria methods consider a set of alternatives and a set of criteria. Each alternative is evaluated according to each criterion. As a result, a best alternative or an alternatives ranking is obtained.

The selection of the appropriate multi-criteria method is an important problem. Two classical multi-criteria methods are TOPSIS and VIKOR which have been proven effective through a large number of applications.

S. Dzitac et al. (Eds.): ICCCC 2022, AISC 1435, pp. 293–302, 2023.
https://doi.org/10.1007/978-3-031-16684-6_25

VIKOR method is a well-known compromise multi-criteria ranking method developed by Opricovic and Tzeng [6]. The notion of compromise solution was introduced in MCDM (Multiple Criteria Decision Making) by [10] and [13]. The method calculated a ranking of alternatives based on an approximation to the "ideal" alternative. The measure for compromise ranking is based on the L_p metric and uses an aggregation function [10, 13]. VIKOR method was used in several areas. A systematic review of the VIKOR method applications domains is presented in the paper [5]. In the recent paper [1] the application domains of VIKOR method are: manufacturing, construction engineering and management, performance evaluation, finance, sustainability and marketing.

The TOPSIS method was first proposed in 1981 by Hwang and Yoon [4]. TOPSIS method is based on the concepts of "ideal" and "anti-ideal" solutions. The method computes, for each alternative, the Euclidean distances from the ideal and anti-ideal solutions. The best alternative should be as close as possible to the ideal solution and also it should be as far as possible away from the anti-ideal solution.

TOPSIS has been widely applied, in the past decades, in different domains [2] and [14]. The important application areas for TOPSIS are cf. [2]: Engineering and Manufacturing Systems, Supply Chain Management and Logistics, Health, Safety and Environment Management, Business and Marketing Management, Energy Management, Water Resources Management and Human Resources Management.

These two methods introduce different metrics for ranking and use different normalization methods to eliminate the criteria units of measure. The VIKOR method uses linear normalization and the TOPSIS method uses vector normalization [6].

The most important difference between VIKOR and TOPSIS is the presence only in the VIKOR method of a user parameter υ, called utility weight. The value of υ can be selected according the decision-makers preferences.

TOPSIS is easy to understand. The implementation process is simple. Both positive and negative ideal solutions are considered. Interaction between criteria is not considered in this method [3].

VIKOR method is a solution when decision-making preferences are not clear. But this method does not work well in conflicting situations. Interaction between criteria is not considered [3].

Sometimes the application of the VIKOR and TOPSIS methods give different alternatives ranking starting from the same input data. In this paper is proposed a new method that is a combination of the two methods. The new method: combined VIKOR-TOPSIS, is applied to evaluate and ranking a set of CSP (Cloud Service Provider) for selecting a best CSP, taking into account the user requirements and Quality of Service (QoS) criteria.

The input data and the algorithm of the combined TOPSIS-VIKOR method are presented in the second section. In the third section an application of the new method to the Cloud Services Provider selection is realized. The implementation is for seven CSPs and ten criteria. An analysis on the ranks obtained with TOPSIS, VIKOR and the combined TOPSIS-VIKOR method is performed. The paper ends with the conclusion section.

2 A Combined VIKOR-TOPSIS Method

2.1 Input Data in the Combined VIKOR-TOPSIS Method

Input data are:

1. V - the set of m alternatives. Denote by $V = \{V_1, V_2, \ldots, V_m\}$.
2. C - the set of n criteria $C = \{C_1, C_2, \ldots, C_n\}$. A criterion C_i from the set C is measured using a measure unit and can be of max (benefit) or min (cost) criterion. A weight can be associated for each criterion. The n-dimensional vector of weights is denoted by $\mathbf{w} = (w_j)$. The weights usually have a numerical value in the range (0,1) and $\sum_{i=1}^{n} w_i = 1$. The weight w_j shows the importance of the criterion j.
3. E - is the evaluation matrix $E = (e_{ij}), i = 1, 2, \ldots, m; j = 1, 2, \ldots, n$.
4. e_{ij} are real numbers and denote the assessment of the i-th alternative for j- th criterion.
5. $\nu \in [0.1]$ (only in the VIKOR method) is a parameter that shows the decision-making strategy between the subjective involvement of experts and the objective evaluation. The case $\nu > 0.5$ is called "Voting by majority rule". The case $\nu = 0.5$ is called "by consensus" and the case $\nu < 0.5$ is called "with a veto".

2.2 The Algorithm of the Combined VIKOR-TOPSIS Method

Below is presented the algorithm of the combined VIKOR-TOPSIS method in steps:

Step 1. Normalization:

Normalization for the VIKOR method:

The following indicators are calculated:

$$e_j^{max} = \begin{cases} \max\limits_i e_{ij} \text{ } if \text{ } C_j \text{ is a max criterion} \\ \min\limits_i e_{ij} \text{ } if \text{ } C_j \text{ is min criterion} \end{cases} \quad (1)$$

$$e_j^{min} = \begin{cases} \min\limits_i e_{ij} \text{ } if \text{ } C_j \text{ is a max criterion} \\ \max\limits_i e_{ij} \text{ } if \text{ } C_j \text{ is a min criterion} \end{cases} \quad (2)$$

The weighted normalized matrix $\overline{\overline{E}}^V = \left(\overline{\overline{e}}_{ij}^V\right)$ is calculated. The arrays of the matrix $\overline{\overline{E}}^V$ are computed as follows:

$$\overline{\overline{e}}_{ij}^V = w_j\left(e_j^{max} - e_{ij}\right)/\left(e_j^{max} - e_j^{min}\right); i = 1, 2, \ldots, m; j = 1, 2, \ldots, n \quad (3)$$

Normalization for the TOPSIS method:

The normalized matrix $\overline{E} = (\overline{e}_{ij})$ is obtained, based on vector normalization:

$$\overline{e}_{ij} = \frac{e_{ij}}{\sqrt{\sum_{k=1}^{m}(e_{kj})^2}} \text{for max criteria} \tag{4}$$

$$\overline{e}_{ij} = 1 - \frac{e_{ij}}{\sqrt{\sum_{k=1}^{m}(e_{kj})^2}} \text{ for min criteria} \tag{5}$$

Then arrays of the weighted normalized matrix $\overline{\overline{E}}^T = \left(\overline{\overline{e}}_{ij}^T\right)$ are calculated as follows:

$$\overline{\overline{e}}_{ij}^T = w_i * \overline{e}_{ij},\ i = 1, 2, \ldots, m; j = 1, 2, \ldots, n \tag{6}$$

Step 2. The distances calculation:

For VIKOR the entries of the vectors $Q = (q_i)$ and $R = (r_i)$ are calculated as follows:

$$q_i = \sum_{j=1}^{n}\left[\overline{\overline{e}}_{ij}^V\right],\quad i = 1, 2, \ldots, m \tag{7}$$

$$r_i = \max_{1 \le j \le n}\left\{\overline{\overline{e}}_{ij}^V\right\},\quad i = 1, 2, \ldots, m \tag{8}$$

q_i represent the overall group benefit and r_i the individual deviation of i-th alternative.

TOPSIS method used the Euclidean distances:

The positive ideal and negative ideal solutions $A^+ = \left(a_i^+\right)$ and $A^- = \left(a_i^-\right), i = 1, 2, \ldots, m$ is calculated as:

$$a_j^+ = \begin{cases} \max\limits_i \overline{\overline{e}}_{ij}^T \ \textit{if}\ C_j\ \textit{is a max criterion} \\ \min\limits_i \overline{\overline{e}}_{ij}^T \ \textit{if}\ C_j\ \textit{is a min criterion} \end{cases}, j = 1, 2, \ldots, n \tag{9}$$

$$a_j^- = \begin{cases} \min\limits_i \overline{\overline{e}}_{ij}^T \ \textit{if}\ C_j\ \textit{is a max criterion} \\ \max\limits_i \overline{\overline{e}}_{ij}^T \ \textit{if}\ \textit{if} C_j\ \textit{is a min criterion} \end{cases}, j = 1, 2, \ldots, n. \tag{10}$$

For each $i = 1, 2, \ldots, m$, are calculated the vectors of the Euclidean distances from the positive and negative ideal solutions $D^+ = \left(d_i^+\right)$ and $D^- = \left(d_i^-\right)$:

$$d_i^+ = \sqrt{\sum_{j=1}^{n}\left|\overline{\overline{e}}_{ij}^T - a_j^+\right|^2} \tag{11}$$

$$d_i^- = \sqrt{\sum_{j=1}^{n}\left|\overline{\overline{e}}_{ij}^T - a_j^-\right|^2} \tag{12}$$

Step 3. The ranking calculation:
For VIKOR method the following indicators are calculated:

$$q^{min} = \min_i q_i\, ;\, q^{max} = \max_i\, q_i;\ i = 1, 2, ..., m \tag{13}$$

$$r^{min} = \min_i r_i;\, r^{max} = \max_i r_i;\, i = 1, 2, ..., m \tag{14}$$

The entries of the vector $S^V = \left(s_i^V\right)$ are calculated as follows:

$$s_i^V = \upsilon \frac{q_i - q^{min}}{q^{max} - q^{min}} + (1 - \upsilon)\frac{r_i - r^{min}}{r^{max} - r^{min}} \tag{15}$$

The choice of parameter $\upsilon \in [0, 1]$ shows the balance between the overall benefit and the maximum individual deviation. Larger values of υ emphasize the group gain, while smaller values of υ emphasize individual deviations.

Alternatives ranking. The entries of vectors S^V, Q and R are ordered in ascending order. Let α, β, γ be permutations of the set {1,2,...,m} such that:

$$s^V_{\alpha(1)} \leq s^V_{\alpha(2)} \leq \cdots \leq s^V_{\alpha(m)}$$

$$q_{\beta(1)} \leq q_{\beta(2)} \leq \cdots \leq q_{\beta(m)}$$

$$r_{\gamma(1)} \leq r_{\gamma(2)} \leq \cdots \leq r_{\gamma(m)}$$

The alternative $\alpha(1)$ is the best ranked (corresponds to the minimum value of the entries of vector S^V) in the VIKOR method if the following two conditions are fulfilled:

Condition1. "Acceptable Advantage":

$$s^V_{\alpha(2)} - s^V_{\alpha(1)} \geq \frac{1}{m-1}$$

where: the alternative $\alpha(2)$ is the second-best ranked alternative in the arrays of vector S^V ranking list.

Condition2. "Acceptable stability in decision making": The alternative $\alpha(1)$ must also be the best ranked in the list of arrays of vector Q or R, that is $\alpha(1) = \beta(1)$ or $\alpha(1) = \gamma(1)$.

If one of the above conditions is not satisfied, then a set of compromise solutions is proposed:

- Alternatives $\alpha(1)$ and $\alpha(2)$ are the best if only Condition2 is not satisfied, or
- Alternatives $\alpha(1), \alpha(2), \ldots, \alpha(k)$ are the best, if the Condition1 is not satisfied, and

$$s^V_{\alpha(i)} - s^V_{\alpha(1)} \leq \tfrac{1}{m-1} \text{ for every i } = 1, 2, \ldots, \text{k and}$$
$$s^V_{\alpha(k+1)} - s^V_{\alpha(1)} > \tfrac{1}{m-1}$$

For TOPSIS the relative closeness to the ideal solutions is computed the following vector $S^T = (s_i^T)$ as follows:

$$s_i^T = \frac{d_i^-}{d_i^+ + d_i^-}; i = 1, 2, \ldots, m \tag{16}$$

The best TOPSIS alternative k is the alternative for which $s_k^T = \max_i s_i^T$.s

For alternatives ranking, the arrays of vector S^T are sorted in descending order.

Step 4. The index of an alternative in the combined VIKOR-TOPSIS method is calculated as follows:

$$s_{max}^V = \max_i s_i^V; s_{min}^V = \min_i s_i^V \tag{17}$$

$$s_{max}^T = \max_i s_i^T; s_{min}^T = \min_i s_i^T \tag{18}$$

$$c_i^V = (s_{max}^V - s_i^V) \Big/ (s_{max}^V - s_{min}^V) \tag{19}$$

$$c_i^T = (s_i^T - s_{min}^T) \Big/ (s_{max}^T - s_{min}^T) \tag{20}$$

The arrays of the vector of the combined VIKOR-TOPSIS method $S^C = (s_i^C)$ are:

$$s_i^C = c_i^V + c_i^T, i = 1, 2, \ldots, m \tag{21}$$

The rank of the best alternative is k that is such that:

$$s_k^C = \max_i s_i^C \tag{22}$$

For alternatives ranking the arrays of the vector $S^C = (s_i^C)$ are sorted in descending order.

3 An Application of the Combined VIKOR-TOPSIS Method for the Cloud Services Provider Selection

One of the applications of multi-criteria methods is for the Cloud Service Provider (CSP) selection, see for example the recent references [7] and [8]. In the following we shall apply the combined VIKOR - TOPSIS method to the CSP selection.

A firm manager wants to identify which IaaS cloud service is suitable for his business requirements and technical environment. A set of 7 IaaS CSPs, S1 to S7 are selected by experts in Cloud computing, based on the user requirements. Ten QoS criteria, for the CSP selection, are chosen by experts: CPU Speed, Memory read, Disk read, Response time, VM cost, Support type, Service Availability, Data security, Platform security, Reputation (Table 1).

Table 1. The criteria selected

Nr. crt.	Criteria	Symbol	Criteria type1	Criteria type2	Measure Unit	Type
1	CPU Speed	C1	Dynamic	Quantitative	GHZ	max
2	Memory read	C2	Dynamic	Quantitative	MB/s	max
3	Disk read	C3	Dynamic	Quantitative	MB/s	max
4	Response time	C4	Dynamic	Quantitative	interval (Sec)	min
5	VM cost	C5	Static	Quantitative	$/Hr or $/M	min
6	Support type	C6	Static	Qualitative	[1, 4] scale	max
7	Service Availability	C7	Dynamic	Quantitative	%	max
8	Data security	C8	Static	Qualitative	[1, 10] scale	max
9	Platform security	C9	Static	Qualitative	[1, 10] scale	max
10	Reputation	C10	Static	Qualitative	[1, 10] scale	max

The values of QoS criteria for these ten CSPs considered for evaluation and analysis are presented in Table 2. This is the matrix E.

Table 2. The matrix E for the selected QoS criteria

Criteria Symbol	S1	S2	S3	S4	S5	S6	S7
C1	10.6	9	11.8	11.9	11.2	10.5	11.2
C2	13	12	15	15	14	12	13
C3	1700	2050	1500	1900	1500	1500	1700
C4	120	100	110	125	100	100	120
C5	0.94	0.96	0.95	0.95	0.96	0.97	0.96
C6	3	1	3	2	4	4	3
C7	99.8	99.9	99.89	99.87	99.89	99.9	99.95
C8	9.5	9.4	9.8	9.8	9.7	9.5	9.7
C9	9.7	9.85	9.8	9.8	9.7	9.85	9.8
C10	9.65	9.5	9.7	9.7	9.7	9.6	9.7

In our case study we shall consider the case when all criteria weights have equal values.

Based on Eqs. (7) and (8) the Q and R are calculated.

Based on Eqs. (11) and (12) the distances D^{+} and D^{-} are calculated (Table 3).

The alternatives ranking obtained with the VIKOR, TOPSIS and with the composed VIKOR-TOPSIS method are presented in Table 4.

Table 3. The VIKOR and TOPSIS distances

CSP	VIKOR		TOPSIS	
	Q	R	D^{+}	D^{-}
S1	0.9636	1.0000	0.0178	0.0262
S2	1.0000	1.0000	0.0398	0.0149
S3	0.0000	1.0000	0.0178	0.0286
S4	0.0963	1.0000	0.0266	0.0203
S5	0.3339	1.0000	0.0128	0.0396
S6	0.7050	1.0000	0.0157	0.0388
S7	0.3454	0.0000	0.0173	0.0268

Table 4. The ranks of the VIKOR, TOPSIS and combined VIKOR - TOPSIS method

CSP	VIKOR S^{V}	VIKOR ranks	TOPSIS S^{T}	TOPSIS ranks	Combined method S^{C}	Combined method ranks
S1	0.9818	6	0.596	5	0.6912	6
S2	1	7	0.2723	7	0.0000	7
S3	0.5	2	0.6166	3	1.3162	3
S4	0.5482	3	0.4321	6	0.8765	5
S5	0.6669	4	0.756	1	1.4026	2
S6	0.8525	5	0.7122	2	1.0877	4
S7	0.1727	1	0.6067	4	1.6913	1

One can easily see that the alternatives order obtained with the VIKOR method differs from the order obtained with the TOPSIS method. The alternative in the first position in VIKOR order is S7 while in TOPSIS order is S5. The common position is for alternative S2. The biggest difference between the alternatives ranks is 3 positions for the S4, S5, S6 and S7 with plus or minus. The combined VIKOR-TOPSIS method provides an order between the VIKOR and TOPSIS orders. The biggest difference between the combined VIKOR-TOPSIS method ranks and the TOPSIS ranks is for the S7 alternative (Fig. 1).

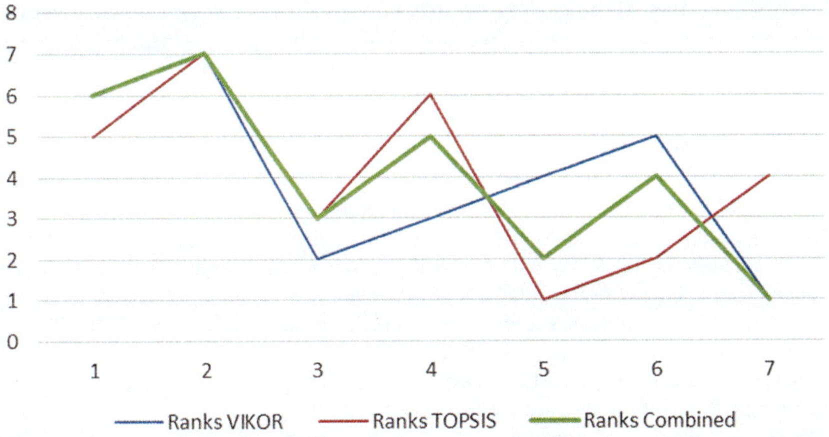

Fig. 1. The differences between the VIKOR, TOPSIS and Combined VIKOR-TOPSIS method rankings

4 Conclusions

This paper has concentrated on the alternatives ranking problem. The aim is to propose a new method which is a combination of VIKOR and TOPSIS methods in order to rank a set of alternatives. The new proposed method is simple, easy-to-use and inherit advantages of TOPSIS and VIKOR methods. This new method integrates the alternatives rankings obtained with the VIKOR and TOPSIS methods.

The VIKOR-TOPSIS method is applied to obtain a ranking of seven CSPs based on ten criteria. A comparative analysis of alternatives rankings obtained with the above methods is realized.

For future research, the proposed method may be generalized in order to be applied for an arbitrary set of multi-criteria methods. Also, a set of criteria weights, that are calculated using weighting methods, will be considered in the VIKOR-TOPSIS method.

Acknowledgements. The research reported in this paper was supported by project PN 19 37 04 01" New solutions for complex problems in current ICT research fields based on modelling and optimization", funded by the Romanian Core Program of the Ministry of Research and Innovation (MCI), 2019–2022.

References

1. Alidrisi, H.: An Innovative job evaluation approach using the VIKOR algorithm. J. Risk Financ. Manag. **14**(6), 271 (2021)
2. Behzadian, M., Otaghsara, S.K., Yazdani, M., Ignatius, J.: A state-of the-art survey of TOPSIS applications. Expert Syst. Appl. **39**(17), 13051–13069 (2012)
3. Effatpanah, S.K., et al.: Comparative analysis of five widely-used multi-criteria decision-making methods to evaluate clean energy technologies: a case study. Sustainability **14**(3), 1403 (2022)

4. Hwang, C.L., Yoon, K.: Methods for multiple attribute decision making. In: Multiple attribute decision making. Lecture Notes in Economics and Mathematical Systems, vol. 186 Springer, Berlin, Heidelberg (1981) https://doi.org/10.1007/978-3-642-48318-9_3
5. Mardani, A., Zavadskas, E.K., Govindan, K., Amat Senin, A., Jusoh, A.: VIKOR technique: a systematic review of the state-of-the-art literature on methodologies and applications. Sustainability **8**(1), 37 (2016)
6. Opricovic, S., Tzeng, G.H.: Compromise solution by MCDM methods: a comparative analysis of VIKOR and TOPSIS. Eur. J. Oper. Res. **156**(2), 445–455 (2004)
7. Radulescu, C.Z., Radulescu, M., Boncea, R., Petre, I., Sandu, I.E., Dumitrache, M.: A multi-criteria framework for cloud service providers selection based on the matter element extension method. Stud. Inform. Control **30**(1), 77–87 (2021)
8. Radulescu, C.Z., Radulescu, M., Filip, F.G.: Cloud provider selection a complex multicriteria problem. Rom. J. Inf. Sci. Technol. (ROMJIST) **24**(4), 337–352 (2021)
9. Radulescu, C.Z., Rădulescu, I.C.: An extended TOPSIS approach for ranking cloud service providers. Stud. Inform. Control **26**(2), 183–192 (2017)
10. Yu, P.L.: A class of solutions for group decision problems. Manage. Sci. **19**(8), 936–946 (1973)
11. Zavadskas, E.K., Stevic, Ž Zenonas, Turskis, U., Tomašević, M.: A novel extended EDAS in Minkowski space (EDAS-M) method for evaluating autonomous vehicles. Stud. Inform. Control **28**(3), 255–264 (2019)
12. Zavadskas, E.K., Bausys, R., Lescauskiene, I., Omran, J.: M-generalised q-neutrosophic MULTIMOORA for Decision Making. Stud. Inform. Control **29**(4), 389–398 (2020)
13. Zopounidis, C., Doumpos, M. (eds.): Multiple Criteria Decision Making. MCDM, Springer, Cham (2017). https://doi.org/10.1007/978-3-319-39292-9
14. Zyoud, S.H., Fuchs-Hanusch, D.: A bibliometric-based survey on AHP and TOPSIS techniques. Expert Syst. Appl. **78**, 158–181 (2017)

Two-Sample Permutation Test for Comparing Marginal Probabilities of Multivariate Bernoulli Distributions

Stefano Bonnini(✉) and Michela Borghesi

University of Ferrara, Ferrara, Italy
{stefano.bonnini,michela.borghesi}@unife.it

Abstract. The contribution of this paper is both methodological and applicative. A combined permutation test for two independent samples is proposed for comparing the vectors of marginal probabilities of two multivariate Bernoulli distributions. The power behavior of the test is analyzed through a Monte Carlo simulation study. The application of the method to real data concerning the propensity towards circular economy of small Italian companies in the sector of metal, contributes to the debates on the effect of firm size to the "circularity" of companies.

Keywords: Bernoulli distribution · Multivariate statistics · Permutation test · Circular economy

1 Introduction

Frequently, empirical studies investigate phenomena that can be represented by multidimensional Bernoulli variables. This is typical of surveys based on a questionnaire that includes simple questions concerning presence or absence of some characteristics or statistical analysis of categorical variables. E.g., in medical studies, the detection of a disease is often based on the presence of symptoms detected through a questionnaire administered to the patient. The customer satisfaction can be represented by a multidimensional variable of which every single component represents the satisfaction (or not) for a specific aspect of the product or service. Other examples of multivariate binary data are gene expression and the binary numeral system in informatics. In general, categorical data can be represented by several dichotomous or binary variables, each of which represents a category.

When a statistical method is based on the assumption that observed data are realizations of random variables, the underlying probability distribution in case of binary variables is the Bernoulli law. This is typical of inferential statistics (for both estimation and testing problems) [1], statistical quality control (e.g. attribute control charts for monitoring the production of defective items) [2], fuzzy set theory [3], neural networks [4], cluster analysis [5], machine learning in general [6] and other fields.

S. Dzitac et al. (Eds.): ICCCC 2022, AISC 1435, pp. 303–312, 2023.
https://doi.org/10.1007/978-3-031-16684-6_26

In the classic univariate two-sample test, where the goal consists of comparing the probabilities of two Bernoulli distributions or the proportions of two populations, sample data can be represented by the 2×2 contingency table whose rows correspond to samples and columns represent the two categories (e.g. $0 =$ failure, $1 =$ success) or vice versa. The classic inferential method to test whether the probability of success is significantly different in the two groups is the well-known chi-square test. When the alternative hypothesis is not two-sided but directional, for large samples, a Z test on the difference between two proportions can be applied whilst for small samples the Fisher exact test based on the permutation approach is a suitable solution [7].

In this paper we deal with the multivariate extension of the two sample test on proportions. In other words, for each statistical unit, a vector of $k > 1$ binary values (0 or 1) is observed and the focus is on the comparison of the vectors of probabilities of success of the multivariate response components of the two populations. When the number of variables k is large, the likelihood approach is not applicable for computational reasons. Another limit of the likelihood based inference and parametric methods in general is the need of taking into account the dependence between the k responses [9]. Hence, a nonparametric solution is required.

When the goal is to test the difference of the two vectors of probabilities without a specific direction in the alternative hypothesis, in other words for two sided tests, a solution based on the multivariate version of the Wald statistic is provided by [8]. This solution can be seen as the equivalent of the Hotelling T^2 test for the comparison of vectors of means, adapted to the case of binary variables. Under the null hypothesis of equal vectors of probabilities, for large samples the test statistic approximately follows a chi-square distribution with k degrees of freedom. An alternative version of this method, based on a simplified computation of the covariance matrix, is proposed by [9]. Given the limits of parametric methods and since asymptotic solutions are not valid because of the presence of sparse data, especially for large k and small samples, [9] identifies the permutation method as the most suitable solution for the problem.

This paper proposes a new permutation solution based on the combination of the partial tests on the marginal probabilities of success. Unlike the solutions proposed by [8] and [9], the test is also valid for directional alternative hypotheses. The paper is organized as follows: in Sect. 2 the testing problem is defined and the permutation solution is presented; Sect. 3 shows the results of a simulation study; Sect. 4 is dedicated to a case study concerning the effect of firm's size on the propensity towards circular economy for small enterprises. Section 5 includes concluding remarks.

2 Multivariate Permutation Test on Proportions

Let us consider the testing problem where two populations are compared with respect to categorical data and that the characteristics observed on the sample statistical units are represented by a multivariate dichotomous variable. For instance, in a sample survey about the propensity of small companies towards circular economy, the statistical units are firms and the variables represent specific activities of circular economy such as the reduction of the use of raw materials (1 = yes, 0 = no), the reduction of emitted waste (1 = yes, 0 = no), etc. Therefore, a vector of values equal to 0 or 1 is observed for each company.

We focus on two sample testing problems and we are interested in comparing two populations with respect to a set of proportions (proportion of companies that reduce the use of raw materials, proportion of companies that reduce the emitted waste, etc.) or, equivalently, population probabilities (probability that a company randomly selected from a population reduces the use of raw materials, probability that a company randomly selected from a population reduces the emitted waste, etc.). It is worth noting that we do not consider problems related to the joint distribution of the response variables but only problems concerning the marginal probabilities. Formally speaking, let us assume that the data refer to two independent samples randomly selected from the two compared populations and let x_{uiv} be the value about the v-th variable observed on the u-th unit of the i-th sample, with $u = 1, \ldots, n_i$, $i = 1, 2$ and $v = 1, \ldots, k$. Assume that, for every u, x_{uiv} is a realization of the Bernoulli random variable X_{iv}. We do not focus on the comparison between the joint probability distribution of $(X_{11}, \ldots, X_{1k})$ and $(X_{21}, \ldots, X_{2k})$ but rather in the differences between $P(X_{11} = 1)$ and $P(X_{21} = 1)$, $P(X_{12} = 1)$ and $P(X_{22} = 1)$, etc. Such a problem is named *simultaneous marginal homogeneity* by [9].

2.1 Testing Problem

Let ξ_{iv} denote the parameter of the Bernoulli random variable X_{iv} i.e. $P(X_{iv} = 1) = \xi_{iv}$ and $P(X_{iv} = 0) = 1 - \xi_{iv}$. Since the one-sided tests, especially in the multivariate case, are more challenging and the extension to the two sided problem trivial, we consider the test with directional alternative hypotheses (not considered by competitors for multivariate Bernoulli). Hence, the null hypothesis of the problem is

$$H_0 : \xi_{1v} = \xi_{2v} \forall v \in \{1, \ldots, k\} \tag{1}$$

whilst the alternative hypothesis is

$$H_1 : \exists v \in \{1, \ldots, k\} \text{ s.t. } \xi_{1v} < \xi_{2v} \tag{2}$$

Under the alternative hypothesis, if the strict inequality is not satisfied for some v, then, for those variables, the equality holds. Such a problem can be defined as a multivariate stochastic dominance for Bernoulli distributions because $\xi_{1v} = \xi_{2v} \iff X_{1v} \overset{d}{=} X_{2v}$ and $\xi_{1v} < \xi_{2v} \iff X_{1v} \overset{d}{<} X_{2v}$.

Some authors define the problem through the following logit model

$$\log\left(\frac{\xi_{iv}}{1 - \xi_{iv}}\right) = \beta_{iv}, \quad i = 1, 2, v = 1, \ldots, k$$

and the hypotheses are $H_0 : \beta_{1v} = \beta_{2v} \forall v$ and $H_1 : \beta_{1v} < \beta_{2v}$ for at least one v. This is a special case of a more general model that includes covariates. We do not follow this approach because it is not functional to the solution proposed in this paper and because we are not interested in predictions of the Bernoulli probabilities as functions of some explanatory variables.

We note that the solutions proposed in the literature assume independence of the distributions with respect to units. The test proposed in this paper does not assume independence but require the milder condition of between-units exchangeability in the null hypothesis. Consequently the $n_1 + n_2$ k-variate Bernoulli random variables can be dependent but the dependence structure between any sub-set of random variables does not change depending on the units considered in the sub-set.

2.2 Permutation Solution

We propose a solution based on a permutation multiple test. Actually the general testing problem can be broken down into k different partial tests, one for each dichotomous response. In the v-th partial test, the null hypothesis is $H_{0,v} : \xi_{1v} = \xi_{2v}$ and the alternative hypothesis is $H_{1,v} : \xi_{1v} < \xi_{2v}$, with $v = 1, \ldots, k$. Such a problem, given the exchangeability under H_0, can be solved with a permutation test for two independent samples, whose test statistic is the difference of the sample proportions

$$T_v = \hat{\xi}_{2v} - \hat{\xi}_{1v} = \sum_{u=1}^{n_2} \frac{x_{u2v}}{n_2} - \sum_{u=1}^{n_1} \frac{x_{u1v}}{n_1}. \tag{3}$$

A suitable statistic for the general problem can be based on the combination of the k partial p-values, provided that a suitable combining function ψ exists. According to [10] and assuming, without loss of generality, that H_0 is rejected for large values of the test statistic, such combining function should satisfy the following simple and common properties: (1) it should be non-decreasing function of the p-values, (2) it should attain its (possible not finite) supremum when at least one p-value tends to 0, (3) for every significance level α, the critical value of the test t_α is finite and less than the supremum, i.e. H_0 is rejected if $T_\psi \geq t_\alpha$ and $t_\alpha < \psi^{\text{sup}}$, where T_ψ is the combined test statistic.

Let $l_1, \ldots, l_k$ be the p-values of the partial tests and $T_\psi = \psi(l_1, \ldots, l_k)$ the combined test statistic. The procedure consists of the following steps:

1. Compute the vector of the observed values of the partial test statistics
$$\mathbf{t_{obs}} = (t_{1,\text{obs}}, \ldots, t_{k,\text{obs}}) = t(\mathbf{X}).$$
2. Generate B random permutations of the rows of the dataset, i.e. generate random assignments of the vectors of values observed on the statistical units to sample 1 and sample 2, B times, obtaining B permuted datasets: $\mathbf{X}_1^*, \ldots, \mathbf{X}_B^*$.
3. For each permuted dataset, compute the corresponding values of the test statistics
$$\mathbf{t_b^*} = (t_{1,b}^*, \ldots, t_{k,b}^*) = t(\mathbf{X_b^*}), \quad b = 1, ..., B$$

4. For each permuted dataset, compute the significance level functions of the partial tests
$$\mathbf{l}_\mathbf{b}^* = (l_{1,b}^*, \ldots, l_{k,b}^*), \quad b = 1, ..., B$$
where $l_{v,b}^* = L(t_{v,b}^*) = \left[\sum_{c=1}^{B} I_{(-\infty, t_{v,c}^*]}(t_{v,b}^*) + 0.5\right] / (B+1)$ is the estimated probability that, under H_0, the v-th partial test statistic, according to the permutation distribution, takes values greater than or equal to $t_{v,b}^*$, formally $\hat{P}(T_v \geq t_{v,b}^* | H_0)$ and $I_A(x) = 1$ if $x \in A$ and 0 if $x \neq A$.
5. Compute the vector of p-values of the partial tests
$$\mathbf{l}_{\mathbf{obs}} = (l_{1,\text{obs}}, \ldots, l_{k,\text{obs}}),$$
where $l_{v,obs} = L(t_{v,\text{obs}}) = \hat{P}(T_v \geq t_{v,\text{obs}} | H_0)$.
6. Compute the permutation distribution of the combined test statistic
$$t_{\psi,b}^* = \psi(l_{1,b}^*, \ldots, l_{k,b}^*) \text{ and } t_{\psi,\text{obs}} = \psi(l_{1,\text{obs}}, \ldots, l_{k,\text{obs}})$$
7. Compute the p-value of the combined test, according to the permutation distribution
$$l_{\psi,\text{obs}} = L(t_{\psi,\text{obs}}) = \hat{P}(T_\psi \geq t_{\psi,\text{obs}} | H_0).$$

As usual, if $l_{\psi,obs} \leq \alpha$ then H_0 must be rejected in favor of H_1 otherwise H_0 is not rejected.

The set of possible combining functions is huge. Two suitable choices are the Fisher combination $T_F = -2\sum_{v=1}^{k} \log(l_v)$ and the Tippett combination $T_T = \max_{v=1,\ldots,k}(1 - l_v)$.

3 Simulation Study

In order to analyze the power behavior of the proposed permutation tests, we carried out a Monte Carlo simulation study. For the implementation of the simulations, specific *R* scripts were created by the authors. For each setting, we simulated 1000 datasets and, for each dataset, we considered 1000 permutations for the computation of p-values. We studied the power as a function of the sample sizes, of the number k of components of the multivariate response and the proportion p of components for which the alternative hypothesis is true, i.e. the proportions of true partial alternative hypotheses among the k of the multiple test. With regard to the number of variables, the two cases $k = 5$ and $k = 10$ were considered.

Data were randomly generated from a multivariate normal distribution and then dichotomized according to the probabilities of success ξ_{1v} and ξ_{2v}. The probability of success for the j-th population ($j = 1, 2$) was assumed to be equal for all the k variables: $\xi_{1v} = \xi_1 \ \forall v$ and $\xi_{2v} = \xi_2 + \delta \ \ \forall v$. For ξ_1, the low value 0.03 was chosen to simulate the challenging situation of sparse data, in the sense of low number of non-zero-values. The values for the "shift" parameter are $\delta = 0$ under H_0 and $\delta = 0.05$ under H_1. In the simulation, the dependence between the k variables was taken into account by assuming correlated normal random variables. We assumed the same value ρ for the Pearson correlation index of any pair of response variables. The results of the simulation study reported in Table 1 and Table 2 of this paper refer to the moderate, non-null and realistic case of $\rho = 0.3$. We took into account balanced samples, i.e. $n_1 = n_2 = n$.

Hence, for each statistical unit, a vector of simulated data $(\tilde{z}_{uj1}, \ldots, \tilde{z}_{ujk})$, $u = 1, \ldots, n_j, j = 1, 2$, was randomly generated from a k-variate normal distribution $Z \sim \mathcal{N}_k(\mathbf{0}, \mathbf{\Sigma})$, where $\mathbf{\Sigma} = [\sigma_{vr}]$ is the covariance matrix, with $\sigma_{vr} = \rho$ if $v \neq r$ and $\sigma_{vr} = 1$ if $v = r$. Then, the z-values were transformed into binary data according to the following rule:

$$\tilde{x}_{ujv} = \begin{cases} 1 & \text{if } \tilde{z}_{ujv} \leq z_{1-\xi_j} \\ 0 & \text{if } \tilde{z}_{ujv} > z_{1-\xi_j} \end{cases}$$

where $z_{1-\xi_j}$ is the quantile of the standard normal distribution with cumulative probability ξ_j.

In Table 1 the rejection rates, in the null hypothesis, of the permutation tests based on Fisher's and Tippett's combination are shown. The probability of wrong rejection of the null hypothesis is less than or very close to the nominal significance α level 0.05. This is true for both the tests and it holds for all the considered sample sizes and for both the cases of 5 and 10 response components. Hence, the tests are well approximated.

Table 1. Rejection rates of the tests for multivariate Bernoulli distributions, with Fisher (T_F) and Tippett (T_T) combination, under H_0 with $\rho = 0.3$, $\xi_1 = 0.3$ and $\delta = 0$.

n	k	p	T_F	T_T	n	k	p	T_F	T_T
20	5	0	0.043	0.050	20	10	0	0.048	0.057
40			0.047	0.046	40			0.049	0.051
60			0.057	0.048	60			0.046	0.043
80			0.046	0.052	80			0.044	0.042
100			0.047	0.044	100			0.048	0.029

In Table 2 the rejection rates under the alternative hypothesis with $\delta = 0.05$ are reported. Since the rejection rates are greater than α, then the tests are unbiased. The results in the table prove that the power is increasing function of the sample sizes, hence the test is consistent. The permutation test based on the combination of Fisher seems to be uniformly more powerful than the permutation test based on the combination of Tippett. Indeed, the rejection rates of the former are greater than those of the latter for all the considered sample sizes, number of variables and proportion of variables under the alternative hypothesis. When the number of responses is 10 the power of the test based on Fisher's combination is slightly greater than in the case of 5 responses. This is not as obvious for the test based on Tippett's combination. For both the permutation tests, the effect of the proportion p of true partial alternative hypotheses on the power is not clear. The expected increasing relationship typical of some combined permutation tests here is not evident.

Table 2. Rejection rates of the tests on multivariate Bernoulli distributions, with Fisher (T_F) and Tippett (T_T) combination, under H_1 with $\rho = 0.3$, $\xi_1 = 0.3$ and $\delta = 0.05$.

n	k	p	T_F	T_T	n	k	p	T_F	T_T
20	5	1.0	0.154	0.139	20	10	1.0	0.147	0.122
40			0.202	0.145	40			0.222	0.156
60			0.269	0.195	60			0.263	0.170
80			0.292	0.194	80			0.355	0.202
100			0.329	0.206	100			0.426	0.270
20		0.8	0.123	0.117	20		0.8	0.159	0.081
40			0.205	0.125	40			0.238	0.143
60			0.268	0.173	60			0.292	0.206
80			0.297	0.220	80			0.347	0.225
100			0.339	0.244	100			0.414	0.243
20		0.6	0.140	0.100	20		0.6	0.176	0.124
40			0.173	0.140	40			0.225	0.147
60			0.206	0.149	60			0.305	0.185
80			0.292	0.205	80			0.354	0.229
100			0.328	0.207	100			0.406	0.230
20		0.4	0.137	0.118	20		0.4	0.156	0.117
40			0.204	0.145	40			0.231	0.152
60			0.260	0.181	60			0.301	0.186
80			0.308	0.211	80			0.362	0.209
100			0.349	0.256	100			0.420	0.264
20		0.2	0.122	0.106	20		0.2	0.148	0.100
40			0.209	0.167	40			0.235	0.129
60			0.249	0.173	60			0.278	0.154
80			0.288	0.203	80			0.331	0.207
100			0.350	0.240	100			0.432	0.269
					20		0.1	0.150	0.096
					40			0.208	0.149
					60			0.291	0.178
					80			0.333	0.218
					100			0.405	0.248

4 Application Problem

Let us consider a case study concerning the propensity of Italian Small and Medium Enterprises (SMEs) towards Circular Economy (CE). In particular, we want to test the effect on such propensity of firm size within population of over six years old enterprises. In fact, young firms (six or less than six years old) seem to be not much involved in activities related to CE [14]. CE is an opportunity and a strategy for providing environmental sustainability, reduction of waste and pollution and environment protection. According to [11], SMEs are approaching circular economy more slowly than others. Some contributions show empirical evidence in favor of the hypothesis that the smaller the firm the lower the propensity to CE [12,13].

We want to contribute to the empirical literature by proposing a new methodological approach based on the presented testing procedure. Empirical literature on CE is not much focused on micro and small enterprises, typical of the economic systems of some developed countries such as Italy. Hence, the goal is to compare micro and small enterprises to test whether the propensity to undertake CE activities of the former is less than that of the latter. In particular we take into account the metal sector, which plays a central role in the recovery and recycling of post-consumer materials. The decision to focus only on older companies in the metal sector, also satisfies the need of comparing homogeneous companies with respect to age and economic sector. This choice allows to overcome a typical limit of the existing empirical literature. Indeed, the confounding effect of age and sector is usually ignored. Another complex aspect of the problem concerns the multivariate nature of the response. To date, a rigorous, unique and objective definition of CE is missing but it is common opinion that CE is characterized by different aspects and therefore it is a multidimensional concept. Each partial aspect can be linked to a specific CE activity and can be represented by a dichotomous variable that takes value 1 if a firm undertakes that activity and 0 otherwise.

Data refer to a sample survey carried out in January 2020. A sample of Italian SMEs were interviewed through the CATI method. The firms considered in this paper consists of a sample of 32 micro enterprises (less than 10 employees) and a sample of 350 small enterprises (form 10 to 49 employees) in the metal production and processing sector and with more than six years of activity. The multivariate response consists of the following six binary variables representing six different CE activities (1 = yes, 0 = no):

1. RED_MAT: reduction of the use of raw materials
2. RED_WASTE_1: reduction of emitted waste (per unit of produced output)
3. RED_WASTE_2: reduction of waste in the production cycle
4. CON_WASTE: conferral of waste to other companies to be used in the production cycle
5. DES_RYC: change in the design of products to maximize their recyclability
6. INV_R&D: investments in R&D aimed at reducing the environmental impact of production

In Table 3, the sample proportions of "circular" companies (companies that under-take the specific listed CE activities) within the groups of micro and small enterprises are reported. The proportions range from 0.000 (micro enterprises that invest in R&D to reduce the environmental impact of production) to 0.200 (small enterprises that decided to reduce the emitted waste in the production output). For every activity, the proportion referred to the micro enterprises is smaller than that referred to the small enterprises. The only exception concerns the decision of changing the design of products to maximize their recyclability.

The permutation test with Fisher combination was applied to test the equality of the proportions versus the hypothesis that the proportion of circular micro enterprises is less than the proportion of circular small enterprises at the significance level $\alpha = 0.10$. The global p-value is equal to 0.068. Hence, the null hypothesis is rejected in favor of the hypothesis that the propensity to CE of micro enterprises is lower. The p-values of the univariate partial tests were adjusted with a closed testing method [10] to control the familywise error rate of the test and avoid the inflation of the I type error rate due to the multiplicity. According to the adjusted p-values, the significance of the global test must be attributed to the difference of the proportions of companies that invest in R&D for reducing the environmental impact of production. For all the other variables, the differences in the proportions are not significant.

Table 3. Sample proportions of circular companies in the groups of micro and small enterprises and p-values of the univariate two-sample tests for proportion comparisons.

Variable	Proportion of circular micro ent.	Proportion of circular small ent.	Unadjusted p-value	Adjusted p-value
RED_MAT	0.063	0.149	0.089	0.260
RED_WASTE_1	0.094	0.200	0.069	0.181
RED_WASTE_2	0.094	0.106	0.435	0.720
CON_WASTE	0.156	0.174	0.416	0.720
DES_RYC	0.031	0.029	0.567	0.720
INV_R&D	0.000	0.094	0.022	**0.088**

5 Conclusions

The combined permutation tests for two independent samples proposed for comparing the vectors of marginal probabilities of two multivariate Bernoulli distributions or simply the multivariate extension of the test on proportions seems to be a valid solution for multivariate problems in presence of binary variables. The test based on Fisher's combination seems to be more powerful than the test based on Tippett's combination. The test is unbiased, consistent and powerful. It doesn't require an asymptotic distribution of the test statistic and is valid also for small sample sizes and large number of variables. According to the results of the case study, micro enterprises seem to have a lower propensity towards circular economy than small enterprises. This is due to the lower propensity to invest in R&D for reducing the environmental impact of production.

Acknowledgments. Authors thank the Italian Ministry of Education, University and Research that funded the departmental development program (DEM - University of Ferrara) for the period 2018–2022, to promote excellence in education and research ("Dipartimenti di Eccellenza")

References

1. Casella, G., Berger, R.L.: Statistical Inference, 2nd edn. Duxbury, Pacific Grove (CA) (2002)
2. Montgomery, D.C., Runger, G.C.: Applied Statistics and Probability for Engineers, 6th edn. Wiley, Hoboken (NJ) (2014)
3. Lofti, A.Z.: Discussion: probability theory and fuzzy logic are complementary rather than competitive. Technometrics **37**, 271–276 (2012)
4. Peters, J.W.T., Welling, M.: Probabilistic binary neural networks. arXiv preprint arXiv:1809.03368 (2018)
5. Ye, M., Zhang, P., Nie, L.: Clustering sparse binary data with hierarchical Bayesian Bernoulli mixture model. Comput. Stat. Data Anal. **123**, 32–49 (2018)
6. Saeed, M., Javed, K., Babri, H.A.: Machine learning using Bernoulli mixture models: clustering, rule extraction and dimensionality reduction. Neurocomputing **119**, 366–374 (2013)
7. Bonnini, S., Corain, L., Marozzi, M., Salmaso, L.: Nonparametric Hypothesis Testing. Rank and Permutation Methods with Applications in R. Wiley, Chichester (2014)
8. Chuang-Stein, C., Mohberg, N.: A unified approach to the analysis of safety data in clinical trials. In: Sogliero-Gilbert, G. (ed.) Drug Safety Assessment in Clinical Trials. Dekker, New York (1993)
9. Agresti, A., Klingenberg, B.: Multivariate tests comparing binomial probabilities, with application to safety studies for drugs. Appl. Stat. **54**(4), 691–706 (2005)
10. Pesarin, F., Salmaso, L.: Permutation Tests for Complex Data. Theory, Applications and Software. Wiley, Chichester (2010)
11. Yadav, N., Gupta, K., Rani, L., Rawat, D.: Drivers of sustainability practices and SMES: a systematic literature review. Eur. J. Sustain. Develop. **7**(4), 531–544 (2018)
12. Bassi, F., Dias, J.G.: Sustainable development of small- and medium-sized enterprises in the European Union: a taxonomy of circular economy practices. Bus. Strateg. Environ. **29**, 2528–2541 (2020)
13. Ghisetti, C., Montresor, S.: On the adoption of circular economy practices by small and medium size enterprises (SMEs): does"financing-as-usual" still matter? J. Evol. Econ. **30**, 559–586 (2020)
14. Bassi, F., Dias, J.G.: The use of circular economy practices in SMEs across the EU. Conserv. Recycl. **146**, 523–533 (2019)

Measuring Consensus in Group Decision-Making Problems Through an Inequality Measure

J. M. Tapia[1(✉)], F. Chiclana[2], M. J. del Moral[3], and E. Herrera–Viedma[4]

[1] Department of Quantitative Methods in Economic and Business, University of Granada, 18071 Granada, Spain
jmtaga@ugr.es

[2] DIGITS, Department of Informatics, Faculty of Technology, De Montfort University, Leicester LE1 9BH, UK
chiclana@dmu.ac.uk

[3] University of Granada, 18071 Granada, Spain
delmoral@ugr.es

[4] Department of Computer Science and A.I, University of Granada, 18071 Granada, Spain
viedma@decsai.ugr.es

Abstract. Gini index, a measure of statistical dispersion intending to represent inequality within a group, used mainly in economics, becomes in this paper a tool to introduce a new index to measure the level of consensus in Group Decision Making problems. An empirical study reveals that the levels of consensus obtained by this index are similar to those derived through the use of a distance function when fuzzy preference relations are considered. The results obtained suggest that this new index can be satisfactorily used to measure the degree of consensus in this framework.

Keywords: Group decision making · Fuzzy preferences · Consensus · Inequality · Gini index · Distance functions

1 Introduction

Faced with a certain problem that requires making a final decision, the group of people who must decide to reach that decision have to choose the best alternative or solution among all the possible ones. This is the idea that underlies the known as Group Decision Making (GDM) problems. It is desirable that these people, usually called experts, agree on their decision and thus reach what is usually called consensus.

Although consensus is often used to represent complete agreement among experts, it is not necessary for all experts to agree on exactly the same alternative to make a decision [1]. This way the concept of soft consensus arises allowing us to measure the degree of agreement as a value between total agreement and null agreement. By using soft consensus, different levels of agreement among experts can be expressed in a decision-making problem. Various measures can be used to calculate this soft consensus.

S. Dzitac et al. (Eds.): ICCCC 2022, AISC 1435, pp. 313–319, 2023.
https://doi.org/10.1007/978-3-031-16684-6_27

The concept of similarity among the preferences of the experts explains the use of different procedures to be able to measure the consensus degree among them. Traditionally, the calculation of consensus levels involves computing and aggregating the distance measures used to represent the proximity among the preferences of each pair of experts over each pair of alternatives [3, 4]. These calculations can be complicated depending on the aggregator used and may require prior construction of the collective preference relation.

Alternative measures based on statistical variability have been used to measure consensus [5, 6]. Inequality measures such as entropy [7] or Gini index [8] have been used in aggregation methods to determine consensus in GDM with linguistic variables and distance-based process among individual assessments. Gini index is a measure of statistical dispersion intended to represent inequality within a group. It was proposed by Corrado Gini [9], who defined this index by using relative mean absolute difference in a given data set. Mainly used in economics, it is customary to define it mathematically in terms of the Lorenz curve, which plots two variables on Cartesian axes.

Following the aforementioned idea, in this paper we introduce a new index of consensus degree that allows us to measure soft consensus among the preferences of the experts in a context of GDM problems with fuzzy preference relations. This new index would allow the calculation of consensus without using distance measures by replacing them with a measure of inequality among individual preferences generated from the concept introduced by Gini. The index does not depend on the different aggregation criteria that could be applied to obtain a collective relation used to find solutions in GDM problems as those introduced above. So, the implementation of this new index could successfully solve, in a straightforward way, problems that have traditionally been resolved by using both distance measures and aggregation operators.

We compare this new consensus measure with a frequently used approach based on an aggregation operator and the Manhattan distance function [4] and interesting results have been found.

This paper is structured as follows: Sect. 2 introduces the theoretical framework in which this study is developed: basic concepts and elements of GDM problems. The proposed index is introduced in Sect. 3. Section 4 is dedicated to the experimental study carried out: the design, the conditions under which it was developed, as well as the results obtained. Finally, the Conclusion section ends this article.

2 Consensus Measurement in GDM Problems

Let us consider a group of experts $E = \{e^1, \ldots, e^n\}$ and a set of possible alternatives $X = \{x_1, \ldots, x_m\}$ $(m, n > 1)$. A GDM problem consists of finding the best alternative from X according to the preferences of the experts E. In this situation it is quite interesting to consider the preference relations as fuzzy preference relations [1, 10].

A fuzzy preference relation on a finite set of alternatives X is defined by a function F such that $F(x_i, x_j) = p_{ij}$ denotes the preference degree of the alternative x_i over x_j given by an expert [3, 5], denoting 0 the minimal or null preference and 1 the maximal or total preference. F verifies the known as reciprocity property, i.e. $p_{ij} + p_{ji} = 1$, i, j in $\{1, \ldots, m\}$. These relations are usually expressed using matrix notation, $P = (p_{ij})$.

In order to calculate the consensus degree, a consensus matrix, $CM = (cm_{ij})$, is introduced by aggregating all similarity matrices previously derived through an Ordered Weighted Averaging (OWA) operator with weighting vector W [11, 12]. Maximum ($W = [1, 0, \ldots, 0]$), Minimum ($W = [0, \ldots, 0,1]$) or Average ($W = [1/n, \ldots, 1/n]$) are some examples of OWA operators. Alternative representations for the concept of fuzzy majority can be found in the literature [12].

Calculating the level of consensus among experts involves measuring the distance among their preference values [13]. In this calculation it is necessary to use a distance function. The Manhattan, Euclidean or Cosine distance functions are frequently used [4, 13]. In Sect. 3 we use the Manhattan distance function as a tool to introduce the new proposed consensus index. Given $A = \{a_1, \ldots, a_n\}$ and $B = \{b_1, \ldots, b_n\}$ two sets of real numbers, the Manhattan distance function for A and B is given by

$$Ma(A, B) = \sum_{i=1}^{n} |a_i - b_i| \tag{1}$$

Any of the distance functions could be used to find the similarity between preference values through the similarity function by setting similarity as $s = 1 - d$ [5].

The similarity matrix for every expert r, $SM^r = (sm^r{}_{ij})$, with $sm^r{}_{ij} = s(p^r{}_{ij}, p_{ij})$ provide an evaluation of the proximity among preference values comparing each expert with the rest. The proximity is obtained for each pair of alternatives (x_i, x_j). Then, the consensus matrix $CM = (cm_{ij})$, with i,j in $\{1,\ldots,m\}$, is obtained as an aggregation of $sm^k{}_{ij}$ with $k = 1,\ldots,n$. CM shows the consensus degree on each pair of alternatives (x_i, x_j) by means of cm_{ij}.

Finally, the consensus degree on the relation, c_r, i.e. the global agreement among all experts, is obtained by aggregating all the consensus degrees at the level of pairs of alternatives.

3 A Gini-Based Consensus Index

There are several measures of statistical dispersion. One of them was proposed by Corrado Gini [9] intending to represent inequality within a group. Mainly used in economics, the so called Gini index is a measure of the variability between a population variable and an income variable inside a given data set, as can be seen in papers on equality/inequality by Czarny [14] or Drobotya et al. [15].

This concept could be used in a GDM context in which fuzzy preference relations are considered, in order to measure agreement or disagreement. Thus, we introduce the Gini-based index for consensus measurement as follows.

Let $\{p_{ij}{}^1,\ldots, p_{ij}{}^n\}$ be the preferences of n experts on a pair of alternatives (x_i, x_j) with i,j in $\{1,\ldots, m\}$. The Gini-based index on each pair of alternatives (x_i, x_j) (IG) is defined as

$$IG(x_i, x_j) = IG_{ij} = \frac{\sum_{k=1}^{n-1} (g_k - q_k)}{\sum_{k=1}^{n-1} g_k} = 1 - \frac{\sum_{k=1}^{n-1} q_k}{\sum_{k=1}^{n-1} g_k}$$

where g_k and q_k a obtained by

$$g_k = \frac{k}{n} \quad \& \quad q_k = \begin{cases} 0, & for \sum_{l=1}^{n} p_{ij}^{l} = 0; \\ \frac{\sum_{r=1}^{k} p_{ij}^{\sigma(r)}}{\sum_{l=1}^{n} p_{ij}^{l}}, & elsewhere. \end{cases}$$

and $\sigma(r)$ is a permutation function such that

$$p_{ij}^{\sigma(s)} \geq p_{ij}^{\sigma(s+1)}, \ \forall s \in \{1, ..., n-1\}$$

To calculate consensus degree on the relation the Gini-based consensus index (C_{IG}) is defined as

$$C_{IG} = \frac{\sum_{i=1}^{m} \sum_{j>i}^{m} CIG_{ij}}{\sum_{k=1}^{m-1} (m-k)} \tag{2}$$

with

$$CIG_{ij} = 1 - Minimun[IG_{ij}, IG_{ji}]$$

4 Comparative Study

In what follows we perform a comparison between a traditional way of measuring the degree of consensus in a GDM problem through a distance function (1) and the way derived from the use of the consensus index (2) proposed above.

To do this, we generated a total of 600 random GDM problems,100 for each of the combinations between alternatives and experts, with 3, 4 and 5 being the alternatives considered and 2 and 4 the number of experts, that is, 100 GDM problems for 3 alternatives and 2 experts, 100 for 4 alternatives and 4 experts, and so on.

Associated with the Manhattan distance function, we have used Average as the OWA aggregation operator, being $w = [1/3, 1/3, 1/3]$ the weighting vector.

This comparative study attempts to answer the question:

> "Can the value obtained by the Gini-based index (C_{IG}) be used to calculate the degree of consensus in a GDM problem instead of the one provided by the use of the Manhattan distance function (Ma)?"

This question is linked in turn to two other questions:

> "Is the value obtained by $C_{IG} - Ma$ a monotonic function, i.e. $C_{IG} > Ma$ ($C_{IG} < Ma$) always or almost always, or does it fluctuate, i.e. sometimes $C_{IG} > Ma$, sometimes $C_{IG} < Ma$, and sometimes $C_{IG} = Ma$?"

and

"In the case of monotonicity, i.e. $C_{IG} > Ma$ or $C_{IG} < Ma$, how much greater is C_{IG} over Ma or vice versa?"

The answer to the first linked question is reflected in Table 1. In almost all cases, the value of the Gini-based index is greater than the value obtained through the use of the Manhattan distance function. In the case of 3 alternatives and 4 experts there are 4% of situations where $C_{IG} < Ma$ and the difference is small (less than 4%). Thus we can assume that the difference is a monotonic function and C_{IG} is greater than Ma.

Table 1. Percentage of C_{IG} values greater than Ma values.

Alt/Exp	2	4
3	100%	94%
4	100%	100%
5	100%	100%

Table 2 shows the approximate difference between C_{IG} and Ma values with $C_{IG} > Ma$, in percentages. It can be seen that the number of alternatives seems to affect this difference very slightly for 2 experts and even more for 4 experts, while the number of experts does seem to influence significantly (10% vs 18%).

Table 2. Increase in the degree of consensus of C_{IG} over Ma (in percentage).

Alt/Exp	2	4
3	11%	16%
4	10%	19%
5	10%	20%
Average	10%	18%

Figures 1 and 2, show the level of consensus (in percentage) achieved in the two cases analysed. The higher the value of the consensus degrees, the higher the global consensus degree. The results show the relative positions of the proposed Gini-based index versus the one derived from the use of the Manhattan distance function.

This lay bare, as occurs in Tables 1 and 2, that this new index could be used as a measure of the degree of consensus in GDM problems.

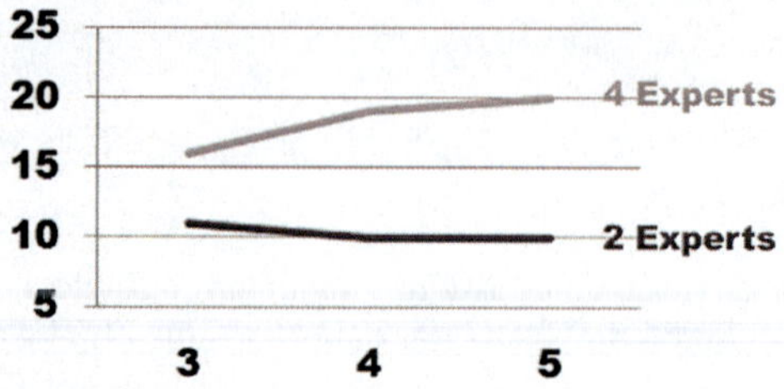

Fig. 1. Increase in the degree of consensus of C_{IG} over *Ma* for 2 and 4 experts (in percentage).

Figure 1 also shows a growth trend in the differences between C_{IG} and *Ma* according to the number of alternatives for 4 experts. Figure 2 clearly shows the indicated difference according to the number of experts considered in the GDM problem.

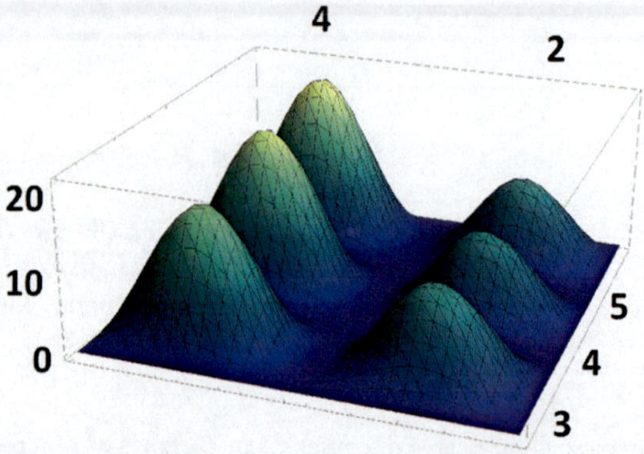

Fig. 2. Increase in the degree of consensus of C_{IG} over *Ma* for 2 and 4 experts (in percentage).

5 Conclusion

In this study we have proposed an index to assess consensus based on a measure of inequality within a group, the Gini-based consensus index. We have compared this new index with the one derived from considering a well-known distance function -the Manhattan distance function-. The empirical study carried out shows interesting results regarding the consensus behavior of the proposed index, being monotonicity one of them. The new C_{IG} index almost always exceeds *Ma*, in an average percentage that depends on the number of experts and alternatives considered, between 10% and 20%, which should be taken into account when interpreting this C_{IG} index.

The possibility of expanding the study with a greater number of experts and alternatives, as well as the comparison with other consensus measures, becomes an interesting question to address in the future.

Acknowledgments. Supported by both the project PID2019-103880RB-I00 funded by MCIN/AEI/https://doi.org/10.13039/501100011033 and the project number P20 00673 funded by the Andalusian Government.

References

1. Herrera-Viedma, E., Cabrerizo, F.J., Kacprzyk, J., Pedrycz, W.: A review of soft consensus models in a fuzzy environment. Inf. Fusion **17**, 4–13 (2014)
2. Del Moral, M.J., Tapia, J.M., Chiclana, F., Al-Hmouz, A., Herrera-Viedma, E.: An analysis of consensus approaches based on different concepts of coincidence. J. Intell. Fuzzy Syst. **34**(4), 2247–2259 (2018)
3. Cabrerizo, F.J., Moreno, J.M., Perez, I.J., Herrera-Viedma, E.: Analyzing consensus approaches in fuzzy group decision making: advantages and drawbacks. Soft. Comput. **14**(5), 451–463 (2010)
4. Del Moral, M.J., Tapia, J.M., Chiclana, F., Herrera-Viedma, E.: Comparing two approaches for consensus computation in group decision making problems. Front. Artif. Intell. Appl. **303**, 312–320 (2018)
5. Akiyama, Y., Nolan, J., Darrah, M., Rahem, M.A., Wang, L.: A method for measuring consensus within groups: an index of disagreement via conditional probability. Inf. Sci. **345**, 116–128 (2016)
6. Del Moral, M.J., Chiclana, F., Tapia, J.M., Herrera-Viedma, E.: An alternative calculation of the consensus degree in group decision making problems. In: Fifth International Conference on Information Technology and Quantitative Management, New Delhi (2017)
7. Lo, C.C., Wang, P.: Using fuzzy distance to evaluate the consensus of group decision-making – an entropy-based approach. In: IEEE International Conference on Fuzzy Systems, Budapest, pp. 1001–1006 (2004)
8. Falco, E., Garcia-Lapresta, J.L., Rosello, L.: Allowing agents to be imprecise: a proposal using multiple linguistic terms. Inf. Sci. **258**, 249–265 (2014)
9. Gini, C.: Measurement of Inequality of Incomes. Econ. J. **31**(121), 124–126 (1921)
10. Pedrycz, W., Parreiras, R., Ekel, P.: Fuzzy Multicriteria Decision-Making. Models, Methods and Applications. Wiley, Chichester (2011)
11. Zadeh, L.A.: A computational approach to fuzzy quantifiers in natural languages. Comput. Math. Appl. **9**(1), 149–184 (1983)
12. Yager, R.R.: On ordered weighted averaging aggregation operators in multicriteria decision making. IEEE Trans. Syst. Man Cybern. **18**(1), 183–190 (1988)
13. Deza, M.M., Deza, E.: Encyclopedia of Distances. Springer, Heidelberg (2009)
14. Czarny, B., Czarny, E.: Efficiency and equity - the Swedish economy in comparison to other countries at the beginning of the 21st century. Int. J. Manag. Econ. **57**(3), 255–267 (2021)
15. Drobotya, Y., Kulinich, T.: Overcoming poverty and social inequality in third world countries (Latin America, Africa). Int. J. Comput. Sci. Netw. Secur. **21**(3), 295–303 (2021)

Some Trends in Fuzzy Decision Making

J. R. Trillo[1(✉)], F. J. Cabrerizo[1], F. Chiclana[2], M. A. Martínez[3], and E. Herrera-Viedma[1]

[1] Department of Computer Science and Artificial Intelligence, Andalusian Research Institute of Data Science and Computational Intelligence, DaSCI, University of Granada, 18071 Granada, Spain
jrtrillo@ugr.esm, {cabrerizo,viedma}@decsai.ugr.es

[2] Faculty of Computing, Engineering and Media, Institute of Artificial Intelligence, Montfort University, Leicester, UK
chiclana@dmu.ac.uk

[3] Department of Social Work and Social Services, University of Granada, 18071 Granada, Spain
mundodesilencio@ugr.es

Abstract. Decision making is a line of research that has been consolidating since its beginnings in the 1960s. The aim of this paper is to show the evolution and future challenges of this line of research, focusing especially on its evaluation and information methods. For this purpose, some issues and trends of the fuzzy decision system are presented. By doing so, it is possible to show which trend fuzzy decision systems will follow and the challenges that may arise.

Keywords: Group decision making system · Linguistic decision making · Multi-criteria decision making system · Fuzzy number assessment · Preference relations

1 Introduction

Decision Making is a process that people go through on a daily basis. This process consists of selecting one option from a set of alternatives. To facilitate this choice, Decision Making systems have been developed. These systems can be grouped according to different environments, as those that process the knowledge of the alternatives to be analysed and the environments that process the uncertainty of the knowledge of the alternatives are evaluated in an approximate and non-probabilistic way. Within the latter type of environment, we find Fuzzy Decision Making, that it analyses the imprecise information of the alternatives. The analysis of fuzzy theory was developed in 1965 by the researcher L.A. Zadeh [41] and he and R.E. Bellman developed a model based on Decision Making in 1970 [3].

From the 1970s s to the present, the Fuzzy Decision Making line of research has gained popularity because it captures the subjectivity that people have when solving a problem. Contributions in this field range from the representation of

S. Dzitac et al. (Eds.): ICCCC 2022, AISC 1435, pp. 320–332, 2023.
https://doi.org/10.1007/978-3-031-16684-6_28

assessment information generated by people, such as fuzzy number-based assessment [10] or linguistic assessment [45], to the creation of systems, such as Group Decision Making (GDM) methods [30] or Multi-criteria Decision Making methods [18].

This article aims to show the advances that have been made in the research on Fuzzy Decision Making systems and some of the future challenges of this line of research. Furthermore, this article will show a broad view of the state of the art of this line of research, being a reference when analysing the existing trend in the field.

This document is divided into the following parts. In Sect. 2, the different forms of representation of assessment information will be presented. Therefore, this section will have a subsection for each of the existing types. In Sect. 3, the different types of models that use information to generate the choice of alternatives will be presented. This section will also contain different subsections, one for each type of model. In Sect. 4, we will outline some of the challenges of the Fuzzy Decision Making research line. Finally, in Sect. 5, conclusions will be drawn.

2 Representation of Assessment Information

In this section, we will elaborate on each of the types of representation of assessment information (see Fig. 1). Section 2.1 develops the representation of linguistic assessment. In Sect. 2.2, the representation of complex linguistic assessment is shown. In Sect. 2.3, we start the numerical representation, namely representation of fuzzy number assessment, where users show their opinions employing the number and not linguistic terms. In Sect. 2.4, the representation within the interval [0, 1] is analysed. In Sect. 2.5 and Sect. 2.6, the representation of Atanassov assessments and representation of Hesitant assessment are developed respectively.

2.1 Representation of Linguistic Assessment

The way humans express themselves is through the use of natural language. Human beings, through this type of language, express their opinions using vague and imprecise expressions. In the Decision Making process, an approach has been considered where experts can express their opinions using words and not numbers. Consequently, a methodology for calculating with words has to be created [15]. This methodology uses linguistic variables, whose value is extracted from the natural language [43,44].

In [42] defines a linguistic variable as an element with five parts, (n,t(n),u,g,p). Each part of this element has the following meaning:

- n: it is name of the variable.
- t(n): it is the collection of values of n.
- u: it is the universe of discourse.

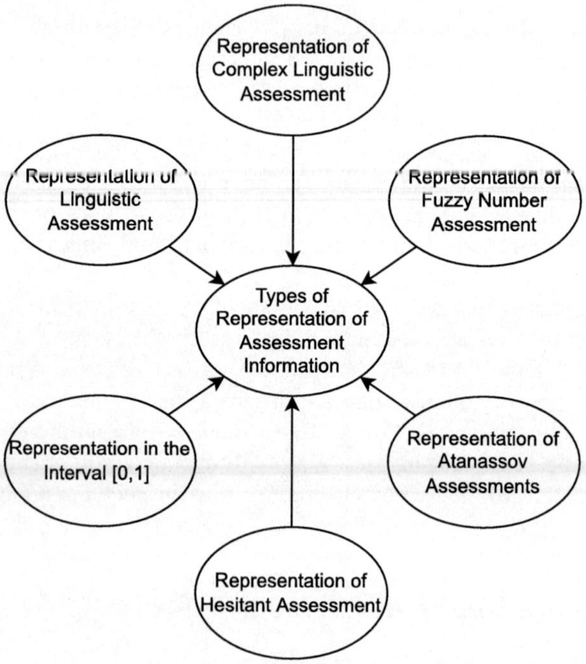

Fig. 1. Types of assessment representation.

- g: it is the set of the syntactic rule generated by the terms of t(n).
- p: it is the semantic rule associated with each linguistic value.

To be able to use the linguistic variables, the linguistic descriptors and their semantics must be determined. Consequently, different approaches are created and classified into two groups. These groups are [48]:

- Groups of approaches based fuzzy number.
- Groups of approaches based on linguistic scale functions.

2.2 Representation of Complex Linguistic Assessment

The approach shown in the previous section shows how experts express their opinions through a single linguistic term chosen from a set. However, depending on the expert's knowledge of the subject, he/she may be more generalist in some terms and more detailed among them [25]. Consequently, the use of a term is sometimes not enough because it restricts the experts' way of expression, as it is they who have to adapt to the set of linguistic terms and not the other way around [36].

To avoid this loss of information, linguistic approaches have to use more than one linguistic term so that experts can correctly express their opinions. This is why complex linguistic expressions are created. These expressions are assessments involving more than one term and are characterised by natural and artificial languages through various elements, such as linguistic hedges [36]. The use of these complex expressions instead of linguistic terms helps experts to express their opinions and thus obtain more accurate expressions from the experts. Nevertheless, the diversity of complex expressions is higher than simple linguistic terms.

The last ten years have seen different approaches to complex expressions [38]. Among the most prominent are approaches based on linguistic distribution [46] and approaches based on hesitant fuzzy linguistic terms [26].

2.3 Representation of Fuzzy Number Assessment

A fuzzy set measures the modelling of an object's membership of a specific class. An example is the satisfaction of a particular property that is measured by using fuzzy terms. Consequently, there is a non-probabilistic uncertainty that establishes the boundary of a class. The definition given by Zadeh in 1965 was the following [41]: Let X be a non-empty universal set. A fuzzy set B, defined a type-1 fuzzy set, over X is defined by a membership function $\mu_B : X \to [0,1]$ which associates to each element of the universal set X, $x \in X$, a degree of membership to B; $\mu_B(x) : B = \{(x, \mu_B(x)) \mid x \in X \wedge \mu_B(x) \in [0,1]\}$.

Using the definition of fuzzy set, fuzzy numbers were introduced that allowed the practical use of imprecise numerical quantities [7]. These generalisations are not limited to a single interval, but to a set of possible values because the number of real values that exist in the interval [0, 1] is infinite.

Such fuzzy sets have different characteristics. The following are shown in [10]:

- B is semi-continuous in $\mathbb{R}$ and therefore $\forall \epsilon > 0\, \exists \delta > 0$; $|x - x_k| < \delta$ and $|B(x) - B(x_k)| < \epsilon$.
- B is compact and therefore the closure of the set is compact. Put formally, $cl = \{x \in \mathbb{R}; B(x) > 0\}$ is compact.
- B is convex and thus, $B(c\cdot x+(1-c)\cdot y) \geq min\{B(x), B(y)\}; \forall c \in [0,1], x, y \in \mathbb{R}$
- B is normal and there exists an $x_k \in X$ such that $\mu(x_k) = 1$.

Finally, when users use fuzzy numbers instead of providing an exact value a real number is used which is practical but also imprecise [35].

2.4 Representation in the Interval [0,1]

Using the fuzzy set definition shown in the previous section, experts can show their opinions using the degree of membership, as in [34] where two alternatives are compared with each other. Moreover, this approach is simpler when creating models, since in this representation of the assessment information the degree of membership is considered as a closed and bounded interval $[0, 1]$. This implies that the models use the uncertain assessment of the experts on the alternatives consisting of using values belonging to the interval [4].

2.5 Representation of Atanassov Assessments

In the theory of fuzzy sets developed in 1960 by Zadeh, the following is defined that the sum of the degree of membership of an element x for the fuzzy set B and the degree of membership of its complement, B^C is equal to 1. Therefore, the degree of membership of a set B^C is equal to $1 - \mu_{B^C}(x)$. This definition put by Zadeh is generalized by the assessment representation of Atanassov [1], where it is shown that $\mu_{B^C}(x) + \mu_B(x) \leq 1$. The difference between Atanassov's representation and Zadeh's is that Atanassov's representation of assessment information allows to better model the validation of inclusion and exclusion of an object [39].

Furthermore, Atanassov defined the hesitancy index as the remainder of the sum of the degrees of membership, i.e., he defined in hesitancy index as $1 - (\mu_B(x) + \mu_{B^C}(x))$. The advantage of considering this index is that you have another degree of freedom that measures the degree of hesitation of the experts [9,32].

2.6 Representation of Hesitant Assessment

This assessment was introduced by Torra in [33]. In this paper, he introduced an extension of fuzzy sets where he proposed to represent the degrees of membership as a subset of $[0, 1]$, these sets were defined as hesitant fuzzy sets. These differ from those sets of the previous section in that it does not give a degree of membership by a margin of error, but by having several among which one would be correct. This situation is usual in Decision Making processes when experts may have several degrees of membership, i.e. $\{0.57, 0.58, 0.55\}$ of an alternative to a specific class [25].

3 Fuzzy Decision Making Methods

This section will review three types of Fuzzy Decision Making methods (see Fig. 2). In Sect. 3.1 looks at the key characteristics of GDM methods. In Sect. 3.2 shows at the characteristics of Multi-criteria Decision Making methods. Finally, in Sect. 3.3, it develops the explanation of Multi-person Multi-criteria Decision Making Decision Making models.

3.1 Group Decision Making Methods

GDM is a process where a group of experts jointly participate in choosing from a set of alternatives. In this process, each expert expresses his or her ideas and preferences individually and it can provide a less biased solution than a single person [16,19]. The complexity of such methods is to efficiently combine the

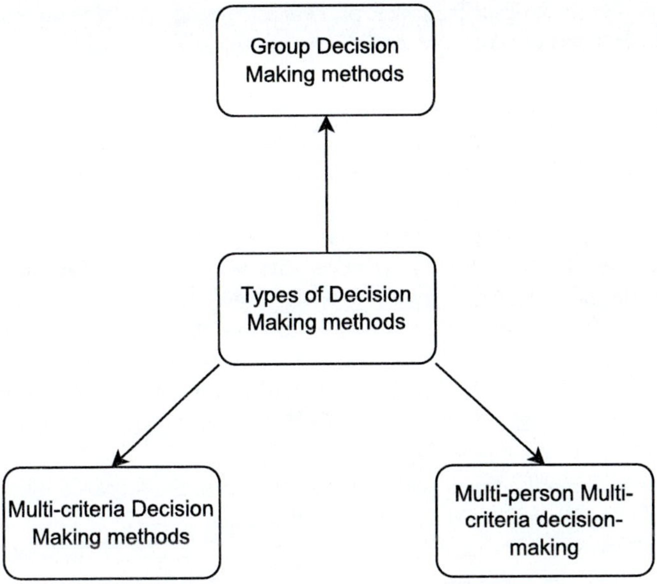

Fig. 2. Types of decision making methods.

information generated by each expert and transform it into a single collective piece of information.

The formal definition of a GDM process is initialised by defining a set of alternatives $X = \{x_1, \ldots, x_m\}$ and a set of experts $E = \{e_1, \ldots, e_n\}$ with $m, n \in \mathbb{N}$; $n, m \geq 2$. In GDM processes, experts evaluate alternatives to find a common solution to the problem. In this type of process, as will be seen in the next section, experts can be assigned weights. These weights indicate the importance of the experts and consequently, the importance of their opinions. Moreover, in GDM problems, each expert usually expresses his or her opinions through comparisons between the available options. Thus, Orlovsky studied fuzzy preference relations in [22]. A fuzzy preference relation is a matrix $m \times m$ where each position is a comparison between two alternatives.

Within these processes, the consensus process is introduced as a fundamental requirement for accessing the alternative selection process within the GDM process [23,47]. The aim of a consensus process is that group decisions are made in a consensual manner among experts. This is done by measuring the consensus among the experts and validating whether the consensus obtained is high enough to be considered acceptable. Nonetheless, if a sufficient value is not reached, measures are taken to bring the experts closer together, as a feedback process. Consensus processes have the following steps: the first step is consensus measurement [8], the second step is consensus control. This step aims to measure whether the level reached is sufficient. The last step is in case a high enough consensus is not reached, a feedback process is carried out to bring the experts' opinions closer together [6].

3.2 Multi-criteria Decision Making Methods

In real life, humans are unlikely to have a single criterion for evaluating an alternative. Therefore, Decision Making processes that evaluate alternatives with a single criterion are not very applicable to real world [27]. Multi-criteria Decision Making methods help to solve this problem so that a decision can be made based on different criteria associated with alternatives [13]. Consequently, a Multi-criteria Decision Making problem can be defined as follows: The set of alternatives is defined as $X = \{x_1, \ldots, x_m\}$; $m \in \mathbb{N}$; $m \geq 2$. The set of experts as $E = \{e_1, \ldots, e_n\}$; $n \in \mathbb{N}$; $n \geq 2$. The set of criteria to be evaluated alternatively as $C = \{c_1, \ldots, c_b\}$; $b \in \mathbb{N}$; $b \geq 2$. Each criterion has an associated weight, denoted as w_q that verifies $\sum_{q=1}^{b} w_q = 1$.

With these elements developed and the information generated by the experts is expressed employing a preference relation matrix with dimension $m \times m$, denoted as M where each position of the matrix is the comparison between two alternatives on a particular criterion c_j. Using weights on criteria implies that in Multi-criteria Decision Making problems, some criteria are more important than others. Moreover, there can be quantitative and qualitative criteria, subjective and objective criteria, and even benefit and cost criteria.

In the real world, there is a lot of diversity in Decision Making. Consequently, there is a great diversity of Multi-criteria Decision Making systems that solve various problems. From an operational point of view [13], there are different approaches to solving these problems:

- Outranking methods are based on the principle of dominance, in this case between alternatives. This type of system performs a pairwise comparison of alternatives. Consequently, the comparison between alternatives is made when measuring a criterion [5,12].
- Pairwise comparison based methods that compare two alternatives against each other. These methods, unlike outranking methods, are used when one alternative is more suitable than another. This type of methods and their extensions allow comparing fuzzy information [29].
- Value measuring methods are used in situations that require the assignment of a single value in order to establish a ranking of the alternative or to find the best alternative among a finite set.
- Reference level methods use reference points, also called ideal solutions, and use metrics to determine alternatives concerning a particular point [17,21], as VIKOR [28] or TOPSIS [31].
- Mixed methods combine aspects of several approaches, such as value measurement and benchmarking mechanisms; this combination is found in Multi-Moora [14]. Furthermore, goal-based programming with multiple objectives is considered [20].

3.3 Multi-person Multi-criteria Decision Making Models

There are processes based on Decision Making that use multiple experts and criteria. As defined above, the set of alternatives is defined as $X = \{x_1, \ldots, x_m\}$; $m \in \mathbb{N}$; $m \geq 2$. The set of experts $E = \{e_1, \ldots, e_n\}$; $n \in \mathbb{N}$; $n \geq 2$. The set of criteria $C = \{c_1, \ldots, c_b\}$; $b \in \mathbb{N}$; $b \geq 2$ [2,40].

Most of these models are under two main approaches: expert aggregation followed by criterion aggregation [37] and criterion aggregation followed by expert aggregation [11]. The first of these two approaches are divided into two phases:

- The data entry phase. Experts provide their opinions through the use of matrices.
- The selection phase. The experts' evaluations are aggregated in such a way that they are aggregated into a single matrix and the selection process of the best alternative is performed with this matrix.

Finally, it is necessary to apply game theory where experts can see that they have links to each other. This implies that there is a process of acceptance and understanding among the experts.

4 Some Trends in Fuzzy Decision Making

In this section, we will develop four challenges facing the Decision Making research line. In Sect. 4.1, we will trend towards the application of a gamification strategy to improve consensus. In Sect. 4.2, we will look at the challenge of some experts' manipulation strategies when using dynamic opinions. In Sect. 4.3, we will present at the trend towards the application of trust models in Decision Making systems. Finally, in Sect. 4.4, we will show at the challenge of applying Artificial Intelligence to Decision Making systems.

4.1 Gamification Strategies

In consensus processes, sometimes the consensus value is not high enough to affirm that the set of experts participating in a Decision Making process agrees with the decision made. Increasing the commitment of experts, one study [24] suggested that there are benefits to be gained by applying gamification, as it increases the commitment of these experts. Consequently, it highlights the need to create ideas that use gamification and game theory as part of the feedback process. This is where the challenge lies, in applying gamification and game theory to the consensus processes of a Decision Making system.

Finally, it is necessary to apply game theory where experts can see that their initially more dispersed opinions have links to each other. This implies that there is a process of acceptance and understanding among the experts.

4.2 Strategic Manipulation

Opinion dynamics are a phenomenon that is widely used in social networks. The use of this type of opinion dynamics affects the consensus among experts and in some cases leads to fragmentation or polarisation of their opinions. This fragmentation can be used to manipulate the experts in order to obtain a previously chosen result. Therefore, the challenge is to detect and create patterns that analyse dynamic opinions and avoid fragmentation and polarisation of experts.

In short, it is necessary to be aware of the comments made by experts, to analyse how their dynamic opinions vary and to observe which experts influence others by fragmenting and polarising them.

4.3 Uses of Trust in Decision Making Models

Trust is a key factor in Decision Making systems because if one expert trusts another, it is more likely that their decisions will be similar. Nevertheless, trust relationships between experts are often difficult to detect, and sometimes there are no initial relationships between experts because they are unknown to each other.

The challenge lies in developing methods that identify expert relationships and analyse how these relationships vary throughout the process, increasing and decreasing the degrees of trust between experts. It is also necessary to analyse the possible uses of trust relationships and how they affect the Decision Making process.

4.4 The Artificial Intelligence Approach to Large Scale Group Decision Making Systems

Artificial Intelligence is a tool that allows extra information to be obtained from data generated by experts. However, there is a tendency to use this tool for a specific type of Decision Making models, Large Scale Group Decision Making system. The application of Artificial Intelligence in this type of system is because they are the models that generate more information and therefore more data can be analysed to generate knowledge.

Artificial Intelligence is very versatile and for this reason, we have the following trends:

- Data integration: one of the challenges is to use the extra information generated by the data to integrate it into the systems belonging to the Large Scale Group Decision Making system. Some of the data that can be integrated is sentiment analysis through the comments generated by the experts.
- Integrating the Internet of Things into Large Scale Group Decision Making systems: the Internet of Things generates information, from devices, sensors, etc. The challenge is to use this information to generate algorithms and integrate them into such systems. To produce this, the set of tools belonging to machine learning can be used.

- Blockchain application: the amount of data generated by Large Scale Group Decision Making systems must be secure, complete and cost-effective, and it eliminates central authority. The Blockchain shows a trend to decentralise systems and improve the traceability of systems.
- Application of Big Data: the more experts involved in the process, the more information is generated. The trend is to analyse, through the use of Big Data tools, the information generated by the experts. Furthermore, Big Data makes it possible to manage the information, through the application of analysis dashboards.

5 Conclusions

In this article, we have seen how Fuzzy Decision Making has evolved over the last fifty years, both in the representation of evaluation information and in Decision Making models. In doing so, we have shown some of the challenges that Decision Making faces today. Moreover, this article allows us to observe how this line of research has been consolidated over time and how the systems related to this line of research are diverse and applicable to everyday situations.

Finally, once some of the challenges of Decision Making have been shown and developed, it is proposed as future work to analyse the current viability of the challenges to be developed by researchers.

Acknowledgement. This work was supported by the project B-TIC-590-UGR20 co-funded by the Programa Operativo FEDER 2014–2020 and the Regional Ministry of Economy, Knowledge, Enterprise and Universities (CECEU) of Andalusia, by the Andalusian Government through the project P20_00673, and by the project PID2019-103880RB-I00 funded by MCIN/AEI/10.13039/501100011033.

References

1. Atanassov, K.T.: Intuitionistic fuzzy sets. Fuzzy Sets Syst. **20**(1), 87–96 (1986)
2. Atanassov, K.T., Pasi, G., Yager, R.R.: Intuitionistic fuzzy interpretations of multi-person multi-criteria decision making. In: Proceedings First International IEEE Symposium Intelligent Systems, Varna, Bulgaria, pp. 115–119, September 2002
3. Bellman, R.E., Zadeh, L.A.: Decision making in a fuzzy environment. Manag. Sci. **17**, 141–164 (1970)
4. Bilgiç, T.: Interval-valued preference structures. Eur. J. Oper. Res. **105**(1), 162–183 (1998)
5. Brans, J.P., Vincke, P., Mareschal, B.: How to select and how to rank projects: the PROMETHEE method. Eur. J. Oper. Res. **24**(2), 228–238 (1986)
6. Cabrerizo, F.J., Ureña, R., Pedrycz, W., Herrera-Viedma, E.: Building consensus in group decision making with an allocation of information granularity. Fuzzy Sets Syst. **255**, 115–127 (2014)
7. Cabrerizo, F.J., Trillo, J.R., Morente-Molinera, J.A., Alonso, S., Herrera-Viedma, E.: A granular consensus model based on intuitionistic reciprocal preference relations and minimum adjustment for multi-criteria group decision making. In: 19th

World Congress of the International Fuzzy Systems Association (IFSA), 12th Conference of the European Society for Fuzzy Logic and Technology (EUSFLAT), and 11th International Summer School on Aggregation Operators (AGOP), pp. 298–305. Atlantis Press (2021)

8. del Moral, M.J., Chiclana, F., Tapia, J.M., Herrera-Viedma, E.: A comparative study on consensus measures in group decision making. Int. J. Intell. Syst. **33**(8), 1624–1638 (2018)
9. Deschrijver, G., Kerre, E.E.: On the relationship between some extensions of fuzzy set theory. Fuzzy Sets Syst. **133**(2), 227–235 (2003)
10. Dijkman, J.G., van Haeringen, H., de Lange, S.J.: Fuzzy numbers. J. Math. Anal. Appl. **92**(2), 301–341 (1983)
11. Gao, J., Guo, F., Ma, Z., Huang, X., Li, X.: Multi-criteria group decision making framework for offshore wind farm site selection based on the intuitionistic linguistic aggregation operators. Energy **204**, 117899 (2020)
12. Govindan, K., Jepsen, M.B.: ELECTRE: a comprehensive literature review on methodologies and applications. Eur. J. Oper. Res. **250**(1), 1–29 (2016)
13. Greco, S., Ehrgott, M., Figueira, J.R. (eds.): Multiple Criteria Decision Analysis: State of the Art Surveys. Springer-Verlag, New York (2016)
14. Hafezalkotob, A., Liao, H., Herrera, F.: An overview of MULTIMOORA for multi-criteria decision making: theory, developments, applications, and challenges. Inf. Fusion **51**, 145–177 (2019)
15. Herrera, F., Alonso, S., Chiclana, F., Herrera-Viedma, E.: Computing with words in decision making: foundations, trends and prospects. Fuzzy Optim. Decis. Making **8**(4), 337–364 (2009). https://doi.org/10.1007/s10700-009-9065-2
16. Herrera, F., Herrera-Viedma, E., Verdegay, J.L.: Direct approach processes in group decision making using linguistic OWA operators. Fuzzy Sets Syst. **79**(2), 175–190 (1996)
17. Hwang, C.-L., Yoon, K.: Multiple Attribute Decision Making: Methods and Applications - A State-of-the-Art Survey. Springer-Verlag, Berlin, Heidelberg (1981)
18. Kacprzyk, J.: Group decision making with a fuzzy linguistic majority. Fuzzy Sets Syst. **18**(2), 105–118 (1986)
19. Morente-Molinera, J.A., Cabrerizo, F.J., Trillo, J.R., Pérez, I.J., Herrera-Viedma, E.: Managing group decision making criteria values using fuzzy ontologies. Procedia Comput. Sci. **199**, 166–173 (2022)
20. Moro, L.M., Ramos, A.: Goal programming approach to maintenance scheduling of generating units in large scale power systems. IEEE Trans. Power Syst. **14**(3), 1021–1028 (1999)
21. Opricovic, S., Tzeng, G.-H.: Compromise solution by MCDM methods: a comparative analysis of VIKOR and TOPSIS. Eur. J. Oper. Res. **156**(2), 445–455 (2004)
22. Orlovsky, S.A.: Decision-making with a fuzzy preference relation. Fuzzy Sets Syst. **1**(3), 155–167 (1978)
23. Palomares, I., Estrella, F.J., Martínez, L., Herrera, F.: Consensus under a fuzzy context: taxonomy, analysis framework AFRYCA and experimental case of study. Inf. Fusion **20**, 252–271 (2014)
24. Pérez, I.J., García-Sánchez, P., Cabrerizo, F.J., Herrera-Viedma, E.: An approach toward a feedback mechanism for consensus reaching processes using gamification to increase the experts' experience. In: Proceedings of the 53rd Hawaii International Conference on System Sciences (HICSS 53), Maui, Hawaii, USA, pp. 1717–1726, January 2020

25. Rodríguez, R.M., et al.: A position and perspective analysis of hesitant fuzzy sets on information fusion in decision making. Towards high quality progress. Inf. Fusion **29**, 89–97 (2016)
26. Rodríguez, R.M., Martínez, L., Herrera, F.: Hesitant fuzzy linguistic term sets for decision making. IEEE Trans. Fuzzy Syst. **20**(1), 109–119 (2012)
27. Roy, B.: Paradigms and challenges. In: Greco, S., Ehrgott, M., Figueira, J.R. (eds.) Multiple Criteria Decision Analysis. ISORMS, vol. 233, pp. 19–39. Springer, New York (2016). https://doi.org/10.1007/978-1-4939-3094-4_2
28. Cristóbal, J.R.S.: Multi-criteria decision-making in the selection of a renewable energy project in Spain: the Vikor method. Renewable Energy **36**(2), 498–502 (2011)
29. Sekitani, K., Takahashi, I.: A unified model and analysis for AHP and ANP. J. Oper. Res. Soc. Jpn. **44**(1), 67–89 (2001)
30. Shendrik, M.G., Tamm, G.B.: An approach to interactive solution of multicriterial optimization problems with linquistic modeling of preferences. Autom. Control. Comput. Sci. **19**(6), 1–7 (1986)
31. Shih, H.-S., Shyur, H.-J., Stanley Lee, E.: An extension of Topsis for group decision making. Math. Comput. Model. **45**(7–8), 801–813 (2007)
32. Szmidt, E., Kacprzyk, J.: A consensus-reaching process under intuitionistic fuzzy preference relations. Int. J. Intell. Syst. **18**(7), 837–852 (2003)
33. Torra, V.: Hesitant fuzzy sets. Int. J. Intell. Syst. **25**(6), 529–539 (2010)
34. Trillo, J.R., Herrera-Viedma, E., Cabrerizo, F.J., Morente-Molinera, J.A.: A multi-criteria group decision making procedure based on a multi-granular linguistic approach for changeable scenarios. In: Fujita, H., Selamat, A., Lin, J.C.-W., Ali, M. (eds.) IEA/AIE 2021. LNCS (LNAI), vol. 12799, pp. 284–295. Springer, Cham (2021). https://doi.org/10.1007/978-3-030-79463-7_24
35. van Laarhoven, P.J.M., Pedrycz, W.: A fuzzy extension of Saaty's priority theory. Fuzzy Sets Syst. **11**(1–3), 229–241 (1983)
36. Wang, H., Xu, Z.S., Zeng, X.-J.: Modeling complex linguistic expressions in qualitative decision making: an overview. Knowl.-Based Syst. **144**, 174–187 (2018)
37. Wen, T.-C., Lai, H.-H., Chang, K.-H.: A new flexible method for solving multi-expert multi-criterion decision-making problems. Appl. Sci. **10**(13), 4582 (2020)
38. Wu, Y., et al.: Distributed linguistic representations in decision making: taxonomy, key elements and applications, and challenges in data science and explainable artificial intelligence. Inf. Fusion **65**, 165–178 (2021)
39. Zeshui, X., Zhao, N.: Information fusion for intuitionistic fuzzy decision making: an overview. Inf. Fusion **28**, 10–23 (2016)
40. Yager, R.R.: Non-numeric multi-criteria multi-person decision making. Group Decis. Negot. **2**, 81–93 (1993). https://doi.org/10.1007/BF01384404
41. Zadeh, L.A.: Fuzzy sets. Inf. Control **5**(3), 338–353 (1965)
42. Zadeh, L.A.: The concept of a linguistic variable and its application to approximate reasoning—I. Inf. Sci. **8**(3), 199–249 (1975)
43. Zadeh, L.A.: The concept of a linguistic variable and its application to approximate reasoning–II. Inf. Sci. **8**(4), 301–357 (1975)
44. Zadeh, L.A.: The concept of a linguistic variable and its application to approximate reasoning–III. Inf. Sci. **9**(1), 43–80 (1975)
45. Zadeh, L.A.: Fuzzy logic = computing with words. IEEE Trans. Fuzzy Syst. **4**(2), 103–111 (1996)
46. Zhang, G.Q., Dong, Y.C., Xu, Y.F.: Consistency and consensus measures for linguistic preference relations based on distribution assessments. Inf. Fusion **17**, 46–55 (2014)

47. Zhang, H., Zhao, S., Kou, G., Li, C.-C., Dong, Y.C., Herrera, F.: An overview on feedback mechanisms with minimum adjustment or cost in consensus reaching in group decision making: Research paradigms and challenges. Inf. Fusion **60**, 65–79 (2020)
48. Zuheros, C., Li, C.-C., Cabrerizo, F.J., Dong, Y.C., Herrera-Viedma, E., Herrera, F.: Computing with words: revisiting the qualitative scale. Internat. J. Uncertain. Fuzziness Knowl.-Based Syst. **26**(Suppl. 2), 127–143 (2018)

Short-Term Covid-19 Incidence Prediction in Countries Using Clustering and Regression Analysis

Fuad Aleskerov[1,2], Sergey Demin[1,2], Alexey Myachin[1,2](✉), and Vyacheslav Yakuba[1,2]

[1] HSE University, 20 Myasnitskaya Street, Moscow 101000, Russia
{alesk,sdemin,amyachin,vyakuba}@hse.ru

[2] Institute of Control Science of Russian Academy of Science, 65 Profsoyuznay Street, Moscow 117997, Russia

Abstract. As of February 07, 2022, more than 395 million cases of COVID-19 had been identified in the world, with 5.74 million deaths. The paper considers methodology for predicting the number of cases in the short term using a preliminary assessment of countries based on three indicators: expert assessments of the law-abiding population, the level of education and restrictive measures taken in the country. The description and composition of the groups obtained are given. An assessment of the accuracy of the forecast results is made. A comparison of the considered models of 2020 with 2022 is given.

Keywords: Covid-19 · Clustering · Pandemic · Incidence prediction · Regression analysis

1 Introduction

In connection with the problems caused by the COVID-19 pandemic that began in 2019, the number of various studies on this topic has grown rapidly. A query on the Google Academy for the keyword "COVID-19" showed about 4,700,000 results, which is a significant number given the relatively short study period. These works deal with the various aspects of the study of this disease: data collection and visualization [3], clinical symptoms [13], and the description of specific cases of patients [11]. Many articles study the impact of COVID-19 on human health, and possible concomitant diseases [5, 7, 12]. However, despite the extensive statistical data, a relatively small number of works, as far as the authors know, apply the quantitative methods for making short and medium term forecasts.

This study focuses on two main aspects. First, the methodology of dividing 66 countries into groups with similar structures according to the following basic system of indicators is proposed: the aggregated value of weekly quarantine measures and restrictions adopted in the country, the general level of education, and expert opinion on the law-abidingness of citizens. Naturally, much attention is paid to the choice of

S. Dzitac et al. (Eds.): ICCCC 2022, AISC 1435, pp. 333–342, 2023.
https://doi.org/10.1007/978-3-031-16684-6_29

the methodology for splitting the sample of countries under study and the basic system of indicators. Second, the compilation of the models for predicting the increase in the number of cases of COVID-19 is performed. Since it is very difficult to draw up a single model for all countries, identifying clusters can significantly increase the final forecast accuracy. Significance of the model is testing took place on data for the period of February 01, 2020–October 01, 2020, and compared with the result for January 2022. In most cases, the determination coefficient exceeds 0.8.

2 Initial Data

Since the ultimate goal of the study is to build a model that allows predicting of the increase in the number of cases in certain countries of the world in the short term, it is assumed that a basic system of indicators is used to determine differences in the behavior of citizens in the presence of restrictive measures, and the trend of the disease itself. To determine the behavior of citizens, it is assumed to use both the general level of education and law-abidingness, and the severity of the restrictive measures adopted in the country. Working with data on education is not particularly difficult: there are publicly available databases on the number of citizens with different levels of education [10]. However, the big question is the operationalization of the concept of "law abiding" in this task and the choice of characterizing indicators. Among the possible options, it is proposed to use various aggregated indices, including [4], or the partial use of indicators from some studies (e.g. Public Trust in Politicians from [9]). However, these indices characterize the level of trust rather than law-abidingness. In this regard, it is decided to switch to expert assessments conducted by specialists from the HSE University and the Trapeznikov Institute of Control Science of Russian Academy of Sciences. The peer review required certain limitations for the original sample, and therefore 66 countries were included. We consider working with the selected scorecard in more detail:

1. Weekly data on the number of cases of COVID-19 in the country, taken from [6]. Since the comparison of the absolute number of values in some cases is not always representative, in this study, a weekly gain is used. All values are logarithms.
2. Aggregate indicator characterizing the level of restrictive measures adopted in the country for the population due to the worsening of the epidemiological situation [2]. The measures taken are normalized on a scale of [0; 1].
3. An aggregate indicator characterizing the level of education, based on the data presented in [10]. The initial data is the percentage of the population with no education, incomplete primary, primary, lower secondary, upper secondary and post secondary. The level of education is normalized on a scale of [0; 1].
4. The general level of law-abidingness in the country. The indicator is evaluated on the basis of expert estimates, in the scale of [0; 10]

3 Finding Clusters

3.1 The Results of 2020

The study assumes an increase in forecast accuracy by compiling not a single model for all countries, but after preliminarily dividing them into clusters. For this purpose, methods of pattern analysis [1, 8], classification and clustering is considered. The results based on pattern analysis is difficult to interpret, and it is very difficult to compose a test sample for classification. In this regard, classical clustering methods are used for further analysis. The big question remains is both the choice of a particular methodology and the validity of this choice. In the study, the selection criterion is the interpretability of the final results and the accuracy of the prediction.

Results that improve the accuracy of predictive models are obtained using the k-means method. The criteria for assessing the quality of clustering to determine the number of clusters are silhouette estimates (in conjunction with the coefficient of the determination of regression models to determine the forecast accuracy). However, due to the use of dynamic data, the resulting clusters may differ in composition and in number, depending on the week studied, in which the measures taken in the country are considered, and a forecast is made about the increase in the number of cases. Consider an example for July 18, 2020. The clustering is shown in Fig. 1.

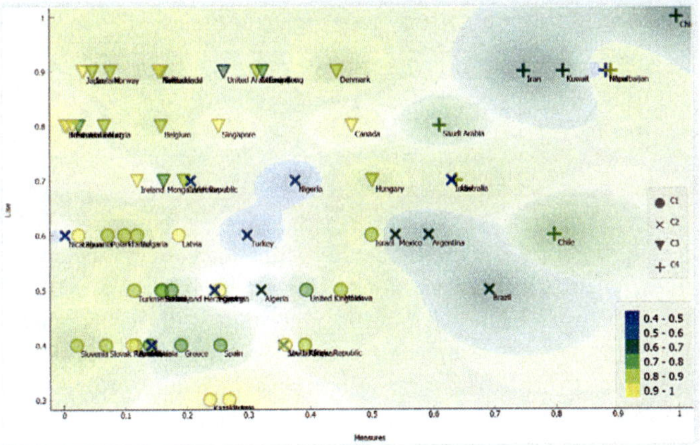

Fig. 1. Splitting the original sample of countries into groups (as of July 18, 2020).

The X axis characterizes the measures taken as of July 18, 2020, the Y axis is the law-abiding population in different countries. The color scale reflects the normalized indicator characterizing the level of education.

The composition of the groups differs somewhat depending on the date in question (although, in some cases, not significantly). For example, consider an alternative breakdown presented on September 19, 2020.

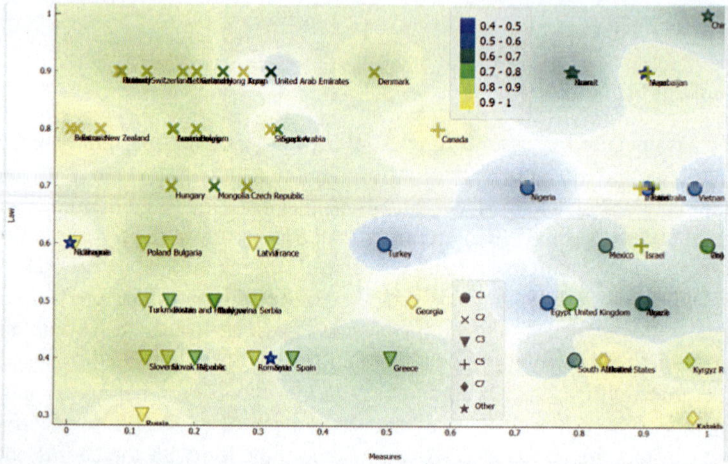

Fig. 2. Splitting the original sample of countries into groups (as of September 19, 2020).

The above figures clearly show the difference between the resulting clusters. In some cases, not only the number of clusters differs, but also their composition. However, such a result does not contradict the original problem, namely, the compilation of regression models to predict the increase in morbidity. This approach, on the contrary, allows an adjustment of the results over time, providing higher accuracy when making short-term forecasts.

3.2 The Results of 2022

The clustering has been carried out for the entire period from the beginning of 2020 to 2022. To compare the results, as well as to demonstrate the effectiveness of the proposed approach, an example of short-term forecasting of the increase in incidence on January 15, 2022 is given. Measures taken are based on data as of January 8, 2022.

Clustering is also based on the k-means algorithm, with the number of clusters adjusted when using silhouette estimates. Note that, for the purpose of comparison, analysis has been also carried out based on other methods, including: ordinal-fixed, ordinal-invariant and diffusion-invariant pattern clustering, hierarchical clustering, density-based spatial clustering of applications with noise (DBSCAN). The efficiency criterion is the interpretability of the final results (in our case, the accuracy of regression analysis models obtained based on clustering for short-term forecasting of the number of cases of COVID-19).

Figure 3 shows the results of the clustering. Designations are similar to Fig. 1 and 2. The X axis characterizes the measures taken as of January 08, 2022, the Y axis is the law-abiding population in different countries. Colors reflect the normalized indicator characterizing the level of education.

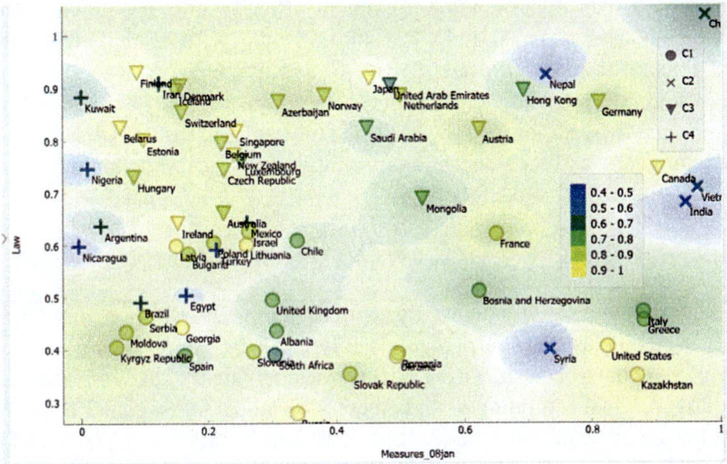

Fig. 3. Splitting the original sample of countries into groups (as of January 08, 2022).

4 The Regression Analysis

4.1 Regression Models for 2020

The next stage of the study is the construction of regression models in order to predict the increase in the number of number of new cases in the short term. Such models cannot take into account all possible factors, but they are very useful in drawing up general recommendations for the population and in making the necessary decisions. As mentioned, the general model gives an insufficiently accurate forecast (the comparison is shown in Table 1), and therefore separate models for each cluster are presented (Fig. 4).

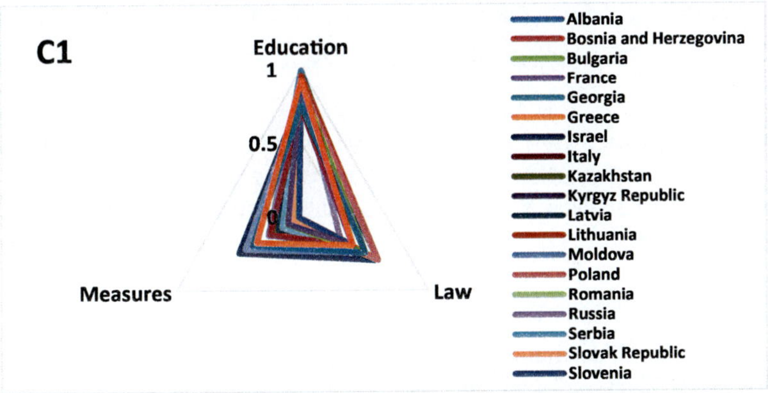

Fig. 4. Visualization of Cluster 1 (as of July 18, 2020).

Cluster 1 consists of 24 countries: Albania, Bosnia and Herzegovina, Bulgaria, France, Georgia, Greece, Israel, Italy, Kazakhstan, Kyrgyz Republic, Latvia, Lithuania, Moldova, Poland, Romania, Russia, Serbia, Slovak Republic, Slovenia, Spain, Turkmenistan, Ukraine, United Kingdom, United States. The relatively average indicators of both the measures taken as of July 18, 2020, and the values of expert assessments regarding law abidingness, and high education indicators are typical. The multiple regression model in this cluster looks like this:

$$N_t(Cluster\ 1) = 0{,}015 + 0{,}523N_{t-1} - 0{,}715N_{t-2} - 0{,}009N_{t-3} + 0{,}868N_{t-4} + 0{,}078N_{t-5}$$

Thus, the increase in the number of incidences of COVID-19 on July 25, 2020 is projected based on clustering carried out on July 18, 2020. For an accurate forecast, the previous 5 weeks have been shown. Moreover, the coefficient of determination is 0.911. For comparison, when compiling a single regression model for 66 countries on the same data (and for the same period), the coefficient of determination is 0.678. Thus, there is a significant improvement in the quality of the forecast due to the division of countries into clusters (a full comparison is shown in Table 1 and 2) (Fig. 5).

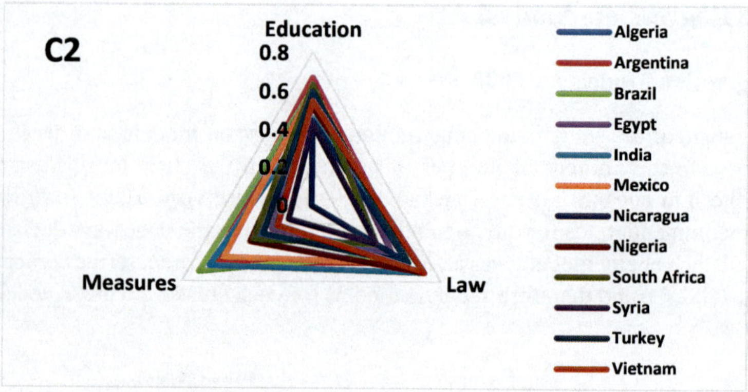

Fig. 5. Visualization of Cluster 2 (as of July 18, 2020).

The second cluster includes 12 countries: Algeria, Argentina, Brazil, Egypt, India, Mexico, Nicaragua, Nigeria, South Africa, Syria, Turkey, Vietnam. In this case, a relatively high indicator of the level of education and law-abidingness, is characteristic. The regression equation is:

$$N_t(Cluster\ 2) = 0{,}036 - 0{,}356N_{t-1} - 0{,}328N_{t-2} - 0{,}461N_{t-3} - 0{,}44N_{t-4} + 1{,}737N_{t-5}$$

The coefficient of determination is the smallest among all the clusters obtained: at 5 weeks it is 0.812. However, in this cluster the R-squared value is relatively high and at 2 weeks, and further changes insignificantly.

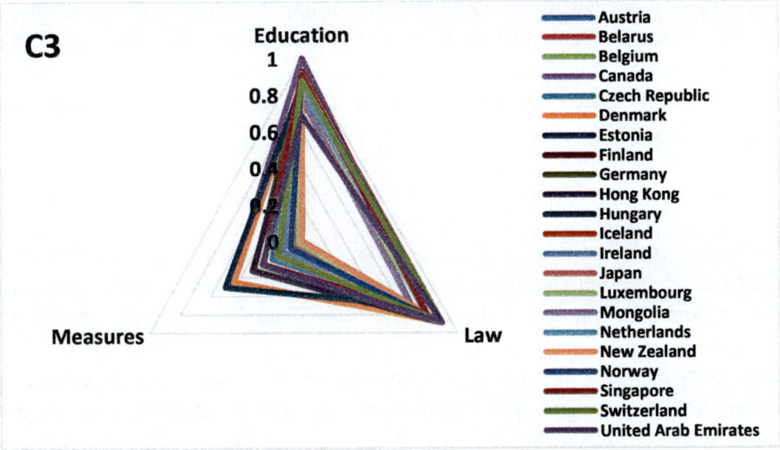

Fig. 6. Visualization of Cluster 3 (as of July 18, 2020).

The third cluster is typical for 22 countries: Austria, Belarus, Belgium, Canada, Czech Republic, Denmark, Estonia, Finland, Germany, Hong Kong, Hungary, Iceland, Ireland, Japan, Luxembourg, Mongolia, Netherland, New Zealand, Norway, Singapore, Switzerland, United Arab Emirates. In this cluster, relatively high indicators of education, the level of law-abidingness, and relatively low values of the aggregate indicator of measures taken as of July 18, 2020, are visible. Regression equation for this cluster is:

$$N_t(Cluster\ 3) = 0{,}001 - 1{,}209N_{t-1} + 0{,}587N_{t-2} - 0{,}923N_{t-3} + 3{,}363N_{t-4} - 0{,}259N_{t-5}$$

An interesting feature of Cluster 3 is the significant increase in the value of the determination coefficient from 1 to 2 weeks: 0.355 and 0.776, respectively. For 5 weeks, R-squared is 0.857, which, despite its high value, only surpasses the value of Cluster 2. However, we note that this value can be increased (although not very significantly) with an increase in the number of periods considered (Fig. 7).

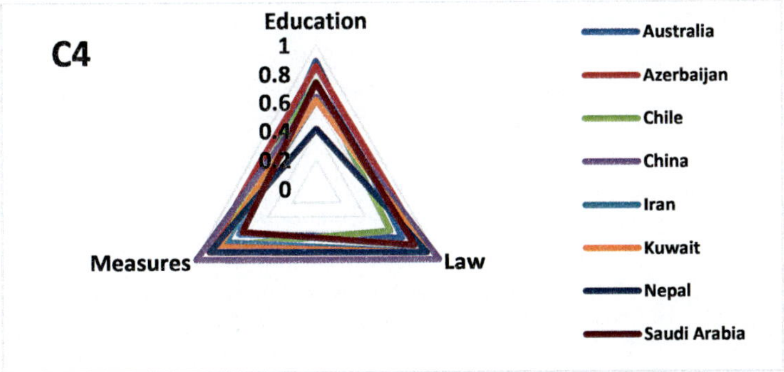

Fig. 7. Visualization of Cluster 4 (as of July 18, 2020).

Cluster 4 is described by 8 countries: Australia, Azerbaijan, Chile, China, Iran, Kuwait, Nepal, Saudi Arabia. Figure 6 shows the features characteristic of this cluster: average and relatively high values for all considered indicators: education level, law-abidingness, and measures taken as of July 18, 2020. The regression equation for the cluster is:

$$N_t(Cluster\ 4) = 0{,}014 + 0{,}048N_{t-1} - 0{,}529N_{t-2} + 0{,}848N_{t-3} - 0{,}922N_{t-4} + 1{,}293N_{t-5}$$

The very high value of R-squared is characteristic, as it is gradual insignificant increase when a larger number of weeks are included in the model: from 0.955 at 2 weeks to 0.991 at 5. Since the value of the determination coefficient in this cluster is initially very high, the increase is the least significant (when compared with other obtained clusters).

Below is a general comparison of the results. As an example, the regression model for Cluster 1 on July 18, 2020 is built followed by the forecast of an increase the number of cases on July 25, 2020. The accuracy of the model, as before, is assessed based on the coefficient of determination. The results are shown in the Table 1.

Table 1. Comparison of the results as of July 25, 2020 for Cluster 1 with the entire sample

Number of weeks	R-squared for cluster 1	R-squared for all countries
1	0.487	0.346
2	0.893	0.644
3	0.896	0.655
4	0.896	0.679
5	0.911	0.678

In Table 1, the advantage of using data from one cluster from the entire sample is obvious, with the exception of the short-term forecast based on one-week data. Next, we present the results of other clusters.

Table 2. Comparison of results as of July 25, 2020 for Clusters 2–4

Number of weeks	Cluster 2	Cluster 3	Cluster 4
1	0.763	0.355	0.883
2	0.763	0.776	0.955
3	0.766	0.832	0.977
4	0.789	0.844	0.985
5	0.812	0.857	0.991

The results for the other periods under consideration shows similar results: with the preliminary division of countries into clusters (for a specific date), the forecast accuracy of increases. The composition of clusters and their number may change over different periods, which is due to the use of a dynamic indicator (measures).

4.2 Regression Models for 2022

To compare the results, we present regression models for predicting the number of cases on January 15, 2022. Models are built for 4 clusters obtained on the basis of k-means.

For all obtained clusters, the coefficient of determination is not lower than 0.95 (for logarithmic data). Below are the resulting models:

$$N_t(\mathit{Cluster}\ 1) = 0{,}001 + 0{,}075N_{t-1} - 0{,}108N_{t-2} + 0{,}248N_{t-3} - 1{,}3N_{t-4} + 1{,}99N_{t-5}$$
$$N_t(\mathit{Cluster}\ 2) = 0{,}025 + 0{,}523N_{t-1} - 1{,}75N_{t-3} + 1{,}25N_{t-5}$$
$$N_t(\mathit{Cluster}\ 3) = 0{,}009 - 0{,}46N_{t-1} + 1{,}59N_{t-2} - 0{,}97N_{t-3} - 1{,}05N_{t-4} + 1{,}68N_{t-5}$$
$$N_t(\mathit{Cluster}\ 4) = 0{,}001 + 0{,}149N_{t-1} - 0{,}239N_{t-2} - 0{,}007N_{t-3} - 0{,}319N_{t-4} + 1{,}217N_{t-5}$$

5 Conclusion

COVID-19 poses a serious threat to the health of the entire world. Restrictive measures have been introduced in many countries, and the consequences of this disease remain to be studied. In this regard, it is highly relevant to create mathematical models for predicting the spread of the disease.

The article provides a methodology for improving the accuracy of predicting the increase in the number of cases in 66 countries of the world with a preliminary division of countries into clusters using cluster analysis methods based on 3 indicators: the level of education in the country, the law-abidingness of citizens and the restrictive measures taken. The use of dynamic data allows an adjustment of the number of clusters and their composition, which also increases the forecast quality.

Acknowledgements. The article was prepared within the framework of the HSE University Basic Research Program.

References

1. Aleskerov, F., Egorova, L., Gokhberg, L., Myachin, A., Sagieva, G.: Pattern analysis in the study of science, education and innovative activity in Russian regions. Proc. Comput. Sci. **17**, 687–694 (2013)
2. Coronavirus Government Response Tracker. https://www.bsg.ox.ac.uk/research/research-projects/coronavirus-government-response-tracker
3. Dong, E., Du, H., Gardner, L.: An interactive web-based dashboard to track COVID-19 in real time. Lancet. Infect. Dis **20**(5), 533–534 (2020)
4. Edelman Trust Barometer (2020). https://www.edelman.com/trustbarometer

5. Fang, L., Karakiulakis, G., Roth, M.: Are patients with hypertension and diabetes mellitus at increased risk for COVID-19 infection? Lancet Respir. Med. **8**(4), e21 (2020)
6. Johns Hopkins University. https://coronavirus.jhu.edu/map.html
7. Li, G., et al.: Coronavirus infections and immune responses. J. Med. Virol. **92**(4), 424–432 (2020)
8. Myachin, A.L.: Pattern analysis in parallel coordinates based on pairwise comparison of parameters. Autom. Remote. Control. **80**(1), 112–123 (2019)
9. The Global Competitiveness Report 2017–2018. http://reports.weforum.org/global-competitiveness-index-2017-2018/
10. The World Bank. Education Statistics – All Indicators. https://databank.worldbank.org/databases/education
11. Zhang, Y., et al.: Coagulopathy and antiphospholipid antibodies in patients with Covid-19. New Engl. J. Med. **382**(17), e38 (2020)
12. Zheng, Y.Y., et al.: COVID-19 and the cardiovascular system. Nat. Rev. Cardiol. **17**(5), 259–260 (2020)
13. Zu, Z.Y., et al.: Coronavirus disease 2019 (COVID-19): a perspective from China. Radiology **296**(2), E15–E25 (2020)

Importance of Flexibility in Design and Organizational Structure

Fabián Barrales[1], Claudia Dúran[2], Diego Fuentealba[3], and Raúl Carrasco[4](✉)

[1] Departamento de Ingeniería Industrial, Universidad de Santiago de Chile, 9170124 Santiago, Chile
[2] Departamento de Ingeniería Industrial, Universidad Tecnológica Metropolitana, 7800002 Santiago, Chile
[3] Departamento de Informática y Computación, Universidad Tecnológica Metropolitana, 7800002 Santiago, Chile
[4] Facultad de Ingeniería y Negocios, Universidad de Las Americas, 7500975 Santiago, Chile
rcarrasco@udla.cl

Abstract. Global changes have influenced organizations generating uncertainty in economic, social, environmental and technological decision-making. To face this situation, it has been necessary to incorporate increasingly the concept of flexibility in the organizational structure of companies. In the present work through a review of the scientific literature of the last 30 years, the key factors related to the new trends, models and areas of application showing as evolved the design and organizational structure is evolved. Successful cases are studied, results and conclusions are delivered.

Keywords: Organizational design · Organizational structure · Flexibility

1 Introduction

Conceptually, we call "structure" the distribution and order of the parts that make up an organization, work, or building, which serves as support and basis for what they form. According to Robbins [33], organizational Structure is the formal distribution of jobs within an organization, which involves decisions on work specialization, departmentalization, chain of command, the span of control, centralization, and formalization. The objective of an adequate organizational structure is to maximize the use of resources, establish a balance between hierarchical control and autonomy, determine different areas of operation, establish job specifications based on tasks and responsibilities of each member of the organization, thus avoiding confusion and conflicts.

There are no components defined as a minimum in organizations because the same people fulfill more than one role in small and medium-sized enterprises

S. Dzitac et al. (Eds.): ICCCC 2022, AISC 1435, pp. 343–352, 2023.
https://doi.org/10.1007/978-3-031-16684-6_30

(SME), being essential elements of authority, control, and communication [14]. However, several factors can negatively affect the organizational structure, such as control over innovation, the technological development of the company, access to financing, the size of the company, and the predominance of multinationals in the marke [17,20].

The constant evolution of the market forces organizations to continuously transform themselves to avoid bankruptcy or default, where additional elements to financial indicators have been proposed, such as flexibility in the face of changes and efficient use of resources in SMEs [7]. It is necessary to have a greater amount of research regarding organizational design to improve operations related to flexibility that deliver added value to organizations [4]. On the other hand, the models that could answer are not based on hierarchical control but recursive structures and living systems, such as the viable system model [13].

For this reason, this paper aims to conduct a literature review of recent scientific articles associated with the design of an organizational structure to propose possible systems that support change. Thus, answering the following questions: 1) What are the new trends regarding organizational design, 2) Have new organizational models emerged? and 3) In what areas has design been applied the most in the last five years?

2 Material and Methods

This work carried out a bibliographic review based on the scientific database "Web of Science", filtering by keywords, years of publication of the different documents, the language of the documents, type of documents, and the number of references. The used keywords were "organizational design" and "organizational structure", between the years 1992 and "2022".

The following search criterion is language. This work selected only English language documents to facilitate the reader from any part of the world to review the references used in this document adequately.

Next, the type of published document was selected as scientific articles, excluding from the search documents such as books or book chapters, editorials,

Table 1. Search Criteria.

Criteria	Description
Topic	TS = (("Organizational design" OR "organizational set-up" OR "organizational setup") AND ("Organizational structure" OR "organization chart" OR "organisational structure" OR "organizational chart" OR "institutional framework" OR "institutional structure" OR "organizational framewor"))
Published year	1992–2022
Document types	Articles, proceding papers, review article, early access & book chapters

letters, and bibliographic reviews. Table 1 shows the search fields, the original fields in English from the Web of Science database, and the criteria used to select the documents.

Finally, we organized the data obtained according to the number of times other works cited each document, and the twenty-five most frequently cited articles were selected for the literature review. This criterion seeks to determine the articles with the most significant impact and those with the most relevance.

3 Results

As a result of this work, 130 articles were found that converge on organizational design and structure. However, Fig. 1 shows that they are contributions from different research areas, focusing mainly on the area of Economics and Business with twenty articles. There are also different articles related to the top ten area of Management (67 articles), Business (31 articles), Operations Research Management Science (12 articles), Economics (11 articles), Engineering Industrial (7 articles), Information Science Library Science (7 articles), Public Administration (5 articles), Computer Science Artificial Intelligence (4 articles), Computer Science Information Systems (4 articles), and Computer Science Interdisciplinary Applications (4 articles). It is important to note that each scientific article can belong to more than one area of research; a clear example would be the article on the structure of the service line and attention in decision making [10] in the health system [1,27], which is a problem in the area of health sciences, but also of organizational structure.

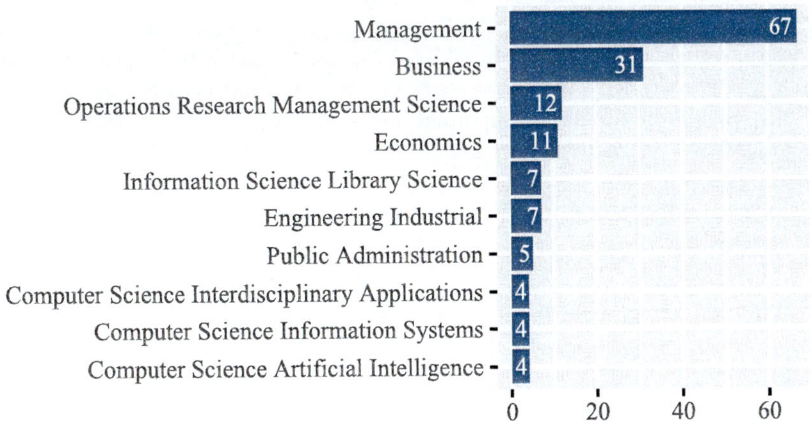

Fig. 1. Top 10 research field.

If we evaluate the findings obtained according to the type of study carried out, we can see that the vast majority corresponds to qualitative research. They used examples of companies in which several studies applied a specific organizational design are presented and, based on this, the factors that could have affected the result were evaluated. Data collection in these studies was done through documentation, press publications, direct observation, and even interviews.

The results obtained are shown in Table 2, where five scientific articles were conceptual analyses based on presenting the information available to date on a specific organizational design, explaining the proposed model, its benefits, and comparisons with other existing models. While 130 articles were observational studies that analyzed the application of these organizational models in various institutions, most of which corresponded to companies.

Table 2. Paper's type.

Type	Number	%
Conceptual analysis	41	31.5%
Observational study	89	68.5%
Total	130	100.0%

According to Chaves and Monzon [6], a country's economy comprises the public economy, the private economy, and the social economy, also known as the third sector. According to this classification, we analyzed which sector of the economy the organizations studied in the selected articles belonged, regardless of whether it was an application of organizational models or the analysis and development of theoretical models for institutions in these sectors. Table 3 shows that 73.1% of the organizations studied belonged to the private economy, 19.2% to the public economy, and 7.7% to the social economy. In addition, some articles do not fall into any of the above classifications, as they develop models or concepts transversely to the type of organization, and even Schwaninger [36] develops a systemic model for sustainability based on the Viable System Model [13], which involves recursive structures that include public, private, and social institutions.

Table 3. Sector of the economy of the studied organizations.

Economy sector	Number	%
Private	95	73.1%
Public	25	19.2%
Social	10	7.7%
Total	130	100.0%

The private sector companies studied present a variety in terms of their characteristics, among which we find large companies in telecommunications

[19,24], fashion [20,25], video games [23], consulting [28], several project-oriented companies [30,38], intelligent manufacturing companies [5], country of origin [34, 40], design companies [15], multidivisional [26], innovative [9], and even private health systems [27]. Regarding the public sector, Di Beningno [8] studied the role of organizational structure in managing conflicts of competing goals in the US military. At the same time, Ingrams [18] analyzed organizational design in the US and UK governments. Finally, cases from the social economy show that models of organizational structures can be applied to any organization, regardless of its purpose or nature. Benkler [2] highlights the influence of peer production structures in computer and open source companies, while McGahan [29] presents a case in the management of a trade association.

Regarding theoretical and conceptual development articles, there is a tendency to analyze the relationship of organizational structure or design with other relevant topics of economics, business, and management, such as strategic entrepreneurship, digital transformation, innovation, competitive strategy, and business initiatives. For example, Simsek [39] develops a model of strategic ventures based on their identity and boundaries. Kretshmer [21], on the other hand, analyzes the effects of digital transformation on organizational design through classification and analysis in the creation of products in companies. On the other hand, Burton, Colombo, Rossi-Lamastra, and Wasserman [4] conducted quantitative studies on the performance of organizational structure designs for business initiatives undertaken in Sweden. On the other hand, Puranam [31] questions the lack of innovation, recommending advancing the organizational science of companies by offering conceptual and technical devices for creating new prototypes of organizational forms and testing them. Finally, Sengul [37] questions the lack of coordination between organizational structure management and strategic management, indicating that the organizational design applied by a company will be reflected in its choices, which will end up affecting its competitive position to its peers. However, the competitive strategy to be adopted is a process independent of the organizational design adopted by companies.

It is possible to highlight the contribution in entrepreneurship pointed out by Burton [4], which exposes that the organizational design of entrepreneurial ventures stands out and can be approached in 3 stages: a) organizational structure: the hierarchy of relationships between organizational actors, the division of labor and specialization of tasks, and the formalization of roles; b) decision systems: the nature of decision-making process and the mechanisms that govern how decisions, once made, are communicated and coordinated; and c) managerial practices: the policies and procedures duration's that determine hiring, firing, training, career progression and intrinsic and extrinsic incentive systems.

4 Discussion

The economic and human resources required to conduct an experimental study are usually considerably more significant than those required for an observational study. This principle translates into the prevalence of theoretical scientific articles rather than analyses of organizational models put into practice. According to Puranam [31], there is enough theory to start putting new models into practice, so the number of case study articles will increase and contribute to enriching the available literature.

Much of the research on organizational models looks at companies as organizations with objectives of profitability and financial efficiency. However, implementing a new model in organizations is a challenge [28], since companies develop

Table 4. Organizational design and organizational structure studies with successful results in their findings.

Study	Author	Findings
By differentiating creative and commercial success, and examines how these outcomes are influenced in different ways by a start-up's hierarchy	[23]	The findings allow the debate to characterize those conditions in which the hierarchy is conducive or detrimental to emerging companies
The study is designed as a single case study of a Norwegian shipping company	[16]	Study analyzes the conflicting elements of the institutional environment that affect the organizational structure and how these conflicts are addressed by the company. In addition, it describes the decoupling mechanism that the company uses to balance these elements to adapt to those of institutional pressures
Study with data from 164 managers of shared service centers in index-listed European companies	[32]	Re-organization capabilities and IT as dynamic capabilities with the organizational structure of activities are the important factors in the successful implementation of shared service centers
Analysis of data collected from 207 Chinese manufacturing firms	[42]	The findings provide an alternative theoretical view by combining resource-based theory and structural contingency theory to explain the successful transformation of firm servitization
Achieve the highest level of management commitment in the process of designing an organizational structure where the organizational design represents a solid and complex architecture	[41]	Managers need more creativity and deeper logic if they want their product to be successful and effective
Research that is addressed with longitudinal data (1998–2010) that link the structure at the corporate level of Fortune 500 companies with their performance	[22]	They show that aligning corporate structure around customers pays off in specific competitive environments
Islamic companies need to learn the art of efficiently transforming strategy into strategic action, and aligning their organizations with their strategies	[35]	Establishing close connections between a company's human resources, organizational design and strategy is essential for a successful strategy
Study on 104 horticultural companies	[3]	The companies with the best economic success depend on the extent that they present a better organizational design suitable for the type of entrepreneur, when the amount of information is reduced

their structure under organizational models as they are created and increasingly incorporate the relationship with society through the social economy [6].

This study found that most of the studies are from the private sector and that companies are looking for organizational models to optimize their operations. According to Sengul [37], organizational design affects the company's decision making and directly influences its competitive position compared to the companies in its environment.

In the literature review, it was determined that the observed models define in their objectives: the increase of profitability [19], the need to improve innovation processes [20], the efficiency and effectiveness of project management that can generate new knowledge [40] and the viability of the organization's subsistence in the face of an uncertain economic, social, technological and environmental environment [11]. The multiple criteria posed by organizations influence decision making, generating risks [12] and questions related to the dynamic adaptation of the strategic definitions of an unpredictable environment in a model of organizational structure.

On the other hand, as shown in Table 4, there are cases of successful companies in the literature in which it can be observed how they adapt to the environment, their organizational structure and their strategic and business definitions.

5 Conclusion

Companies regardless of the sector to which they belong and the services/products offered to the market, must have an organizational structure that allows them to operate and achieve the objectives that arise. If there is no design of the defined organization, it may increase the risk of its survival in an increasingly uncertain environment influenced by the permanent social, environmental and technological changes that globality contributes. It is important to have organizational models applicable to each particular case, which to define strategies and carry them out through concrete actions.

In this work, it is observed that bibliography is lacking that the institutions with proposals of new organizational structures, which support companies to adapt to potential markets and generate more innovation with digital transformations to increase competitiveness.

Acknowledgements. This research aknowledge to the government of Chile, the project FONDEF ID21I10191 and Centre for Multidisciplinary Research in Telecommunication Technologies (CIMTT).

References

1. Alvial, X., Rojas, A., Carrasco, R., Duran, C., Fernandez-Campusano, C.: Overuse of health care in the emergency services in Chile. Int. J. Environ. Res. Public Health **18**(6), 3082 (2021). https://doi.org/10.3390/ijerph18063082
2. Benkler, Y.: Peer production, the commons, and the future of the firm. Strateg. Organ. **15**(2), 264–274 (2017). https://doi.org/10.1177/1476127016652606
3. Bockelmann, S., Schenk, E.W.: Type of entrepreneur and organizational-structure. Gartenbauwissenschaft **58**(2), 49–54 (1993)
4. Burton, M.D., Colombo, M.G., Rossi-Lamastra, C., Wasserman, N.: The organizational design of entrepreneurial ventures. Strategic Entrep. J. **13**(3, SI), 243–255 (2019). https://doi.org/10.1002/sej.1332
5. Cagliano, R., Canterino, F., Longoni, A., Bartezzaghi, E.: The interplay between smart manufacturing technologies and work organization: the role of technological complexity. In. J. Oper. Prod. Manag. **39**(6/7/8, SI), 913–934 (2019). https://doi.org/10.1108/IJOPM-01-2019-0093
6. Chaves, R., Monzón, J.L.: Beyond the crisis: the social economy, prop of a new model of sustainable economic development. Service Bus. **6**(1), 5–26 (2012). https://doi.org/10.1007/s11628-011-0125-7
7. Ciampi, F., Giannozzi, A., Marzi, G., Altman, E.I.: Rethinking SME default prediction: a systematic literature review and future perspectives. Scientometrics **126**, 2141–2188 (2021). https://doi.org/10.1007/s11192-020-03856-0
8. DiBenigno, J.: Anchored personalization in managing goal conflict between professional groups: the case of us army mental health care. Adm. Sci. Q. **63**(3), 526–569 (2018). https://doi.org/10.1177/0001839217714024
9. Escrig, E.D., Broch, F.F.M., Alcami, R.L., Gomez, R.C.: How to enhance radical innovation? The importance of organizational design and generative learning. Rev. Manag. Sci. **14**(5), 1101–1122 (2020). https://doi.org/10.1007/s11846-019-00326-7
10. Duran, C., Palominos, F., Carrasco, R., Carrillo, E.: Influence of strategic interrelationships and decision-making in chilean port networks on their degree of sustainability. Sustainability **13**(7), 3959 (2021). https://doi.org/10.3390/su13073959
11. Durán, C., Carrasco, R., Sepúlveda, J.M.: Model of decision for the management of technology and risk in a port community. Decis. Sci. Lett. **7**(3), 211–224 (2018). https://doi.org/10.5267/j.dsl.2017.10.002
12. Durán, C., Sepulveda, J., Carrasco, R.: Determination of technological risk influences in a port system using DEMATEL. Decis. Sci. Lett. **7**(1), 1–12 (2018). https://doi.org/10.5267/j.dsl.2017.5.002
13. Espejo, R., Bowling, D., Hoverstadt, P.: The viable system model and the Viplan software. Kybernetes **28**(6–7), 661–678 (1999). https://doi.org/10.1108/03684929910282944
14. Fuentealba, D., Liu, K., Li, W.: Organisational responsiveness through signs. In: Baranauskas, M.C.C., Liu, K., Sun, L., Neris, V.P.A., Bonacin, R., Nakata, K. (eds.) Socially Aware Organisations and Technologies. Impact and Challenges. ICISO 2016. IFIP Advances in Information and Communication Technology, vol 477. Springer, Cham. https://doi.org/10.1007/978-3-319-42102-5_13
15. Furniss, L.: Beyond discipline: evolving design practice and design education in the twenty-first century. Archit. Mps **18**(1, SI) (2020). https://doi.org/10.14324/111.444.amps.2020v18i1.004
16. Henk, A., Fallmyr, T.: Navigating through institutional complexity: adoption of a process view in functional organizations. Bus. Process. Manag. J. **26**(2), 593–612 (2020). https://doi.org/10.1108/BPMJ-01-2019-0038

17. Herreros, S., Inoue, K., Mulder, N.: Innovation and SME internationalization in Korea and Latin America and the Caribbean. United Nations publication (2018)
18. Ingrams, A.: Organizational design in open government: two cases from the United Kingdom and the United States. Public Perform. Manag. Rev. **43**(3), 636–661 (2020). https://doi.org/10.1080/15309576.2018.1562947
19. Joseph, J., Wilson, A.J.: The growth of the firm: an attention-based view. Strategic manag. j. **39**(6, SI), 1779–1800 (2018). https://doi.org/10.1002/smj.2715
20. Keum, D.D., See, K.E.: The influence of hierarchy on idea generation and selection in the innovation process. Organ. Sci. **28**(4), 653–669 (2017). https://doi.org/10.1287/orsc.2017.1142
21. Kretschmer, T., Khashabi, P.: Digital transformation and organization design: an integrated approach. Calif. Manag. Rev. **62**(4, SI), 86–104 (2020). https://doi.org/10.1177/0008125620940296
22. Lee, J.Y., Sridhar, S., Henderson, C.M., Palmatier, R.W.: Effect of customer-centric structure on long-term financial performance. Mark. Sci. **34**(2), 250–268 (2015). https://doi.org/10.1287/mksc.2014.0878
23. Lee, S.R.: The myth of the flat start-up: reconsidering the organizational structure of start-ups. Strategic Manag. J. **43**(1), 58–92 (2022). https://doi.org/10.1002/smj.3333
24. Leyer, M., Iren, D., Aysolmaz, B.: Identification and analysis of handovers in organisations using process model repositories. Bus. Process. Manag. J. **26**(6), 1599–1617 (2020). https://doi.org/10.1108/BPMJ-01-2019-0041
25. Li, P., Wu, C., Zheng, J., Chen, J.: Consumer-centered collaborative design of fashion clothing brands: a communication and organizational structure study. J. Text. Inst. **111**(3), 405–415 (2020). https://doi.org/10.1080/00405000.2019.1642698
26. Li, X., Holsapple, C.W., Goldsby, T.J.: The structural impact of supply chain management teams Supply chain agility development in multidivisional firms. Manag. Res. Rev. **42**(2), 290–310 (2019). https://doi.org/10.1108/MRR-04-2018-0163
27. Louis, C.J., Clark, J.R., Gray, B., Brannon, D., Parker, V.: Service line structure and decision-maker attention in three health systems: implications for patient-centered care. Health Care Manag. Rev. **44**(1), 41–56 (2019). https://doi.org/10.1097/HMR.0000000000000172
28. Magistretti, S., Trabucchi, D., Dell'Era, C., Buganza, T.: A new path toward a hybrid model insights from PwC's Italian experience centre a dedicated agile unit can help proliferate change across the entire organization. Res. Technol. Manag. **62**(5), 30–37 (2019). https://doi.org/10.1080/08956308.2019.1638223
29. McGahan, A.M.: My presidency of the academy of management: moral responsibility, leadership, governance, organizational change, and strategy. J. Manag. Inquiry **28**(3), 251–267 (2019). https://doi.org/10.1177/1056492619850900
30. Miterev, M., Turner, J.R., Mancini, M.: The organization design perspective on the project-based organization: a structured review. Int. J. Manag. Proj. Bus. **10**(3), 527–549 (2017). https://doi.org/10.1108/IJMPB-06-2016-0048
31. Puranam, P.: When will we stop studying innovations in organizing, and start creating them? Innov.-Organ. Manag. **19**(1), 5–10 (2017). https://doi.org/10.1080/14479338.2016.1272420
32. Richter, P.C., Bruhl, R.: Ahead of the game: antecedents for the success of shared service centers. Eur. Manag. J. **38**(3), 477–488 (2020). https://doi.org/10.1016/j.emj.2019.10.006
33. Robbins, S., Coulter, M.: Management, 8th edn. Pearson Prentice Hall, New Jersey (2005)

34. Rodriguez-Gutierrez, P., Correa, C., Larrinaga, C.: Is integrated reporting transformative? An exploratory study of non-financial reporting archetypes. Sustain. Account. Manag. Policy J. **10**(3), 617–644 (2019). https://doi.org/10.1108/SAMPJ-12-2017-0156
35. Saeed, M.M., Shah, S.Z., Qureshi, T.M.: Organizing and executing a strategy for firms operating in and from the Islamic World. Afr. J. Bus. Manage. **4**(13), 2899–2902 (2010)
36. Schwaninger, M.: Systemic design for sustainability. Sustain. Sci. **13**(5), 1225–1234 (2018). https://doi.org/10.1007/s11625-018-0538-5
37. Sengul, M.: Organization design and competitive strategy: an application to the case of divisionalization. In: Joseph, K., Baumann, J., Burton, O., Srikanth, R. (eds.) Organization Design. Advances in Strategic Management, vol. 40, pp. 207–228. Emerald publishing Ltd., Bingley (2019). https://doi.org/10.1108/S0742-332220180000040007
38. Simard, M., Aubry, M., Laberge, D.: The utopia of order versus chaos: a conceptual framework for governance, organizational design and governmentality in projects. Int. J. Project Manage. **36**(3), 460–473 (2018). https://doi.org/10.1016/j.ijproman.2018.01.003
39. Simsek, Z., Heavey, C., Fox, B.C.: (Meta-)framing strategic entrepreneurship. Strategic Organ. **15**(4), 504–518 (2017). https://doi.org/10.1177/1476127017711720
40. Stojanovic-Aleksic, V., Nielsen, J.E., Boskovic, A.: Organizational prerequisites for knowledge creation and sharing: empirical evidence from Serbia. J. Knowl. Manag. **23**(8), 1543–1565 (2019). https://doi.org/10.1108/JKM-05-2018-0286
41. Tripunoski, M., Ferati, F., Sibinovski, B.: Product design in the service of quality international marketing. Casopis Za Ekon. Trzisne Komunikacije **7**(1), 109–120 (2017). https://doi.org/10.7251/EMC1701109T
42. Yan, K.N., Li, G., Cheng, T.C.E.: The impact of service-oriented organizational design factors on firm performance: the moderating role of service-oriented corporate culture. Int. J. Prod. Econ. **228**, 107745 (2020). https://doi.org/10.1016/j.ijpe.2020.107745

Decision Support Systems for Integrated Land Management and Transport Infrastructure in Support of Climate Change Mitigation and Pandemic

Maria Vişan[1](✉) and Sorin Lenus Negrea[2]

[1] Department of Engineering, Mechanics, Computers, School of Advanced Studies of the Romanian Academy (SCOSAAR), 125 Calea Victoriei, 010071 Bucharest, Romania
maria.visan@ingr.ro

[2] Founding Member of the Romanian Association for Smart City and Mobility, Tudor Vladimirescu Boulevard 22, 050883 Bucharest, Romania

Abstract. The third sector is important in the climate change race for the "reduction of carbon emission effects", severely affected by the pandemic and under pressure to implement drastic measures, the transportation from *urban agglomerations* (UA) requires a sustainable transformation solution. An important point in this major effort is the understanding of the cognitive, institutional, strategic and regulatory barriers that may hinder the achievement of the objectives set by international climatic changes agreements. An integrated territory and infrastructure management center equipped with decision support systems is a solution for this challenge. Dynamic spatial planning contributes through the regulatory framework to the operationalization of this solution in the "new normality" after the pandemic. But no matter how agile the adoption of technologies could be at the UA level (through the collaboration of all actors involved in the process, namely public administration, business environment, citizens), in a society where the "new normal" is disruption, the key to the change remains the vision and how the transformation is conducted. However, the challenge remains on the integrated management of the territory and operation under hybrid roles, and the development of collaborative workflows between all actors involved in the workflows processes, between organizations and the development of the relationship with civil society and the public or private business environment.

Keywords: Climatic changes · Multicriterial decision support system · Intelligent agents · Big data · Data analytics · Transportation

1 Introduction

Unlike past climate change, today this one is not due to the natural causes but mostly the effect of the destructive human interventions in the last century. Over time, the effects have been intensified, with the development of climate change now having a much faster rate than in the Agreement past and continuing to increase in intensity and

S. Dzitac et al. (Eds.): ICCCC 2022, AISC 1435, pp. 353–365, 2023.
https://doi.org/10.1007/978-3-031-16684-6_31

duration of manifestation, taking place with increasing frequencies and in more and more areas, affecting all continents. The global manifestation of these changes exacerbates a package of particularly serious effects, some even irreversible, as stated in the latest Intergovernmental Panel on Climate Change (IPCC) report [1]. It is already clear that if severe measures to stabilize or eliminate emissions are applied starting from today, some of the effects will continue to manifest themselves, remaining active for a long time to come.

Climate change targets set by the Paris Agreement [2] at the pre-industrial level means a 50% reduction in CO_2 emissions by 2030. This means a fundamental global change achieved through a global economic and social transformation by adopting the available scientific results, generated with the effort of all nations.

Although almost all states signed the Paris Agreement, the emissions and CO_2 levels in the atmosphere continued to rise. At the United Nations Climate Change Conference [3], all the commitments made by the signatory countries failed to reach the proposed target of pre-industrial 1,5-degrees Celsius.

The *European Environment Agency* (EEA) has estimated the economic impact of climate change in the EU-27 Member States for the period 1980 and 2020 at € 487 billion, stating that losses could be even higher in the coming years if no action is taken [4]. For example, the Swiss Re Institute [5] estimates an impact on global GDP by 2050 of more than 4% if the terms of the Paris Agreement are met, and rise to 18% and the temperature will continue to rise if no global action is taken. Throughout history, massive reductions in CO_2 emissions have been generated only during major global events (wars, global recessions, etc.), but by far the highest value was recorded during the pandemic crisis, as evidenced by Statista, Fig. 1 [6] (Fig. 2).

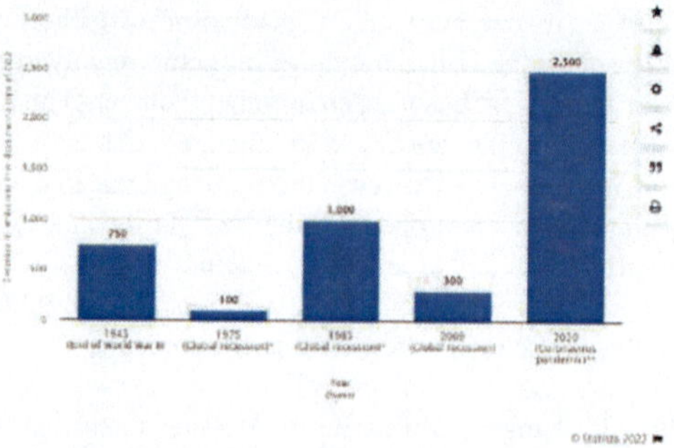

Fig. 1. Decrease in carbon dioxide emissions worldwide between 1945 and 2020, by major historical event, https://www.statista.com/statistics/1111452/co2-emissions-decrease-due-to-majorhistorical-events-globally/ [6]

However, with the passage of the crisis periods, global annual *dioxide carbon* (CO_2) emissions have continued to increase amplified by population growth, increased energy demand and mobility [7].

Under the Paris Agreement, each country must regularly set, plan, and report on its contributions (changes meant to reduce global effects) to achieving zero emissions

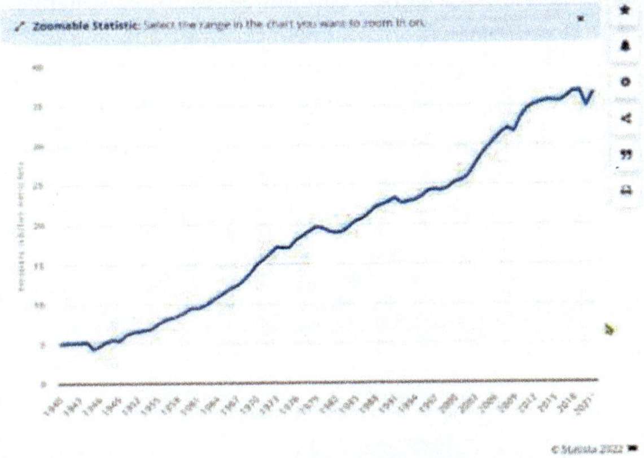

Fig. 2. Annual CO_2 emissions worldwide from 1940 to 2020 (in billion metric tons) [7]

targets. Of course, these changes also require sufficient budget allocation. This is where the gaps appear. There are no restrictions imposed, each country can set its targets, but over time they should be more and more ambitious. The perceived gap is that there are no differences between developed and developing countries, as the latter lack sufficient financial resources. As a result, developing countries will feel the greatest pressure in the race to reduce emissions, but they will be the biggest beneficiaries of these changes.

The article aims to address by 2050 the path of radical transformation in the transportation sector by and the corresponding contribution of Information and communications technology (ICT). We will assist in the gradual replacement of everything we know today, and the emergence of another transportation world based on other business models aimed at new infrastructure and new modes of operation. The article is structured into 6 sections, as follows: Sect. 1 presented the gravity of the climate change situation; Sect. 2 identifies the impact of climate change on transport infrastructure and operations, highlighting the climate changes that have already occurred; Sect. 3 presents the analysis of the uncertainties that can constitute blockages of the transformation; Sect. 4 presents a vision on the future of transport in UA; Sect. 5 highlights the role of multi-criteria and multi-agent decision support systems as mandatory tools for collaboration at the level of an integrated command center, and Sect. 6 highlight the conclusions for the transportation sector.

2 The Impact of Climate Change and the Pandemic on Transportation Infrastructure and Operations

The impact of climate change on transportation infrastructure and operations is no longer an unknown topic, and multiple studies have already been conducted [8], many books highlighting these issues have been written for different countries [Potential Impacts of

Climate change on U.S. Transportation. Committee on climate change and U.S. Transportation. Transportation Research Board Division and Earth and life Studies], or professional studies conducted by various experts, Deloitte [9], PCW [10], McKinsey [11], and much more examples can continue also naming the academic research environment.

Transport is one of the key accelerator sectors for the reduction of CO_2. According to Statista, it ranks third (7.29) in terms of energy (13.13) and is similar to other combustion industries (7.82 billion metric tons of carbon dioxide) [12] (Fig. 3).

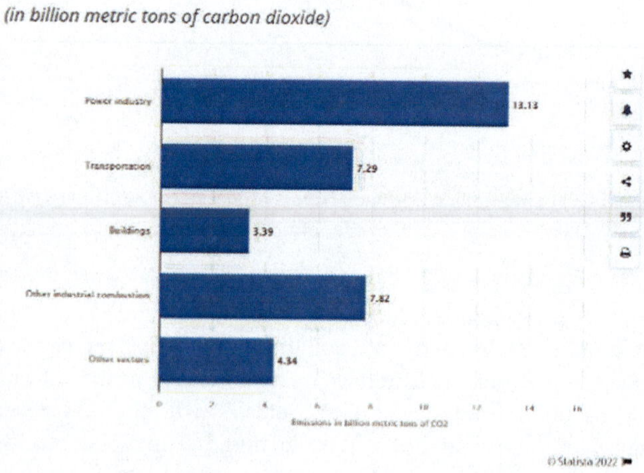

Fig. 3. Global carbon dioxide emissions in 2020, by sector [12]

Achieving the 1,5-degree temperature reduction target, one of the drastic measures applicable to the transport sector, involves the gradual reduction of internal combustion vehicles to half by 2030 and the replacement with electric vehicles by 2050 [13]. Transport professionals will have to adapt to the consequences of these changes. The major impact of these measures will affect all categories of "actors" involved in the planning, design, construction, operation and maintenance of infrastructure, but also other industries affected by these changes, such as financial institutions (banks, investment firms, insurance and reinsurance), manufacturers and distributors in the supply chain, to name just a few of them.

Five climate changes of particular importance for the transport sector have been identified since 2007 [14] and the probability of their occurrence in the 21st century has been estimated, which has happened (Fig. 4):

In addition, the emergence of new classes of meteorological and climatic extremes (in terms of size and frequency), such as record rainfall and record heatwaves, not been experienced so far, leads to a review of historical climate patterns used regionally by transportation planners to draw up plans, future guidance and mobility operations and investments.

Transportation management with older models and methods is not an option today. UA congestion requires innovation, and the task no longer belongs only to public

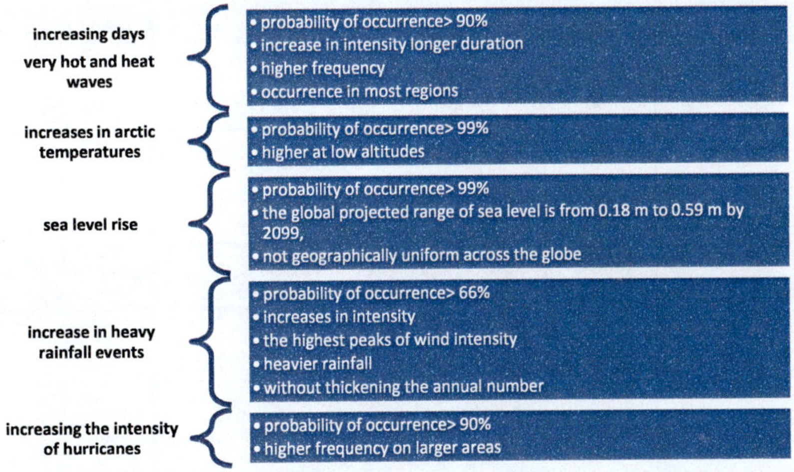

Fig. 4. 2007 Intergovernmental Panel for climate change summit for policy Makers on the Pasca Science Basis (Fourth Assessment Report), The Petersson et all. 2006, see Appendix C [14]

administrations and central regulators, but to all actors in civil society and the business environment.

Adding the disruption caused by the digital transformation that shifts the transportation sector from a physical to a digital set of orders, it also generates new activities, unexpected before the pandemic, in which big data become a current reality. The information-driven decisions and the generation of new services through data analytics and processing take place with a much higher absorption rate than before the pandemic (pandemic has accelerated the development of digital transformation in transportation and affects the field of transport).

In a world full of variables and uncertainty, with major changes and continuous innovation, the transport sector needs more and more monitoring and action to meet the goals set by international climate agreements.

3 Uncertainty Management

Understanding the obstacles that can hinder the achievement of the goals set by international climate agreements goes beyond identifying sustainable plans to adapt transport to climate change. Starting from the approach to uncertainty management introduced by Lidskog and Löfmarck [15] to the components of the decision-making process in the management of social and scientific constructs, perceptions and attitudes, we identify 4 categories of obstacles (Fig. 5):

- **cognitive**: driven by the lack, of inadequate or uncertain knowledge, this obstacle generates inappropriate interventions. Cognitive barriers have often been highlighted as crucial for climate change decision making, and generate an increase in uncertainty [16]

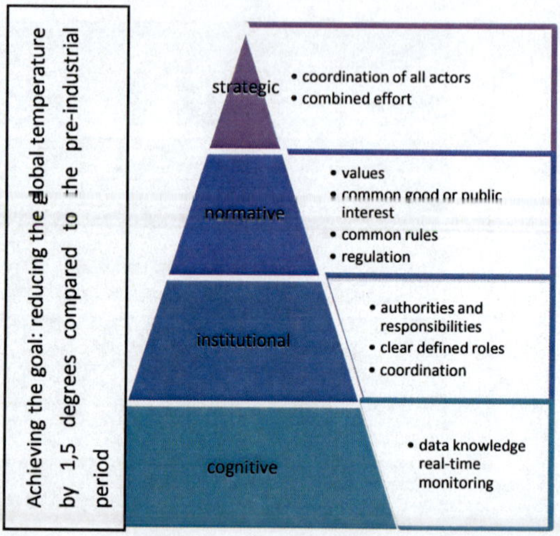

Fig. 5. Uncertainty management

- **institutional**: generated by unsuitable or undefined authority/ responsibility and coordination or when a cause is linked to a person, this obstacle generates atomized decisions or the decision is blocked [17]
- **normative**: generated by the lack of same understanding of common values, norms or difficulties in setting their importance to affect the decision-making and draw a different interpretation of common good and different levels of commitments
- **strategic**: generated by unclear interpretations or by the lack of correct understanding and assessment of a problem or hesitation in formulating answers as a result of different interpretations of other participants in the process, can lead to a lack of decision making, delayed or unnecessary decisions, constituting obstacles even for those who are determined to act. The results are even more serious when some actions require coordination to be effective.

Real-time data collection allows institutions to be organized for a common good by roles (execution, regulation, standardization and monitoring). Adoption of rules and standards enables the communication between institutions with responsibilities in achieving the reducing emission targets. The adoption of strategies allows the coordinating of all participants (actants) for gaining the goal of 1.5 degrees compared to the pre-industrial period.

4 Vision on Transport in the Urban Agglomerations of the Future

Given the unprecedented rate of urban growth, Urban Agglomerations (UAs) require measures related to climate change and measurements of their impact on the transport sector.

A good lesson for developing countries is that models of cities in other countries cannot be copied to drive their development in the right direction when it comes to transportation, because even these big models suffer and need to be adapted themselves to the new challenges.

There are several challenges [18], manifested in UAs as **disruptive innovation**, that should be taken into consideration when an intelligent urban mobility system is designed:

- **Connected and Automated Vehicles (CAVs)** even there are still open challenges related to safety that should be addressed, and Control Barrier Function (CBF) to avoid collission
- **Traffic control**, based on traffic control schemes for traffic flow regulation, and control strategies, including ecorouting and eco-drive [18]
- **Ultra-high-speed travel options, hyperloop** [19], that revolutionize the people commututing between cities, or high-priority goods procurement (such as medical supplies and food). These will affect real estate markets along the routes. Large economic zones can emerge, including cities connected by ultra-fast transport solutions
- **Underground ultra-fast transport alternatives** that leaving room for autonomous or individual transport. For the main needs of travelling inside or outside the AU, there will be associated services (considering the elimination by 2040 of vehicles powered by petrol and diesel), by sharing/renting any type of vehicle (preferably electric) and by planning and communicating the request to an integrated territory and infrastructure management center
- **Zonal Supply Hub for freight transport** [20] designed for underground and surface stations with efficiently sized (autonomous and electric) vehicles. Travel is no longer required for daily/regular supplies, everything can be ordered online and containerized transport can provide all functions (security and temperature conditions) - similar to the concepts implemented by the OCADO hypermarket chain (robotics for general merchandise), vertical farming (JFC, Infinite Acres), automated meal prep (Karakuri) and 3D printing
- **Pop prototype. UP Next| Audi Nanaimo** [21], drone capsule for passengers designed as a result of cooperation between Audi, Airbus, and Ital design and presented at Geneva Autosalon (8 to 18 March 2018)
- **Green power supply as mixed alternatives sources of fuel – solar energy** by implementing **photovoltaic panel fields** (on the lake's surface or on mountain slopes where there is no vegetation, etc.) and **hydrogen** ones proposed to support carbon emissions. Examples of good practice are the investments made by ALSTOM following the Nordic land, water and air transport model [22] or the first hydrogen train [23] made in France
- **High-capacity lithium-ion batteries** since 2017 [24, 25]
- **Digital twins** [26] **of UA**, that support the command center of integrated land and infrastructure management
- **Mobility *as a Service* (MaaS)** [37], the concept with the best suited perspective to the needs and alternative forms of multi-modal transport infrastructures. Workable information obtained by real-time data aggregation is a great example of reinvention, as in the case of the Finnish company MaaS Global, which has developed its business model based not on operating infrastructure, as Uber does, but on the aggregation and interpretation of data and service offers of public and private transport operators [27]. This helps UA increase the efficiency of its existing business. An extension in the

future may be the aggregation of congestion levels. Service frequency and cleanliness of public transport generate estimations or risks of delays

- **Dynamic urban planning and regulations modifications** [35] based on new smart technology, changing work habits (from home) investments development, while maintaining the volume of services designed adapted to the urban fabric, regulated function and 15 min walking solutions for reducing travel needs. This allow the UA adaptation to the current situation, without existing static jam. Implementing dynamic urban planning and, along with digital landscaping brings new options for integrated multi-sectorial analysis. The pandemic has generated opportunities to resettle UA and transform it into a digital, green and more stable one based on principles in terms of consequences for events similar to the pandemic that can cause global destabilization.

All analysts predict that future traffic is based on autonomous vehicles, which stop only for recharging and service. This could make the use of cars more efficient, as most of them are currently parked 96% of the time. The emergence of autonomous machines also leads to the implementation of centralized decision-making systems, capable of monitoring and controlling the entire transport system to avoid the conflict of different AI algorithms of different manufacturers, until the standardization of all interactions.

Meta-applications in the central decision making system based on data centre [18] must exist to prevent accidents. The proliferation of self-driving cars and sharing services greatly reduces the need for parking and can free up urban space for other purposes such as residential neighborhoods and parks.

Integrated public and private mobility networks could start to grow in demand, responding to the changing needs of passenger preferences.

The passenger economy, combined with the latest technological innovations, will potentially open up a wide range of new business opportunities that could lead to a completely new range of onboard services.

Although the known effects of the pandemic have been a terrible way to go about this stage, we can say that we have managed to do the impossible in many ways and it has revealed capabilities that we did not know we had and made us rethink what was impossible.

All of these scenarios and innovations below require planning, procedure, a major investment and long-term continuity, and decision-making system architectures capable of storing, processing, analyzing, and reporting large volumes of data in real-time and communicating disaster response plans, roles and specific actions on each type of intervention. In support of the proposed solution, come the 2021 predictions of IDC [28] which show that by 2025 most organizations (80%) will use products, goods and own digital twins to share data and information with other participants in the industrial ecosystem.

Putting into practice, the vision although it began in some parts of the world and requires a global approach, new innovative solutions to support the achievement of climate goals. The implementation of decision support systems can support UA in the integrated management of the territory and infrastructures.

5 Multi-criteria and Multi-agent Decision Support Systems for Integrated Land and Infrastructure Management

The implemented multi-criteria in dynamically changing environments use complex programs called Intelligent Agents. These agents interact and cooperate in carrying out decision-making tasks. They extract and integrate data to gain knowledge. Through collaborative action, Intelligent Agents allow *Decision Support Systems* (DSS) users to make more informed and accurate decisions [29]. The autonomous execution and the dynamic nature of Intelligent Agents increase the potential, efficiency and effectiveness of decision-making. Intelligent agents are the key technology at the DSS level, ensuring data distribution, interconnection and openness. The proposed environment allows autonomous planning and action monitoring. The intelligent agents have flexibility and intelligence in highly unpredictable situations, bringing significant improvements in the quality and software sophistication [30]. In the context of multi-agent reinforcement learning (RL) and an extended actor-critic (AC) algorithm implementation, there have been demontrated a significant efficiency improvement and transportation cost reduction. Using these methods, urban transportation vehicles became capable to make decision by themslves, improving service cycle in real time, in the same time reducing the requirements for communication cost [31].

For the transportation sector, DSS functional modules include the components described in [33]. New technologies are, and require a multi-disciplinary approach to support collaborative work [33, 35]. To meet the challenges of climate change and the pandemic, DSS monitoring virtually any type of sensor (which collects traditional geospatial content, satellite imagery or IoT data), unstructured data, real-time or near data) in real-time. The dynamics of urban development depend on the maturity of the UA, respectively on stages of evolution: at the level of data, workflows or strategy. Although challenging, the decision-making process is adapted to maintain an optimal level of services designed for the population in terms of recommended indicators of quality of life for each regulated functional area. Dynamic urban planning allows the dynamic remodeling of the functional area according to the investments made, maintaining the function and services adapted to the 21st century, implementing the measures and monitoring the fulfilment of the objectives of combating climate change without destroying the habitat of the blue/green planet.

According to the CIO's agenda, decision support systems will need to include three components, described below (Fig. 6).

- integrative technologies - digital platforms and *Big data* volumes

  - DSS manage live information into the continuous decision-making flow and metadata of all autonomous transport systems; real-time data collection from sensor monitor the successful implementation of the climate strategy
  - the major technological areas: Big Data, Data Analytics, Cloud, Geospatial, Robotics, AI, ML, Social Media, IoT/IoE, Sharing Economy, Digital Gemini for UA territory, and others, as described in [31, 33]. New technologies are constantly being added as digital transformation approaches and support [32, 34]

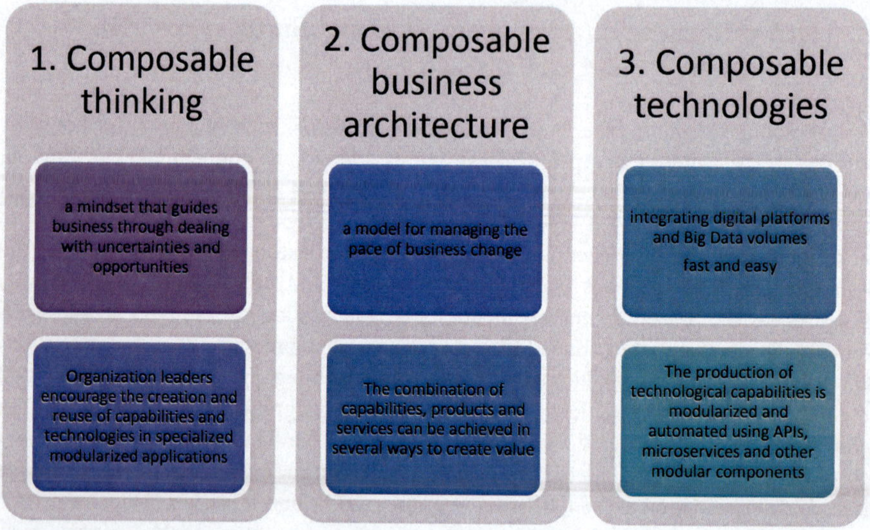

Fig. 6. Adapted from Gartner, 2022 CIO Agenda: Create an Action Plan to Master Business Composability, 2022-CIO-agenda-ebook.pdf (gcom.cloud) [36]

- Data Analytics engines correlate information at different levels of the multi-criteria and multi-institutional cooperation workflows. The proposed decision-making system will incorporate the new services for citizens such as Maas [37]
- local digital transport planning tools become a subset of regional and national planning:
  - mobility plans with standardized framework content, measurable indicators
  - transport models adapted to climatic and post-pandemic transformations
  - monitoring of transport to assess dynamic reorganization within the territory's functions, while maintaining the level of services designed

– a business model, which involves combining capabilities, products and services can be achieved in several ways to create value. Transforming the business model means much more because it becomes the digital orchestrator of the transport industry o remodeling and automating basic processes to make them more connected, more dynamic, with more data than ever before

- Transforming individual systems into a network of interoperable systems of systems allow a free flows of information and opens new opportunities for intelligent control at a very different level
- connected and dynamic operations
- dashboards for decisions based on actionable data
- care for the experience of the employee responsible for the execution of operations, has become as important as the customers

– a mentality to guide business by dealing with uncertainties and opportunities oorganization leaders encourage the creation and reuse of capabilities and technologies in specialized modularized applications

- hybrid models (fluid workforce) bring a flow of innovation that proves agility and contributes to the improvement of components through the participation of independent experts in their development.

6 Conclusions

Transport has always been one of the main engines of the economy and it requires continuous and special attention, analysis and understanding of the digital transformation obstacles due to its influence on many other areas. The transportation sector was highly affected in the last two years by pandemic quarantine and global restrictions. Other factors related to new habits, such as work and distance learning, social distancing or rising prices for supplies and services (airfare), and banning events and conferences have led to a reduction in passenger/visitor/travel flows. Fears of infection continue to manifest themselves for a long time. Returning to what was before the pandemic and "the new normal" is a matter of reinvention or disappearance. The economic disruptions caused by the drastic measures reduce the effects of climate change adopted by the Paris Agreement.

The losses are greater if we ignore the effects of climate change. Traditional transportation is no longer a universally valid option and requires a different approach. The solution for this challenge is dynamic urban planning that changes the traffic flows and continuous monitoring in the integrated management center of the territory and infrastructures. Decision Support Systems that use Intelligent Agents can (re)plan the transport system, the new means of transport. DSS can also treat metadata of all autonomous transport systems, communicate with users and notify them of information or avoidance of accidents, integrate different mobility systems, assist land and infrastructure managers in decision-making or offer services based on data that enhance the activities of business participants in a UA in terms of infrastructure and the operation of the transport sector.

No matter how powerful the technology is, although it retains the tireless and very fast-changing nature of all disruptive digital transformations, organizations are still adopting it much more slowly. Over the last years, technology was not a problem. The uncertainty management and adapting to the hybrid model of work have been a challenge. Change is slower or faster depending on the AU's ability, through its organizations, to visualize the right direction of progress through the uncertainties and to drive this change. The key to transformation is the vision and how is conducted the way to the transformation. Although Covid was a very bad thing, it helped us to move faster, and think differently. Incorporated at the territorial level between organizations on collaborative workflows, the priorities and measures adopted by each country depend on AU maturity. Reflecting on the assumptions from which we started today, the steps are faster and the decision support systems come to assist and support the development of volume and speed through intelligent agents capable of collaboration, communication and autonomy. In the context of the "new normal" disruption, the organizations that have begun the transformation will not only survive but also thrive.

References

1. AR6 Climate Change 2021: The Physical Science Basis (2021). https://www.ipcc.ch/report/ar6/wg1/
2. The Paris Agreement (2015). https://unfccc.int/process-and-meetings/the-paris-agreement/the-paris-agreement%20
3. Climate change COP 26 in Glasgow 2021, November, Politico (2021). https://www.politico.com/cop26-climate-conference-glasgow-live-updates
4. Economic losses from climate-related extremes in Europe (2022). https://www.eea.europa.eu/ims/economic-losses-from-climate-related#:~:text=Between%201980%20and%202020%2C%20weather,EUR%2011.9%20billion%20per%20year
5. The economics of climate change: no action not an option (2021). https://www.swissre.com/dam/jcr:e73ee7c3-7f83-4c17-a2b8-8ef23a8d3312/swissre-institute-expertise-publication-economics-of-climate-change.pdf
6. Decrease in carbon dioxide emissions worldwide between 1945 and 2020, by a major historical event, Statista (2022). https://www.statista.com/statistics/1111452/co2-emissions-decrease-due-tomajor-historical-events-globally/
7. Annual CO_2 emissions worldwide from 1940 to 2020 (in billion metric tons), Statista (2022). https://www.statista.com/statistics/276629/global-co2-emissions/
8. Biesbroek, R., Wright, S.J., Eguren, S.K., Bonotto, A., Athanasiadis, N.I.: Policy attention to climate change impacts, adaptation and vulnerability: a global assessment of National Communications (1994–2019). Climate Policy **22**(1), 97–111 (2022). https://www.tandfonline.com/doi/full/10.1080/14693062.2021.2018986
9. The United States' turning point on climate change, Deloitte (2022). https://www2.deloitte.com/content/dam/Deloitte/global/Documents/gx-global-turning-point-report.pdf
10. Digital Auto Report 2021 Accelerating towards the "new normal", PCW (2022). Digital Auto Report 2021 Vol. 1|Strategy& (pwc.com)
11. The net-zero transition: What it would cost, what it could bring, McKinsey (2022). https://www.mckinsey.com/business-functions/sustainability/our-insights/the-net-zero-transition-what-it-would-cost-what-it-could-bring
12. Global carbon dioxide emissions in 2020, by sector, Statista (2022). https://www.statista.com/statistics/276480/world-carbon-dioxide-emissions-bysector/
13. McKinsey's Scary Climate Math, McKinsey (2020). https://medium.com/clean-revolution/mckinseys-scary-climate-math-c08f0a56070d
14. 2007 Intergovernmental Panel for climate change summit for policy Makers on the Pasca Science Basis (Fourth Assessment Report), The Petersson et al., Appendix C (2006)
15. Lindskog, R., Löfmarck, E.: Managing uncertainty: forest professionals' claim and epistemic authority in the face of societal and climate change. Risk Manag. **17**, 145–164 (2015). https://doi.org/10.1057/rm.2015.10
16. van Bueren, E., Klijn, E.-H., Koppenjan, J.F.M.: Dealing with wicked problems in networks: analyzing an environmental debate from a network perspective. J. Public Adm. Res. Theory **13**(2), 193–212 (2003)
17. Betsill, M.M., Bulkeley, H.: Transnational networks and global environmental governance: the cities for a climate protection program. Int. Stud. Q. **48**, 471–493 (2004)
18. Qing-Shan, J., Hervé, P., Macchi, M., Siri, S., Weichhart, G., et al.: Control for smart systems: challenges and trends in smart cities. Ann. Rev. Control **53**, 358–369 (2022). https://doi.org/10.1016/j.arcontrol.2022.04.010
19. Motwani, S., Gupta, A.: Experiencing hyperloops: the transit of the future. Computer **54**(7), 37–45 (2021). https://doi.org/10.1109/MC.2021.3074006

20. Ocado Group-Annual Report 2020-Strategic-Report.pdf (2021). https://www.ocadogroup.com/media/eagb12qy/ocado-group-annual-report-2021-strategic.pdf
21. Pop Prototype: UP Next, prototype capsule with passengers taken over by an Audi Pop drone. Up Next|Audi Nanaimo (2018). https://www.audinanaimo.com/news-2018-audi-pop-up-next
22. Huhtamaa, H., Fredrik, C.L.: The climate in Nordic historical research – a research review and future perspectives, Scand. J. Hist. **46**(5), 665–695 (2021). https://doi.org/10.1080/03468755.2021.1929455
23. Hydrogen-acquisition-Helion-hydrogen-power (2021). https://www.alstom.com/press-releases-news/2021/4/alstom-enhances-its-expertise-hydrogen-acquisition-helion-hydrogen-power
24. SA 100MW battery storage tender won by Tesla and Neoen (2017). https://reneweconomy.com.au/sa-100mw-battery-storage-tender-won-by-tesla-and-neoen-65874/
25. National blueprint for lithium batteries 2021–2030 (2021). https://www.energy.gov/sites/default/files/2021-06/FCAB%20National%20Blueprint%20Lithium%20Batteries%200621_0.pdf
26. Rudskoy, A.., Ilin, I., Prokhorov, A.: Digital twins in the intelligent transport systems. Transp. Res. Proc. **54**, 927–935 (2021). https://doi.org/10.1016/j.trpro.2021.02.152. ISSN 2352-1465
27. Smith, G.: Making Mobility-as-a-Service: Towards Governance Principles and Pathways (2020). https://ses.library.usyd.edu.au/bitstream/handle/2123/21881/ITLS-WP-20-04.pdf?sequence=1
28. IDC FutureScape: Top 10 Predictions for the Future of Industry Ecosystems, IDC, (2021). https://www.idc.com/getdoc.jsp?containerId=prUS48441121
29. Thi Thi Tun, T.: Agent based decision support system for travelers. Int. J. Sci. Res. Sci. Eng. Technol. (IJSRSET) **7**(2), 293–298 (2020). https://doi.org/10.32628/IJSRSET207258. https://ijsrset.com/PDF.php?pid=6298&v=7&i=2&y=2020&m=March-April. Online ISSN 2394-4099. Print ISSN 2395–1990
30. Wooldridge, M., Nicholas, R.J.: Intelligent agents: theory and practice. Knowl. Eng. Rev. **10**, 115–152 (1995)
31. Wei, Q., Yan-Ning, S., Zi-Long, Z., Zhi-Yao, L., Yao-Ming, Z.: Multi-agent reinforcement learning-based dynamic task assignment for vehicles in urban transportation system. Int. J. Prod. Econ. **40**, 108251 (2021). https://doi.org/10.1016/j.ijpe.2021.108251. ISSN 0925-5273
32. Visan, M., Negrea, S.L., Firicel, M.: Towards intelligent public transport systems in Smart Cities. Collaborative decisions to be made. Proc. Comput. Sci. **199**, 1221–1228 (2022)
33. Filip, F.G., Zamfirescu, C.B., Ciurea, C.: Computer Supported Collaborative Decision-Making. Springer Cham (2017). https://doi.org/10.1007/978-3-319-47221-8
34. Filip, F.G.: Collaborative decision-making: concepts and supporting information and communication technology tools and systems. Int. J. Comput. Commun. Control [S.l.] **17**(2) (2022). ISSN 1841-9844. https://doi.org/10.15837/ijccc.2022.2.4732
35. Visan, M.: Spatial and territorial development planning: digital challenge and reinvention using a multi-disciplinary approach to support collaborative work. In: 7th International Conference on Information Technology and Quantitative Management (ITQM 2019) (2019). Proc. Comput. Sci. **162**, 795–802
36. 2022 CIO Agenda: Create an Action Plan to Master Business Composability (2022). 2022-cio-agenda-ebook.pdf (gcom.cloud)
37. Mladenović, M.N.: Mobility as a service. In: International Encyclopedia of Transportation, pp. 12–18 (2021). ISBN 978-0-08-102672-4. S2CID 242133419. https://doi.org/10.1016/B978-0-08-102671-7.106074

Research on Early-Warning of Financial Risks of Logistics Enterprises

Yi Song and Dan Chang(✉)

Beijing Jiaotong University, Beijing 10091, China
{19113051,chang}@bjtu.edu.cn

Abstract. The logistics industry is a composite service industry integrating transportation, warehousing, freight forwarding, and information industries. It is an important part of the national economy. Therefore, the research on the early-warning of the financial risks of logistics enterprises and whether they can be developed continuously are also necessary. Through the study of enterprise risk early-warning, doing a good job in financial risk management and control is the most important issue that must be solved to thoroughly evaluate, prevent, and resolve the financial risks of logistics companies so that they can develop in a sound and orderly manner. This paper builds 22 financial early-warning indicators from the six aspects of each share index, growth capability index, profitability index, profit quality index, operation quality index, risk capability index, etc., using PCA to reduce and optimize the latitude, and build a logistics enterprise Financial Risk Early-warning Model, which is based on BP neutral network.

Keywords: Logistics enterprise · Financial risk · BP neural network

1 Introduction

With the increasing development of information technology and the integration of the global economy, market competition is becoming increasingly fierce, and the social division of labor has been further refined. Logistics has begun to separate from traditional corporate functions and gradually form an independent industry. In recent years, as an emerging industry with rapid development, logistics companies have continuously expanded their scale and provided diversified services in the supply chain. Moreover, as the scale of the supply chain has expanded globally and the structure has become increasingly complex, the instability of the supply chain has increased. The management level of logistics enterprises also lags behind the pace of their scale development, all of which determines that it must face huge financial risks, and financial issues such as investment and financing have also attracted more and more attention from all walks of life.

In this context, paying attention to the financial risks faced by listed companies in the logistics industry and predicting their risks will play a key role in their healthy development and are also an important part of establishing a perfect modern logistics service

S. Dzitac et al. (Eds.): ICCCC 2022, AISC 1435, pp. 366–383, 2023.
https://doi.org/10.1007/978-3-031-16684-6_32

system. This paper uses BP neural network as a research tool for financial risk early-warning model, and conducts an empirical study on the financial risks of 78 logistics sample companies, and finally puts forward corresponding suggestions on how to further reduce the financial risks of China's logistics industry.

2 Research Status and Purpose

2.1 Financial Risk Early-Warning Model Research Study

An effective financial risk early-warning system should reflect the value of 'pre', be able to timely discover the signs and causes of financial risks, prevent the spread of crises, and reduce the expansion of losses. Financial risk early-warning can be roughly divided into univariate analysis, multiple discriminant analysis, logistic regression analysis, artificial neural network method, support vector machine method, etc.

Fitzpatrick (1932) first used the net profit/shareholders' equity and shareholder's equity/liability to analyze the difference between bankrupt and non-bankrupt companies for financial risk warning; Altman (1968) [5] Z-score model included unit asset working capital and unit asset retention 5 financial indicators: earnings, earnings before interest and taxes on unit assets, the market value of shareholders' equity/book value of liabilities, asset turnover rate; Zhou Shouhua et al. (1996) [5] added cash flow indicators to transform the Z-score model; Martin (1977) [5], James (1980) [5] He Qian Aimin and others (2008) [5] used logistic regression for financial risk early-warning analysis. This method is based on the cumulative probability function, with high accuracy and strong applicability; Odom and Sharda (1990) [5], Arzum and Yalcin (2007) [5], Yang Baoan et al. (2001) [5] and Lu Changjiang et al. (2004) [5] applied the artificial neural network method to financial risk early-warning. The biggest advantage of this method is that it has strong error correction and learning, but the method is based on brain nerve analog, constantly need to artificially debug, predicted strong randomness; Kyung-shikin et al. (2005) [5] and Yanjuan Juan et al. (2006) [5] using the branch. Vector machine for financial risk prediction, the results show that the SVM model is better than an artificial neural network. Peng Jing et al. (2008) [5] proposed a particle algorithm - support vector machine (PSO-SVM) model, which simultaneously optimizes the feature set and kernel function parameters, thereby improving the prediction results of the SVM model. Wu Dongmei (2010) [1] based on the screening of financial indicators and corporate governance variables, using a support vector machine method for empirical analysis, the results show that the model has a stronger predictive ability. Jia Wei Ying, etc. (2013) [5] was constructed using PCA of the financial crisis early-warning model, and the back-generation test and test samples in this test. The research results show that the prediction accuracy rate of the early-warning model for the test sample exceeds 76%, but it is highly professional and the analysis process is complex. It requires professionals to perform prediction and analysis, which is difficult for companies to use themselves and has low applicability.

2.2 Research on Financial Risks of Logistics Companies

Kentang Hong (2010) [2] believes that the financial risks of logistics companies exist in the overall supply chain of logistics companies, whether upstream, midstream, or downstream may be affected by external environmental factors or internal operating factors: when a specific debtor does not perform the contract. At this time, it will cause risks to the entire supply chain where logistics companies are located, causing financial losses between organizations.

Caihong Wang (2010) [3] pointed out that as an extension of the logistics business model and the supply chain financing supervision business of the logistics value-added business, it not only brings new profit growth points for traditional logistics companies but also promotes its transformation into a modern logistics company. At the same time, it also analyzed the financial risks of the supply chain financing supervision business, pointed out that the core of its business operations is risk control, and proposed some measures to prevent risks.

Qi Yujuan (2018) [6] analyzed the financial risks of the logistics industry from the perspective of the financial internal control of logistics enterprises. She believes that under the new situation, there are several outstanding problems in the financial internal control of logistics enterprises: a. The internal control environment is poor. Logistics companies have been under extensive management for a long period, and there are many shortcomings in the governance structure. The basic quality of internal control-related staff is low, making the internal control work of logistics companies unable to be fully carried out. At the same time, a large number of logistics companies are not paying enough attention to the modeling of logistics enterprise culture in the actual operation process, which will also be extremely detrimental to the internal control of logistics companies. b. There are outstanding financial risks in the operation of logistics enterprises. Interest rate risk, foreign exchange risk, commodity price risk, postponement risk, etc. are common business risks in the business activities of modern logistics companies. For example, due to the impact of tight monetary policy, modern logistics companies are facing huge financing difficulties, which not only hinders the effective flow of logistics companies' funds but also logistics companies will pay higher capital costs for this. Some modern logistics companies undertake cross-border projects. With constant exchange rate changes, logistics companies will face financial risks on the foreign exchange level. c. Risk control execution is low. The focus of the existing problems is that the risk identification of logistics companies does not set up a special department, and there is no very precise risk identification method and path. Besides, from a risk perspective, the professionalism of the risk management team of logistics companies is generally low, and their experience is not very rich.

Other research results are mostly from the perspective of the supply chain to discuss the financial risks of logistics companies, but there is still a lack of scientific and systematic in-depth research on the financial risks of logistics companies themselves, and the selection of financial indicators is relatively old.

This article will use the BP neural network to build a financial early-warning model. The construction of this forward multi-layer neural network is based on the neural operation principle of the human brain. Information processing is divided into forwarding propagation and backward learning. Repeated learning and training to reduce the error makes the actual output close to the expected output, to predict the future financial situation.

Compared with other methods, BP neural network has a high degree of fault tolerance and strong nonlinear mapping ability, and the requirements for data distribution and completeness are not very strict, so that more data samples are feasible for analysis; The outstanding learning ability makes the model more flexible to adapt and analyze the financial data representing different financial environments, and make more accurate early-warning.

This article will use PCA to optimize the selected financial early-warning indicators, and use the optimized early-warning indicators to establish a logistics enterprise financial early-warning model based on the BP neural network model. On the one hand, it can further explore more scientific and effective logistics enterprise financial early-warning Model; on the other hand, it also enriches the application of BP neural network in logistics enterprise financial early-warning model.

3 Empirical Research Design and Results

3.1 Definition of the Financial Crisis

Foreign countries usually use the company's bankruptcy as a criterion for determining whether the company is in financial crisis, while most domestic standards for determining financial crisis are whether listed companies are specially treated by the stock exchange and are given the title of "*ST". However, the number of logistics companies in China that have been specially treated so far is very small, only four, so this method is unreasonable. This article will continue to follow Youhong Xing (2011) [4] defined methods, EVA is positive indicates an increase enterprise value of value, classified as finance normal (EVA $>$ 0); EVA is negative it means reducing the corporate value of that business into a financial crisis (EVA $<$ 0). The EVA value selected in this paper is directly extracted from CSMAR, and some enterprises that are missing in some years are calculated according to the following formula: EVA = NOPAT - adjusted capital × KWACC. Which, NOPAT = Net profit + (interest expense + research and development expenses adjustments) × (1 –25%), adjusted capital = Average owner's equity + average interest-bearing liabilities Total - average in construction; KWACC = KD × average Interest-bearing liabilities/(average interest-bearing liabilities + average owner's equity) × (1–25%) + KS × average owner's equity/(average interest-bearing liabilities + average owner's equity). Among them, NOPAT means net profit after tax, KWACC means the average cost of capital, KD means debt capital cost, and KS means equity capital cost.

3.2 Sample Selection and Data Sources

After excluding B-share companies, ST companies, * ST companies, companies with incomplete financial data, and companies with a short establishment time, a total of 468 samples were selected from 78 logistics companies, 327 samples were used for training, and 141 samples were used for training. Inspection, annual interception is selected from the data of the six-year annual report of 2013–2018. The selection principle is that the enterprise data set during the six years is the most complete and the number of companies that exist is the largest. All data selected in this article come from eastmoney.com and CSMAR. The application software includes SPSS 20.0 and Python 3.7.

3.3 Selection and Optimization of Forecast Indicators

Primary Selection of Financial Indicators

This article selects 22 sub-indicators that are complete, conducive to analysis, and more relevant to the operating characteristics of logistics enterprises under the six main indicators, and constructs all financial indicators required by the financial risk carly-warning indicator system (Table 1).

Table 1. Financial indicators.

Financial dimension	Serial number	Index
Index per share	X 1	Basic earnings per share (yuan)
	X 2	Non-earnings per share (yuan)
	X 3	Net assets per share (yuan)
	X 4	Provident fund per share (yuan)
	X 5	Undistributed profit per share (yuan)
	X 6	Operating cash flow per share (yuan)
Growth capability index	X 7	Total operating income (yuan)
	X 8	Gross profit (yuan)
	X 9	Attributable net profit (yuan)
Profitability index	X 10	Weighted return on net assets (%)
	X 11	Diluted return on net assets (%)
	X 12	Diluted return on total assets (%)
	X 13	Gross margin (%)
	X 14	Net interest rate (%)
Profit quality index	X 15	Sales cash flow/operating income
	X 16	Operating cash flow/operating income
Operating capability index	X 17	Turnover of total assets (times)
	X 18	Receivables turnover days (days)
Financial risk indicators	X 19	Assets and liabilities (%)
	X 20	Current liabilities/total liabilities (%)
	X 21	Current ratio
	X 22	Quick ratio

Optimization of Early-Warning Indicators

Too many input nodes of the neural network will cause the convergence and stability of the neural network to deteriorate, and may even show an overfitting phenomenon. Therefore, the PCA method is used for dimensionality reduction.

KMO Test and Bartlett Sphericity Test

Import the obtained data into SPSS22.0, input the 22 indicators as variables into the factor analysis model for PCA, and obtain the KMO and Buffett test values of 0.642 > 0.5, the significance is 0.000 < 0.01, At a significance level of 1%, this data is suitable for factor analysis (Table 2).

Table 2. KMO and Bartlett inspection.

KMO sampling suitability measure		.642
Bartlett's sphericity test	Chi-square degrees of freedom last read Distinctiveness	9899.552
		231
		0.000

Selection of Principal Components

Through factor analysis, the total variance interpretation table is obtained, and the feature values greater than 1 are extracted as the principal components. From the table below, it can be seen that 7 principal components have been extracted.

- Factor 1 can explain 24.232% of the original variable information;
- Factor 2 can explain 14.856% of the original variable information;
- Factor 3 can explain 13.078% of the original variable information;
- Factor 4 can explain 8.059% of the original variable information;
- Factor 5 can explain 7.324% of the original variable information;
- Factor 6 can explain 5.380% of the original variable information;
- Factor 7 can explain 4.762% of the original variable information.

The accumulation of all common factors can explain 78.493% of the original variables.

Seven principal components are rotated by factor loading to obtain a component matrix table. In principal component 1, the absolute value of the coefficient of basic earnings per share and deduction of non-earnings per share is relatively large, which mainly measures the company's per share index; The absolute value of the coefficient of interest rate is large, which mainly measures the profitability of the company; the absolute value of the coefficient of the quick freezing ratio in the main component 3 is large, which can be interpreted as a measure of the financial risk index of the enterprise; the absolute value of the coefficient of gross profit in the main component 5 is large, you can interpreted as a measure of business and growth capacity indicators; principal components factor 6 sales cash flow/revenue, it can be interpreted as a measure of earnings quality indicators; in the measure of business operating capability index coefficient has an absolute value of the index is the largest principal components factor 7 the main components in addition to the 7 diluted ROA coefficient larger absolute value, the absolute value as well as cash flow from operations per share in the index is also large, comprehensive to look at the main components 7 can be used as a measure of business operating capacity indicators representative. Principal component 4 can be interpreted as the principal component of other indicators.

On the whole, the initial screening business indicators also from share index, growth capacity indicators, profitability indicators, profitability indicators of quality, operating capacity indicators, financial business risk index for screening. This shows that the component factors obtained by the PCA are representative (Tables 3, 4 and 5).

Table 3. Total variance explained (Extraction method: PCA).

Components	Initial eigenvalue			Extract the sum of squared loads		
	Total	Variance percentage	Cumulative %	Total	Variance percentage	Cumulative %
1	5.331	24.232	24.232	5.331	24.232	24.232
2	3.268	14.856	39.088	3.268	14.856	39.088
3	2.877	13.078	52.166	2.877	13.078	52.166
4	1.949	8.859	61.025	1.949	8.859	61.025
5	1.611	7.324	68.350	1.611	7.324	68.350
6	1.184	5.380	73.730	1.184	5.380	73.730
7	1.048	4.762	78.493	1.048	4.762	78.493
8	.972	4.419	82.912			
9	.793	3.603	86.514			
10	.726	3.298	89.813			
11	.538	2.447	92.260			
12	.348	1.583	93.843			
13	.328	1.490	95.333			
14	.306	1.389	96.722			
15	.217	.985	97.707			
16	.189	.861	98.567			
17	.120	.545	99.113			
18	.062	.284	99.396			
19	.057	.260	99.656			
20	.038	.174	99.830			
21	.025	.113	99.944			
22	.012	.056	100.000			

Table 4. Rotation component matrix a (Extraction method: PCA).

	Components						
	1	2	3	4	5	6	7
Basic earnings per share (yuan)	.934	.154	– .034	.019	– .160	.012	.083
Non-earnings per share (yuan)	.896	.109	– .065	.013	– .173	– .016	.064
Net assets per share (yuan)	.454	.309	– .036	.742	– .317	.030	.069
Provident fund per share (yuan)	– .007	.305	.115	.632	– .086	– .094	– .348
Undistributed profit per share (yuan)	.587	.225	– .103	.406	– .326	.107	.352
Operating cash flow per share (yuan)	.243	– .270	– .213	.238	– .283	.063	– .603
Camp industry gross income into the (yuan)	.192	.527	– .435	.053	.472	.238	.141
Gross profit (yuan)	.416	.256	– .582	.152	.572	– .094	– .022
Attributable net profit (yuan)	.518	.076	– .458	.112	.491	– .252	– .063
Weighted return on net assets (%)	.829	– .028	.072	– .430	– .108	.009	– .066
Diluted return on net assets (%)	.818	– .064	.022	– .393	– .144	– .006	– .013
Spread a thin return on total assets (%)	.832	– .072	.249	– .320	– .011	– .045	– .183
Gross margin (%)	.231	– .829	– .080	.067	.064	– .008	.115
Net interest rate (%)	.470	– .653	.002	.047	.103	– .004	.207
Sales cash flow/operating income	– .085	– .177	.006	.083	– .050	.804	.127
Operating cash flow/operating income	.142	– .769	– .283	.185	.016	.185	– .047
Total assets week transfer rate (times)	.133	.705	.121	– .282	.009	.306	– .103
Receivables turnover days (days)	– .081	.053	.367	.121	– .149	– .430	.459
Assets and liabilities (%)	– .258	.206	– .614	– .206	– .242	.138	.240
Flow movable liabilities/Total liabilities (%)	.068	.522	.462	– .152	– .089	.006	– .069
Flow dynamic ratio	.250	– .083	.753	.214	.424	.189	.054
Quick ratio	.246	– .064	.776	.204	.389	.113	.006

Table 5. Component score coefficient matrix (Extraction method: PCA).

	Components						
	1	2	3	4	5	6	7
Basic earnings per share (yuan)	.175	.047	–.012	.010	–.099	.010	.079
Non-earnings per share (yuan)	.168	.033	–.022	.006	–.107	–.014	.061
Net assets per share (yuan)	.085	.094	–.012	.381	–.197	.025	.065
Each share fund (yuan)	–.001	.093	.040	.325	–.053	–.080	–.332
Undistributed profit per share (yuan)	.110	.069	–.036	.208	–.203	.090	.336
Operating cash flow per share (yuan)	.046	–.083	–.074	.122	–.175	.053	–.575
Total operating income (yuan)	.036	.161	–.151	.027	.293	.201	.134
Gross profit (yuan)	.078	.078	–.202	.078	.355	–.079	–.021
Attributable net profit (yuan)	.097	.023	–.159	.057	.305	–.213	–.060
Weighted return on net assets (%)	.156	–.008	.025	–.221	–.067	.007	–.063
Diluted return on net assets (%)	.153	–.020	.008	–.202	–.089	–.005	–.012
Diluted return on total assets (%)	.156	–.022	.086	–.164	–.007	–.038	–.175
Gross margin (%)	.043	–.254	–.028	.035	.040	–.007	.110
Net interest rate (%)	.088	–.200	.001	.024	.064	–.003	.197
Sales cash flow/operating income	–.016	–.054	.002	.043	–.031	.679	.121
Operating cash flow/operating income	.027	–.235	–.098	.095	.010	.156	–.045
Turnover of total assets (times)	.025	.216	.042	–.145	.006	.259	–.098
Accounts receivable turnover days (days)	–.015	.016	.128	.062	–.092	–.363	.438

(*continued*)

Table 5. (*continued*)

	Components						
	1	2	3	4	5	6	7
Asset-liability ratio (%)	–.048	.063	–.213	–.106	–.150	.117	.229
Current liabilities/total liabilities (%)	.013	.160	.160	–.078	–.055	.005	–.066
Current ratio	.047	–.025	.262	.110	.263	.159	.052
Quick ratio	.046	–.020	.270	.105	.242	.096	.006

The principal component coefficient matrix is obtained, and the seven principal component scores are calculated based on the standardized values of the 22 selected indicators.

$$\begin{aligned}
F1 = {} & 0.175X_1 + 0.168X_2 + 0.085X_3 - 0.001X_4 \\
& +0.110X_5 + 0.046X_6 + 0.036X_7 + 0.078X_8 \\
& +0.097X_9 + 0.156X_{10} + 0.153X_{11} + 0.156X_{12} \\
& +0.043X_{13} + 0.088X_{14} - 0.016X_{15} + 0.027X_{16} \\
& +0.025X_{17} - 0.015X_{18} - 0.048X_{19} + 0.013X_{20} \\
& +0.047X_{21} + 0.046X_{22}
\end{aligned}$$

$$\begin{aligned}
F2 = {} & 0.047X_1 + 0.033X_2 + 0.094X_3 + 0.093X_4 \\
& +0.069X_5 - 0.083X_6 + 0.161X_7 + 0.078X_8 \\
& +0.023X_9 - 0.008X_{10} - 0.020X_{11} - 0.022X_{12} \\
& -0.254X_{13} - 0.200X_{14} - 0.054X_{15} - 0.235X_{16} \\
& +0.216X_{17} + 0.016X_{18} + 0.063X_{19} + 0.160X_{20} \\
& -0.025X_{21} - 0.020X_{22}
\end{aligned}$$

$$\begin{aligned}
F3 = {} & -0.012X_1 - 0.022X_2 - 0.012X_3 + 0.04X_4 \\
& -0.036X_5 - 0.074X_6 - 0.151X_7 - 0.202X_8 \\
& -0.159X_9 + 0.025X_{10} + 0.008X_{11} + 0.086X_{12} \\
& -0.028X_{13} + 0.001X_{14} + 0.002X_{15} - 0.098X_{16} \\
& +0.042X_{17} + 0.128X_{18} - 0.213X_{19} + 0.160X_{20} \\
& +0.262X_{21} - 0.270X_{22}
\end{aligned}$$

$$\begin{aligned}
F4 = {} & 0.010X_1 + 0.006X_2 + 0.381X_3 + 0.325X_4 \\
& +0.208X_5 + 0.122X_6 + 0.027X_7 + 0.078X_8 \\
& +0.057X_9 - 0.221X_{10} - 0.202X_{11} - 0.164X_{12} \\
& +0.035 + 0.024X_{14} + 0.043X_{15} + 0.095X_{16} \\
& -0.145X_{17} + 0.062X_{18} - 0.106X_{19} - 0.078X_{20} \\
& +0.110X_{21} + 0.105X_{22}
\end{aligned}$$

$$F5 = -0.099X_1 - 0.107X_2 - 0.197X_3 - 0.053X_4 \\ -0.203X_5 - 0.175X_6 + 0.293X_7 + 0.355X_8 \\ +0.305X_9 - 0.067X_{10} - 0.089X_{11} - 0.007X_{12} \\ +0.040X_{13} + 0.064X_{14} - 0.031X_{15} + 0.010X_{16} \\ +0.006X_{17} - 0.092X_{18} - 0.150X_{19} - 0.055X_{20} \\ +0.263X_{21} + 0.242X_{22}$$

$$F6 = 0.010X_1 - 0.014X_2 + 0.025X_3 - 0.080X_4 \\ +0.090X_5 + 0.053X_6 + 0.201X_7 - 0.079X_8 \\ -0.213X_9 + 0.007X_{10} - 0.005X_{11} - 0.038X_{12} \\ -0.007X_{13} - 0.003X_{14} + 0.679X_{15} + 0.156X_{16} \\ +0.259X_{17} - 0.363X_{18} + 0.117X_{19} + 0.005X_{20} \\ +0.159X_{21} + 0.096X_{22}$$

$$F7 = 0.079X_1 + 0.061X_2 + 0.065X_3 - 0.332X_4 \\ +0.336X_5 - 0.575X_6 + 0.134X_7 - 0.021X_8 \\ -0.060X_9 - 0.063X_{10} - 0.012X_{11} - 0.175X_{12} \\ +0.110X_{13} + 0.197X_{14} + 0.121X_{15} - 0.045X_{16} \\ -0.098X_{17} + 0.438X_{18} + 0.229X_{19} - 0.066X_{20} \\ +0.052X_{21} + 0.006X_{22}$$

The seven principal components F1–F7 are used as input layer variables of the BP neural network to train and test the financial early-warning model.

BP Neural Network Design

In this paper, a total of 468 samples are used, 327 samples are used for training, and 141 samples are used for detection (Table 6).

Table 6. Actual value and predictive value of training samples.

Sample name	Actual value	Predictive value	Sample name	Actual value	Predictive value
Oriental Fashion 2013	1	1	Yichang Delivery 2013	1	1
Xiamen Xiangyu 2016	0	0	City Development Environment 2016	1	1
Beiba Media 2017	1	1	Qingdao Port 2018	0	0

(*continued*)

Table 6. (*continued*)

Sample name	Actual value	Predictive value	Sample name	Actual value	Predictive value
STO Express 2016	1	1	Longjiang Transportation 2017	1	1
Hyundai Investment 2014	1	1	Fulin Transport 2017	1	1
Qingdao Port 2013	1	1	Shanxi Road and Bridge 2014	1	1
Dongfang Jiasheng 2016	1	0	City Development Environment 2017	1	1
Bohai Ferry 2013	1	1	Jinzhou Port 2014	1	1
Daqin Railway 2018	0	0	Tielong Logistics 2015	1	0
Tianshun shares 2014	1	1	Hunan Investment 2016	1	1
China Merchants Port 2017	0	1	Wuzhou Transportation 2016	0	1
Huaihe Energy 2015	1	1	STO Express 2018	0	0
Haiqi Group in 2015	1	1	Zhongyuan Expressway 2017	1	1
Nanjing Port 2013	1	1	Bohai Ferry 2015	1	1
Qingdao Port 2014	1	1	Flyida 2013	1	1
China Reserve Holdings 2018	0	0	Hyundai Investment 2013	1	1
Mass Transportation 2015	1	1	C & D shares 2014	1	0
Yingkou Port 2014	1	1	Sinotrans Development 2014	1	0

(*continued*)

Table 6. (*continued*)

Sample name	Actual value	Predictive value	Sample name	Actual value	Predictive value
Guangzhou Port 2014	1	1	Jinjiang Investment 2014	1	1
Fulin Transport 2015	1	1	SF Holdings 2018	0	0
Western Venture 2013	1	1	Dexin Delivery 2018	1	1
Huamao Logistics 2015	1	1	Xiamen Xiangyu 2014	0	0
Oriental Fashion 2018	0	1	Zhuhai Port 2016	1	1
Shenzhen Expressway 2014	1	1	Dalian Port 2013	1	1
Wuzhou Transportation 2018	1	1	Shentong Subway 2018	1	1
Longjiang Transportation 2014	1	1	Xinning Logistics 2017	1	1
Hua Pengfei 2017	1	1	Beibu Gulf Port 2014	1	1
China Merchants Port 2018	0	0	Qingdao Port 2015	1	1
Longjiang Transportation 2016	1	1	Jiangxi Changyun 2017	1	1
Sichuan Chengyu 2016	1	1	Beibu Gulf Port 2016	1	1
Fujian Expressway 2013	1	1	Hua Pengfei 2016	1	1
Hengtong shares 2018	1	0	Rui Maotong 2016	0	0
Zhuhai Port 2015	1	1	Lianyungang 2018	1	1

(*continued*)

Table 6. (*continued*)

Sample name	Actual value	Predictive value	Sample name	Actual value	Predictive value
Fulin Transport Industry 2014	1	1	Yichang Delivery 2017	1	1
Xinning Logistics 2016	1	1	Changjiu Logistics 2014	1	1
Jiangxi Changyun 2014	1	1	SF Holdings 2017	0	0
Yuanshang shares 2017	1	1	Tianshun shares 2017	1	1
Beibu Gulf Port 2018	0	1	Dexin Delivery 2014	1	1
Rizhao Port 2013	1	1	Henderson Group 2014	1	1
Hengtong shares 2016	1	1	Yinfei Storage 2016	1	1
Zhongyuan Expressway 2013	0	1	Flytec 2018	1	1
Yinfei Storage 2014	0	1	Henderson Daxin 2017	1	1
Changlian Shares 2013	1	1	Hyundai Investment 2016	1	1
Jinzhou Port 2015	1	1	Jiangxi Changyun 2018	1	1
Sichuan Chengyu 2017	1	1	Haiqi Group 2014	1	1
China Merchants Port 2016	1	1	Kason International 2018	1	1
Beiba Media 2013	1	1	Fulin Transport 2016	1	1
Guangzhou Port 2013	1	1	Dalian Port 2016	0	1
COSCO Hite 2014	1	0	Kason International 2014	1	1

(*continued*)

Table 6. (*continued*)

Sample name	Actual value	Predictive value	Sample name	Actual value	Predictive value
Aoyang Shunohang 2017	1	1	Aoyang Shunohang 2018	1	1
Nanjing Port 2015	1	1	China Reserve Holdings 2014	0	0
Fujian Expressway 2015	1	1	Fujian Expressway 2018	1	1
Tianshun shares 2016	0	1	Dalian Port 2017	1	1
STO Express 2015	1	1	Oriental Fashion 2016	0	1
Changlian Co., Ltd. 2015	1	1	Tielong Logistics 2014	1	0
COSCO Hite 2018	1	1	Ningbo Shipping 2015	1	1
Lianyungang 2013	1	1	Shenzhen Expressway 2013	1	1
Henderson Daxin 2016	1	1	Tianshun shares 2018	1	1
Guangzhou Port 2018	0	0	Hunan Investment 2017	1	1
Hengtong shares in 2014	1	1	Rizhao Port 2017	1	1
Chuanhua Zhilian 2017	0	0	Tangshan Port 2014	1	1
Yichang Delivery 2014	1	1	Changlian Shares 2017	1	1
City Development Environment 2015	1	1	Changlian shares 2018	1	1
YTO Express 2015	0	0	Yichang Delivery 2018	0	1

(*continued*)

Table 6. (*continued*)

Sample name	Actual value	Predictive value	Sample name	Actual value	Predictive value
COSCO SHIPPING 2016	0	0	Xiamen Xiangyu 2013	0	0
Haiqi Group 2018	1	1	Jiangxi Changyun 2015	1	1
COSCO SHIPPING 2015	0	0	Dexin Delivery 2017	1	1
Henderson Daxin 2018	1	1	STO Express 2013	1	1
Beiba Media 2016	1	1	Chongqing Gangjiu 2017	1	0
City Development Environment 2013	1	1	Western Venture 2015	1	1
			Sinotrans Development 2013	1	0

The first type of misjudgment result in the model is to mistakenly judge 0 "company with no financial risk or normal financial status" as 1 "company with the high-risk financial crisis". The data of this test are as follows: Oriental Fashion 2018, The actual value is 0, and the misjudgment is 1; the second misjudgment result is to judge 1 "the company with high-risk financial crisis" as 0 "the company without financial risk or financial normal". The data of this test is Dongfang Jiasheng in 2016, the actual value was 1 and the misjudgment was 0. Among the 141 test samples, the accuracy rate of the first type of misjudgment results is 34%, but the result of the second type of misjudgment rate is only 9%. Overall, the accuracy rate of the neural network model is as high as 87% (Table 7).

Table 7. Actual value and predictive value of training samples.

0	0.65	0.63	0.64	27
1	0.91	0.92	0.92	114
Accuracy			0.87	141

(*continued*)

Table 7. (*continued*)

0	0.65	0.63	0.64	27
Macro avg	0.78	0.78	0.78	141
Weighted avg	0.86	0.87	0.86	141

4 Conclusion

This paper constructs a logistics enterprise financial early-warning model based on the BP neural network model. Through the analysis and testing of 78 logistics enterprises, the scientific validity of the logistics enterprise financial early-warning model of this model is verified.

First, dimensionality reduction and optimization of primary selection indicators. On the one hand, the selection of financial indicators is more comprehensive, on the other hand, it avoids the excessive burden of the BP neural network; Second, the BP neural network can be repeatedly trained and learned, and the error is propagated backward from the input layer to the output layer, and the value is corrected Small errors can get higher accuracy; Third, the model is not a mechanical input and output. With the update of training samples and detection samples, the network parameters of the model are constantly changing. New training and learning are performed. The adjustable network parameters make the model adaptable to more variable financial environment of logistics enterprises.

Although the BP neural network model is more accurate in the early-warning of financial risks of logistics companies, the managers of logistics companies should still maintain their focus on financial risk management, improve the risk management system and risk monitoring system, and avoid falling into financial crises, making logistics companies continue healthy development.

The shortcoming of this article is that there are currently few samples of logistics companies that can be used, and the impact of text information on financial risk early-warning is not considered. With the increase of samples, the BP neural network model may be further optimized. If the influence of text information is included and sentiment analysis is performed, the accuracy of risk warning may be higher.

Acknowledgement. This work is supported by Beijing Logistics Informatics Research Base.

References

1. Wu, D., Zhu, J., Zhuang, X., Yang, L.: Support Vector Machine based financial crisis early-warning model. J. Northeast. Univ. (Nat. Sci. Ed.) **31**(04), 601–604 (2010)
2. Hong, K.: Classification of supply chain risks and research on risk sources—Based on literature review method. Logist. Technol. **29**(08), 97–100 (2010)
3. Wang, C.: An analysis of financial risks in supply chain financing supervision business. Transp. Fi-nance Account. (03), 14–17 (2010)

4. Xing, Y., Zhang, J.: Empirical research on airline fi-nancial early-warning. Account. Newslett. (36), 80–82 (2011)
5. Zhang, X., Hu, A.: Literature review of empirical re-search on financial risk early-warning models at home and abroad. China Secur. Futures (03), 109–111 (2013)
6. Qi, Y.: Analysis of the problems and countermeas-ures in the financial internal control of logistics enter-prises under the new situation. Finance Econ. (Acad. Ed.) (13), 77–78 (2018)

Trustworthy and Green Design

Bridging Reliability to Efficiency Consecutive Elegant and Simple Design

Valeriu Beiu[1], Sorin-Horațiu Hoară[2], and Roxana-Mariana Beiu[1](✉)

[1] "Aurel Vlaicu" University of Arad, 2-4 Elena Drăgoi Str., 310330 Arad, Romania
{valeriu.beiu,roxana.beiu}@uav.ro

[2] Faculty of Automation and Computers, Politehnica University of Timișoara, 2 Vasile Pârvan Blvd., 300223 Timișoara, Romania
sorin-horatiu.hoara@student.upt.ro

Abstract. An acute problem when moving into the few nanometers highly advanced CMOS technologies is represented by the inexorable cost-yield balance, lately clearly tilted by the hefty costs. Amongst the options worth investigating (for reducing costs), designing for (enhanced) reliability has still not gained traction, as implicitly considered a power-/energy-hungry solution (due to redundancy), as well as a complex design alternative (hence risky). From the biological side, neurons are prime examples of highly efficient designs reaching outstanding communication and computation reliabilities, although relying on random devices (known as ion channels or, more fittingly, single ion transistors). In this paper, bridging from biology to circuits, we will show how kind of overlooked mathematical results (about consecutive systems) together with novel Binet-equivalent formulas (for Fibonnaci numbers of higher orders) can be pieced together for completely avoiding reliability calculations for consecutive systems. Finally, using such results, in combination with freshly defined cost functions (for reliability), we are able to present a trivial design scheme for consecutive systems which is balancing reliability and efficiency.

Keywords: Reliability · Consecutive system · Efficient/optimal design

1 Introduction and Motivations

By far the most famed development of the 19th century was the computer! It is amazing that it was conceived two centuries ago (around 1820 by Charles Babbage, whose ideas on computation were rewarded in 1824 by the Gold Medal of the Royal Astronomical Society *"for his invention of an engine for calculating mathematical and astronomical tables"*). The brilliant concepts put forward lay dormant for over a century, but a detailed recount was presented in a review from 1834 [1]. It relied on documents from 1822–1826 as well as on a report from 1829 (of a committee appointed by the Council of the Royal Society). As our interest in this paper is entirely on efficient and reliable designs we are going to include a few quotes from [1] which will make it clear that ***reliability was a concern even before the first computer was built***, and that the concept of modular redundancy mentioned by John von Neumann in [2] was known and used. Here are two quotes from [1] on the making of numerical tables.

S. Dzitac et al. (Eds.): ICCCC 2022, AISC 1435, pp. 387–400, 2023.
https://doi.org/10.1007/978-3-031-16684-6_33

- *"Each number was calculated by two or more distinct and independent computers, and its truth and accuracy determined by the coincidence of the results thus obtained."* Obviously, this is (at least) duplication followed by checking (if the results are identical).
- *"[T]hose numbers in which they are found to agree must be correct; inasmuch as the chances are supposed to be very considerable against two or more independent computers falling into precisely the same errors."* This means that, as $p_1, p_2, \ldots$ are (reasonably) small, then $p_1 \times p_2 \ldots$ is even smaller (negligible).

Clearly, computer reliability was seen as a prerequisite, and reliability enhancements based on redundancy were envisaged from the very early beginning. Additionally, Charles Babbage was aware of tolerances and variations as well as mechanical wear and tear (something the semiconductor industry is now calling "ageing"), and his designs have included particular (mechanical) methods to deal with these. With respect to the mechanical technology (of his time), his conclusion was that ***"numerical correctness is altogether unattainable until some more perfect method be discovered,"*** and suggested that *"the most certain and effectual check upon errors which arise in the process of computation, is to cause the same computations to be made by separate and independent computers; and this check is rendered still more decisive if they make their computations by different methods."*

Nowadays, computers are relying on chips built using tiny CMOS transistors (currently 5 nm process being in production, with 3 nm in the pipeline) connected by a complex mesh of criss-crossing wires. Scaling and new processing steps (e.g., EUV lithography) are only complicating the fabrication process and affecting yield, hiking up prices which are already high [3]. Obviously, efficiency is of paramount importance as money (related to area, hence number of transistors) should translate into high speeds (short delays on wires) and low power/energy (a cost function capturing both the number of transistors and the length of the wires). The other parameter to consider is yield, and that is why reliability (of computations and communications) is gaining importance. Still, designing for reliability has not seen much progress although a highly efficient example exists: the brain! Explanation normally fall on the line of *"it is a different technology,"* but we still do not have a clear understanding of how the brain is able to achieve outstanding reliability (through redundancy) while simultaneously dissipating $10^{3 \ldots 4}$ times less than the most advanced CMOS.

In the following we shall focus on a very particular type of network (consecutive), and present an extremely simple design approach for *linear consecutive systems*. The scheme can be used to optimize such networks for either communications or computations. The approach we are going to present here is entirely avoiding the time consuming calculations of the associated reliability polynomial (or of its approximations), being comparable in spirit to computing the simplest bounds.

The paper is structured as follows. Section 2 will go briefly over the state-of-the-art of network reliability and consecutive systems, while mentioning consecutive systems in biology, particularly neurons. Section 3 will link the reliability of linear consecutive systems at $p = 0.5$ to Fibonacci numbers of higher orders—which can be represented using a Bine-like formula. Section 4 will show how to approximate k when n is 'large,' either for communications or for computations (the optimization criteria being slightly different), and follow with supporting simulations. Conclusions and further directions for research are going to end the paper.

2 State-of-the-Art

2.1 Network Reliability

Network reliability was introduced by Moore and Shannon [4]. Conceptually they abstracted a network N as a graph of nodes connected by edges, and assumed that the nodes always work perfectly, while only the edges fail independently. More precisely, each edge works (sends information, hence it is 'closed') with a probability p, and fails (is not able to send information as it is 'open') with probability $q = 1-p$. This gives rise to a reliability polynomial associated to each network $Rel(N; p)$. The opposite concept when edges are perfect and only nodes are failing somehow matches the schemes put forward by von Neumann [2], if the nodes represent Boolean logic gates (a concept also advanced by Pierce [5]).

In the following, we shall rely on the classical interpretation which implies that a network N fails if edges have failed such as information cannot travel through the network (with the caveat that one could consider 2 nodes, k nodes, or *all* nodes).

Optimality has obviously been a major concern (see [6]), while the interested reader is referred to two fresh reviews [7, 8] which mention quite a few open problems (see also [9]). Many of the open problems have to do with the fact that calculating the reliability polynomial $Rel(N; p)$ can be reasonable simple for some (particular) cases, while for others it can be an extremely difficult (#P-complete) task [10, 11].

2.2 Consecutive Systems

Consecutive systems were introduced by Kontoleon in 1980 as "r-successive-out-of-n:F" systems [12]. This concept was a departure from, and generalization of the "r-out-of-n:F" systems, while the two concepts perfectly overlap for two cases: when $r = 1$ (leading to a series system) and $r = n$ (leading to a parallel system). Naming them consecutive-k-out-of-n:F was done by Chiang and Niu one year later [13].

Since the 1980s quite a few variations, besides linear, have been proposed: circular, 2D, cylindrical, and others. Consecutive systems have also been use, early examples being represented by street lights (linear consecutive-3-out-of-n:F) and gas/oil pipelines, recent examples including telecommunications and reaction chambers [14–22]. They are appealing as they span from series to parallel, and are ordered when holding one parameter constant and varying the other one.

Lately it was found that some of the equivalent statistical problems were known and solved by de Moivre [23], but dealing with large n and finding economical solutions (minimizing k) is nontrivial.

Further we shall consider linear consecutive-k-out-of-16 ($Lin/Con/k/16$).

The coefficients of the reliability polynomials associated to $Lin/Con/k/16$ for $k = 1, 2, \ldots, 16$ are presented in Table 1, and simulations are shown in Fig. 2.

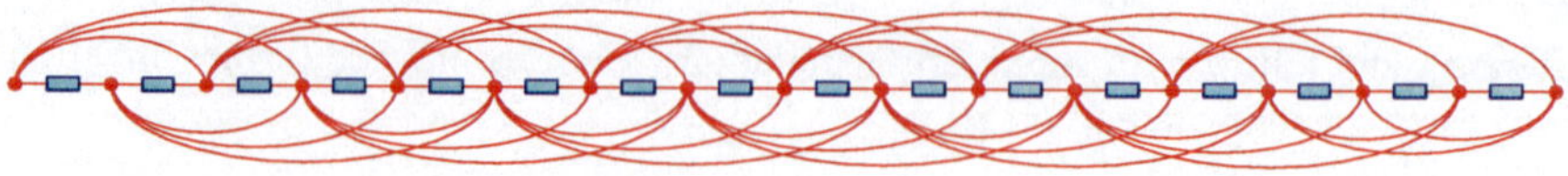

Fig. 1. A particular linear consecutive-4-out-of-16:F system ($Lin/Con/4/16$).

Table 1. The coefficients of $Rel(Lin/Con/k/16; p)$ for $k = 1 \dots 16$.

k	p^0 q^{16}	p^1 q^{15}	p^2 q^{14}	p^3 q^{13}	p^4 q^{12}	p^5 q^{11}	p^6 q^{10}	p^7 q^9	p^8 q^8	p^9 q^7	p^{10} q^6	p^{11} q^5	p^{12} q^4	p^{13} q^3	p^{14} q^2	p^{15} q^1	p^{16} q^0
1	0	0	0	0	0	0	0	0	0	0	0	0	0	0	0	0	1
2	0	0	0	0	0	0	0	0	9	120	462	792	715	364	105	16	1
3	0	0	0	0	0	6	161	1016	2907	4740	4917	3432	1651	546	120	16	1
4	0	0	0	0	35	456	2128	5328	8451	9240	7282	4224	1807	560	120	16	1
5	0	0	0	20	320	1686	4795	8800	11385	10890	7887	4356	1820	560	120	16	1
6	0	0	3	104	780	2856	6538	10480	12465	11340	7997	4368	1820	560	120	16	1
7	0	0	15	224	1190	3612	7420	11152	12789	11430	8008	4368	1820	560	120	16	1
8	0	0	36	336	1470	4032	7812	11376	12861	11440	8008	4368	1820	560	120	16	1
9	0	2	57	420	1645	4242	7959	11432	12870	11440	8008	4368	1820	560	120	16	1
10	0	4	75	480	1745	4332	8001	11440	12870	11440	8008	4368	1820	560	120	16	1
11	0	6	90	520	1795	4362	8008	11440	12870	11440	8008	4368	1820	560	120	16	1
12	0	8	102	544	1815	4368	8008	11440	12870	11440	8008	4368	1820	560	120	16	1
13	0	10	111	556	1820	4368	8008	11440	12870	11440	8008	4368	1820	560	120	16	1
14	0	12	117	560	1820	4368	8008	11440	12870	11440	8008	4368	1820	560	120	16	1
15	0	14	120	560	1820	4368	8008	11440	12870	11440	8008	4368	1820	560	120	16	1
16	0	16	120	560	1820	4368	8008	11440	12870	11440	8008	4368	1820	560	120	16	1

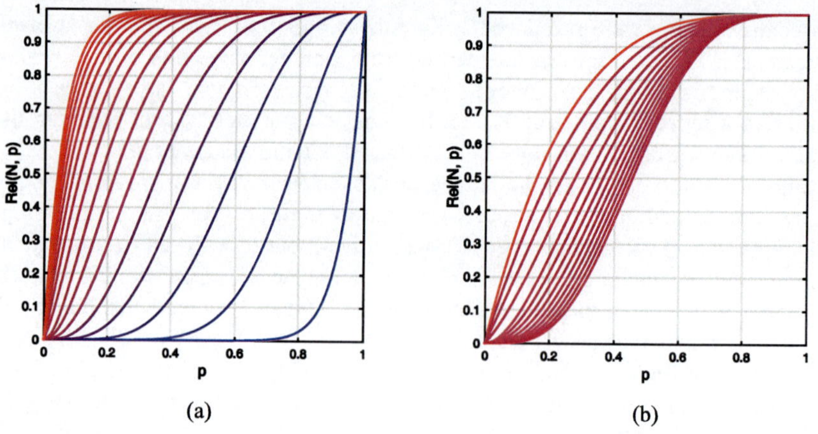

Fig. 2. Reliability of linear consecutive-k-out-of-n:F systems $Rel(Lin/Con/k/n; p)$ plotted for: (a) constant $n = 16$ and $k = 1 \dots 16$; (b) constant $k = 4$ and $n = 4 \dots 16$.

2.3 Consecutive Systems in Nature

Unfortunately, it looks like consecutive systems are not used for electrical circuits (in spite of the fact that they have been used for street lights), a possible reason being that they are kind of wire-demanding (both in the number of wires and with respect to their total length). Still, consecutive systems have been used for fluidic solutions in general (e.g., oil and gas pipelines), as well as biological nano-fluidics, a fact mentioned and argued about as early as 2014 [21].

The highly regular organization, observed for the first time in 2013 on the axons using STORM imaging [24], has actin-adducin forming circular structures (on the circumference) which are evenly distributed at about 180–190 nm from one another. This was also observed for spectrin which is scattered similarly just that in antiphase (i.e., shifted with about 90 nm). Additionally, the sodium channels seem to follow this regular distribution pattern of the actin-spectrin cytoskeleton. These results were confirmed by [25, 26], and a flurry of publications followed. In 2015 it was shown that these regular structures are in fact ubiquitous in neurons being found not only on axons (as originally revealed in [24]), but also on dendrites and even on the nodes of Ranvier [27]. Evidently, plenty of questions were raised on how these link to function [28, 29]. Complementing the picture, in 2016 it was shown that periodic F-actin structures exist even on the neck of dendritic spines [30]. The major goal of identifying and linking these regular structures to particular functions kept unabated but with slow progress, and the interested reader should consult [31–44] (a very small sample of articles on this topic).

Being arranged on the surface of cylinders/pipelines, biology seems to rely on 2.5D structures (see [21, 22]), while in this paper we are aiming to make only a step forward with respect to 1D structures, i.e., networks N like the one in Fig. 1, known as linear consecutive-k-out-of-n systems, or $Lin/Con/k/n$.

3 Fibonacci and Consecutive

3.1 Linear Consecutive at $p = 0.5$

Fibonacci numbers form a very well-known sequence [45–48]. They were generalized by d'Ocagne in 1884 [49–51], while those generalizations were rediscovered independently ten years later, in 1894, by Schlegel [52]. They were re-rediscovered in 1960 by Miles [53] (see [54, 55], with [56–66] as examples).

Similarly, although $Lin/Con/k/n$ were introduced in 1980 [12], a plethora of related statistical results were developed since 1738 when the successive coin tossing problem was stated and solved by de Moivre [23]. Of interest are [67] and [68], while linking Fibonacci to $Lin/Con/k/n$ was presented in 1985 (see [69]) as:

$$Rel(Lin/Con/k/n; 0.5) = \frac{F_{n+2}^{(k)}}{2^n}. \tag{1}$$

This basically means that the sum of all the coefficients of the reliability polynomial of a $Lin/Con/k/n$ network is $F_{n+2}^{(k)}$.

3.2 Fibonacci Numbers of Higher Orders

As already stated above, Fibonacci numbers of higher orders $F_n^{(k)}$ were introduced and studied by d'Ocagne in 1884 [49–51] and rediscovered by Schlegel in 1894 [52]. It looks like they were forgotten, being re-rediscovered by Miles in 1960 [53].

The recurrence relation these numbers are built upon is:

$$F_n^{(k)} = F_{n-1}^{(k)} + F_{n-2}^{(k)} + \cdots + F_{n-k}^{(k)} \tag{2}$$

with the associated characteristic equation [70] being:

$$x^k - x^{k-1} - x^{k-2} - \cdots - x^2 - x - 1 = 0 \tag{3}$$

having roots $\alpha_1, \ldots, \alpha_k$.

Theorem (1998 Wolfram [71]): Let $2(1 - \varepsilon_k)$ be the positive root α_k of the characteristic equation (i.e., $\alpha_k = 2 - 2\varepsilon_k$). Then

$$\varepsilon_k = \sum\nolimits_{i \geq 1} \frac{1}{i} \binom{(k+1)i - 2}{i - 1} \frac{1}{2^{(k+1)i}}. \tag{4}$$

This was rediscovered by Hare et al. in 2014 [72], and used (together with results reported by Forsyth et al. 2013 [73]) to establish upper and lower bounds

$$2 - \frac{1}{2^{k-1}} < 2 - \frac{1}{2^{k - \frac{k}{2} \frac{k^2}{2^k}}} < \alpha_k < 2 - \frac{1}{2^{k - \frac{k}{2}}} < 2 - \frac{1}{2^k}. \tag{5}$$

Table 2 presents numerical results for α_k (from Eq. (3)) and bounds from Eq. (5).

3.3 Binet-Like Formula

The first Binet-like formula for $F_n^{(k)}$ is due to Spickerman and Joyner [70]. Their efforts were continued [74–76].

The following theorems/formulas have been firstly discussed by Zhao Hui Du in a Math R&D forum (https://bbs.emath.ac.cn/thread-667-4-1.html#pid9145) in 2008, while one year later (2009) they were included in a technical report posted independently on arXiv [77] by Gregory Dresden. Subsequently we are going to refer to their joint paper published in 2014 [78].

Theorem (2014 Dresden & Du [78]): The nth k-generalized Fibonacci number $F_n^{(k)}$ can be computed as

$$F_n^{(k)} = \sum\nolimits_{i=1}^{k} \frac{\alpha_i - 1}{2 + (k+1)(\alpha_i - 2)} \alpha_i^{n-1} \tag{6}$$

where α_k are the roots of $x^k - x^{k-1} - \cdots - 1 = 0$.

Dresden and Du also brought up the fact that for n sufficiently 'large' Eq. (6) ends up in the much simpler form

$$F_n^{(k)} \cong \frac{\alpha_k - 1}{2 + (k+1)(\alpha_k - 2)} \alpha_k^{n-1} \tag{7}$$

as Eq. (3) has only one root $|\alpha_k| > 1$, while all the other roots are inside the unit circle, hence their contributions quickly fade away.

Table 2. Positive roots of the characteristic equation (aka *"generalized golden ratios"* [72]).

k	α_k	$2-\frac{1}{2^{k-1}}$	$2-\frac{1}{2^{k-\frac{k}{2}-\frac{k^2}{2^k}}}$	$2-\frac{1}{2^{k-\frac{k}{2}}}$	$2-\frac{1}{2^k}$
1	1.0000000000000000000000000000	1.0000000000000	1.0000000000000	1.3333333333333	1.5000000000000
2	1.6180339887498948482045868344	1.5000000000000	1.5000000000000	1.6666666666666	1.7500000000000
3	1.8392867552141611325518525646	1.7500000000000	1.8139534883720	1.8461538461538	1.8750000000000
4	1.9275619754829253042619058615	1.8750000000000	1.9230769230769	1.9285714285714	1.9375000000000
5	1.9659482366454853371899373754	1.9375000000000	1.9651795429815	1.9661016949152	1.9687500000000
6	1.9835828434243263303856292932	1.9687500000000	1.9834539813857	1.9836065573770	1.9843750000000
7	1.9919641966050350210977417542	1.9843750000000	1.9919430981305	1.9919678714859	1.9921875000000
8	1.9960311797354145898153179014	1.9921875000000	1.9960278053624	1.9960317460317	1.9960937500000
9	1.9980294702622866986623314668	1.9960937500000	1.9980289422118	1.9980295566502	1.9980468750000
10	1.9990186327101011386634092391	1.9980468750000	1.9990185516736	1.9990186457311	1.9990234375000
11	1.9995104019782854914400439566	1.9990234375000	1.9995103897541	1.9995104039167	1.9995117187500
12	1.9997555009373175366974267623	1.9995117187500	1.9997554991208	1.9997555012224	1.9997558593750
13	1.9998778327115455400399468997	1.9997558593750	1.9998778324451	1.9998778327530	1.9998779296875
14	1.9999389387495946484978450597	1.9998779296875	1.9999389387109	1.9999389387555	1.9999389648437
15	1.9999694754345032923601892011	1.9999389648437	1.9999694754289	1.9999694754353	1.9999694824218
16	1.9999847393479441072890596901	1.9999694824218	1.9999847393471	1.9999847393480	1.9999847412109

Theorem (2014 Dresden & Du [78]): The nth k-generalized Fibonacci number $F_n^{(k)}$ can be computed as

$$F_n^{(k)} = \text{rnd}\left[\frac{\alpha_k - 1}{2+(k+1)(\alpha_k-2)}\alpha_k^{n-1}\right] \tag{8}$$

for all $n \geq 2-k$ and for α_k the unique positive root of $x^k - x^{k-1} - \cdots - 1 = 0$.

4 A Simple Design Scheme

4.1 Solving k for 'Large' n

We can use Eqs. (1) and (7) to calculate $Rel(Lin/Con/k/n; 0.5)$ as

$$\frac{F_{n+2}^{(k)}}{2^n} \cong \frac{\frac{(\alpha_k-1)}{2+(k+1)(\alpha_k-2)}\alpha_k^{n+1}}{2^n} = \frac{(\alpha_k-1)\alpha_k}{2-(2-\alpha_k)(k+1)} \times \left(\frac{\alpha_k}{2}\right)^n = f(k) \times g(k,n). \tag{9}$$

First of all, let us look at the second function $g(k,n)$, and approximate it by replacing α_k with one of the simpler bounds from Eq. (5)

$$g(k,n) = \left(\frac{\alpha_k}{2}\right)^n < \left(\frac{2-\frac{1}{2^k}}{2}\right)^n = \left(1-\frac{1}{2^{k+1}}\right)^n = \left[1+\frac{\left(-\frac{1}{2}\right)}{2^k}\right]^n. \tag{10}$$

Due to the 2^k term it is quite natural to expect k being in the vicinity of $\log_2 n$, but our previous simulations have also suggested this value [79]. That is why we have decided to consider $k = \log_2 n + \delta$, which gives

$$g\left(\log_2 n + \delta, n\right) = \left[1 + \frac{\left(-\frac{1}{2}\right)}{2^{\log_2 n + \delta}}\right]^n = \left[1 + \frac{\left(-\frac{1}{2^{1+\delta}}\right)}{n}\right]^n. \tag{11}$$

For large values of n this converges to

$$\lim_{n\to\infty} g\left(\log_2 n + \delta, n\right) = e^{-1/2^{1+\delta}}. \tag{12}$$

Now let us check the other function $f(k)$

$$f(k) \cong \frac{\left(2 - \frac{1}{2^k}\right)\left(2 - \frac{1}{2^k} - 1\right)}{2 - \left(2 - 2 + \frac{1}{2^k}\right)(k+1)} = \frac{\left(2 - \frac{1}{2^k}\right)\left(1 - \frac{1}{2^k}\right)}{2 - \frac{k+1}{2^k}} = \frac{2 - \frac{2}{2^k} - \frac{1}{2^k} + \frac{1}{2^{2k}}}{2 - \frac{k+1}{2^k}}$$
$$\cong \frac{2 - \frac{3}{2^k}}{2 - \frac{k+1}{2^k}} = \frac{2^{k+1} - 3}{2^{k+1} - (k+1)}. \tag{13}$$

For the same $k = \log_2 n + \delta$ we have

$$f\left(\log_2 n + \delta\right) \cong \frac{2^{1+\delta} n - 3}{2^{1+\delta} n - \left(\log_2 n + \delta + 1\right)} \tag{14}$$

which converges to 1 as n grows

$$\lim_{n\to\infty} f\left(\log_2 n + \delta\right) = \lim_{n\to\infty} \frac{2^{1+\delta} n - 3}{2^{1+\delta} n - \left(\log_2 n + \delta + 1\right)} = 1. \tag{15}$$

Putting Eqs. (1), (9), (12) and (15) together we have that

$$\lim_{n\to\infty} Rel\left(Lin/Con/\log_2 n + \delta/n; 0.5\right) = \lim_{n\to\infty} \frac{F_{n+2}^{(\log_2 n + \delta)}}{2^n} = e^{-1/2^{1+\delta}}. \tag{16}$$

Table 3 presents both Eq. (16) for $k = \log_2 n + \delta$ by varying $\delta = -2, -1, 0, 1, 2, 3$, versus exact simulation results obtained using an algorithm based on the Markov chain method suggested by Fu [17].

4.2 Simulation Results

For benchmarking the results reported above we have performed plenty of exact simulations, only a small portion of which will be presented here.

First and foremost we need to stress that there are two different design options for reliability: one intended for *communications* (the bulk of research on network reliability,

Table 3. Theoretical and exact reliability of $Lin/Con/k/n : F$ at $p = 0.5$.

k	Theoretical (eq. (16))	Numerical estimates	Exact (simulations)
$\log_2 n - 2$	e^{-2}	0.135335283236613	0.135521348936266
$\log_2 n - 1$	e^{-1}	0.367879441171442	0.368621154368109
$\log_2 n$	$e^{-1/2}$	0.606530659712633	0.607246063637544
$\log_2 n + 1$	$e^{-1/4}$	0.778800783071405	0.780024776838979
$\log_2 n + 2$	$e^{-1/8}$	0.882496902584595	0.883741786178675
$\log_2 n + 3$	$e^{-1/16}$	0.939413062813476	0.941082001454758

see [7, 8]), and another one aimed at *computations* (the original goal of Moore and Shannon [4], and inline with [2, 5] but using a different model). As we have already shown (see [80] for details), the two different design options are separated by the optimum being either the classical Heaviside function (for communications), or a version of the Heaviside function shifted at $p_0 = 0.5$ (for computations). It follows that for communications $Rel(N; 0.5)$ should be as close as possible to 1 (say > 0.9, i.e., larger than 90%), while for computations $Rel(N; 0.5)$ should be as close as possible to 0.5. For a better understanding, exact simulations of $Rel(Lin/Con/k/16; p)$ are presented in Fig. 3 (similar to those in Fig. 2(a)), with the optimal solutions highlighted for communication in Fig. 3(a), and for computations in Fig. 3(b).

Because Eq. (16) was obtained using approximations and limits, it seems natural to ask oneself *"How large might n be?"* This is an important question for practical design purposes, as we should be aware of the smallest value of n for which a design method relying on Eq. (16) will still be holding.

For identifying the range of values of n we have computed $Rel\big(Lin/Con/\log_2 n + \delta/n; 0.5\big)$ by varying δ as $-2, -1, 0, 1, 2, 3$ (see Table 3), and n in the range 2^2 to 2^{50} (!). The exact results reported were obtained using a matrix-based Markov chain method [17]. The full range of simulation results are plotted in Fig. 4(a), where the horizontal axis is logarithmic representing *exp* (the exponent), which makes $n = 2^{exp}$. It is clear that, for all the δ we have considered the plots are asymptotic (Eq. (16)), clearly saturating for $n > 2^{10}$ (outside the yellow rectangle).

The minimum value of n is still unclear, so a zoom in is presented in Fig. 4(b). This reveals that things are way better than $n > 1024$ suggested by Fig. 4(a)). That yellow rectangle has morphed into a yellow ellipse in Fig. 4(b)—uneven scaling on the axes makes it look more like a circle. Surprisingly, the approximations are confirmed to be outstandingly good, valid even for the smallest value of $n = 4$:

- $Rel\big(Lin/Con/\log_2 n/n; 0.5\big)$ being constantly the closest to 0.5;
- $Rel\big(Lin/Con/\log_2 n + 2/n; 0.5\big)$ being constantly above 0.9.

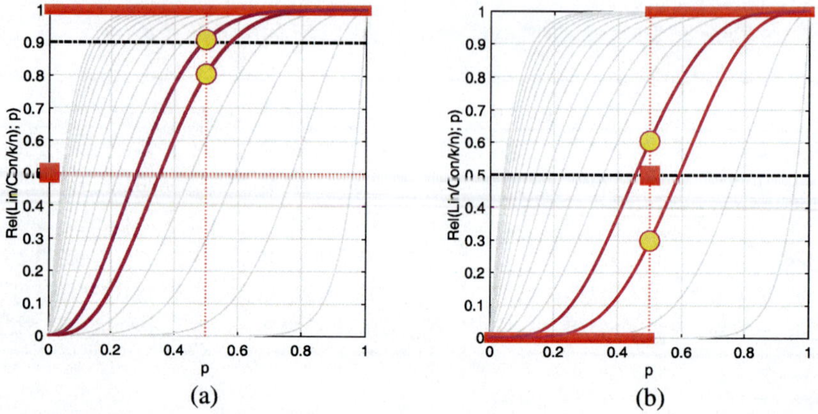

Fig. 3. Optimal design for linear consecutive-k-out-of-n: F systems when $n = 16$ (identified by the yellow dots): (a) for communication $k = 6$, corresponding to $\log_2 n + 2$; (b) for computation $k = 4$, corresponding to $\log_2 n$.

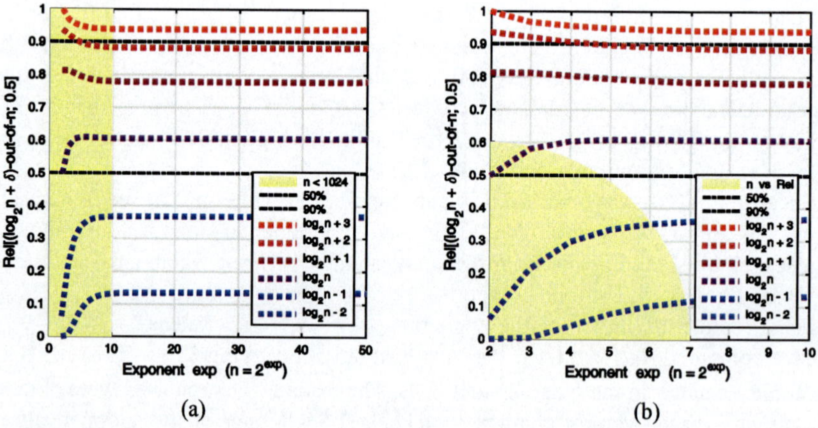

Fig. 4. Reliability of consecutive-k-out-of-n:F systems (exact simulations) at $p = 0.5$ for $k = \log_2 n + \delta$ (with $\delta = -2, -1, 0, 1, 2, 3$) and variable n in the range: (a) $2^2 \ldots 2^{50}$ (large); (b) $2^2 \ldots 2^{10}$ (small).

5 Conclusions

This paper has suggested a fresh and extremely simple approach for the optimal design of linear consecutive systems. The approach we have taken has used Binet-like formulas (allowing to compute Fibonacci numbers in $O(1)$), followed by quite simple approximations. Additionally, thoroughly detailed and exact simulations have shown that the approximations we have used are valid for any value of $n \geq 4$.

The quick optimal design for $Lin/Con/k/n$ (minimizing k) based on the results presented in this paper goes as follows:

- If used for *communication*, one should start by checking n versus 32 and selecting either $k = \log_2 n + 2$ or $k = \log_2 n + 3$ (which guarantees over 90% reliability);
- If used for *computation*, select $k = \log_2 n$. The problem here is a bit more nuanced as networks performing better for computations are known (see [80]), so we should rely on one of the better ones, like, e.g., hammocks (basically, consecutive systems should not be used for computations).

We also argue that this is (most probably) the simplest possible design for linear consecutive systems as the computations required are: (*i*) a test of n versus 32; (*ii*) computing $\log_2 n$; and (*iii*) an addition (eventually).

Acknowledgements. This research was partly funded by a grant of the Romanian Ministry of Education and Research, CNCS-UEFISCDI, project no. PN-III-P4-ID-PCE-2020-2495, within PNCDI III (ThUNDER^2 = *Techniques for Unconventional Nano-Designing in the Energy-Reliability Realm*).

References

1. Babbage, C.: Babbage's calculating engine. Edinb. Rev. **59**(120), 263–327 (1834). https://en.wikisource.org/w/index.php?title=Edinburgh_Review/Volume_59/Babbage%27s_Calculating_Engine&oldid=5473361. Accessed 06 June 2022
2. von Neumann, J.: Probabilistic logics and the synthesis of reliable organisms from unreliable components. In: Shannon, C.E., McCarthy, J. (eds.): Automata Studies, pp. 43–98. Princeton University Press, Princeton (1956). Lectures at Caltech, 4–15 January 1952. https://archive.org/details/vonNeumann_Prob_Logics_Rel_Org_Unrel_Comp_Caltech_1952/mode/2up. Accessed 06 June 2022
3. European Commission: European Chips Act, Brussels, Belgium, 8 February 2022. https://ec.europa.eu/commission/presscorner/detail/en/ip_22_729. see also U. von der Leyend, 15 March 2022. https://twitter.com/vonderleyen/status/1503723837501935618. Accessed 06 June 2022
4. Moore, E.F., Shannon, C.E.: Reliable circuits using less reliable relays – Part I. J. Franklin Inst. **262**(3), 191–208 (1956). Preliminary version: Reliable circuits using crummy relays. Tech. Rep. Memo. 54-114-42. Bell Labs, Murray Hill, NJ, USA, 29 November 1954
5. Pierce, W.H.: Interwoven redundant logic. J. Frankl. Inst. **277**(1), 55–85 (1964)
6. Deng, H., Chen, J., Li, Q., Li, R., Gao, Q.: On the construction of most reliable networks. Discrete Appl. Math. **140**(1–3), 19–33 (2004)
7. Brown, J.I., Colbourn, C.J., Cox, D., Graves, C., Mol, L.: Network reliability: Heading out on the highway. Networks **77**(1), 146–160 (2020)
8. Romero, P.: Uniformly optimally reliable graphs: A survey. Networks (2021). Art. 22085. https://doi.org/10.1002/net.22085. Accessed 06 June 2022
9. Drăgoi, V.-F., Beiu, V.: Fast reliability ranking of matchstick minimal networks. Networks **79**(4), 479–500 (2022)
10. Valiant, L.G.: The complexity of computing the permanent. Theoret. Comput. Sci. **8**(2), 189–201 (1979)
11. Valiant, L.G.: The complexity of enumeration and reliability problems. SIAM J. Comput. **8**(3), 410–421 (1979)
12. Kontoleon, J.M.: Reliability determination of a r-successive-out-of-n:F system. IEEE Trans. Reliab. **R-29**(5), 437 (1980)

13. Chiang, D.T., Niu, S.-C.: Reliability of consecutive-k-out-of n:F system. IEEE Trans. Reliab. **R-30**(1), 87–89 (1981)
14. Chao, M.T., Lin, G.D.: Economical design of large consecutive-k-out-of n:F systems. IEEE Trans. Reliab. **R-32**(5), 411–413 (1984)
15. Fu, J.C.: Reliability of a large consecutive-k-out-of n:F system. IEEE Trans. Reliab. **R-34**(2), 127–130 (1985)
16. Fu, J.C.: Bounds for reliability of large consecutive-k-out-of n:F systems with unequal component reliability. IEEE Trans. Reliab. **R-35**(3), 316–319 (1986)
17. Fu, J.C.: Reliability of consecutive-k-out-of n:F systems with $(k-1)$-step Markov dependence. IEEE Trans. Reliab. **R-35**(5), 602–606 (1986)
18. Fu, J.C., Hu, B.: On reliability of a large consecutive-k-out-of n:F systems with $(k-1)$-step Markov dependence. IEEE Trans. Reliab. **R-36**(1), 75–77 (1987)
19. Makri, F.S., Psillakis, Z.M.: On success runs of length exceeding a threshold. Methodol. Comput. Appl. Probab. **13**(2), 269–305 (2011)
20. Beiu, V., Dăuş, L.: Review of reliability bounds for consecutive-k-out-of-n systems. In: Proceedings of the International Conference on Nanotechnology (IEEE-NANO), Toronto, Canada, pp. 302–307 (2014)
21. Beiu, V., Dăuş, L.: Deciphering the reliability scheme of the neurons – one ion channel at a time. In: Proceedings of the International Conference on Bio-Inspired ICT (BICT), Boston, MA, USA, pp. 182–187 (2014)
22. Beiu, V., Dăuş, L.: Reliability bounds for two dimensional consecutive systems. Nano Commun. Netw. **6**(3), 145–152 (2015)
23. de Moivre, A.: The Doctrine of Chances, 2nd edn. H. Woodfall, London (1738). https://archive.org/details/doctrineofchance00moiv. Accessed 06 June 2022. Links to the 3rd edn.
24. Xu, K., Zhong, G., Zhuang, X.: Actin, spectrin and associated proteins form a periodic cytoskeletal structure in axons. Science **339**(6118), 452–456 (2013)
25. Lukinavičius, G., et al.: Fluorogenic probes for live-cell imaging of the cytoskeleton. Nat. Methods **11**(7), 731–733 (2014)
26. Zhong, G., et al.: Developmental mechanism of the periodic membrane skeleton in axons. elife **3**, e04581(1–21) (2014)
27. D'Este, E., Kamin, D., Göttfert, F., El-Hady, A., Hell, S.W.: STED nanoscopy reveals the ubiquity of subcortical cytoskeleton periodicity in living neurons. Cell Rep. **10**(8), 1246–1251 (2015)
28. Kevenaar, J.T., Hoogenraad, C.C.: The axonal cytoskeleton: Fom organization to function. Front. Mol. Neurosci. **8**, 44(1–12) (2015)
29. Stewart, E., Shen, K.: STORMing towards a clear picture of the cytoskeleton in neurons. eLife **4**, e06235(1–3) (2015)
30. Bär, J., Kobler, O., van Bommel, B., Mikhaylova, M.: Periodic F-actin structures shape the neck of dendritic spines. Sci. Rep. **6**, 37136(1–9) (2016)
31. D'Este, E., Kamin, D., Balzarotti, F., Hell, S.W.: Ultrastructural anatomy of nodes of Ranvier in the peripheral nervous system as revealed by STED microscopy. PNAS **114**(2), 191–199 (2016)
32. D'Este, E., Kamin, D., Velte, C., Göttfert, F., Simons, M., Hell, S.W.: Subcortical cytoskeleton periodicity throughout the nervous system. Sci. Rep. **6**, 22741(1–8) (2016)
33. He, J., et al.: Prevalent presence of periodic actin−spectrin-based membrane skeleton in a broad range of neuronal cell types and animal species. PNAS **113**(21), 6029–6034 (2016)
34. Leite, S.C., Sousa, M.M.: The neuronal and actin commitment: Why do neurons need rings? Cytoskeleton **73**(9), 424–434 (2016)
35. Roy, S.: Waves, rings, and trails: The scenic landscape of axonal actin. J Cell Biol. **212**(2), 131–134 (2016)

36. Sidenstein, S.C., D'Este, E., Böhm, M.J., Danzl, J.G., Belov, V.N., Hell, S.W.: Multicolour multilevel STED nanoscopy of actin/spectrin organization at synapses. Sci. Rep. **6**, 26725(1–8) (2016)
37. Han, B., Zhou, R., Xia, C., Zhuang, X.: Structural organization of the actin-spectrin–based membrane skeleton in dendrites and soma of neurons. PNAS **114**(32), 6678–6685 (2017)
38. Leterrier, C., Dubey, P., Roy, S.: The nano-architecture of the axonal cytoskeleton. Nat. Rev. Neurosci. **18**(12), 713–726 (2017)
39. Hauser, M., Yan, R., Li, W., Repina, N.A., Schaffer, D.V., Xu, K.: The spectrin-actin-based periodic cytoskeleton as a conserved nanoscale scaffold and ruler of the neural stem cell lineage. Cell Rep. **24**(6), 1512–1522 (2018)
40. Vassilopoulos, S., Gibaud, S., Jimenez, A., Caillol, G., Leterrier, C.: Ultrastructure of the axonal periodic scaffold reveals a braid-like organization of actin rings. Nat. Commun. **10**, 5803(1–13) (2019)
41. Lorenzo, D.N.: Cargo hold and delivery: Ankyrins, spectrins, and their functional patterning of neurons. Cytoskeleton **77**(3–4), 129–148 (2020)
42. Chu, C., Zhong, G., Li, H.: Structure and function of subcortical periodic cytoskeleton throughout the nervous system. STEMedicine **1**(1), e9(1–8) (2020)
43. Leterrier, C.: Putting the axonal periodic scaffold in order. Curr. Opin. Neurobiol. **69**, 33–40 (2021)
44. Zahavi, E.E., Hoogenraad, C.C.: Multiple layers of spatial regulation coordinate axonal cargo transport. Curr. Opin. Neurobiol. **69**, 241–246 (2021)
45. Pisano/Fibonacci, L.: Liber Abaci (Book of Calculations), Codice Magliabechiano, Conventi Sopressi C. 1, 2616, Biblioteca Nazionale Centrale di Firenze, Florence, Italy (1202). https://bibdig.museogalileo.it/Teca/Viewer?an=1072400. Accessed 06 June 2022
46. Sigler, L.E.: Fibonacci's Liber Abaci – A Translation into Modern English of Leonardo Pisano's Book of Calculation. Springer, New York (2002). https://doi.org/10.1007/978-1-4613-0079-3
47. Wilson, R., Watkins, J.J. (eds.): Combinatorics: Ancient and Modern. Oxford University Press, Oxford (2015)
48. OEIS®: Fibonacci numbers. https://oeis.org/A000045. Accessed 06 June 2022
49. d'Ocagne, P.M.: Sur un algorithme algébrique (On an algebraic algorithm). Nouv. Ann. Math. **2**, 220–226 (1883). http://www.numdam.org/item/?id=NAM_1883_3_2__220_0. Accessed 06 June 2022
50. d'Ocagne, P.M.: Théorie élémentaire des séries récurrentes (Elementary theory of recurring series). Nouv. Ann. Math. **3**, 65–90 (1884). http://www.numdam.org/item?id=NAM_1884_3_3__65_0. Accessed 06 June 2022
51. d'Ocagne, P.M.: Sur une suite récurrente (On a recurring series). Bull. Soc. Math. France **14**, 20–41 (1886). http://www.numdam.org/item/?id=BSMF_1886__14__20_1. Accessed 06 June 2022
52. Schlegel, V.: Petites observations mathématiques (Minor mathematical remarks). El Progreso Matemático **4**(5), 169–177 (1894). http://hemerotecadigital.bne.es/issue.vm?id=0004532418. Accessed 06 June 2022
53. Miles, E.P., Jr.: Generalized Fibonacci numbers and associated matrices. Am. Math. Mon. **67**(8), 745–752 (1960)
54. Ferguson, D.E.: An expression for generalized Fibonacci numbers. Fibonacci Quart. **4**(3), 270–273 (1966)
55. Miller, M.D.: On generalized Fibonacci numbers. Am. Math. Mon. **78**(10), 1108–1109 (1971)
56. OEIS®: Tribonacci numbers. https://oeis.org/A000073. Accessed 06 June 2022
57. OEIS®: Tetranacci numbers. https://oeis.org/A000078. Accessed 06 June 2022
58. OEIS®: Pentanacci numbers. https://oeis.org/A001591. Accessed 06 June 2022

59. OEIS®: Hexanacci numbers. https://oeis.org/A001592. Accessed 06 June 2022
60. OEIS®: Heptanacci numbers. https://oeis.org/A122189. Accessed 06 June 2022
61. OEIS®: Octanacci numbers. https://oeis.org/A079262. Accessed 06 June 2022
62. OEIS®: Enneanacci numbers. https://oeis.org/A104144. Accessed 06 June 2022
63. OEIS®: Decanacci numbers. https://oeis.org/A122265. Accessed 06 June 2022
64. OEIS®: 11-nacci numbers. https://oeis.org/A168082. Accessed 06 June 2022
65. OEIS®: Dodecanacci numbers. https://oeis.org/A207539. Accessed 06 June 2022
66. OEIS®: 13-nacci numbers. https://oeis.org/A163551. Accessed 06 June 2022
67. Shane, H.D.: A Fibonacci probability function. Fibonacci Quart. **11**(5), 517–522 (1973)
68. Turner, S.J.: Probability via the nth-order Fibonacci-T sequence. Fibonacci Quart. **17**(1), 23–28 (1979)
69. Philippou, A.N., Georghiou, C., Philippou, G.N.: Fibonacci-type polynomials of order k with probability applications. Fibonacci Quart. **23**(2), 100–105 (1985)
70. Spickerman, W.R., Joyner, R.N.: Binet's formula for the recursive sequence of order k. Fibonacci Quart. **22**(4), 327–331 (1984)
71. Wolfram, D.A.: Solving generalized Fibonacci recurrences. Fibonacci Quart. **36**(2), 129–145 (1998)
72. Hare, K., Prodinger, H., Shallit, J.: Three series for the generalized golden mean. Fibonacci Quart. **52**(4), 307–313 (2014)
73. Forsyth, M., Jayakumar, A., Shallit, J.: Remarks on privileged words. Int. J. Found. Comput. Sci. **27**(4), 431–442 (2016)
74. Lee, G.-Y., Lee, S.-G., Kim, J.-S., Shin, H.-K.: The Binet formula and representations of k-generalized Fibonacci numbers. Fibonacci Quart. **39**(2), 158–164 (2001)
75. Kessler, D., Schiff, J.: A combinatoric proof and generalization of Ferguson's formula for k-generalized Fibonacci numbers. Fibonacci Quart. **42**(3), 266–273 (2004)
76. Howard, F.T., Cooper, C.: Some identities for r-Fibonacci numbers. Fibonacci Quart. **49**(3), 231–242 (2004)
77. Dresden, G.P.B.: A simplified Binet formula for k-generalized Fibonacci numbers. arXiv Number Theory (math.NT) 0905.0304 (2009). https://arxiv.org/abs/0905.0304. Accessed 06 June 2022
78. Dresden, G.P.B., Du, Z.: A simplified Binet formula for k-generalized Fibonacci numbers. J. Integer Seq. **17**(4) (2014). Art. 14.4.7
79. Beiu, R.-M., Hoară, S.-H., Drăgoi, V.-F., Beiu, V.: 3D hammocks and 2.5D consecutive – Biology fine balancing. In: Proceedings of the International Workshop Soft Computing Applications (SOFA), Arad, Romania, 27–29 November 2020 (2023, in press)
80. Beiu, V., Drăgoi, V.-F., Beiu, R.-M.: Why reliability for computing needs rethinking. In: Proceedings of the IEEE International Conference on Rebooting Computing (ICRC), Atlanta, GA, USA, pp. 16–25 (2020)

On the Roots of Certain Reliability Polynomials

Leonard Dăuş[1](✉), Vlad-Florin Drăgoi[2], Marilena Jianu[1], Dominic Bucerzan[2], and Valeriu Beiu[2]

[1] Technical University of Civil Engineering of Bucharest, 020396 Bucharest, Romania
{leonard.daus,marilena.jianu}@utcb.ro

[2] "Aurel Vlaicu" University of Arad, 310032 Arad, Romania
{vlad.dragoi,dominic.bucerzan,valeriu.beiu}@uav.ro

Abstract. In this paper we assume that a network is a probabilistic graph where the vertices do not fail while the edges do, being operational independently with probability p. The two most common probabilistic models of network reliability are the *all-terminal* and the *two-terminal reliability* ones, with *k-terminal reliability* modeling the whole spectrum. For any of these models, the reliability of a network is represented by a polynomial in p. While the roots of all-terminal reliability polynomials have been extensively studied, only a small fraction of papers has converged on the roots of two-terminal reliability polynomials. A particular sub-class of two-terminal networks is represented by *composition of series and parallel*. Obviously, networks formed by composition are expected to be easier to understand, exhibiting kind of simpler/regular reliability polynomials. Additionally, compositions of series and parallel networks have been shown to be highly competitive on improving the reliability of networks doing computations—even when compared to hammock networks (a particular sub-class of two-terminal networks introduced by Moore and Shannon in 1956). Neither the roots associated to the reliability polynomials of hammock networks, nor those of composition of series and parallel networks have yet been investigated. In this paper we are going to start filling such voids by studying the roots of the two-terminal reliability polynomials associated to series and parallel compositions. This is an early stage investigation which will lead to several conjectures while also suggesting further directions for research.

Keywords: Network reliability · Composition of series and parallel · Polynomial roots

1 Introduction

Network reliability has become a well-recognized research topic. Its origins are deeply rooted in electrical circuit theory [32], while it is considered as being properly established in the middle of the 20th century through the seminal work of Moore and Shannon [35,36]. In those early days, much effort was allocated

S. Dzitac et al. (Eds.): ICCCC 2022, AISC 1435, pp. 401–414, 2023.
https://doi.org/10.1007/978-3-031-16684-6_34

towards determining the complexity classes of several computational problems, such as, determining the reliability polynomial of a graph/network, computing the coefficients of the reliability polynomial, etc. Such problems turned to be difficult [37,39,44], and hence, research re-focused on finding practical algorithms for bounding or approximating the reliability polynomials [2,17,20–23,31]. Not all networks fall into this worst-case scenario, therefore, identifying sub-classes of networks—for which efficient algorithms for computing the reliability polynomial are known—became of interest. One such sub-class is represented by compositions of series and parallel, on which this article will be focusing.

A recent survey [6] recalls some of the most challenging problems in network reliability, as well as pointing out new directions *"heading out on the highway"*. Among the new avenues for network reliability, a particular one is slowly taking shape: *analytical properties of the reliability polynomials.* This is a topic which covers various aspects from convexity and shape properties [4,24,30,33], to inflection points [13,14,29], through searching for the real/complex roots of the reliability polynomials [3,5,7–12,25]. The expectation is that, having a clear understanding of the complex roots of the reliability polynomial could help answering other fundamental questions, like how optimal those networks are, e.g., with respect to the concept of uniformly most/least reliable. In a series of papers published starting from the early 1990's [3–12], Brown et al. have provided an extensive and well-organized study of the roots of reliability polynomials for different networks. The first remark worth mentioning was about the location of these roots, which do not spread much out in the complex plane. In particular, the roots of all-terminal reliability polynomials were conjectured to belong to the unit disk centered at $z = 1$ in the complex plane [3], a fact disproved in [40], but only by shy margins. In [9] and [11], the roots of independence polynomial were investigated, proving that the real roots are dense in $(-\infty, 0]$, while complex roots are dense in $\mathbb{C}$. Recently, the nature and location of the roots of two-terminal reliability polynomials have also been under scrutiny [7]. It is an open question if the roots of two-terminal reliability polynomials have arbitrary modulus, while the results reported by Tanguy for several two-terminal networks [41–43] show that all the roots are lying in a rectangular region of the complex plane (see also [16]).

Our Contribution. In this article, an algorithm for computing the set of complex roots of the reliability polynomial of a binary composition of series and parallel is provided. It is characterized by the following features.

- The algorithm has time complexity $O(n)$, where $n = 2^m$ denotes the size of the network;
- The algorithm does not require the computation of the reliability polynomial, a fact which is not true in general (normally, one needs to first of all compute the reliability polynomial, and only afterwards to find its roots);
- The algorithm presents a recursive feature, in the sense that, when one requires the roots for all the compositions of series and parallel (for a fixed value of n), for at least half of the reliability polynomials the roots can easily be derived from the previous set of networks of size $n/2$.

We have implemented the algorithm and present graphical visualizations of the complex roots for all the compositions obtained for small values of $n = 2^m$ (i.e., 16, 32, 64, 128). We expect that studying the structural properties of these roots could reveal (some of the) fundamental properties of compositions. Here we identify one particular characteristic of such complex roots: symmetry. In particular, if the last operator of the composition is "in series", then the roots are located symmetrically with respect to the origin. By duality, when the last composition is "in parallel", the point $z = 1$ is the center of symmetry of the set of roots. Simulations for increasing $n = 2^m$ reveal a limit shape of the set of all complex roots, for both compositions ending either "in parallel" or "in series".

2 Matchstick and Series and Parallel Networks

The focus of this article is on two-terminal networks, in particular on matchstick minimal networks. This type of networks was introduced more than half a century ago by Moore and Shannon [35,36], being closely scrutinized lately [1,20,22–24,26,27].

On one hand, a subfamily for which nice formulae exists (e.g., the reliability polynomial, the properties of the underlying circuit, ordering reliability with respect to figures-of-merit) is represented by the set of compositions of series and parallel [1,26,27]. On the other hand, one can view compositions of series and parallel as an important subclass of two-terminal series-parallel networks, which were introduced and firstly studied by MacMahon as early as 1892 [34]. Half a century later, Riordan and Shannon [38] provided a generating function from which the number of series-parallel networks could be computed. We have to point out a significant paper by Duffin [28], where several important results on series-parallel networks were developed. A few years later, Brylawski [15] used completely different techniques (based on combinatorial pregeometries) to study series-parallel networks.

2.1 Compositions of Series and Parallel

Binary compositions of series and parallel are a particular type of matchstick minimal networks. We shall use the existing notations and definitions from the literature [26,27]. If $\boldsymbol{N_1}$ and $\boldsymbol{N_2}$ are two matchstick minimal networks, then the composition of $\boldsymbol{N_1}$ and $\boldsymbol{N_2}$, denoted $\boldsymbol{N_1} \bullet \boldsymbol{N_2}$, is obtained by replacing each device in $\boldsymbol{N_1}$ by a copy of $\boldsymbol{N_2}$. Let us denote two devices in series $\boldsymbol{C}^{(0)}$, and two devices in parallel $\boldsymbol{C}^{(1)}$. In a natural way we can generalize compositions of $\boldsymbol{C}^{(0)}$ and $\boldsymbol{C}^{(1)}$ to any m-length binary vector $\boldsymbol{u} = (u_1, \ldots, u_m) \in \{0,1\}^m$ as

$$\boldsymbol{C}^{\boldsymbol{u}} = \boldsymbol{C}^{(u_1)} \bullet \cdots \bullet \boldsymbol{C}^{(u_m)}, \qquad (1)$$

where $\bullet$ denotes the compositions of two-terminal networks (see [26] for more details). Thus, for any $\boldsymbol{u} \in \{0,1\}^m$, $\boldsymbol{C}^{\boldsymbol{u}}$ is a network from $\mathcal{C}_{2^m}$, the set of all 2^m-size compositions of $\boldsymbol{C}^{(0)}$ and $\boldsymbol{C}^{(1)}$. The composition is in general non-commutative. In Fig. 1 we illustrate the composition of two networks $\boldsymbol{C}^{(0,1)}$ and $\boldsymbol{C}^{(0,1,1,0)}$.

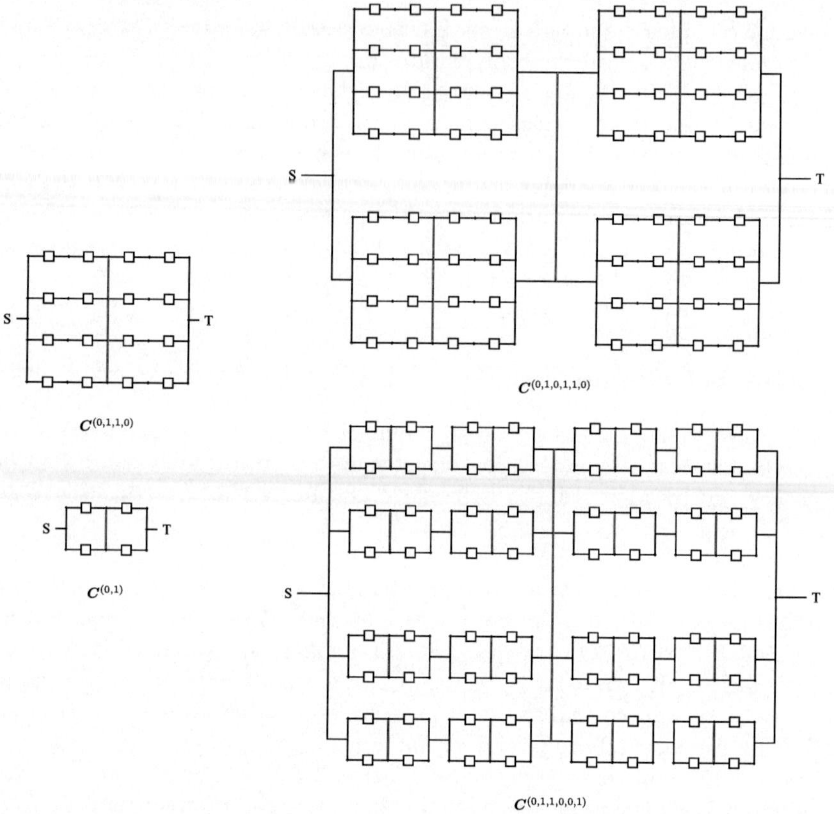

Fig. 1. The two composition of network $\boldsymbol{C}^{(0,1)}$ and $\boldsymbol{C}^{(0,1,1,0)}$.

2.2 Reliability Polynomials

The reliability of a two-terminal network is defined as the probability that the source S and the terminus T are connected (also known as s,t-connectedness) [18], where $p \in [0,1]$ is the probability that a device works (is operational). The reliability polynomial associated to a network $\boldsymbol{N}$ will be $\mathrm{Rel}(\boldsymbol{N};p)$, and we shall rely on the following form:

$$\mathrm{Rel}(\boldsymbol{N};p) = \sum_{k=0}^{n} N_k(\boldsymbol{N})\, p^k (1-p)^{n-k}, \tag{2}$$

where the coefficients $N_k(\boldsymbol{N})$ are integers representing the number of length k pathsets from S to T, satisfying $0 \leq N_k(\boldsymbol{N}) \leq \binom{n}{k}$ for any n-size network $\boldsymbol{N}$ (see [18,19]).

Now, taking $\boldsymbol{N} = \boldsymbol{C}$ we have $\mathrm{Rel}(\boldsymbol{C};p)$, which can be computed using the following theorem.

Theorem 1 ([26,27]). *Let m be a positive integer. For any $\boldsymbol{u} = (u_1, \ldots, u_m) \in \{0,1\}^m$ we have*

$$\mathrm{Rel}(\boldsymbol{C}^{\boldsymbol{u}}; p) = \mathrm{Rel}(\boldsymbol{C}^{(u_1)}; p) \circ \cdots \circ \mathrm{Rel}(\boldsymbol{C}^{(u_m)}; p), \tag{3}$$

where $\mathrm{Rel}(\boldsymbol{C}^{(0)}; p) = p^2$ *and* $\mathrm{Rel}(\boldsymbol{C}^{(1)}; p) = 1 - (1-p)^2$.

3 Roots of Series and Parallel Compositions

3.1 Generating the Roots Efficiently

Since there is an efficient recursive formula for the reliability polynomial of composition of series and parallel, it is natural to expect that efficient procedures for finding the complex roots of reliability polynomials of such networks might exist. We do propose an algorithm that computes directly all the complex roots of the reliability polynomial of $\boldsymbol{C}^{\boldsymbol{u}}$, i.e., without calculating the reliability polynomial itself. We will also show that this recursive algorithm is efficient. The following theorem stands at the core of our method.

Theorem 2. *If $P(x) = \prod_{k=1}^{l}(x - \alpha_k)$ is a polynomial of degree l, having the complex roots α_k, $k = 1, \ldots, l$, then*

- $P(\mathrm{Rel}(\boldsymbol{C}^{(0)}; p))$ *has $2l$ complex roots* $\{\sqrt{\alpha_k}, -\sqrt{\alpha_k}\}_{1 \le k \le l}$.
- $P(\mathrm{Rel}(\boldsymbol{C}^{(1)}; p))$ *has $2l$ complex roots* $\{1 - \sqrt{1-\alpha_k}, 1 + \sqrt{1-\alpha_k}\}_{1 \le k \le l}$.

Proof. $P(\mathrm{Rel}(\boldsymbol{C}^{(0)}; p)) = \prod_{k=1}^{l}(p^2 - \alpha_k) = \prod_{k=1}^{l}(p - \sqrt{\alpha_k})(p + \sqrt{\alpha_k})$.
$P(\mathrm{Rel}(\boldsymbol{C}^{(1)}; p)) = \prod_{k=1}^{l}(1 - (1-p)^2 - \alpha_k) = \prod_{k=1}^{l}(-p^2 + 2p - \alpha_k)$.

Remark 1. For $m = 1$ we can see that

- $\mathrm{Rel}(\boldsymbol{C}^{(0)}; p) = p^2$ has 0 as double root;
- $\mathrm{Rel}(\boldsymbol{C}^{(1)}; p) = 1 - (1-p)^2$ has the simple roots $0, 2$.

From Theorem 2 we can deduce an algorithm that computes the list of all 2^m complex roots of the polynomial $\mathrm{Rel}(\boldsymbol{C}^{\boldsymbol{u}}; p)$ for any vector $\boldsymbol{u} \in \{0,1\}^m$. Our algorithm uses the following functions:

- $\mathrm{Append}(S, x)$: appends an element x to a list S;
- $\mathrm{Elem}(S)$: extracts all the elements from the list S;
- $\mathrm{Eval}(f, x)$: evaluates a function/a list of functions f at the value x.

The algorithm works as follows. It starts by initializing the parameters, i.e., considers the two lists $[\sqrt{x}, -\sqrt{x}]$, $[1 - \sqrt{1-x}, 1 + \sqrt{1-x}]$. It also considers the "network" formed by a single device, which has a single root $x = 0$. Then it runs through two loops: an outer loop for all the values u_k, and an inner loop that updates the list of complex roots for each level k. The inner loop applies either the operator $[\sqrt{x}, -\sqrt{x}]$ or $[1 - \sqrt{1-x}, 1 + \sqrt{1-x}]$ (depending on the value of u_k) on each existing complex root. The new set of roots is thus updated with respect to the value u_k.

Algorithm 1. Generating the roots

Require: $\boldsymbol{u} \in \{0,1\}^m$
Ensure: The list L of all complex roots of $\mathrm{Rel}(\boldsymbol{C}^{\boldsymbol{u}};p)$.
  $g_0 = [\sqrt{x}, -\sqrt{x}], g_1 = [1-\sqrt{1-x}, 1+\sqrt{1-x}]$
  $L = [0]$ {Initialization phase}
  $S = []$
  for $k \leftarrow 1, m$ **do**
    for $j \leftarrow 1, |L|$ **do**
      $S = \mathrm{Append}(S, \mathrm{Elem}(\mathrm{Eval}(g_{u_l}, L[j])))$
    end for
    $L = S$
  end for
  return L

Proposition 1. *Let m be a positive integer and $\boldsymbol{u} \in \{0,1\}^m$. Algorithm 1 outputs the list of all complex roots of* $\mathrm{Rel}(\boldsymbol{C}^{\boldsymbol{u}};p)$.

Proof. Use Theorem 2 and Theorem 1 in order to check that L contains all the complex roots of $\mathrm{Rel}(\boldsymbol{C}^{\boldsymbol{u}};p)$. Also, the size of the list L returned by Algorithm 1 equals 2^m, which is exactly the number of complex roots or $\mathrm{Rel}(\boldsymbol{C}^{\boldsymbol{u}};p)$.

Proposition 2. *The time complexity of Algorithm 1 is $O(n)$.*

Proof. We notice that Algorithm 1 creates a binary tree of elements/complex roots. The tree has depth m (given by the outer loop) and width 2^m (the width of the last level, where the leaves of tree are situated, given by the size of L once the two loops are finished). Hence, it has to compute a total of $\sum_{l=1}^{m} 2^l = 2^{m+1} - 2$ complex numbers.

3.2 When m is Small

For verification, the reliability polynomials for compositions can either be computed using the formula from Theorem 1, or taken directly from [1] (for $m = 4$).

1. For $m = 1$ we have $\mathrm{Eval}(g_k, 0) = \begin{cases} \{0,0\}, & k = 0 \\ \{0,2\}, & k = 1 \end{cases}$.
   These values correspond to the roots of $\mathrm{Rel}(\boldsymbol{C}^{(0)};p)$ and $\mathrm{Rel}(\boldsymbol{C}^{(1)};p)$.
2. For $m = 2$ we need first to compute $\mathrm{Eval}(g_k, 2) = \begin{cases} \{\sqrt{2}, -\sqrt{2}\}, & k = 0 \\ \{1-i, 1+i\}, & k = 1 \end{cases}$.
   Now, we can determine the roots of all four (2^2) compositions.

$$\begin{cases} \{0,0,0,0\}, & \boldsymbol{u} = (0,0) \\ \{0,0,\sqrt{2},-\sqrt{2}\}, & \boldsymbol{u} = (1,0) \\ \{0,0,2,2\}, & \boldsymbol{u} = (0,1) \\ \{0,2,1-i,1+i\}, & \boldsymbol{u} = (1,1) \end{cases}.$$

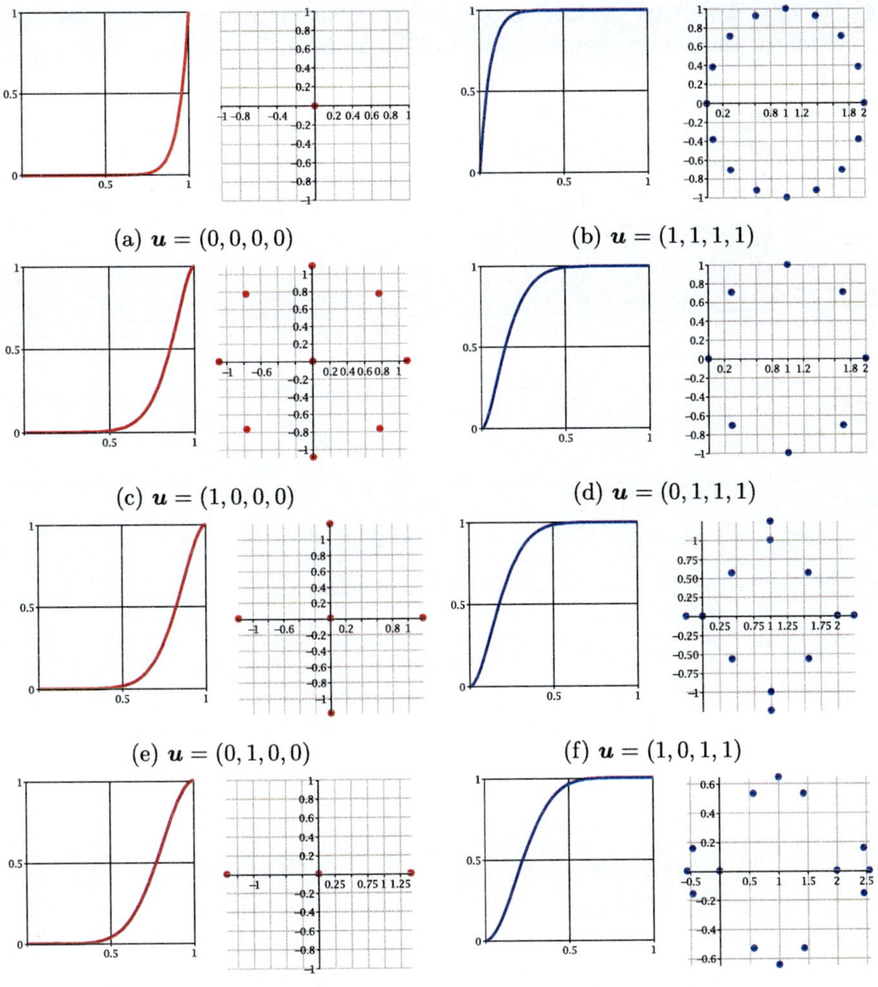

(a) $\boldsymbol{u} = (0,0,0,0)$ (b) $\boldsymbol{u} = (1,1,1,1)$

(c) $\boldsymbol{u} = (1,0,0,0)$ (d) $\boldsymbol{u} = (0,1,1,1)$

(e) $\boldsymbol{u} = (0,1,0,0)$ (f) $\boldsymbol{u} = (1,0,1,1)$

(g) $\boldsymbol{u} = (0,0,1,0)$ (h) $\boldsymbol{u} = (1,1,0,1)$

Fig. 2. Reliability polynomials and their complex roots for $\boldsymbol{C^u}$ with $\boldsymbol{u} \in \{0,1\}^4$.

3. For $m = 3$ we have

$$\begin{cases} \{0,0,0,0,0,0,0,0\}, & \boldsymbol{u} = (0,0,0) \\ \left\{0,0,0,0,\sqrt{2},\sqrt{2},-\sqrt{2},-\sqrt{2}\right\}, & \boldsymbol{u} = (0,1,0) \\ \{0,0,0,0,2,2,2,2\}, & \boldsymbol{u} = (0,0,1) \\ \{0,0,2,2,1-i,1-i,1+i,1+i\}, & \boldsymbol{u} = (0,1,1) \\ \left\{0,0,0,0,\sqrt[4]{2},-\sqrt[4]{2},i\sqrt[4]{2},-i\sqrt[4]{2}\right\}, & \boldsymbol{u} = (1,0,0) \\ \left\{0,0,2,2,1 \pm i\sqrt{\sqrt{2}-1}, 1 \pm \sqrt{\sqrt{2}+1}\right\}, & \boldsymbol{u} = (1,0,1) \\ \left\{0,0,\sqrt{2},-\sqrt{2},\sqrt{\frac{\sqrt{2}+1}{2}} \pm i\sqrt{\frac{\sqrt{2}-1}{2}}, -\sqrt{\frac{\sqrt{2}+1}{2}} \pm i\sqrt{\frac{\sqrt{2}-1}{2}}\right\}, & \boldsymbol{u} = (1,1,0) \\ \left\{0,2,1-i,1+i,1 \pm \frac{1}{\sqrt{2}}(1-i), 1 \pm \frac{1}{\sqrt{2}}(1+i)\right\}, & \boldsymbol{u} = (1,1,1) \end{cases}.$$

4. For $m = 4$ we have decided to plot the reliability polynomials together with their corresponding complex roots in Figs. 2 and 3.

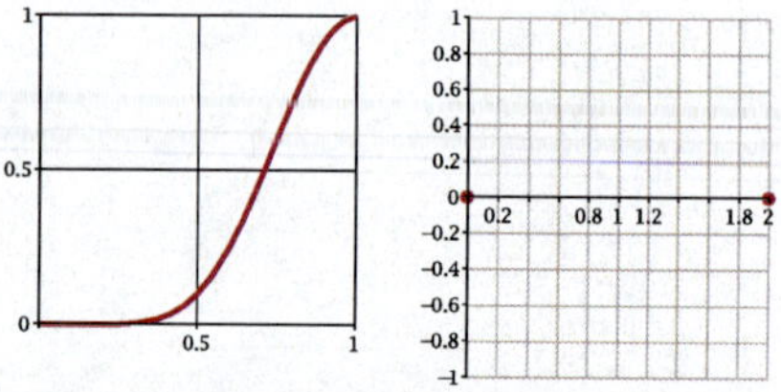

(a) $\boldsymbol{u} = (0, 0, 0, 1)$

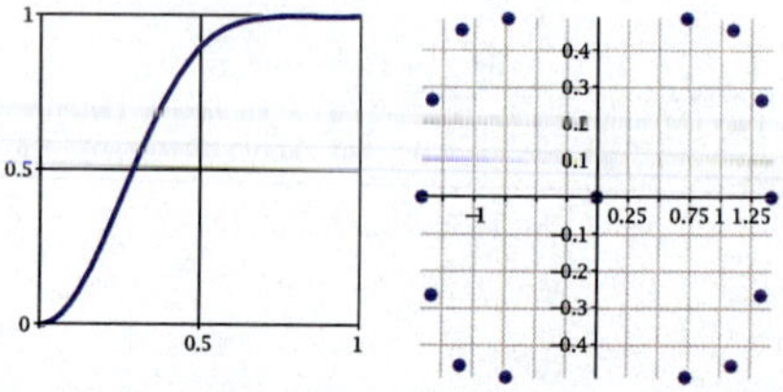

(b) $\boldsymbol{u} = (1, 1, 1, 0)$

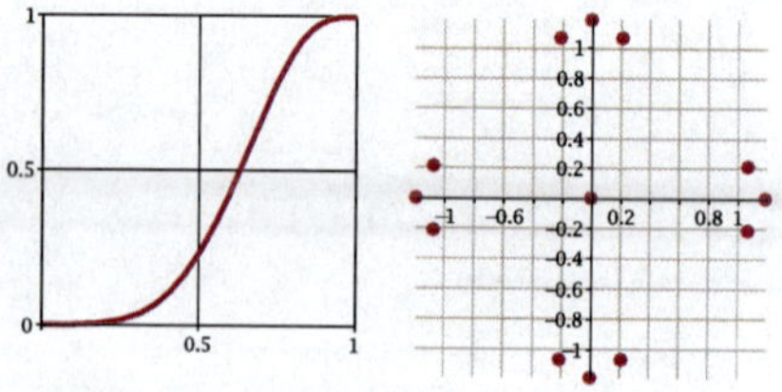

(c) $\boldsymbol{u} = (1, 1, 0, 0)$

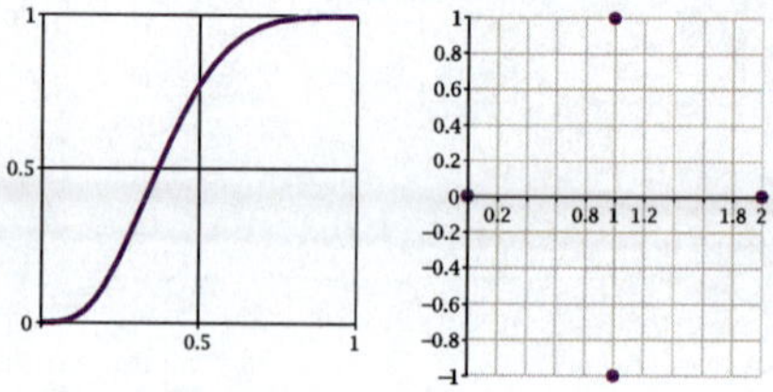

(d) $\boldsymbol{u} = (0, 0, 1, 1)$

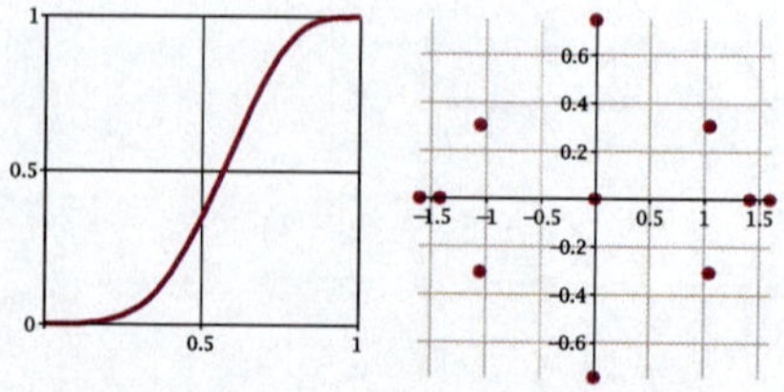

(e) $\boldsymbol{u} = (1, 0, 1, 0)$

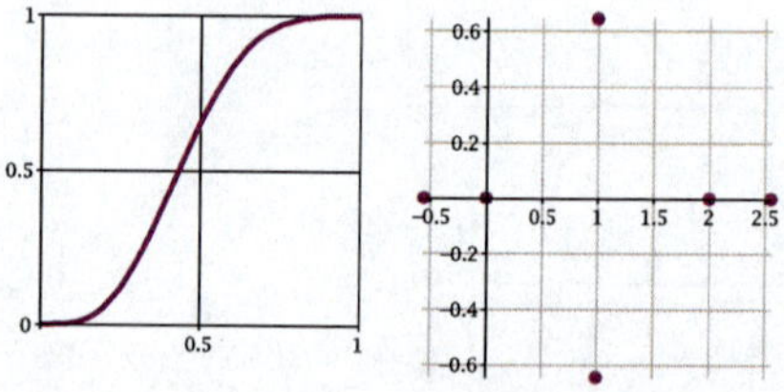

(f) $\boldsymbol{u} = (0, 1, 0, 1)$

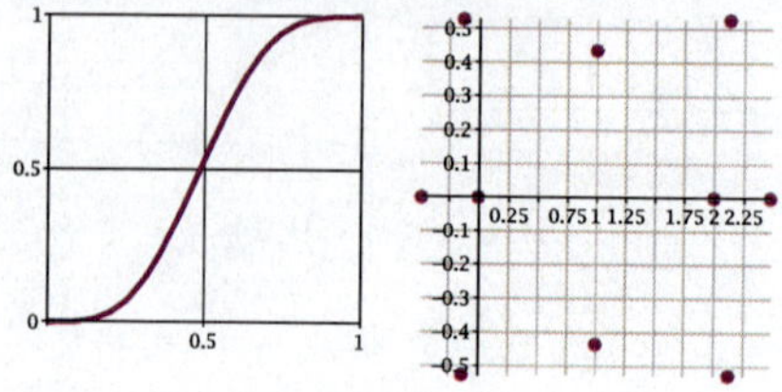

(g) $\boldsymbol{u} = (1, 0, 0, 1)$

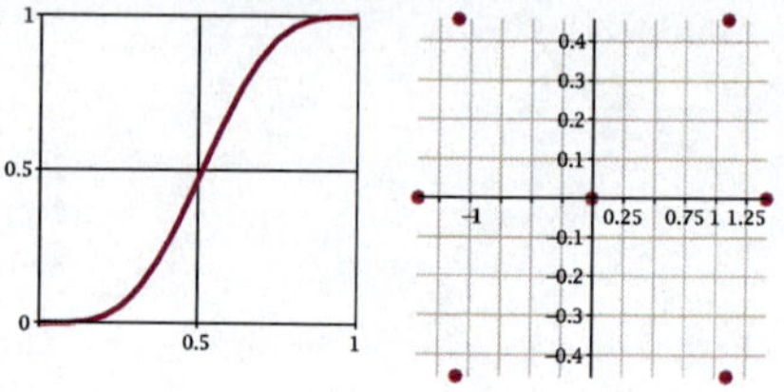

(h) $\boldsymbol{u} = (0, 1, 1, 0)$

Fig. 3. Reliability polynomials and their complex roots for $\boldsymbol{C}^{\boldsymbol{u}}$ with $\boldsymbol{u} \in \{0, 1\}^4$.

3.3 Geometric Features of the Set of Roots

Computing the Roots for all the Compositions. When m is incremented, computations are required for only half of the networks. In particular, we have the following (simple) optimization result.

Proposition 3. *Let* $\boldsymbol{u} \in \{0,1\}^m$ *with* $u_1 = 0$. *Then* $\mathrm{Rel}(\boldsymbol{C}^{\boldsymbol{u}}; p)$ *has the same roots as* $\mathrm{Rel}(\boldsymbol{C}^{(u_2,\ldots,u_m)}; p)$, *with doubled multiplicity.*

Proof. Since $\mathrm{Rel}(\boldsymbol{C}^{(0)}; p) = p^2$, for $\boldsymbol{u} \in \{0,1\}^m$, $\boldsymbol{u} = (0, u_2, \ldots, u_m)$, we have:

$$\mathrm{Rel}(\boldsymbol{C}^{\boldsymbol{u}}; p) = \mathrm{Rel}(\boldsymbol{C}^{(0)}; p) \circ \mathrm{Rel}(\boldsymbol{C}^{(u_2,\ldots,u_m)}; p) = \left[\mathrm{Rel}(\boldsymbol{C}^{(u_2,\ldots,u_m)}; p)\right]^2.$$

This implies that, when keeping track of the existing lists, at least for half of the networks no computations are needed.

We also remark that the network $\boldsymbol{C}^{(0,0,\ldots,0)}$ obtained only by series compositions, has 0 as multiple root (of order n), while the network $\boldsymbol{C}^{(1,1,\ldots,1)}$ obtained only by parallel compositions has the roots $1+\omega_k$, where $\{\omega_k, k = 0, 1, \ldots, n-1\}$ are the roots of order n of the unity.

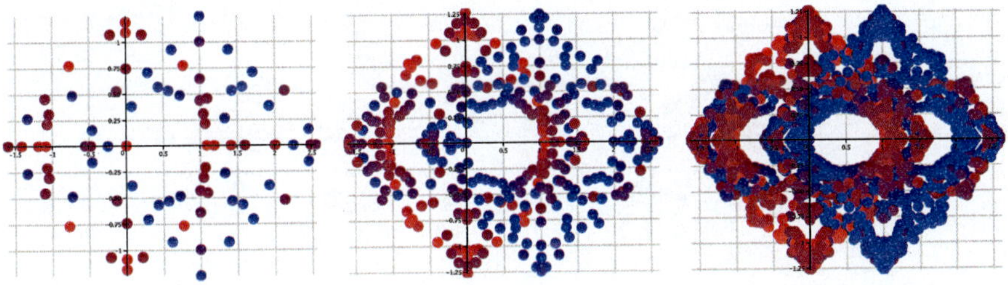

(a) $m = 4$ (b) $m = 5$ (c) $m = 6$

Fig. 4. The complex roots of $\mathrm{Rel}(\boldsymbol{C}^{\boldsymbol{u}}; p)$ for all $\boldsymbol{u} \in \{0,1\}^m$.

Symmetries. The first symmetry comes from the fact that all the coefficients of a reliability polynomial are real numbers (in fact, integers), so, if it has a complex root then the conjugate of this root is also a root of the reliability polynomial. Hence, the symmetry of the set of roots with respect to the x-axis is easily inferred.

The second type of symmetry comes from the composition law.

Proposition 4. *Let* $\boldsymbol{u} \in \{0,1\}^m$.

i) If $u_m = 0$ *then the origin* $z = 0$ *is the center of symmetry of the set of roots of the polynomial* $\mathrm{Rel}(\boldsymbol{C}^{\boldsymbol{u}}; p)$.

ii) If $u_m = 1$ *then the point* $z = 1$ *is the center of symmetry of the set of roots of the polynomial* $\mathrm{Rel}(\boldsymbol{C}^{\boldsymbol{u}}; p)$.

Proof. By Theorem 2, if $u_m = 0$ and α is a root of $\mathrm{Rel}(\boldsymbol{C}^{\boldsymbol{u}}; p)$, then $-\alpha$ is also a root, hence the set of roots is symmetric with respect to the origin.

Similarly, if $u_m = 1$ and α is a root of $\mathrm{Rel}(\boldsymbol{C}^{\boldsymbol{u}}; p)$, then $2 - \alpha$ is also a root, hence the set of roots is symmetric with respect to the point $z = 1$.

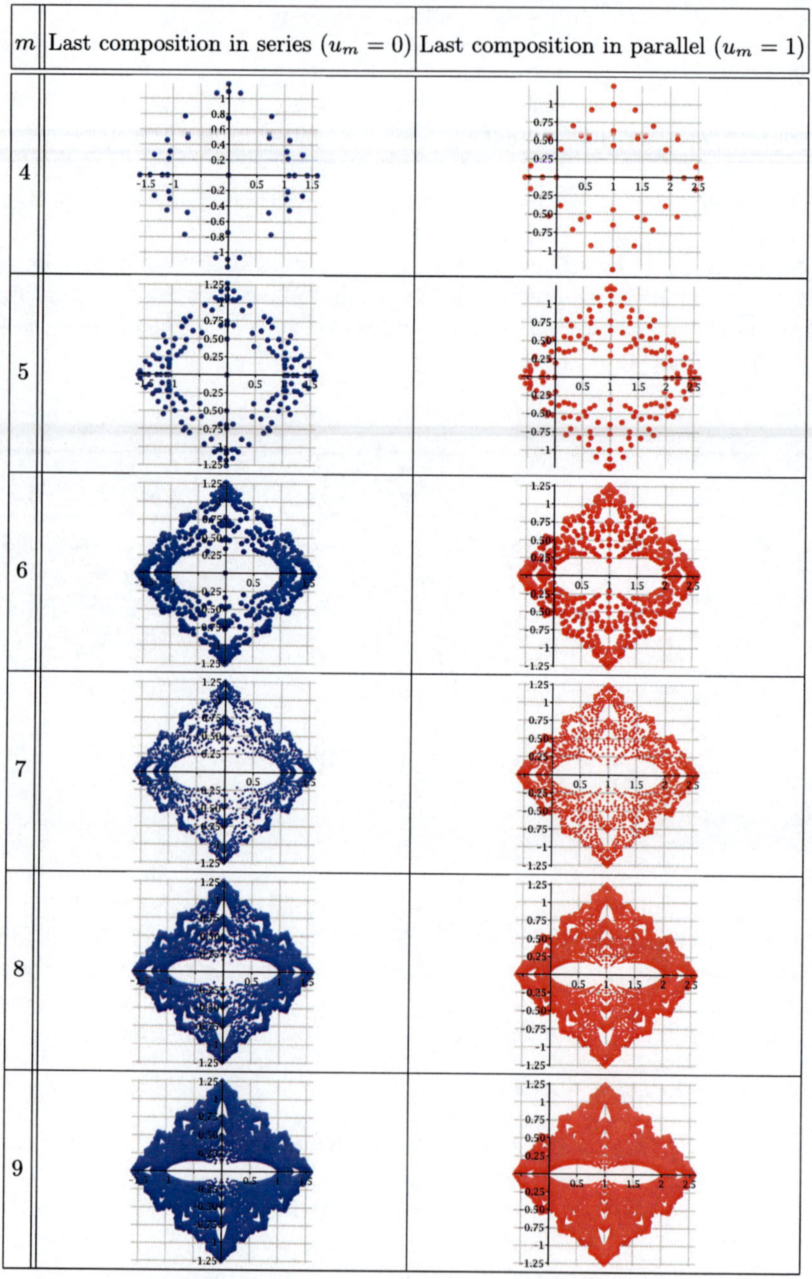

Fig. 5. The complex roots of $\mathrm{Rel}(\boldsymbol{C}^{\boldsymbol{u}}; p)$ for all $\boldsymbol{u} \in \{0, 1\}^m$.

By Proposition 4, and using the symmetry with respect to the x-axis, the next corollary follows.

Corollary 1. *Let* $\boldsymbol{u} \in \{0,1\}^m$.

i) If $u_m = 0$ *then the roots of* $\mathrm{Rel}(\boldsymbol{C^u}; p)$ *are located symmetrically with respect to the* y*-axis.*
ii) If $u_m = 1$ *then the roots of* $\mathrm{Rel}(\boldsymbol{C^u}; p)$ *are located symmetrically with respect to the vertical line at* $z = 1$.

We plot in Fig. 4 the roots for all the compositions for $m \in \{4, 5, 6\}$. Looking at this figure the following aspects are to be mentioned:

- two distinct overlapping shapes (set of points) are noticeable;
- one of them has the symmetry center at the origin;
- the other one has the symmetry center at the point $z = 1$;
- the two shapes are similar but not identical.

These properties are illustrated in Fig. 5 where the roots of the networks of size 2^m with $m \in \{4, \ldots, 9\}$ are represented.

4 Conclusions

This paper has presented an analysis of the roots of the two-terminal reliability polynomials associated to *binary composition of series and parallel*, which, as far as we know, have never been properly investigated before (except for suggestions/hints in a very fresh presentation [16]).

An $O(n)$ time complexity algorithm for computing the set of complex roots of the reliability polynomial of a composition of series and parallel was introduced. A particular feature of this algorithm is that it does not require computing the reliability polynomials themselves. Additionally, the algorithm exhibits recursivity in the sense that for (at least) half of the compositions of series and parallel networks of size n (2^m) the roots of the associated reliability polynomials follow directly from those corresponding to networks of size $n/2$ (i.e., 2^{m-1}). Using this algorithm we have reported here preliminary results for all the compositions of series and parallel networks of sizes $n = 2^m$, with $m \in \{4, \ldots, 9\}$.

Lastly, the roots of the reliability polynomials of compositions of series and parallel networks are symmetric either with respect to the origin (if the last composition is "in series"), or with respect to $z = 1$ (if the last composition is "in parallel").

Acknowledgment. L. Dăuş, M. Jianu, and V. Beiu were supported by a grant of the Romanian Ministry of Education and Research, CNCS-UEFISCDI, project number PN-III-P4-ID-PCE-2020-2495, within PNCDI III (ThUNDER^2 = *Techniques for Unconventional Nano-Designing in the Energy-Reliability Realm*).

V.-F. Drăgoi was supported by a grant of the Ministry of Research, Innovation and Digitization, CNCS/CCCDI-UEFISCDI, project number PN-III-P1-1.1-PD-2019-0285, within PNCDI III.

References

1. Beiu, V., Drăgoi V.-F., Beiu, R.-M.: Why reliability for computing needs rethinking. In: Proceedings of International Conference on Rebooting Computing (ICRC2020), pp. 16–25. IEEE, Piscataway (2020)
2. Brecht, T.B., Colbourn, C.J.: Lower bounds on two-terminal network reliability. Discrete Appl. Math. **21**(3), 185–198 (1988)
3. Brown, J.I., Colbourn, C.J.: Roots of the reliability polynomial. SIAM J. Discrete Math. **5**(4), 571–585 (1992)
4. Brown, J.I., Colbourn, C.J.: On the log concavity of reliability and matroidal sequences. Adv. Appl. Math. **15**, 114–127 (1994)
5. Brown, J.I., Cox, D.: The closure of the set of roots of strongly connected reliability polynomials is the entire complex plane. Discrete Math. **309**(16), 5043–5047 (2009)
6. Brown, J.I., Colbourn, C.J., Cox, D., Graves, C., Mol, L.: Network reliability: Heading out on the highway. Networks **77**(1), 146–160 (2021)
7. Brown, J.I., DeGagné, C.D.C.: Roots of two-terminal reliability polynomials. Networks **78**(2), 153–163 (2021)
8. Brown, J.I., Dilcher, K.: On the roots of strongly connected reliability polynomials. Networks **54**(2), 108–116 (2009)
9. Brown, J.I., Dilcher, K., Nowakowski, R.J.: Roots of independence polynomials of well covered graphs. J. Algebraic Combin. **11**(3), 197–210 (2000)
10. Brown, J.I., Hickman, C.A., Sokal, A.D., Wagner, D.G.: On the chromatic roots of generalized theta graphs. J. Comb. Theory Ser. A **83**(2), 272–297 (2001)
11. Brown, J.I., Hickman, C.A., Nowakowski, R.J.: On the location of roots of independence polynomials. J. Algebraic Combin. **19**(3), 273–282 (2004)
12. Brown, J.I., Mol, L.: On the roots of all-terminal reliability polynomials. Discrete Math. **340**(6), 1287–1299 (2017)
13. Brown, J.I., Koç, Y., Kooij, R.E.: Reliability polynomials crossing more than twice. In: Proceedings of International Congress on Ultra Modern Telecommunications and Control Systems and Workshops (ICUMT'11), Budapest, Hungary, art. 6078860 (1–6). IEEE, Piscataway (2011)
14. Brown, J.I., Koç, Y., Kooij, R.E.: Inflection points for network reliability. Telecomm. Syst. **56**(1), 79–84 (2014)
15. Brylawski, T.H.: A combinatorial model for series-parallel networks. Trans. Amer. Math. Soc. **154**, 1–22 (1971)
16. Buys, P.: Zero-locus and activity-locus of the two-terminal reliability polynomial. Presentation at the Atlantic Graph Theory Seminar Series, 9 March 2022. https://mathstat.dal.ca/~brown/sound/AARMS/AARMSGTS-03-09-22-buys.pdf. Accessed 11 June 2022
17. Chari, K.M., Colbourn, C.J.: Reliability polynomials: A survey. J. Combin. Inf. Syst. Sci. **22**(3–4), 177–193 (1997)
18. Colbourn, C.J.: The Combinatorics of Network Reliability. Oxford University Press, New York (1987)
19. Colbourn, C.J.: Combinatorial aspects of network reliability. Ann. Oper. Res. **33**(1), 1–15 (1991)
20. Cowell, S.R., Hoară, S., Beiu, V.: Experimenting with beta distributions for approximating hammocks' reliability. In: Dzitac, I., Dzitac, S., Filip, F.G., Kacprzyk, J., Manolescu, M.-J., Oros, H. (eds.) ICCCC 2020. AISC, vol. 1243, pp. 70–81. Springer, Cham (2021). https://doi.org/10.1007/978-3-030-53651-0_6

21. Cristescu, G., Drăgoi, V.-F.: Cubic spline approximation of the reliability polynomials of two dual hammock networks. Transylvanian J. Math. Mech. **11**(1/2), 77–90 (2019)
22. Cristescu, G., Drăgoi, V.-F.: Efficient approximation of two-terminal networks reliability polynomials using cubic splines. IEEE Trans. Reliab. **70**(3), 1193–1203 (2021)
23. Dăuş, L., Jianu, M.: Full Hermite interpolation of the reliability of a hammock network. Appl. Anal. Discrete Math. **14**(1), 198–220 (2020)
24. Dăuş, L., Jianu, M.: The shape of the reliability polynomial of a hammock network. In: Dzitac, I., Dzitac, S., Filip, F.G., Kacprzyk, J., Manolescu, M.-J., Oros, H. (eds.) ICCCC 2020. AISC, vol. 1243, pp. 93–105. Springer, Cham (2021). https://doi.org/10.1007/978-3-030-53651-0_8
25. DeGagné, C.D.C.: Network reliability, simplicial complexes, and polynomial roots. Ph.D. thesis, Dalhousie Univ., Halifax, Nova Scotia, Canada (2020). https://dalspace.library.dal.ca/handle/10222/78281. Accessed 5 May 2022
26. Drăgoi, V.-F., Beiu, V.: Fast reliability ranking of matchstick minimal networks. Networks **79**(4), 479–500 (2022)
27. Drăgoi, V.-F., Cowell, S.R., Beiu, V., Hoară, S., Gaşpar, P.: How reliable are compositions of series and parallel networks compared with hammocks? Intl. J. Comput. Commun. Control **13**(5), 772–791 (2018)
28. Duffin, R.J.: Topology of series-parallel networks. J. Math. Anal. Appl. **10**(2), 303–318 (1965)
29. Graves, C.: Inflection points of coherent reliability polynomials. Austr. J. Combin. **49**, 111–126 (2011)
30. Huh, J.: h-vectors of matroids and logarithmic concavity. Adv. Math. **270**, 49–59 (2015)
31. Jianu, M., Ciuiu, D., Dăuş, L., Jianu, M.: Markov chain method for computing the reliability of hammock networks. Probab. Eng. Inf. Sci. **36**(2), 276–293 (2022)
32. Kirchhoff, G.R.: Ueber die Auflösung der Gleichungen, auf welche man bei der Untersuchung der linearen Vertheilung galvanischer Ströme geführt wird. Annalen der Physik **148**(12), 497–508 (1847). [Translated by O'Toole, J.B.: On the solution of the equations obtained from the investigation of the linear distribution of galvanic currents. IRE Trans. Circ. Theory **5**(1), 4–7 (1958)]
33. Lenz, M.: The f-vector of a representable-matroid complex is log-concave. Adv. Appl. Math. **51**(5), 543–545 (2013)
34. MacMahon, P.A.: The combination of resistances. Electrician **28**, 601–602 (1892). [Reprinted in Discr. Appl. Maths. **54**(2–3), 225–228 (1994)]
35. Moore, E.F., Shannon, C.E.: Reliable circuits using less reliable relays - part I. J. Frankl. Inst. **262**(3), 191–208 (1956)
36. Moore, E.F., Shannon, C.E.: Reliable circuits using less reliable relays - part II. J. Frankl. Inst. **262**(4), 281–297 (1956)
37. Provan, J.S., Ball, M.O.: The complexity of counting cuts and of computing the probability that a graph is connected. SIAM J. Comput. **12**(4), 777–788 (1983)
38. Riordan, J., Shannon, C.E.: The number of two-terminal series-parallel networks. J. Math. Phys. **21**(1–4), 83–92 (1942)
39. Rosenthal, A.: Computing the reliability of complex networks. SIAM J. Appl. Math. **32**(2), 384–393 (1977)
40. Royle, G., Sokal, A.D.: The Brown-Colbourn conjecture on zeros of reliability polynomials is false. J. Combin. Theory Ser. B **91**(2), 345–360 (2004)

41. Tanguy, C.: Exact solutions for the two- and all-terminal reliabilities of a simple ladder network. Technical report arXiv:cs/0612143 [cs.PF] (2006). https://doi.org/10.48550/arXiv.cs/0612143
42. Tanguy, C.: Exact solutions for the two- and all-terminal reliabilities of the Brecht-Colbourn ladder and the generalized fan. Technical report arXiv:cs/0701005 [cs.PF] (2006). https://doi.org/10.48550/arXiv.cs/0701005
43. Tanguy, C.: Exact two-terminal reliability of some directed networks. In: IEEE International Workshop Design & Reliable Communication Networks (DRCN 2007), La Rochelle, France, art. 4762273, pp. 1–8. IEEE, Piscataway (2007)
44. Valiant, L.G.: The complexity of enumeration and reliability problems. SIAM J. Comput. **8**(3), 410–421 (1979)

Using Delta-Wye Transformations for Estimating Networks' Reliability

Marilena Jianu[1(✉)], Leonard Dăuş[1], Sorin-Horaţiu Hoară[2], and Valeriu Beiu[3]

[1] Department of Mathematics and Computer Science, Technical University of Civil Engineering Bucharest, 020396 Bucharest, Romania
{marilena.jianu,leonard.daus}@utcb.ro

[2] Faculty of Automation and Computers, Politehnica University of Timişoara, 300223 Timişoara, Romania
sorin-horatiu.hoara@student.upt.ro

[3] Faculty of Exact Sciences, "Aurel Vlaicu" University of Arad, 310330 Arad, Romania
valeriu.beiu@uav.ro

Abstract. It is well-known that finding the exact reliability polynomial of a given two-terminal network in general is a highly demanding computational task (belonging to the $\#P$-complete class of problems), while for particular networks this process might get to be significantly simpler (e.g., of polynomial complexity). This statement is especially true in the case of (very) large networks which cannot be reduced to compositions of simpler (e.g., series and parallel) networks, e.g., those containing complex bridge-type sub-networks. A promising, and not much explored approach, is to borrow concepts and methods from electrical circuits, in particular the delta-wye transformation which has long been established and used for computing exactly the equivalent resistance of a resistor network. We shall review how such concepts (from electronics) should be adapted and applied to reliability evaluations, showing that (as opposed to the case of electrical circuits) these are not always exact, hence sometimes leading to reliability estimates. We shall exemplify approximations of the reliability polynomials when using delta-wye transformations, and show that this approach is able to significantly reduce complexity. Finally, we will apply the delta-wye transformation method to the Moore-Shannon hammock networks (for the first time ever), and show how hammock networks can be transformed into ladder networks/graphs whose reliability can be easily computed by a recurrence formula. Illustrative examples revealing the accuracy of this type of approximation will also be presented.

Keywords: Delta-wye transformation · Hammock network · Ladder graph

S. Dzitac et al. (Eds.): ICCCC 2022, AISC 1435, pp. 415–426, 2023.
https://doi.org/10.1007/978-3-031-16684-6_35

1 Introduction

The two-terminal reliability of a network with perfect nodes and failing edges is the probability that two specified vertices (normally called source S and terminus T) are connected by a path of operational edges. Valiant [20] proved that the problem of calculating two-terminal reliability belongs to the class of $\#P$-complete problems (being at least as difficult a the NP-complete ones).

Hammock networks (or brick-wall networks) are a regular type of networks introduced and analyzed by Moore and Shannon [17] more than sixty years ago, their aim being to design reliable circuits using less reliable relays. In spite of their regularity, a reliability formula for such networks is still an open problem. We will only briefly mention here the steps made until now in this direction. Cowell et al. [3] have calculated the exact reliability polynomials for the 29 hammock networks of small dimensions ($w, l \geq 5$) presented by Moore and Shannon in their original paper [17]. Drăgoi et al. [12] have compared compositions of series and parallel networks of two devices to size-equivalent hammock networks, showing that compositions of series and parallel networks are not surpassing hammock networks of the same size in terms of reliability. Dăuş and Jianu [9] have calculated the first and second non-zero coefficients of the reliability polynomials, while also approximating the reliability polynomials by full Hermite interpolation. Recently, Cowell et al. [5] have used the beta distribution to approximate reliability polynomials. In a fresh paper [14], Jianu et al. have developed an approximation method based on Markov chains, the method providing the *exact* reliability for networks with $w \leq 5$ and arbitrary length l.

In this paper we will use the delta-wye transformation aiming to reduce a hammock network to a ladder graph—whose reliability polynomial can be calculated in polynomial time. The accuracy of the approximation will be verified for hammock networks of width $w = 3$, using the exact reliability computed using the Markov chains method (see [14]).

2 Two-Terminal Reliability

The mathematical model of a network is a probabilistic graph [2], $\mathbf{G} = (V, E)$, where V is the set of nodes (vertices) and E is the set of (undirected) edges. Each edge e_i operates (is closed) with probability p_i and fails (is open) with probability $q_i = 1 - p_i$. We assume that nodes do not fail, hence the failure of the network is always a consequence of edge failures, which are supposed to be independent.

A two-terminal network has two special nodes (normally called terminals): S (source/input) and T (terminus/output). The reliability of such a network is the probability that the terminals S and T are connected by a path (a sequence of adjacent edges) made of operational (i.e., closed) edges. A two-terminal network may also have a *set* of source/input nodes $S_1, \ldots, S_s$ and a *set* of terminus/output nodes $T_1, \ldots, T_t$, respectively. By adding the fictitious nodes S and T and the fictitious edges SS_i, TT_j—all of probability 1—the classical two-terminal network is obtained.

A *pathset* P of a network/graph $\mathbf{G}$ is a subset of E that contains a path between the source node S and the terminus node T:

$$P = \{e_{k_1}, e_{k_2}, \ldots, e_{k_m}\}.$$

Let $\mathcal{P}$ be the set of all the pathsets of $\mathbf{G}$. If $n = |E|$ is the size of the graph $\mathbf{G}$, then the reliability of the two-terminal network is the multivariate polynomial:

$$Rel_{\mathbf{G}}(p_1, \ldots, p_n) = \sum_{P \in \mathcal{P}} \prod_{e_i \in P} p_i \prod_{e_j \notin P} (1 - p_j). \tag{1}$$

If all the edges have the same probability, i.e., $p_1 = p_2 = \ldots = p_n = p$, then the expression of the reliability polynomial becomes:

$$h(p) = \sum_{P \in \mathcal{P}} p^{|P|} q^{n-|P|} = \sum_{i=1}^{n} N_i p^i (1 - p)^{n-i}, \tag{2}$$

where N_i is the number of pathsets of exactly i edges.

The initial network/graph can be reduced to a simpler one by using the following five transformations, which preserve the reliability of the network (see Fig. 1):

(1) *Loop reduction*: An edge with identical endpoints, i.e., $e = VV$, can be removed.
(2) *Pendant edge reduction*: If an edge $e = UV$ has an end point V of degree 1, and V is not a terminal node, then e and V can be removed.
(3) *Series reduction*: If $e_1 = U_1V_1$ and $e_2 = U_2V_2$ are two adjacent edges such that $V_1 = U_2$ is a non-terminal node of degree 2, then V_1 can be removed and e_1, e_2 can be replaced by the edge $e_{12} = U_1V_2$ with the probability:

$$p_{12}^{(s)} = p_1 p_2. \tag{3}$$

(4) *Parallel reduction*: Two edges $e_1 = U_1V_1$ and $e_2 = U_2V_2$ are parallel edges if they have the same endpoints ($U_1 = U_2$ and $V_1 = V_2$), or if they have a common endpoint $U_1 = U_2$ and the others are both either source nodes or terminal nodes. If this is the case, the two parallel edges can be replaced by a single edge $e_{12} = U_1V_1$, having the probability:

$$p_{12}^{(p)} = 1 - (1 - p_1)(1 - p_2) = p_1 + p_2 - p_1 p_2. \tag{4}$$

(5) *Sure-edge reduction*: If an edge $e = UV$ has the probability $p(e) = 1$, then it can be deleted and its endpoints can be replaced by a single node: $U = V = W$.

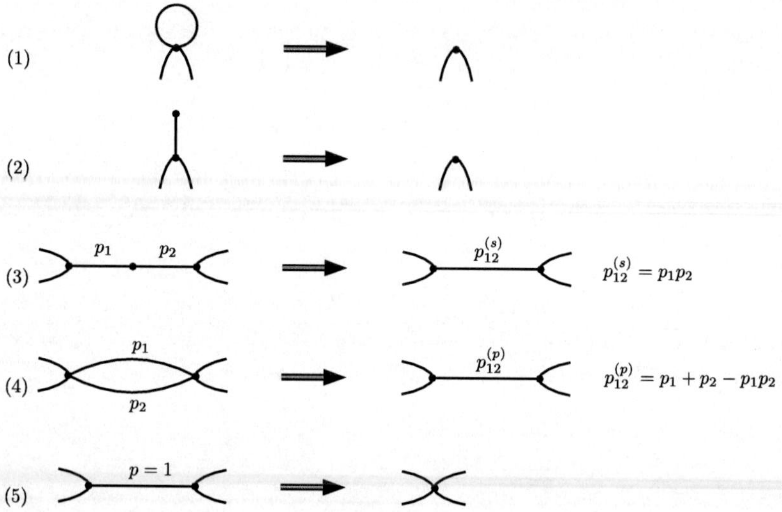

Fig. 1. Reductions preserving the reliability of a network.

3 Ladder Networks

A ladder network of size n (see Fig. 2) is a graph $L(n) = (V_n, E_n)$ with the set of vertices:

$$V_n = \{A_0, A_1, \dots, A_n, B_0, B_1, \dots, B_n\},$$

where A_0, B_0 are source nodes, A_n, B_n are terminus nodes, connected by the set of edges:

$$E_n = \left\{e_i = A_{i-1}A_i, f_i = B_{i-1}B_i, i = \overline{1,n}, g_j = A_jB_j, j = \overline{1,n-1}\right\}.$$

We use the following notation for the probability assigned to each edge:

$$\begin{aligned} p(e_i) &= x_i,\ i = 1, 2, \dots, n,\\ p(f_i) &= y_i,\ i = 1, 2, \dots, n,\\ p(g_i) &= z_i,\ i = 1, 2, \dots, n-1. \end{aligned}$$

We say that a node of a network/graph is *operational* if it is connected to a source node. We denote, for every $k = 1, 2, \dots, n$:

$$\begin{aligned} P_{A_k} &= \text{the probability that } A_k \text{ is operational;}\\ P_{B_k} &= \text{the probability that } B_k \text{ is operational;}\\ P_{A_kB_k} &= \text{the probability that both } A_k \text{ and } B_k \text{ are operational;}\\ P_k &= \text{the probability that at least one of } A_k \text{ and } B_k \text{ is operational.} \end{aligned}$$

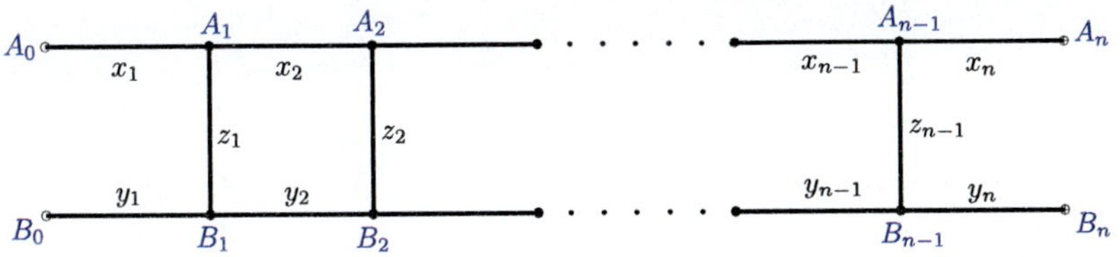

Fig. 2. Ladder network $L(n)$.

We notice that P_n is the reliability of the ladder network $L(n)$.

For $n = 1$, we have:

$$P_{A_1} = x_1,\ P_{B_1} = y_1,\ P_{A_1 B_1} = x_1 y_1,\ P_1 = x_1 + y_1 - x_1 y_1.$$

The reliability of a ladder $L(n)$ can be computed in polynomial time using the following recurrence relations:

$$\begin{cases} P_{A_n} = x_n \left[P_{A_{n-1}} + z_{n-1}(P_{B_{n-1}} - P_{A_{n-1}B_{n-1}})\right], \\ P_{B_n} = y_n \left[P_{B_{n-1}} + z_{n-1}(P_{A_{n-1}} - P_{A_{n-1}B_{n-1}})\right], \\ P_{A_n B_n} = x_n y_n \left[(1 - z_{n-1})P_{A_{n-1}B_{n-1}} + z_{n-1}P_{n-1}\right], \\ P_n = P_{A_n} + P_{B_n} - P_{A_n B_n}. \end{cases} \tag{5}$$

4 Hammock Networks

Hammock networks, which are also known as brick-wall networks, were introduced and analyzed by Moore and Shannon [17], as a solution for designing highly reliable circuits using significantly less reliable relays (probably the best option for computing—an aspect which is still under debate/investigations). A detailed presentation of these networks and of their duality properties can be found in [10], while here we shall describe them briefly.

A hammock/brick-wall network of length l and width w is formed by $l \times w$ identical relays (nowadays this should be transistors) disposed in w rows, each row consisting of l relays/transistors connected in series. Beside these horizontal connections, there exist also alternate interleaved vertical connections between consecutive rows that create the brick-wall pattern (see dotted lines in Fig. 3 (a)). The relays/transistors in the leftmost column are connected to the source node, S, while the relays/transistors in the rightmost column are connected to the terminus node, T. We suppose that each relay/transistor works with probability p (and fails with the probability $q = 1-p$). Additionally, the vertical connections between adjacent rows are supposed to have probability $p = 1$. As we have mentioned above (transformation (5)), each one of these edges with $p = 1$ can be reduced to a node, and that is how the rectangular "bricks" become rhombs, explaining why brick-wall networks are also hammock networks. By splitting the terminal nodes S and T into several source $S_1, S_2, \ldots, S_k$ and terminus

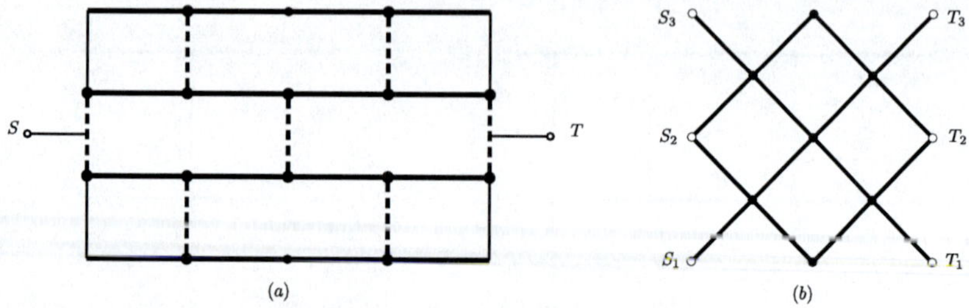

Fig. 3. The classical Moore-Shannon network of dimensions $w = l = 4$ when drawn as: (a) brick-wall network; (b) hammock network.

$T_1, T_2, \ldots, T_{k'}$ nodes, respectively, we obtain a regular, symmetric network, as can be seen in Fig. 3 (b).

For a clear understanding, Fig. 4 shows how the hammock network of dimensions $w = l = 3$ can be transformed into the ladder network $L(2)$ using the *series reduction* (3) and *parallel reduction* (4). Thus, using Eq. (5), the exact expression of the reliability polynomial can be obtained.

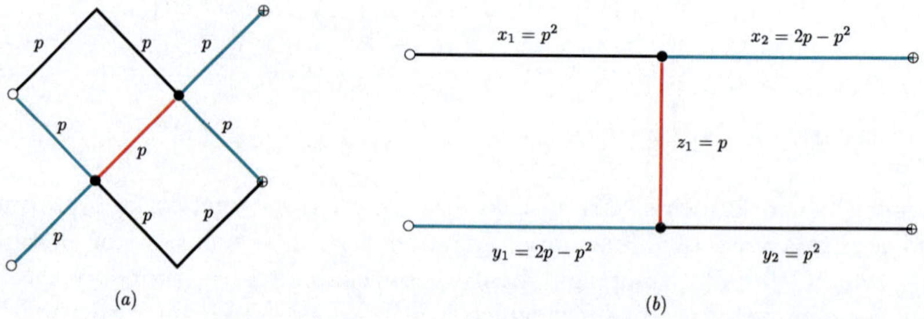

Fig. 4. The classical Moore-Shannon network of dimensions $w = l = 3$ represented as: (a) brick-wall network; (b) equivalent ladder network of dimensions $L(2)$.

5 Delta-Wye Reduction

Delta-wye and wye-delta transformations have long been used for analyzing electrical networks (since at least 1899 [15]).

For analyzing reliability, delta-wye and wye-delta transformations were used for the first time by Lehman [16], who pointed out that the method based on delta-wye and wye-delta reductions does not calculate the *exact* reliability of a network, and it can only *approximate* the reliability polynomial. Lehman also conjectured that any planar two-terminal network $G = (V, E)$ can be reduced to a single edge connecting the terminals $\{S, T\}$ by a finite sequence of transformations of the following types:

- loop reduction;
- pendant edge reduction;
- series reduction;
- parallel reduction;
- delta-wye reduction;
- wye-delta reduction.

Truemper [19], as well as Feo and Provan [13], proved that this can be performed in polynomial time: $O(n^2)$, where $n = |V|$.

Since solving the two-terminal reliability in general is a #P-complete problem, it follows that sometimes this cannot be answered *exactly* by using such reductions (as it *cannot* be done in polynomial time). While the first four transformations do not affect the reliability of the network, delta-wye and wye-delta reductions are the ones which do not always preserve the exact reliability of the network.

A delta-wye transformation (see Fig. 5) takes a triangle (Δ) formed by the vertices V_1, V_2, V_3 and transforms it into an Y by deleting the edges $e_1 = V_2V_3$, $e_2 = V_3V_1$, $e_3 = V_1V_2$, and introducing a new vertex U and three new edges (adjacent in U): $e'_1 = UV_1$, $e'_2 = UV_2$, $e'_3 = UV_3$. Conversely, a wye-delta transformation takes a three-degree non-terminal vertex U and deletes it together with its three adjacent edges $e'_1 = UV_1$, $e'_2 = UV_2$, $e'_3 = UV_3$, introducing three new edges $e_1 = V_2V_3$, $e_2 = V_3V_1$, $e_3 = V_1V_2$.

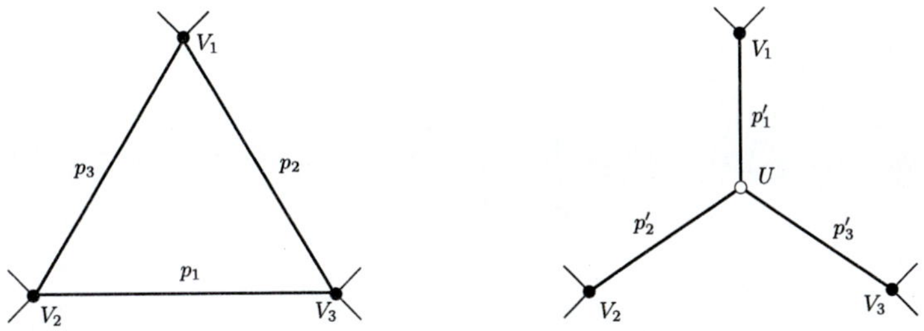

Fig. 5. Delta-wye reduction.

For each $i = 1, 2, 3$ we denote by $p_i = p(e_i)$ the probabilities of delta's edges and by $p'_i = p(e'_i)$, the probabilities of wye's edges. The probabilities p_1, p_2, p_3 are related to the probabilities p'_1, p'_2, p'_3 by two set of equations [1,16]: *open-circuit* Eq. (6), which equates the probabilities that each pair of vertices $\{V_1, V_2\}$, $\{V_2, V_3\}$ and $\{V_3, V_1\}$ are connected, and the *short-circuit* eq. (7), which equates the probabilities that each vertex V_1, V_2 and V_3 is connected to at least one other vertex.

The probability that the vertices V_1 and V_2 are connected is $p'_1p'_2$ in the Y network, and $p_3 + (1 - p_3)p_1p_2$ in the Δ-network, so the *open-circuit* equations are:

$$\begin{cases} p_1'p_2' = p_3 + p_1p_2 - p_1p_2p_3, \\ p_2'p_3' = p_1 + p_2p_3 - p_1p_2p_3, \\ p_3'p_1' = p_2 + p_3p_1 - p_1p_2p_3. \end{cases} \tag{6}$$

The probability that the vertex V_1 is connected to at least one of the vertices V_2 and V_3 is $1-(1-p_2)(1-p_3)$ in the Δ-network, and $p_1'\,[1-(1-p_2')(1-p_3')]$ in the Y-network, so the *short-circuit* system is written:

$$\begin{cases} p_1'(p_2' + p_3' - p_2'p_3') = p_2 + p_3 - p_2p_3, \\ p_2'(p_3' + p_1' - p_3'p_1') = p_3 + p_1 - p_3p_1, \\ p_3'(p_1' + p_2' - p_1'p_2') = p_1 + p_2 - p_1p_2. \end{cases} \tag{7}$$

Suppose now that we make a Δ-Y transformation, that is, the probabilities p_1, p_2, p_3 are known, and we want to find the probabilities p_1', p_2', p_3' of the new-added edges e_1', e_2', e_3'. Obviously, since we have only three unknowns, they cannot be the solution of a system of six equations, so we choose to solve the open-circuit system (6) for p_1', p_2', p_3'. Thus, if we denote:

$$\begin{cases} \delta_1 = p_1 + p_2p_3 - p_1p_2p_3, \\ \delta_2 = p_2 + p_3p_1 - p_1p_2p_3, \\ \delta_3 = p_3 + p_1p_2 - p_1p_2p_3, \end{cases} \tag{8}$$

then, from (6) we find that:

$$p_1'p_2'p_3' = \sqrt{\delta_1\delta_2\delta_3},$$

and we obtain the following expressions for p_1', p_2', p_3':

$$\begin{cases} p_1' = \sqrt{\frac{\delta_2\delta_3}{\delta_1}}, \\ p_2' = \sqrt{\frac{\delta_3\delta_1}{\delta_2}}, \\ p_3' = \sqrt{\frac{\delta_1\delta_2}{\delta_3}}. \end{cases} \tag{9}$$

6 Delta-Wye Reduction of Hammock Networks

Let us consider, as an example, the hammock network of dimensions $w = 3$ and $l = 4$ (see Fig. 6 (a)).

First, we apply the series reduction (and obtain the edges S_1A, AT_1 and BC, each one of probability p^2), and parallel reduction (getting the edges S_2B and CT_2, each one of probability $2p - p^2$). Then, we apply the delta-wye reduction for the triangle ABC with $p(AB) = p(AC) = p$ and $p(BC) = p^2$ (see Fig. 6 (b)). By Eq. (9) we find that the probabilities of the new edges are:

$$\begin{cases} p(OB) = p(OC) = P_1(p) = p\sqrt{2-p^2}, \\ \qquad\qquad p(OA) = P_2(p) = \dfrac{1+p^2-p^3}{\sqrt{2-p^2}}. \end{cases} \tag{10}$$

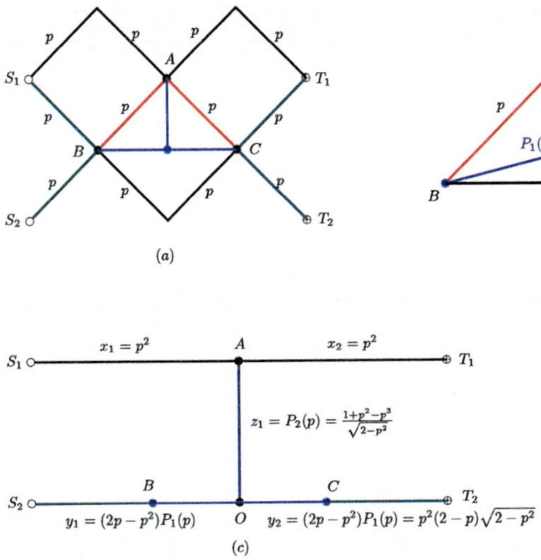

Fig. 6. (*a*) The hammock network of dimensions $w = 3$, $l = 4$; (*b*) Δ-Y reduction applied to the triangle ABC; (*c*) the ladder network $L(2)$ obtained by Δ-Y reduction.

Lastly, we apply once again the series reduction to obtain the ladder network $L(2)$ which can be seen in Fig. 6 (*c*). Its reliability approximates the reliability of the hammock network of dimensions $w = 3$ and $l = 4$.

The same process can be applied to a larger hammock network of dimensions $w = 3$ and $l = 5$, leading to a ladder network $L(3)$ (see Fig. 7), and to an even larger hammock network of dimensions $w = 3$ and $l = 6$, obtaining a ladder network $L(4)$ (see Fig. 8).

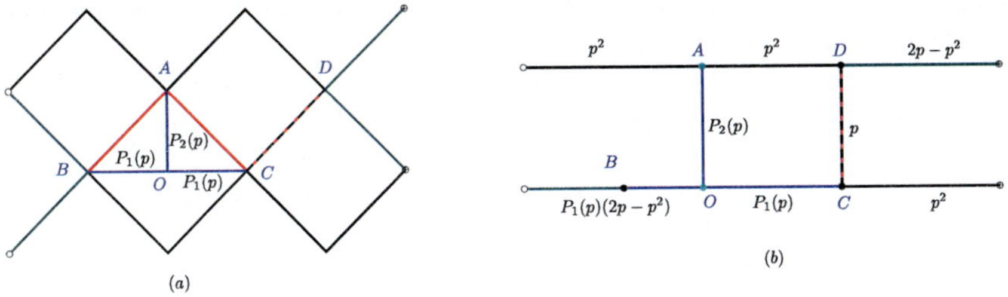

Fig. 7. (*a*) The hammock network of dimensions $w = 3$, $l = 5$; (*b*); the ladder network $L(3)$ obtained from the hammock network by the Δ-Y reduction applied to the triangle ABC.

For understanding/appreciating the accuracies such transformations are able to achieve we have computed exactly the errors, and they can be seen in Fig. 9.

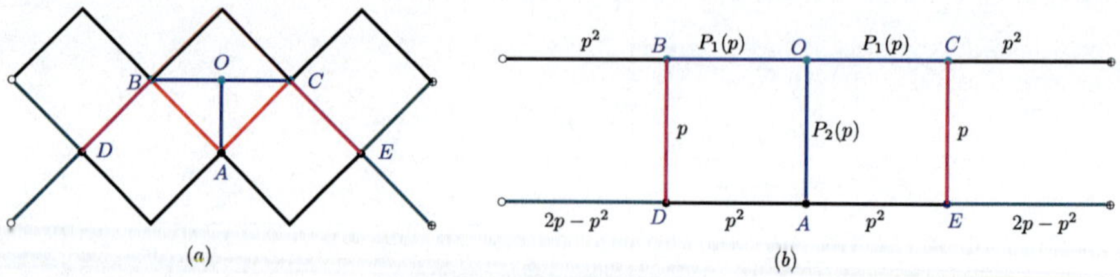

Fig. 8. (*a*) The hammock network of dimensions $w = 3$, $l = 6$; (*b*) the ladder network $L(4)$ obtained from the hammock network by the Δ-Y reduction applied to the triangle *ABC*.

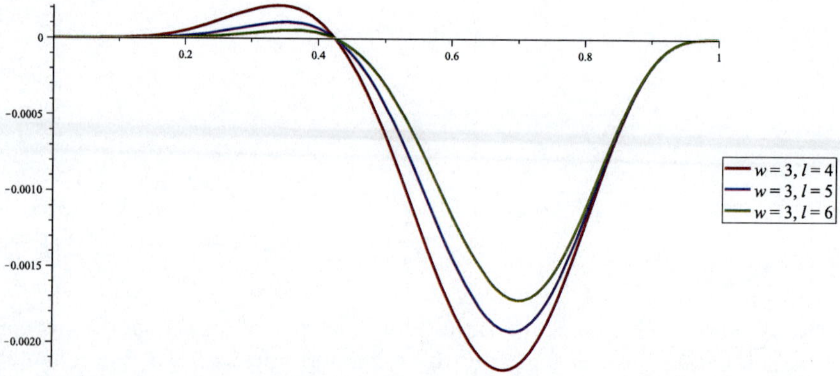

Fig. 9. Errors when estimating the reliability of hammock networks of $w = 3$ and $l = 4, 5, 6$ when transformed into ladder networks ($L(2)$, $L(3)$ and $L(4)$, respectively) using Δ-Y transformations.

We remark that, in all these three cases we have analysed here, the Δ-Y transformations are exact (i.e., the errors are 0) in the case when the short-circuit equations (7) are also satisfied, that is when:

$$P_2(p)\left[2P_1(p) - P_1(p)^2\right] = 2p - p^2$$

and

$$P_1(p)\left[P_1(p) + P_2(p) - P_1(p)P_2(p)\right] = p + p^2 - p^3.$$

Since $P_1(p)$ and $P_2(p)$ are given by Eq. (10), we find that $p = p_0$ is the unique solution between 0 and 1 of the equation:

$$\frac{1 + 2p - 2p^2}{1 + p^2 - p^3} = \sqrt{2 - p^2}. \tag{11}$$

It follows that the transformations are exact for $p_0 = 0.42579$, and also that as l is increased the reliability estimates are getting better and better, i.e., the errors are getting smaller and smaller (e.g., the largest absolute error of 0.0022 is for $l = 4$).

7 Conclusions

In this paper the difficult problem of finding the exact reliability polynomial of a two-terminal network was approached using electrical engineering methods, in particular the delta-wye transformation widely established for computing the equivalent resistance of a resistor network. This has led to reliability estimates, i.e., approximate solutions, which can be obtained in polynomial time.

In particular, we have used the delta-wye transformation to estimate the reliability of Moore-Shannon hammock networks, and have shown that particular hammock networks can be transformed into ladder networks/graphs (whose reliability polynomials can be computed by a recurrence formula). Preliminary examples on these lines have also presented the accuracy achieved by using delta-wye transformations.

In the end, it is worth mentioning that for the particular small hammocks considered here the reliability estimates are getting better and better (i.e., the errors are getting smaller and smaller) as l is increasing, an aspect which clearly deserves further investigations.

Acknowledgment. This research was partly supported by a grant of the Romanian Ministry of Education and Research, CNCS-UEFISCDI, project number PN-III-P4-ID-PCE-2020-2495, within PNCDI III (ThUNDER^2 = *Techniques for Unconventional Nano-Designing in the Energy-Reliability Realm*).

References

1. Chari, M.K., Feo, T.A., Provan, J.S.: The delta-wye approximation procedure for two-terminal reliability. Oper. Res. **44**(5), 745–757 (1996)
2. Colbourn, C.J.: The Combinatorics of Network Reliability. Oxford Univ. Press, Oxford, UK (1987)
3. Cowell, S.R., Beiu, V., Dăuş, L., Poulin, P.: On the exact reliability enhancements of small hammock networks. IEEE Access **6**, 25411–25426 (2018)
4. Cowell, S. R., Drăgoi, V.-F., Rohatinovici, N.-C., Beiu, V.: Effective conductances of Moore-Shannon hammocks. In: Proceedings of IEEE International Conference on Nanotech. (IEEE-NANO 2018), pp. 1–4 (art. 8626295), Cork, Ireland (2018)
5. Cowell, S.R., Hoară, S., Beiu, V.: Experimenting with beta distributions for approximating hammocks' reliability. In: Dzitac, I., Dzitac, S., Filip, F.G., Kacprzyk, J., Manolescu, M.-J., Oros, H. (eds.) ICCCC 2020. AISC, vol. 1243, pp. 70–81. Springer, Cham (2021). https://doi.org/10.1007/978-3-030-53651-0_6
6. Cristescu, G., Drăgoi, V.-F.: Efficient approximation of two-terminal networks reliability polynomials using cubic splines. IEEE Trans. Reliab. **70**(3), 1193–1203 (2021)
7. Cristescu, G., Drăgoi, V.-F., Hoară, S. H.: Generalized convexity properties and shape-based approximation in networks reliability. Mathematics **9**(24), art. 3182, 1–21 (2021)
8. Dăuş, L., Beiu, V., Cowell, S. R., Poulin, P.: Brick-wall lattice paths and applications. Tech. Rep. arXiv (math.CO), 1–16 (2018). https://arxiv.org/abs/1804.05277

9. Dăuş, L., Jianu, M.: Full Hermite interpolation of the reliability of a hammock network. Appl. Anal. Discr. Math. **14**(1), 198–220 (2020)
10. Dăuş, L., Jianu, M.: The shape of the reliability polynomial of a hammock network. In: Dzitac, I., Dzitac, S., Filip, F.G., Kacprzyk, J., Manolescu, M.-J., Oros, H. (eds.) ICCCC 2020. AISC, vol. 1243, pp. 93–105. Springer, Cham (2021). https://doi.org/10.1007/978-3-030-53651-0_8
11. Drăgoi, V.-F., Beiu, V.: Fast reliability ranking of matchstick minimal networks. Networks **79**(4), 479–500 (2022)
12. Drăgoi, V.-F., Cowell, S.R., Beiu, V., Hoară, S.H., Gaşpar, P.: How reliable are compositions of series and parallel networks compared with hammocks? Intl. J. Comp. Comm. Ctrl. **13**(5), 772–791 (2018)
13. Feo, T.A., Provan, J.S.: Delta-wye transformations and the efficient reduction of two-terminal planar graphs. Oper. Res. **41**(3), 572–582 (1993)
14. Jianu, M., Ciuiu, D., Dăuş, L., Jianu, M.: Markov chain method for computing the reliability of hammock networks. Probab. Eng. Inf. Sci. **36**(2), 276–293 (2022)
15. Kennelly, A.E.: The equivalence of triangles and three-pointed stars in conducting networks. Electr. World Eng. **34**(12), 413–414 (1899)
16. Lehman, A.B.: Wye-delta transformations in probabilistic networks. J. SIAM **11**(3), 773–805 (1963)
17. Moore, E.F., Shannon, C.E.: Reliable circuits using less reliable relays - Part I. J. Frankl. Inst. **262**(3), 191–208 (1956)
18. von Neumann, J.: Probabilistic logics and the synthesis of reliable organisms from unreliable components. In: Shannon, C. E., McCarthy, J. (eds.): Automata Studies, Princeton University Press, Princeton, NJ, USA, pp. 43–98 (1956). Based on five lectures given at Caltech during 4–15 January 1952. https://archive.org/details/vonNeumann_Prob_Logics_Rel_Org_Unrel_Comp_Caltech_1952/mode/2up
19. Truemper, K.: On the delta-wye reduction for planar graphs. J. Graph Theory **13**(2), 141–148 (1989)
20. Valiant, L.: The complexity of enumeration and reliability problems. SIAM J. Comp. **8**(3), 410–421 (1979)

Extension of HyperAlgebra to SuperHyperAlgebra and Neutrosophic SuperHyperAlgebra (Revisited)

Florentin Smarandache(✉)

Mathematics, Physics, and Natural Science Division, The University of New Mexico, 705 Gurley Ave., Gallup, NM 87301, USA
smarand@unm.edu

Abstract. This is a review paper. The nth-Powerset of a Set, and the concepts built on it such as SuperHyperOperation, SuperHyperAxiom, SuperHyperAlgebra, and their corresponding Neutrosophic SuperHyperOperation, Neutrosophic SuperHyperAxiom and Neutrosophic SuperHyperAlgebra are recalled and then prolonged to the Neutrosophic SuperHyperStructures {or more accurately Neutrosophic (m,n)-SuperHyperStructures} .

Keywords: n^{th}-Powerset of a Set · HyperOperation · HyperAxiom · HyperAlgebra · SuperHyperOperation · SuperHyperAxiom · SuperHyperAlgebra · Neutrosophic SuperHyperAlgebra · SuperHyperStructure · Neutrosophic SuperHyperStructure

1 Introduction

F. Marty [1] has introduced in 1934 the HyperAlgebra that is based on HyperOperations and consequently on HyperAxioms. One presents them below.

Further on we revisit the SuperHyperAgebra and Neutrosophic SuperHyperAlgebra introduced and developed by Smarandache [2–4] between 2016–2022.

2 Definition of Classical HyperOperations

Let U be a universe of discourse and H a non-empty set, $H \subset U$.

A **classical Binary HyperOperation** $\circ_2^*$ is defined as follows:

$$\circ_2^* : H^2 \to P_*(H)$$

where H is a discrete or continuous set, and $P_*(H)$ is the powerset of H without the empty-set ϕ or $P_*(H) = P(H) - \{\phi\}$.

A **classical m-ary HyperOperation** $\circ_m^*$ is defined as:

$$\circ_m^* : H^m \to P_*(H)$$

for integer $m \geq 1$. For $m = 1$ one gets a **Unary HyperOperation**.

The **classical HyperStructures** are structures endowed with classical HyperOperations and classical HyperAxioms.

S. Dzitac et al. (Eds.): ICCCC 2022, AISC 1435, pp. 427–432, 2023.
https://doi.org/10.1007/978-3-031-16684-6_36

3 Definition of the nth-Powerset of a Set [2]

The nth-Powerset of a Set was introduced in [2–4] in the following way:

$P^n(H)$, as the nth-Powerset of the Set H, for integer $n \geq 1$, is recursively defined as:

$$P^2(H) = P(P(H)),\ P^3(H) = P\left(P^2(H)\right) = P(P(P(H))),\ \ldots,$$

$$P^n(H) = P\left(P^{n-1}(H)\right),$$

where $P^{\circ}(H) \stackrel{\text{def}}{=} H$, and $P^1(H) \stackrel{\text{def}}{=} P(H)$.

The nth-Powerset of a Set better reflects our complex reality, since a set H (that may represent a group, a society, a country, a continent, etc.) of elements (such as: people, objects, and in general any items) is organized onto subsets *P(H),* and these subsets are again organized onto subsets of subsets *P(P(H)),* and so on. That's our world.

4 Neutrosophic HyperOperation and Neutrosophic HyperStructures [1, 2]

In the classical HyperOperation and classical HyperStructures, the empty-set ϕ does not belong to the power set, or $P_*(H) = P(H) - \{\phi\}$.

However, in the real world we encounter many situations when a HyperOperation $\circ$ is:

- *indeterminate*, for example $a \circ b = \phi$ (unknown, or undefined),
- or *partially indeterminate*, for example: $c \circ d = \{[0.2, 0.3],\ \phi\}$.

In our everyday life, there are many more operations and laws that have some degrees of indeterminacy (vagueness, unclearness, unknowingness, contradiction, etc.), than those that are totally determinate.

That's why in 2016 we have extended the classical HyperOperation to the Neutrosophic HyperOperation, by taking the whole power *P(H)* (that includes the empty-set ϕ as well), instead of $P_*(H)$ (that does not include the empty-set ϕ), as follow.

4.1 Definition of Neutrosophic HyperOperation

Let U be a universe of discourse and H a non-empty set, $H \subset U$.

A **Neutrosophic Binary HyperOperation** $\circ_2$ is defined as follows:

$$\circ_2 : H^2 \to P(H)$$

where H is a discrete or continuous set; *P(H)* is the powerset of H that includes the empty-set ϕ.

A **Neutrosophic m-ary HyperOperation** $\circ_m$ is defined as:

$$\circ_m : H^m \to P(H)$$

for integer $m \geq 1$. Similarly, for $m = 1$ one gets a **Neutrosophic Unary HyperOperation**.

4.2 Neutrosophic HyperStructures

A Neutrosophic HyperStructure is a structured endowed with Neutrosophic HyperOperations.

5 Definition of SuperHyperOperations

We recall our 2016 concepts of SuperHyperOperation, SuperHyperAxiom, SuperHyperAlgebra, and their corresponding Neutrosophic SuperHyperOperation Neutrosophic SuperHyperAxiom and Neutrosophic SuperHyperAlgebra [2].

Let $P_*^n(H)$ be the nth-powerset of the set H such that none of $P(H)$, $P^2(H)$, ..., $P^n(H)$ contain the empty set ϕ.

Also, let $P^n(H)$ be the nth-powerset of the set H such that at least one of the $P^2(H)$, ..., $P^n(H)$ contain the empty set ϕ.

The SuperHyperOperations are operations whose codomain is either $P_*^n(H)$ and in this case one has **classical-type SuperHyperOperations**, or $P^n(H)$ and in this case one has **Neutrosophic SuperHyperOperations**, for integer $n \geq 2$.

5.1 Classical-Type Binary SuperHyperOperation

A **classical-type Binary SuperHyperOperation** $\circ_{(2,n)}^*$ is defined as follows:

$$\circ_{(2,n)}^* : H^2 \to P_*^n(H)$$

where $P_*^n(H)$ is the nth-powerset of the set H, with no empty-set.

5.2 Examples of Classical-Type Binary SuperHyperOperation

1) Let $H = \{a, b\}$ be a finite discrete set; then its power set, without the empty-set ϕ, is (Table 1):

$$P(H) = \{a, b, \{a, b\}\}, \text{ and :}$$
$$P^2(H) = P(P(H)) = P(\{a, b, \{a, b\}\}) =$$
$$\{a, b, \{a, b\}, \{a, \{a, b\}\}, \{b, \{a, b\}\}, \{a, b, \{a, b\}\}\}.$$

$$\circ_{(2,2)}^* : H^2 \to P_*^2(H)$$

Table 1. Example 1 of classical-type Binary SuperHyperOperation.

$\circ_{(2,2)}^*$	a	b
a	$\{a, \{a, b\}\}$	$\{b, \{a, b\}\}$
b	a	$\{a, b, \{a, b\}\}$

2) Let $H = [0, 2]$ be a continuous set.
$P(H) = P([0, 2]) = \{A|A \subseteq [0, 2], A = subset\}$,
$P^2(H) = P(P([0, 2]))$.
Let $c, d \in H$ (Table 2).

$$\circ^*_{(2,2)} : H^2 \rightarrow P^2_*(H)$$

Table 2. Example 2 of classical-type Binary SuperHyperOperation.

$\circ^*_{(2,2)}$	c	d
c	{[0, 0.5], [1, 2]}	{0.7, 0.9, 1.8}
d	{2.5}	{(0.3, 0.6), {0.4, 1.9}, 2}

5.3 Classical-Type m-ary SuperHyperOperation {or More Accurate Denomination (m, n)-SuperHyperOperation}

Let U be a universe of discourse and a non-empty set $H, H \subset U$. Then:

$$\circ^*_{(m,n)} : H^m \rightarrow P^n_*(H)$$

where the integers $m, n \geq 1$,

$$H^m = \underbrace{H \times H \times \ldots \times H}_{m\ times},$$

and $P^n_*(H)$ is the nth-powerset of the set H that includes the empty-set.

This SuperHyperOperation is a m-ary operation defined from the set H to the nth-powerset of the set H.

5.4 Neutrosophic m-ary SuperHyperOperation {or More Accurate Denomination Neutrosophic (m, n)-SuperHyperOperation}

Let U be a universe of discourse and a non-empty set $H, H \subset U$. Then:

$$\circ_{(m,n)} : H^m \rightarrow P^n(H)$$

where the integers $m, n \geq 1$; $P^n(H)$- the n-th powerset of the set H that includes the empty-set.

6 SuperHyperAxiom

A **classical-type SuperHyperAxiom** or more accurately a ***(m, n)*-SuperHyperAxiom** is an axiom based on classical-type SuperHyperOperations.

Similarly, a **Neutrosophic SuperHyperAxiom** {or Neutrosphic (m, n)-SuperHyperAxiom} is an axiom based on Neutrosophic SuperHyperOperations.

There are:

- **Strong SuperHyperAxioms**, when the left-hand side is equal to the right-hand side as in non-hyper axioms,
- and **Week SuperHyperAxioms**, when the intersection between the left-hand side and the right-hand side is non-empty.

For examples, one has:

- Strong SuperHyperAssociativity, when $(x \circ y) \circ z = x \circ (y \circ z)$, for all $x, y, z \in H^m$, where the law $\circ^*_{(m,n)} : H^m \to P^n_*(H)$;
- and Week SuperHyperAssociativity, when $\left[(x \circ y) \circ z\right] \cap \left[x \circ (y \circ z)\right] \neq \phi$, for all $x, y, z \in H^m$.

7 SuperHyperAlgebra and SuperHyperStructure

A **SuperHyperAlgebra** or more accurately ***(m-n)*-SuperHyperAlgebra** is an algebra dealing with SuperHyperOperations and SuperHyperAxioms.

Again, a **Neutrosophic SuperHyperAlgebra** {or Neutrosphic (m, n)-SuperHyperAlgebra} is an algebra dealing with Neutrosophic SuperHyperOperations and Neutrosophic SuperHyperOperations.

In general, we have **SuperHyperStructures** {or *(m-n)*-SuperHyperStructures}, and corresponding **Neutrosophic SuperHyperStructures**.

For example, there are SuperHyperGrupoid, SuperHyperSemigroup, SuperHyper-Group, SuperHyperRing, SuperHyperVectorSpace, etc [5–9].

8 Distinction Between SuperHyperAlgebra vs. Neutrosophic SuperHyperAlgebra

i. If none of the power sets $P^k(H)$, $1 \leq k \leq n$, do not include the empty set ϕ, then one has a classical-type SuperHyperAlgebra;
ii. If at least one power set, $P^k(H)$, $1 \leq k \leq n$, includes the empty set ϕ, then one has a Neutrosophic SuperHyperAlgebra.

9 Conclusion

A set *H* (that may represent a group, a society, a country, a continent, etc.) of elements (such as: people, objects, and in general any items) is organized onto subsets *P(H),* and these subsets in their turn are again organized onto subsets of subsets *P(P(H)),* and so on, the nth-PowerSet of a Set [2] was introduced to better reflect our world.

The most general form of algebras, which is based on the nth-PowerSet of a Set, called SuperHyperAlgebra {or more accurate denomination (m, n)-SuperHyperAlgebra} and the Neutrososophic SuperHyperAlgebra, and their extensions to SuperHyperStructures and respectively Neutrosophic SuperHyperAlgebra in any field of knowledge are recalled.

References

1. Marty, F.: Sur une généralisation de la Notion de Groupe. In: 8th Congress Math. Scandinaves, Stockholm, Sweden, pp. 45–49 (1934)
2. Smarandache, F.: SuperHyperAlgebra and Neutrosophic SuperHyperAlgebra. In: Section into the Author's Book Nidus Idearum. Scilogs, II: de rerum consectatione, 2nd edn., pp. 107–108 (2016)
3. Smarandache, F.: Introduction to the n-SuperHyperGraph - the most general form of graph today. In: Neutrosophic Sets and Systems, vol. 48, pp. 483–485 (2022). http://fs.unm.edu/NSS/n-SuperHyperGraph.pdf
4. Smarandache, F.: Extension of HyperGraph to n-SuperHyperGraph and to Plithogenic n-SuperHyperGraph, and Extension of HyperAlgebra to n-ary (Classical-/Neutro-/Anti-) HyperAlgebra. Neutrosophic Sets and Systems, vol. 33, pp. 290–296 (2020). https://doi.org/10.5281/zenodo.3783103. http://fs.unm.edu/NSS/n-SuperHyperGraph-n-HyperAlgebra.pdf
5. Rezaei, A., Smarandache, F., Mirvakili, S.: Applications of (neutro/anti)sophications to SemiHyperGroups. J. Math. 1–7 (2021)
6. Ibrahim, M.A., Agboola, A.A.A.: Introduction to NeutroHyperGroups. Neutrosophic Sets and Systems, vol. 38, pp. 15–32 (2020). https://doi.org/10.5281/zenodo.4300363. http://fs.unm.edu/NSS/IntroductionToNeutroHyperGroups2.pdf
7. Al-Tahan, M., Davvaz, B., Smarandache, F., Anis, O.: On some NeutroHyperstructures. Symmetry **13**, 535, 12 (2021). https://doi.org/10.3390/sym13040535. http://fs.unm.edu/NeutroHyperstructure.pdf
8. Agboola, A.A.A., Davvaz, B.: On neutrosophic canonical hypergroups and neutrosophic hyperrings. In: Neutrosophic Sets and Systems, vol. 2, pp. 34–41 (2014)
9. Nawaz, S., Gulistan, M., Khan, S.: Weak LA-hypergroups; neutrosophy, enumeration and redox reaction. In: Neutrosophic Sets and Systems, vol. 36, pp. 352–367. https://doi.org/10.5281/zenodo.4065464

Green AI from Kirchhoff to Shannon

Mihai Tache[1,2], Sorin-Horațiu Hoară[3], Vlad-Florin Drăgoi[2,4], and Valeriu Beiu[2(✉)]

[1] Faculty of Electronics, Telecommunications and Information Technology, University Politehnica of Bucharest, Iuliu Maniu 1-3, 061071 Bucharest, Romania
mihai.tache@uav.ro

[2] Faculty of Exact Sciences, "Aurel Vlaicu" University of Arad, Elena Drăgoi 2-4, 310330 Arad, Romania
{vlad.dragoi,valeriu.beiu}@uav.ro

[3] Faculty of Automation and Computers, Politehnica University of Timișoara, Vasile Pârvan 2, 300223 Timișoara, Romania
sorin-horatiu.hoara@student.upt.ro

[4] LITIS, University of Rouen Normandie, Avenue de l'Université, 76800 Saint-Etienne-du-Rouvray, France

Abstract. From the earliest beginnings, communications (theorized by Shannon) and computations (abstracted by Turing and von Neumann) have relied on schemes able to augment reliability. One was proposed by von Neumann and the other one was introduced by Moore and Shannon. For computations the gate-level method of von Neumann was embraced, while another one at the device-level was crafted by Moore and Shannon laying the foundation of network reliability. This device-level scheme was also intended for enhancing the reliability of computations, but was kind of sidetracked and ended up being employed for communications. Since then, steady advances towards single digit nanometer transistors have been constant. Lately, advanced technologies are in high demand for supporting the growing computational needs on which AI is blooming (as well as IoT and G5). As advance CMOS are sensitive to variations, it seems normal to analyze networks for identifying optimal ones, and the Moore and Shannon concept is a perfect fit for transistors/devices built and organized as arrays: FinFET, MBCFET, GAAFET, etc. Still, defining optimality in this context is a non-trivial task, as the conflicting power consumption and reliability requirements need to be analyzed and optimized simultaneously. For a better understanding we decided to start performing very detailed simulations of 3×3 resistor/transistor networks including variations (relying on the free ASAP7 PDK). We are reporting and discussing a few of those simulations here. Their statistical processing will reveal subtle and unexpected links between classical Kirchhoff theorems (pertinent for power consumption) and Moore and Shannon reliability concepts (theoretical). Plenty of additional investigations will be needed. Still, the main conclusion (based on the preliminary simulations we shall present) is that identifying optimal networks for low-power reliable CMOS designs (for computations) is a complex but doable task.

Keywords: Reliability · Power · Networks · Computations · Advanced technology · Variations · ASAP7 PDK

S. Dzitac et al. (Eds.): ICCCC 2022, AISC 1435, pp. 433–443, 2023.
https://doi.org/10.1007/978-3-031-16684-6_37

1 Introduction

Offering high-quality products has been crucial for both communications and computations. Quality has been a concern for transatlantic cables (communication) as well as vacuum tube computers (computations), both in need of enhanced reliability. That is why in the early 50's John von Neumann [1] together with Edward F. Moore and Claude E. Shannon [2], have pioneered two different schemes for building reliable systems from unreliable components. Still, [2] has also established a probabilistic network model where nodes are perfectly reliable, while edges fail independently with a probability q ($p = 1 - q$). The fundamental problem was (and still is) that of determining the probability that two (or more) nodes are *connected*, the solution being represented by the now well-known reliability polynomial.

Currently, for advanced and beyond CMOS technologies reliability issues can be seen both as a sleeping giant as well as a skeleton in the closet. Papers like [3] and [4] have restarted the discussions on issues raised in the 50's, but it took another few years till they resurfaced for memory [5]. Additionally, serious safety concerns have been raised for automotive applications [6, 7], as well as for the booming IoT field [8]. The latest publications [9, 10] are anticipating the trends we are going to face in the "angstrom era" (i.e., a few nanometers).

Amazingly, the problem is kind of the same as in the 50's, namely how to design systems with large numbers of unreliable (nanoscale) devices, but for a different technology. Without redundancy schemes such complex systems will either fail, or their (very complex) design and (ultra-precise) fabrication will require huge costs. This means that, an unconventional solution to atomically precise fabrication is to use some form of redundant design for offsetting imperfections, variations and noises. Currently, the hottest examples here come from quantum computing where, irrespective of the technology used, the current gate fidelities mandate for error correction schemes [11–13].

Obviously, the theoretical results from [2] are a foundation on which designs and analyses of more reliable computing schemes should be built. It is to be mentioned that [2] has proposed a practical redundancy scheme for computations, which unfortunately was not embraced by circuit designers.

Here, after analyzing the signal to noise margins (*SMNs*) of two types of SRAM cells (with and without redundancy), we shall examine (as a small example) all the seven 3-by-3 matchstick minimal networks (*MMNs*) implemented in ASAP7 PDK [14–16] (see also Figs. 1 and 2). These simulations will reveal reliability rankings reinforcing the fact that hammocks are prime contenders for doing reliable *computations*. Our particular choice of 3-by-3 minimal networks is due to the fact that 2-by-2 networks are kind of too simple (and have already been investigated [17]), while the more complex 4-by-4 networks have also been analyzed (preliminary results reported in [18]).

Trying to understand and better motivate reliable designs for computations this paper will briefly go over networks (Sect. 2) and their associated reliability polynomials (Sect. 3). Detailed analyses, simulations and comparisons will be presented in Sect. 4, while conclusions and future directions of research are going to end the paper.

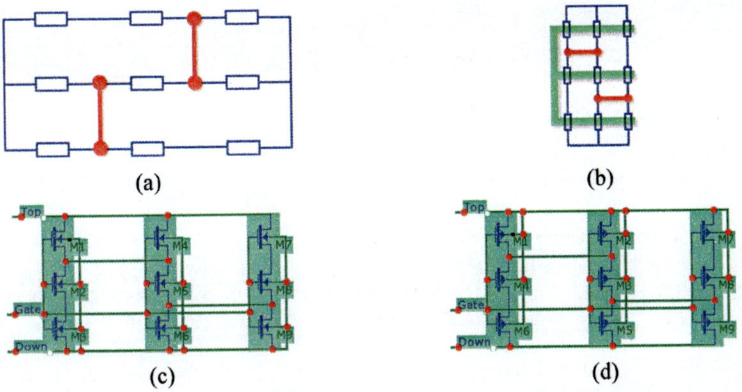

Fig. 1. $H_{3,3}$ (or MMN_{2a}): (a) classical representation; (b) rotated 90° and showing the connections to the gates of the devices/transistors (green); (c) nMOS finFET; (d) pMOS finFET.

2 Networks

A particular type of networks is known as hammocks. This network was introduced by Moore and Shannon [2], where it was shown how one could use such networks to design more reliable circuits (see [19–22]). For additional details on hammocks the interested reader should also check [23], while it is worth mentioning that lately several extensions have been suggested: cylindrical [24], 3D [25], as well as compositions [26]. Still, hammocks represent nothing else but a particular matchstick minimal network [27], being equivalent to a hexagonal lattice (heavily used in quantum computing). An interesting alternative is represented by small optimal sorting networks which have been shown to perform extremely well [28, 29].

The standard notation in the literature is to denote a network as $\boldsymbol{N}$. To any such network $\boldsymbol{N}$ we are associating a size n (which represents the number of devices/transistors), as well as a length l (which is the size of a minimum pathset), and a width w (which is the size of a minimum cutset). Additional explanations and details, including examples, can be found in [30–32].

Obviously, for any $\boldsymbol{N}$ we have $n \geq lw$ (Theorem 3 in [2]). The particular case when $n = lw$ corresponds to nets which are known as *minimal networks*. These can be divided into planar and non-planar, with the planar ones including *matchstick minimal networks* (*MMNs*), i.e., networks that can be represented/identified by their matchstick incidence matrix (see [27] and examples in Figs. 1 and 2).

3 Reliability of Networks

The reliability of a network $\boldsymbol{N}$ is always estimated from the associated reliability polynomial $Rel(\boldsymbol{N}; p)$. This is a polynomial in p (where p is the probability that a device is closed), and corresponds to the probability of connectedness of the network $\boldsymbol{N}$. Implicitly, the original definition from [2] accounts for the disconnectedness of $\boldsymbol{N}$ as well. One

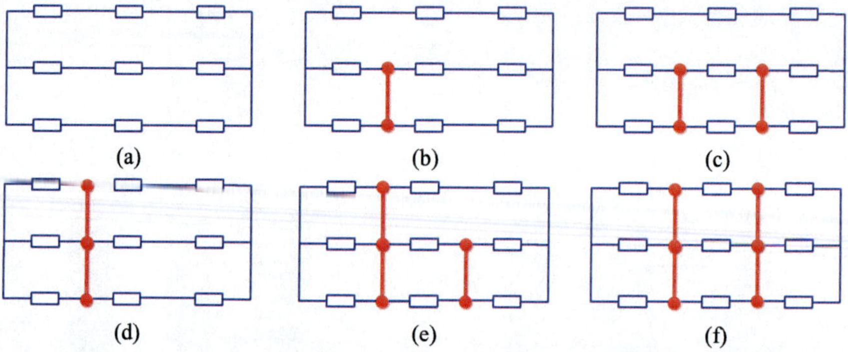

Fig. 2. All the 3 × 3 *MMNs* except for the hammock (already detailed in Fig. 1): (a) *PoS* (no matchsticks, hence MMN_0); (b) one matchstick, MMN_1; (c) two matchsticks MMN_{2b}; (d) two matchsticks MMN_{2c}; (e) three matchsticks MMN_3; (f) *SoP* (four matchsticks, hence MMN_4).

of the forms used to represent the polynomial $Rel(N; p)$ is the so-called N-form [30, 31]:

$$Rel(N;p) = \sum_{i=0}^{n} c_i(N)p^i(1-p)^{n-i} \tag{1}$$

where $c_i(N) = \beta_i C_n^i$, with C_n^i being the binomial coefficient, and with $\beta_i \in [0, 1]$ sub unitary coefficients, making this a Bernstein representation [33].

If designing for communication the polynomial $Rel(N; p)$ should be as close as possible to the Heaviside function, i.e., as close as possible to 1 for any value of p (ideally being 1 for $p \in (0, 1]$):

$$\theta_0(p) = \begin{cases} 0, & p < 0 \\ 0.5, & p = 0 \\ 1, & p > 0 \end{cases}. \tag{2}$$

In the opposite case, when N should be disconnected, i.e., 0 for $p \in (0, 1]$, the ideal polynomial corresponds to $\theta_1(p) = \theta_0(p-1)$. Finally, for computation the optimum is in-between, the ideal being $\theta_{0.5}(p) = \theta_0(p-0.5)$. Because $Rel(N; p)$ is in Bernstein form (1), it is possible to determine approximations of the Heaviside function by means of Bernstein operators (see [18]).

It follows that reliable *communications* and reliable *computations* are requiring quite different optimization conditions. In particular, computation wants both connectedness (sometimes) as well as disconnectedness (some other times). That is why the optimality of a network N intended for *communications* or for *computations* cannot be evaluated in the same way. This implies that the figures of merit (*FoMs*) used for evaluating network reliability for *computations* are different from the *FoMs* which are (commonly) used for networks intended and designed for *communications*. As networks intended for *communications* have received lots of attention, their evaluations have led to several concepts like *uniformly most reliable* [34], *average reliability* [35], or *domination reliability* [36].

On the other side, *FoMs* for networks intended for *computations* have been introduced and analyzed only recently [18]. This is unfortunate, as Moore and Shannon were clearly aiming for *computations* [2], as supported by their comparisons with [1]. While big strides were being made on the *reliability of communications*, the *reliability of computations*, although the goal stated by [2], was kind of forgotten and overlooked.

4 Simulation Results

For getting a clear understanding we have decided to analyze very thoroughly all 3-by-3 *MMNs*, both theoretically as well as through simulations (using ASAP7 PDK [14–16]). The hammock $H_{3,3}$ has 2 matchsticks, so it is also MMN_{2a} (already detailed in Fig. 1 (a)). Figures 1 (c) and (d) show two practical implementations of $H_{3,3}$ as an nMOS finFET 3-by-3 array, and respectively a pMOS finFET 3-by-3 arrays.

The other six *MMNs* of 3-by-3 were presented in Fig. 2, which shows them by increasing number of matchsticks (short red vertical connections) as:

- no (zero) matchsticks MMN_0 (also known as "parallel of series," or *PoS*);
- one matchstick MMN_1;
- two (more) versions with two matchsticks, MMN_{2b} and MMN_{2c};
- three matchsticks (MMN_3); and
- four matchsticks (MMN_4, also known as "series of parallel," or *SoP*).

4.1 Theoretical Analysis

For these seven *MMNs* of 3-by-3 we have computed the coefficients $c_i(N)$ of $Rel(N;p)$. They are presented in Table 1. Using these coefficients we have evaluated $Rel(N;p)$ for all the *MMNs* under consideration. These theoretical results are detailed in Fig. 3. It can be seen that the three *MMNs* with two matchsticks are the closest ones to $\theta_{0.5}(p) = \theta_0(p-0.5)$, hence being better fitted for *computations*.

Table 1. The coefficients of the reliability polynomials of the *MMNs* under investigation.

MMN		c_0	c_1	c_2	c_3	c_4	c_5	c_6	c_7	c_8	c_9
MMN_0	*PoS*	0	0	0	3	18	45	57	36	9	1
MMN_1		0	0	0	5	28	63	69	36	9	1
MMN_{2a}	$H_{3,3}$	0	0	0	8	42	84	76	36	9	1
MMN_{2b}		0	0	0	9	42	81	75	36	9	1
MMN_{2c}		0	0	0	9	45	84	75	36	9	1
MMN_3		0	0	0	15	63	98	79	36	9	1
MMN_4	*SoP*	0	0	0	27	81	108	81	36	9	1

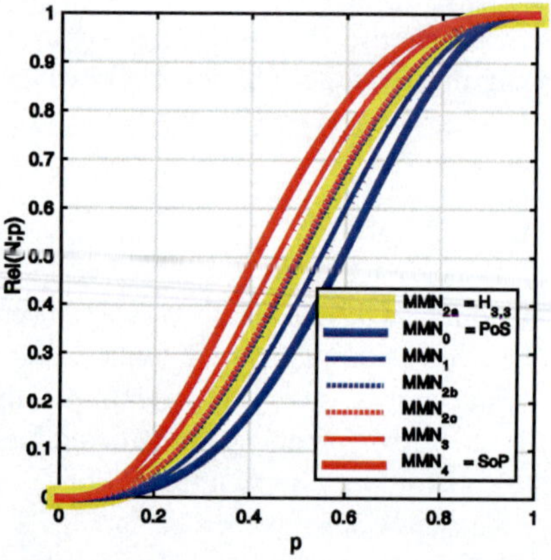

Fig. 3. Plot of $Rel(N; p)$ for N one of the seven *MMNs* under investigation.

4.2 Monte Carlo Simulations

The next step was to implement $H_{3,3}$ using nMOS and pMOS transistors as presented in Fig. 1. For all the following simulations we have used ASAP7 [14–16]. This PDK has several different finFETs, and for each of them one can selected from T (typical), F (fast), and S (slow) versions.

As a first test we decided to use two inverters forming a classical SRAM cell, and to evaluate their *SNMs* (Signal to Noise Margins). One SRAM cell has transistors with a single fin (1×1), while another SRAM cell has finFETs connected as $H_{3,3}$ (as in Figs. 1 (c) and (d)). In combinations with T/F/S we have generated many different versions of the two SRAM cells in a Monte Carlo like fashion. We have selected 25 different SRAM cells. The results of the simulations are presented in a compact form in Fig. 4. The blue bars show *SNMs* achieved using 1×1 finFETs, while the red bars show the corresponding ones achieved by using 3×3 finFETs ($H_{3,3}$) instead.

These results show that the *SNMs* obtained when using 3×3 finFETs are systematically larger than those achieved by cells using 1×1 finFETs, so redundancy does matter, while unfortunately it is not possible to have a meaningful statistical understanding/interpretation of when and by how much.

That is why for the next set of simulations we have decided to start the analysis from the transistors themselves by performing Monte Carlo like simulations accounting for line edge roughness (LER), which is expected to be the largest part of variations in advanced technologies. We decided to have the simplest possible approach and emulate them as networks of resistors on which later we will introduce variations corresponding to (particular points) T, F, S. This approach is used to tackle variations, but it also links to power/energy which can be calculated from Kirchhoff laws [37]. The resistor networks corresponding to the networks in Fig. 2 are presented in Fig. 5, while their equivalent resistances are as follows:

$$R_{MMN_0} = \frac{(R_1+R_2+R_3)(R_4+R_5+R_6)(R_7+R_8+R_9)}{(R_1+R_2+R_3)(R_4+R_5+R_6)+(R_1+R_2+R_3)(R_7+R_8+R_9)+(R_4+R_5+R_6)(R_7+R_8+R_9)} \quad (3)$$

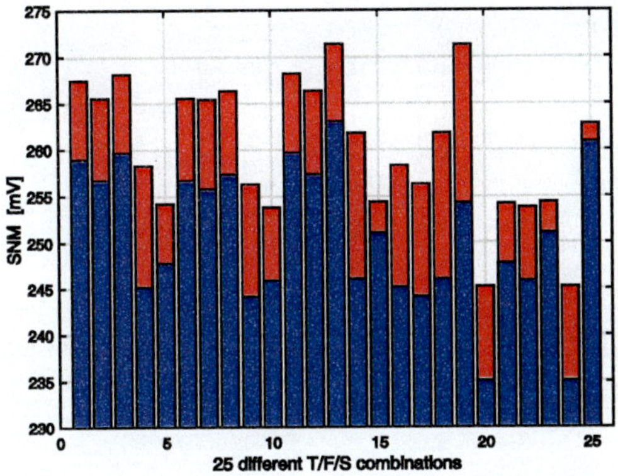

Fig. 4. *SNMs* of a classical SRAM cell implemented using different combinations of T/F/S for 1×1 finFETs transistors (blue), and, respectively, 3×3 finFETs transistors (blue).

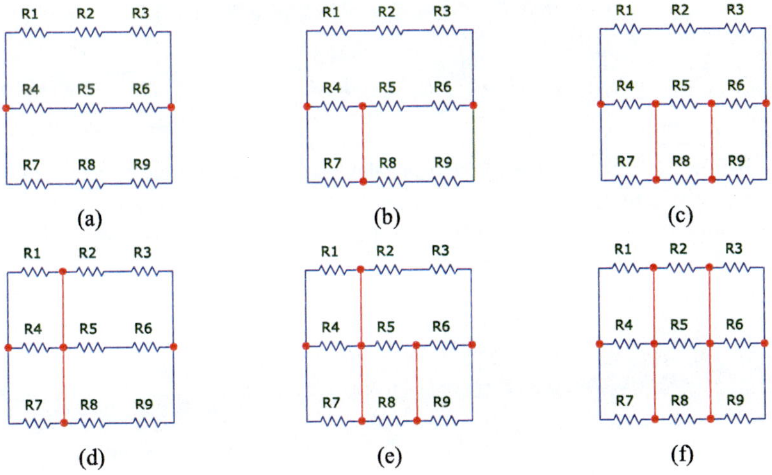

Fig. 5. All the 3×3 *MMNs*, except for the hammock $H_{3,3}$ (MMN_{2a}), as resistor networks: (a) no matchsticks (MMN_0); (b) one matchstick (MMN_1); (c) two matchsticks (MMN_{2b}); (d) two matchsticks (MMN_{2c}); (e) three matchsticks (MMN_3); (f) four matchsticks (MMN_4).

$$R_{MMN_1} = \frac{(R_1 + R_2 + R_3) \cdot \frac{(R_5+R_6)(R_8+R_9)}{R_5+R_6+R_8+R_9} + \frac{R_4R_7}{R_4+R_7}}{R_1 + R_2 + R_3 + \frac{(R_5+R_6)(R_8+R_9)}{R_5+R_6+R_8+R_9} + \frac{R_4R_7}{R_4+R_7}} \quad (4)$$

$$R_{MMN_{2b}} = \frac{(R_1 + R_2 + R_3)\left(\frac{R_4R_7}{R_4+R_7} + \frac{R_5R_8}{R_5+R_8} + \frac{R_6R_9}{R_6+R_9}\right)}{R_1 + R_2 + R_3 + \frac{R_4R_7}{R_4+R_7} + \frac{R_5R_8}{R_5+R_8} + \frac{R_6R_9}{R_6+R_9}} \quad (5)$$

$$R_{MMN_{2c}} = \frac{R_1R_4R_7}{R_1R_4 + R_1R_7 + R_4R_7} + \frac{(R_2 + R_3)(R_5 + R_6)(R_8 + R_9)}{(R_2 + R_3)(R_5 + R_6) + (R_2 + R_3)(R_8 + R_9) + (R_5 + R_6)(R_8 + R_9)} \quad (6)$$

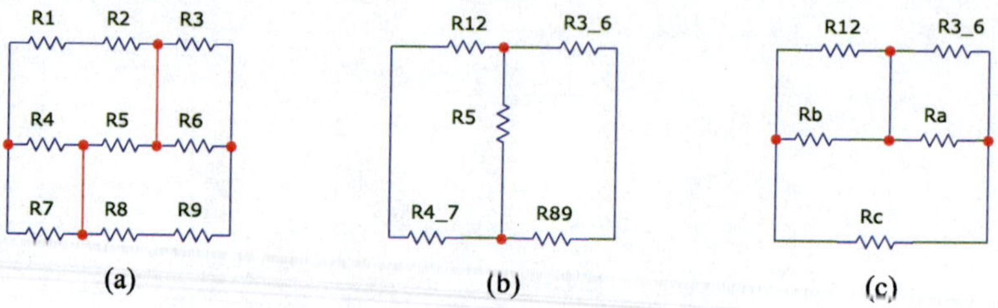

Fig. 6. The hammock $H_{3,3}$ (MMN_{2a}): (a) as a resistor network; (b) after two series ($R1 + R2$, $R8 + R9$) and two parallel transformations ($R3||R6$, $R4||R7$); (c) after a start-triangle (also known as wye-delta) transformation.

$$R_{MMN_3} = \frac{(R_2 + R_3)\left(\frac{R_5R_8}{R_5+R_8} + \frac{R_6R_9}{R_6+R_9}\right)}{R_2 + R_3 + \frac{R_5R_8}{R_5+R_8} + \frac{R_6R_9}{R_6+R_9} + \frac{R_1R_4R_7}{R_1R_4+R_1R_7+R_4R_7}} \tag{7}$$

$$R_{MMN_4} = \frac{R_1R_4R_7}{(R_1R_4 + R_1R_7 + R_4R_7)} + \frac{R_2R_5R_8}{(R_2R_5 + R_2R_8 + R_5R_8)} + \frac{R_3R_6R_9}{(R_3R_6 + R_3R_9 + R_6R_9)} \tag{8}$$

For the hammock network $H_{3,3}$ (MMN_{2a}) calculating the equivalent resistance is no so straightforward as for the other six networks which have required only series and parallel transformations. The school method relies on a start-triangle (also known as wye-delta) transformation, and is presented in Fig. 6. It leads to:

$$R_{MMN_{2a}} = \frac{\Omega\left[\frac{R_1+R_2}{\Omega+(R_1+R_2)(R_8+R_9)} + \frac{\frac{R_3R_6}{R_3+R_6}\cdot\frac{R_4+R_7}{R_4R_7}}{\frac{R_3R_6}{R_3+R_6}+\Omega\frac{R_4+R_7}{R_4R_7}}\right]}{1 + R_5\left[\frac{R_1+R_2}{\Omega+(R_1+R_2)(R_8+R_9)} + \frac{\frac{R_3R_6}{R_3+R_6}\cdot\frac{R_4+R_7}{R_4R_7}}{\frac{R_3R_6}{R_3+R_6}+\Omega\frac{R_4+R_7}{R_4R_7}}\right]} \tag{9}$$

where

$$\Omega = R_5(R_8 + R_9) + R_5\frac{R_4R_7}{R_4 + R_7} + (R_8 + R_9)\frac{R_4R_7}{R_4 + R_7} \tag{10}$$

Alternatively, one can use matrix methods (easier to program), or rely on EDA tools.

Each of these seven formulas (one for each of the seven different networks) has been used to evaluate the equivalent resistance for 19, 683 cases (i.e., 3^9), as each network has 9 resistors, and each resistor can have one of the following three values: $R - \Delta$ (corresponding to F), R (corresponding to T), and $R + \Delta$ (corresponding to S). Obviously, the number of different equivalent resistances is smaller than the total number of simulations ($137, 781 = 7 \times 19, 683$) as the same value appears more than once (an aspect related to Farey sequence). All the simulations have been performed numerically in Matlab and symbolically in Maple for making sure that we have corrected all numerical errors (due to propagation and accumulation).

A detailed statistical analysis of these equivalent resistances has been performed by binning the results in 50, 20 and 10 groups. The results are presented both as PDF (bar graphs) and CDF (interpolated curves). An amazing similarity can be observed between the CDF obtained (especially for 20 and 10 bins) and $Rel(N; p)$ (Fig. 7).

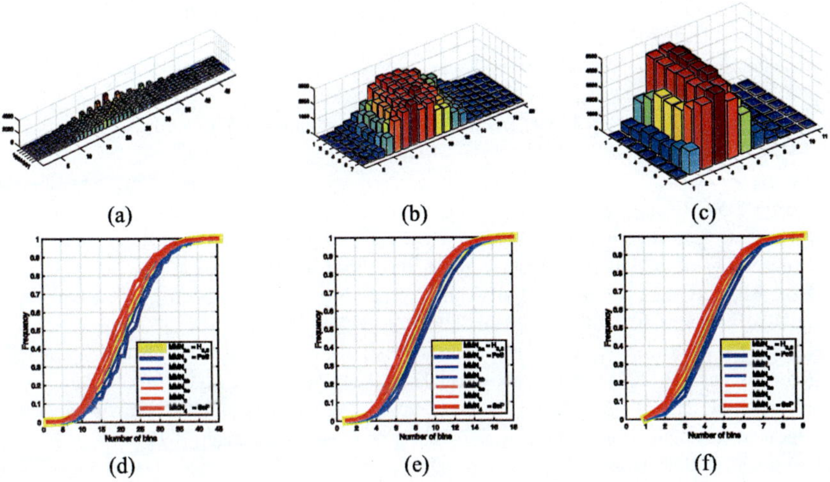

Fig. 7. PDF (bar graphs) for: (a) 50 bins; (b) 20 bins; (c) 10 bins. CDF (interpolated curves) for: (d) 50 bins; (e) 20 bins; (e) 10 bins.

5 Conclusions

This is an early stage investigation of the advantages and disadvantages of using network-based design in advanced CMOS technologies. It aims to bridge reliability (Moore and Shannon style) to power and energy (classical Kirchhoff laws). We plan to follow with more detailed statistical analyses [38], and simulations of SRAM cells mixing the *MMNs* we have investigated here. We also plan to carefully evaluate the power and energy consumption of such SRAM cells (see [39, 40]), and investigate how such SRAM cells are behaving in sub-threshold [41].

Acknowledgements. This research was partly funded by a grant of the Romanian Ministry of Education and Research, CNCS-UEFISCDI, project no. PN-III-P4-ID-PCE-2020-2495, within PNCDI III (ThUNDER^2 = *Techniques for Unconventional Nano-Designing in the Energy-Reliability Realm*).

References

1. von Neumann, J.: Probabilistic logics and the synthesis of reliable organisms from unreliable components. In: Shannon, C.E., McCarthy, J. (eds.): Automata Studies,

pp. 43–98. Princeton University Press, Princeton (1956). Lectures at Caltech, 4–15 January 1952. https://archive.org/details/vonNeumann_Prob_Logics_Rel_Org_Unrel_Comp_Caltech_1952/mode/2up. Accessed 06 June 2022
2. Moore, E.F., Shannon, C.E.: Reliable circuits using less reliable relays – Part I. J. Frankl. Inst. **262**(3), 191–208 (1956). Preliminary version: Reliable circuits using crummy relays. Tech. Rep. Memo. 54-114-42. Bell Labs, Murray Hill, NJ, USA, 29 November 1954
3. Visweswariah, C.: Death, taxes and failing chips. In: Proceedings of the DAC, Anaheim, CA, pp. 343−347. IEEE (2003)
4. Constantinescu, C.: Trends and challenges in VLSI circuit reliability. IEEE Micro **23**(4), 14–19 (2003)
5. Schroeder, B., et al.: DRAM errors in the wild: A large-scale field study. Commun. ACM **54**(2), 100–107 (2011)
6. Iturbe, X., et al.: Soft error vulnerability assessment of the real-time safety-related ARM Cortex-R5 CPU. In: Proceedings of DFT, Storrs, CT, pp. 91−96. IEEE (2016)
7. Abraham, J.A.: Resiliency demands on next generation critical embedded systems. In: Proceedings of the IOLTS, Rhodes, Greece, pp. 135–138. IEEE (2019)
8. Szentesi, S.-G., et al.: Internet of Things, challenges and perspectives in Romania: A qualitative research. Amfiteatru Econ. **23**(57), 448–464 (2021)
9. Dixit, H.D., et al.: Silent data corruptions at scale. Technical Report. Hardware Architecture (cs.AR), pp. 1–8 (2021). https://arxiv.org/abs/2102.11245. Accessed 06 June 2022
10. Hochschild, P.H., et al.: Cores that don't count. In: Proceedings of the Workshop Hot Topics Operating System (HotOS 2021), pp. 9–16. ACM Press, Ann Arbor (2021)
11. Boixo, S., et al.: Characterizing quantum supremacy in near-term devices. Nature Phys. **14**(6), 595–600 (2018)
12. Arute, F., et al.: Quantum supremacy using a programmable superconducting processor. Nature **574**(7779), 505–510 (2019)
13. Google Quantum: Exponential suppression of bit or phase errors with cyclic error correction. Nature **595**(7867), 383–387 (2021)
14. Clark, L.T., et al.: ASAP7: A 7-nm finFET predictive process design kit. Microelectr. J. **53**, 105–115 (2016)
15. Arizona State Univ. (in partnership with ARM): ASAP7/ASU 7nm predictive PDK. (2022). https://asap.asu.edu. Accessed 20 June 2022
16. OpenROAD: ASAP7 7nm predictive PDK. GitHub, 15 December 2020. https://github.com/The-OpenROAD-Project/asap7. Accessed 20 June 2022
17. Robinett, W., et al.: Defect tolerance based on coding and series replication in transistor-logic demultiplexer circuits. IEEE Trans. Circ. Syst. I **54**(11), 2410–2421 (2007)
18. Beiu, V., et al.: Why reliability for computing needs rethinking. In: Proceedings of ICRC (Virtual), pp. 16–25. IEEE (2020)
19. Creveling, C.J.: Increasing the reliability of electronic equipment by the use of redundant circuits. Proc. IRE **44**(4), 509–515 (1956)
20. Lipp, J.P.: Topology of switching elements vs. reliability. IRE Trans. Reliab. Qual. Control **10**, 21–33 (1957)
21. Sorensen, A.A.: Digital-circuit reliability through redundancy. Electr. Reliab. **1**(1), 27–37 (1962)
22. Suran, J.J.: Use of circuit redundancy to increase system reliability. In: Proceedings of ISSCC, Philadelphia, PA, pp. 82–83. IEEE (1964)
23. Cowell, S.R., et al.: On the exact reliability enhancements of small hammock networks. IEEE Access **6**, 25411–25426 (2018)
24. Cowell, S.R., et al.: On cylindrical hammock networks. In: Proceedings of IEEE-NANO, Pittsburgh, PA, pp. 185–188. IEEE (2017)

25. Beiu, V., et al.: Transport reliability on axonal cytoskeleton. In: Proceedings of EMES, Oradea, Romania, pp. 160–163. IEEE (2017)
26. Beiu, V., et al.: Hammocks versus hammock. In: Proceedings of ICCCC, Oradea, Romania, pp. 119–123. IEEE (2018)
27. Drăgoi, V.-F., Beiu, V.: Fast reliability ranking of matchstick minimal networks. Network **79**(4), 479–500 (2022)
28. Nagy, M., et al.: Employing sorting nets for designing reliable computing nets. In: Proceedings of IEEE-NANO (Virtual), pp. 370–375. IEEE (2020)
29. Drăgoi, V.-F., Cowell, S.R., Beiu, V.: Four input sorter good, larger ones not so good. IEEE Trans. Nanotech. **20**, 775–783 (2021)
30. Colbourn, C.J.: The Combinatorics of Network Reliability. Oxford Univ. Press, New York (1987)
31. Chari, M., Colbourn, C.J.: Reliability polynomials: A survey. J. Comb. Inf. Syst. Sci. **22**(3–4), 177–193 (1997)
32. Pérez-Rosés, H.: Sixty years of network reliability. Math. Comput. Sci. **12**(3), 275–293 (2018)
33. Bernstein, S.N.: Démonstration du théorème de Weierstrass fondée sur le calcul des probabilities. Commun. Kharkov Math. Soc. **13**, 1–2 (1912/1913)
34. Deng, H., et al.: On the construction of most reliable networks. Discrete Appl. Math. **140**(1–3), 19–33 (2004)
35. Brown, J.I., et al.: Network reliability: Heading out on the highway. Networks **77**(1), 146–160 (2021)
36. Dohmen, K., Tittmann, P.: Domination reliability. Electr. J. Combin. **19**(1), 1–14 (2012). Art P15
37. Kirchhoff, G.: Ueber die Auflösung der Gleichungen, auf welche man bei der Untersuchung der linearen Vertheilung galvanischer Ströme geführt wird. Annalen der Physik und Chemie **148**(12), 497–508 (1847). [Translated by O'Toole, J. B.: On the solution of the equations obtained from the investigation of the linear distribution of galvanic currents. IRE Trans. Circ. Theory **5**(1), 4–7 (1958)]
38. Sevüktekin, N.C., et al.: Linear noisy networks with stochastic components. In: Proceedings of CDC, Nice, France, pp. 5386–5391. IEEE (2019)
39. International Roadmap for Devices and Systems (IRDS™), 2021 Edition. IEEE, Piscataway. https://irds.ieee.org/editions/2021. Accessed 20 Sep 2022
40. Horowitz, M.: Computing's energy problem (and what we can do about it). In: Proceedings of ISSCC, San Francisco, CA, pp. 10–14. IEEE (2014)
41. Rezaei, E., et al.: Fundamental thermal limits on data retention in low-voltage CMOS latches and SRAM. IEEE Trans. Dev. Mater. Reliab. **20**(3), 488–497 (2020)